AN ESSAY ON THE FOUNDATIONS
OF OUR KNOWLEDGE

Antoine Augustin Cournot

AN ESSAY
ON THE FOUNDATIONS
OF OUR KNOWLEDGE

Translated with an Introduction by
MERRITT H. MOORE
Professor of Philosophy, The University of Tennessee

1956
THE LIBERAL ARTS PRESS
NEW YORK

To JEAN, ROSELLA, MINER and ARTHUR,
whose bearing and forbearing have made
this book possible.

TRANSLATOR'S PREFACE

The publication of this translation of the *Essai sur les fonde-ments de nos connaissances* by Antoine Augustin Cournot makes available to English readers one of the major contributions of this important nineteenth-century French mathematician, economist, and philosopher. Only one other of Cournot's studies in these areas, his *Recherches sur les principes mathématiques de la théorie des richesses,* has previously been available in English, although some have been translated into other languages, particularly German and Italian. Long before the translation of the *Recherches* was made, Cournot's position as a creative and significant contributor to the general field of mathematical economics had been thoroughly established. This fact makes even more pointed the neglect which has been accorded his equally valuable work in philosophy.

My interest in lessening this neglect stems from a conversation I had over twenty years ago with Professor Arthur E. Murphy. At his suggestion I read the *Essai.* Being impressed by the range of Cournot's knowledge, by his concern to carry forward a careful, penetrating and constructive analysis of certain persistent philosophical problems, and by the ingenuity and prescience with which he addressed himself to this work, I decided to undertake a translation of it.

The first draft of the translation was completed in the early 1930's. Difficulties in getting the work published were almost insurmountable, but, among others, Professors Lucien Lévy-Bruhl, Morris R. Cohen, Curt J. Ducasse, George Boas, Frank Knight, and Arthur E. Murphy urged me to persist in my efforts.

The onset of the Second World War brought the whole project to a halt for over ten years. In the spring of 1950, Mr. Oskar Piest, Editor of the Liberal Arts Press, expressed an interest in adding the *Essai* to the listings of this Press. Dr. Sharvy G. Umbeck, President of Knox College (where I was teaching at that time), strongly urged me to go ahead with arrangements which Mr. Piest and I had tentatively worked out. More important, through the good offices of President Umbeck I received generous assistance from Knox

College and was able to enter into a definite agreement with the
Liberal Arts Press.

It was my good fortune to be selected to go to France during the
academic year 1951–52 under the auspices of the Fulbright program. Mr. Piest immediately suggested that I take the completed
manuscript with me and make improvements in it that might result
from working with Cournot material not available in the United
States. This was both a wise and generous suggestion. In Paris
the assistance of numerous French and American friends enabled
me to improve both the manuscript and my knowledge of Cournot.

Shortly after reaching Paris I met Professor Jean Cournot, a
grandnephew of A. A. Cournot, and a member of the Faculty of the
Conservatoire des Arts et Métiers. M. Cournot befriended me and
assisted me in numerous ways. He introduced me to four of the
most eminent French experts on A. A. Cournot: Monseigneur E. P.
Bottinelli, the editor of Cournot's *Souvenirs* and the author of
A. Cournot: Métaphysician de la connaissance; Messieurs René
Roy, an Inspector-General in the Ministry of Public Works; François Divisia, professor at the Ecole Polytechnique, and, finally,
Georges Lutfalla, Assistant Director of the National Credit Bank
for the Departments and Districts, and Actuary in the Ministry of
Labor. With the exception of Monseigneur Bottinelli, these men
are primarily interested in Cournot as an economist; yet each gave
me valuable insights into various aspects of Cournot's thought.

As technical editor of the manuscript, Professor Joseph L. Blau
of Columbia University gave unstintingly of his time and effort, and
did much to improve the work. His comments gave convincing
evidence of the meticulous care with which he read the whole
manuscript. For his work, I owe Professor Blau a debt which I can
neither express in words nor hope to repay. Much of the value and
accuracy of the finished work must be credited to him. Special
credit is also due to the editorial staff of the publishers for their
co-operation which exceeded by far the usual routine editorial work
of publishing houses. A number of errors have been eliminated by
their competent editing and checking of the translation, but I must
assume the sole responsibility for such errors and inadequacies as
may remain.

My greatest debt is to the men mentioned above. But numerous other persons gave invaluable assistance in many ways. Among

these it is possible for me to single out only three for particular mention: Professors Charles J. Adamec and Lucius W. Elder, colleagues and friends at Knox College, prepared translations of the numerous Greek and Latin quotations and footnotes used by Cournot; my wife has given me unstinting help of all sorts at all stages as the project developed, in spite of her continuous responsibilities as wife and mother. To these and to all others who have helped me, I wish to express my heartfelt thanks.

It is also incumbent upon me, and a pleasure, to acknowledge permissions generously given by several publishing houses to use in the translation quotations from the following copyrighted material:

Clarendon Press of the Oxford University Presses for Aristotle, *Works*, translated by W. D. Ross and others;
    for Gaius, *Commentaries on the Institutes*, with a translation and commentary by Edward Poste. Second edition, 1884.
Columbia University Press for Macrobius, *Commentary on the Dream of Scipio*, translated by William Harris Stahl.
E. P. Dutton and Company for Spinoza, *Ethics*, "Everyman's Library."
Harvard University Press for Plato, *Republic; Theaetetus;* and *Timaeus*, "The Loeb Classical Library."
A. A. Knopf for De Tocqueville, *Democracy in Action*. The Henry Reeve text revised by Francis Bowen and further corrected by Phillips Bradley.
Macmillan and Company, Ltd., for Kant, *The Critique of Pure Reason*, translated by Norman Kemp Smith.
Methuen and Company, Ltd., for Gibbon, *The History of the Decline and Fall of the Roman Empire*.
Charles Scribner's Sons for *Bacon Selections*, edited by Matthew T. McClure. From "The Modern Student's Library";
    for *Leibniz Selections*, edited by Philip P. Wiener. From "The Modern Student's Library."

Specific acknowledgments are made in the translation at the points where quotations from the above sources appear.

Since the original form of the *Essai* included no Index, I have prepared one for the translation. To increase its usefulness, important references to items included in it have been set in italics.

M. H. M.

CONTENTS

AN ESSAY ON THE FOUNDATIONS OF OUR KNOWLEDGE

INTRODUCTION

This translation makes one of Antoine Augustin Cournot's philosophical works available in English for the first time. He is already well known and highly regarded by economists;[1] his philosophical writings are known by only relatively few persons, yet these few have a high regard for their merit. It is the purpose of this translation to increase, if possible, the number of persons who will read Cournot as a philosopher: it is the purpose of this Introduction to help acquaint such possible readers with Cournot as a thinker and as a man.

I

It is difficult for those who know his writings to understand why Cournot's philosophical works have not had a wider audience. The reason can certainly not be that his writings are in French. Many French works are widely read, either in the original or in numerous translations, by persons for whom French is not a native tongue. This is true of many of Cournot's contemporaries, some of whom certainly made a less significant contribution to the philosophical literature of our Western tradition than did he.

Cournot died in 1877; yet it was not until 1905 that any con-

[1] A cursory examination of the development of economic thought in the nineteenth century is all that is needed to convince one of the value and impact of Cournot's pioneering work in mathematical economics. Long before his *Recherches sur les principes mathématiques de la théorie des richesses* was translated in 1897 by Nathaniel Bacon, many of the important economists of the century—Jevons, Marshall, Edgeworth, Walras, Pareto, Irving Fisher, and others —had specifically acknowledged their indebtedness to him. In 1938 one session of the Econometric Society was dedicated to the commemoration of the hundredth anniversary of the publication of the *Recherches*. This session is reported in the journal *Econometrica*. Two recent books by A. Schumpeter attest to the originality and the worth of Cournot as an economist. These are *Economic Doctrine and Method* and *History of Economic Analysis*. Both were published by the Oxford University Press in 1954.

See also " L'Oeuvre économique de Cournot " by Albert Aupetit, *Revue de métaphysique et de morale*, XIII, 3 (1905), pp. 377ff. and especially pp. 392f.

certed effort was made to bring his philosophical work to the attention even of his own countrymen. In May of that year a memorial to Cournot appeared in the form of a number of the *Revue de métaphysique et de morale* [1] which was given over entirely to articles about him. Nearly two decades later Professor Lévy-Bruhl [2] of the University of Paris was responsible for the appearance of a second edition of the *Essai*. At the present time Cournot's *Matérialisme, vitalisme, rationalisme* is on the list of books prepared by the French Ministry of Public Instruction for students who complete the program of the secondary schools.

The neglect of Cournot in no way alters the fact that while his contemporaries were often concerning themselves with the tur-

---

[1] This number contains twelve articles on various aspects of Cournot's work. It is interesting to notice not only that each writer acknowledges the importance and originality of Cournot's thought but also that each finds it difficult to explain the general neglect of Cournot.

[2] In his *History of Modern Philosophy in France*, which was published in 1899 and reprinted without change in 1924, Professor Lévy-Bruhl himself gives scant attention to Cournot (pp. 457-459 and 470) and did not deem Cournot's philosophical writings of sufficient importance to be included in his rather extensive bibliography of modern French philosophers (pp. 483–494). Yet when as a graduate student I met and talked with Professor Lévy-Bruhl, while he was a visiting professor at the University of Chicago, he not only urged me to continue my study of Cournot but also to come to France to continue this work after I had completed my degree. By that time Professor Lévy-Bruhl seemed to have modified radically his estimate of Cournot as a philosopher. Certainly philosophical questions being considered then and later involved matters that Cournot had examined between fifty and a hundred years earlier, long before others seemed to be aware of them or, at least, of their significance.

It is also interesting to note that in his *History* Professor Lévy-Bruhl attempts to account for the neglect of Cournot. " May it not be due to its extreme cautiousness that Cournot's doctrine owns the relative obscurity in which, *in spite of its rare value,* it has remained? " (Page 459, italics mine.) Professor Lévy-Bruhl implies a tentative answer in putting a second question: " Can a philosophy exist that dare not assert itself as a philosophy? " (*Ibid.*) It is hard to understand how anyone who has read Cournot's *Exposition de la théorie des chances et des probabilités; Essai sur les fondements de nos connaissances; Traité des idées fondamentales*; the *Matérialisme, vitalisme, rationalisme,* and his *Considérations sur la marche des idées*, to say nothing of his books on mathematics and economics, can seriously suggest that they do not individually and together assert a philosophy. See my article, " The Place of A. A. Cournot in the History of Philosophy," *The Philosophical Review*, Vol. XLIII, No. 4 (July, 1934).

bulent eddies that developed in Western philosophical thought
during the nineteenth century, the problems to which Cournot addressed himself lie in the main stream of our tradition. Local and
passing turbulences always tend to excite us and to distract our
attention even though we may escape being engulfed in them; but
their transient life usually has little if any lasting effect on the
bearing of the main stream. Cournot's work is worth attending
to if for no other reason than that he was one of relatively few of
his contemporaries who seems to have been fruitfully and un-
distractedly concerned with the persistent rather than the ephemeral problems with which men concern themselves.

But there are other reasons for being interested in Cournot.
Perhaps the most important is that in each of the three areas to
which he made a contribution—mathematics, economics, and
philosophy—his work is creative in the double sense of containing
novel elements and of indicating new modes of inquiry whose implications others may fruitfully explore. He leaves us with surprisingly few paradoxes and dead ends. In mathematics, in addition to his work on algebra, geometry and the calculus, he
contributed to the development of the theory of probability: this
work is referred to by practically all later writers on probability
theory whether they agree or disagree with Cournot's assumptions
and conclusions. In economics, he was among the first to attempt
at all, and is commonly regarded as the first to make a successful
attempt, to apply the instruments of mathematical probability to
the examination of problems in the physical sciences and the
social studies: he is regarded by economists from Walras, Jevons
and Marshall on as the effective founder of mathematical economics or econometrics. In philosophy, like Kant, he proposed to
examine the antagonism between classical rationalism and classical
empiricism to see if he might discover some median position which
would enable us to limit the stultifying skepticism or the various
forms of romantic escapism implicit in a radical and logically consistent empiricism and at the same time avoid the dogmatism and
obscurantism of unmitigated rationalism: his critical rationalism
makes articulate those self-corrective features of certain rational-
empirical procedures which are so much a development of modern
thought and which are commonly, if erroneously, identified as
" scientific method "—erroneously identified as such principally

because these features are applicable in the investigation of any field to the degree that it may be organized into a rational study.

Claiming nothing more than probable truth, the coincident use of Cournot's metaphysical and epistemological principles frequently brings us to conclusions whose probability is of such an order that any reasonable being must accept them as practical certainties concerning both the nature of things and the nature and conditions of knowledge. This is more than is claimed and more than can be attained by any consistent empiricism; it is less than is claimed but more than can be certainly established by any consistent rationalism. The temperate skepticism of Cournot's philosophy makes unnecessary the excessive consequences, often of a mystical or romantic sort, which are the result of a dogmatic skepticism; its temperate rationalism, its probabilistic metaphysics and epistemology give an examinable basis for operational and other pragmatic techniques and make it possible to consider pragmatism as it should be considered—as a methodology, not as a total philosophy.

II

Whether or not a writer is noticed by his contemporaries depends, probably to a large degree, upon whether or not his writings do or do not reflect the issues and "tone" characteristic of the period in which he writes. The attention given one is often a function of the vigor with which one either embraces and exploits or rejects and opposes the issues which excite his time. This is likely to be particularly true during periods when wars and revolutions make the fortunes of man and, indeed, of life itself uncertain.

The first estimates of a writer's potential significance are always given by his contemporaries. Unfortunately, in all too many cases later histories tend to be written largely from these contemporary appraisals and from the earliest histories of the period. Little if any effort is made to re-examine periodically and systematically the period about which the later historian proposes to write, to try to find out whether or not, in the light of new information and new grounds for evaluation, someone of real importance may not have been overlooked earlier. Thus, he who is overlooked at first tends to be overlooked at last, in spite of the fact that his work may be a contribution of more lasting significance than that of some of those whom fortune favored with original attention.

Cournot certainly ran such a risk. Partly from choice, partly as a result of increasingly poor eyesight, and partly because of other circumstances, his direct participation in the lively political life of his times could scarcely have been less. By 1859 his eyesight had failed to such an extent that he felt that he would be forced into retirement. In that year he prepared his *Souvenirs* and assumed that his days as a writer were over. This fortunately was not true; as a matter of fact, a number of his most interesting and valuable books were written after this date with the help of various amanuenses.[1]

During Cournot's lifetime France was almost continuously beset with great troubles. These were times in which men tended to find their strength or their solace in pamphlets or slogans rather than in philosophy. Only at long last, if at all, does philosophy yield security; yet in nineteenth-century France as in our twentieth-century world " security " came to be almost synonymous with man's highest hopes and aspirations. Philosophies which attract notice and win a following are likely to be such that more quiet and orderly periods would regard as too simple or too violent, as panacean or utopian, or as superficial and ephemeral. When men are exposed to forces that threaten to destroy their accustomed world, most of them have two needs: one, to protect such cherished things as may be salvaged and carried with them as they seek to move with their lares and penates to some spot deemed safe from the encompassing storm; the other, to create and to foster a great hope and vision for the future—a hope and vision which often turns out to be an idealized resurrection of the past or a messianic ideal impossible of actualization. At such times men tend to heed those who claim to know the road to either supposed haven. Those who can and will remain relatively detached from and undisturbed by immediate concerns, who have the rare gift of finding their own

[1] The *Recherches*, which appeared as early as 1838, contains numerous arithmetical errors of a sort one would not expect to find in the work of a mathematician of Cournot's undoubted skill and power. It is highly probable that at least most of these errors may be attributed to the fact that Cournot had trouble reading the proof of the book.

peace in times of storm and stress and who, therefore, can continue the pursuit of activities congenial to times of greater peace are likely to be little regarded. Men who think and move apart from their fellows at such times tend to be ignored if they are not scorned and condemned: with life itself seemingly in the balance, what man of parts can view the passing scene with apparent detachment and with neutral and dispassionate purpose set himself the task of attempting to view and evaluate his present against a background of more fundamental issues? Such a person almost asks to be neglected.

French nineteenth-century philosophy inevitably bears the mark of distraught times. Two dominant themes developed in it, each having several variations. One group proposed to build a philosophy on the foundation of a psychological sensationalism or of a scientific positivism. In the case of some writers this represents a return to British empiricism and particularly to Locke; in the case of others, it appears as a continuation of the radical sensationalism developed by Condillac in the eighteenth century; still others, of whom Auguste Comte is the outstanding representative, purported to take their cue from science, and proposed to ignore all data and ideas not attested to by scientific procedures and criteria as they understood these procedures and criteria. On the other hand were groups who proposed spiritual, mystical and romantic solutions to man's problems. Joseph de Maistre and the Viscount Louis de Bonald sought a return to Roman Catholic orthodoxy, a suggested return to the past as romantic as Sir Walter Scott's recapture of the Middle Ages and as inappropriate to the temper of nineteenth-century France as anything one can imagine. Others in France, as in Germany, spun romantic philosophies from the extruded stuff of their own egos and read the meaning of all things, and particularly of history, in the projection and objectification of their autobiographies. Victor Cousin, like Schelling, Hegel, Marx, and Nietzsche, calls to mind this group.

III

By no stretch of the imagination could Cournot be identified with any extreme movement, whether political or philosophical. He had great respect for and acknowledged the influence of Kant; but his philosophy has nothing in common with that of those who find an Absolute implicit in a transcendental synthetical unity of apperception. Cournot's critical rationalism accepts the principle of the relativity of our knowledge, a relativity which so far as we can foresee can never be overcome. But the structure of our knowledge is not determined by the structure of our minds; it rests upon our apprehension and progressive approximation of structures objectively inherent in the nature of things, processes, and events whether we deal with these " things " at the level of naive, uncritical, or pragmatic behavior or at the level of our most sophisticated, penetrating, comprehensive, and reasonable understanding. Cournot will have no part of either the epistemological or the metaphysical inferences that Kant drew from his "Copernican Revolution."

Like Comte, Cournot was interested in " positive " knowledge, in the role of sensory experience and of empirical verification in our efforts to grasp the nature of things; but he clearly points out that the use of such experience is circumscribed and that our knowledge simply is not limited to, and is much less determined by, the nature or content of its empirical and positive elements. These elements, by themselves, could not possibly constitute either an adequate science or an adequate philosophy. Like Comte again, Cournot was interested in the classification of the various disciplines and branches of learning, but he argues that the principle of such classification must be logical rather than chronological or historical if the resulting classification is to have value. In the *Essai* and in his last two books on economics—the *Principes de la théorie des richesses* (1863) and the *Revue sommaire des doctrines économiques* (1877)—Cournot undertakes the twofold task of criticizing the classificatory systems of other authors from Francis Bacon to his friend André Marie Ampère, and to work out a more satisfactory system of his own. Finally, like Comte, Cournot was interested and creatively helpful in finding ways through which what are still frequently referred to as the social " studies " might be given at least some of the structured character of the accepted " sciences."

In an early article [1] and particularly in the *Recherches,* Cournot undertook successfully to apply the methods of mathematics to the

---

[1] "Application de la théorie des chances à la statistique judiciaire et à la probabilité," *Journal des mathématiques pures et appliquées,* IV (Louisville, 1838), 257–334.

analysis of social and economic questions. Cournot's interest in the rational concatenation of knowledge led him to propose no such absolute and non-positivistic principle as Comte's law of the three stages; his application of mathematical methods to the examination and possible solution of social and economic matters is a far cry from Comte's ultimate, romantic and completely amazing Religion of Humanity with Reason enthroned at the altar to be served—one is tempted, irreverently perhaps, to say serviced—by "scientists" as its priests. Yet in his last philosophical study, *Materialisme, vitalisme, rationalisme,* Cournot shows himself to be in fundamental agreement with Comte in viewing both mechanism and materialism as inadequate as well as untrue philosophies, however useful they may have been as hypotheses in examining limited problems among the orders of natural phenomena.

Cournot's philosophical position was median. Thus despite the merits of his work, despite the fact that he was well served both by his friends and by events which would seem to have favored his embarking upon a career of some note, and despite unusual practical and speculative abilities and attainments, Cournot's name and work did not catch the public eye and fancy. The quiet and retiring manner in which he lived—partly as a result of choice, partly of necessity—his impeccable honesty, his dislike of excess in action or in thought, his self-discipline which put him at odds with his undisciplined time, his refusal to have any truck with pamphleteers whose wares were slogans and clichés, his avoidance of intrigues of any sort, and his adamant refusal to use his friends to gain notice for himself, even when they themselves almost forced positions and honors upon him—all these worked against the possibility of his attaining any great fame during his lifetime. It is very likely that his defense of probabilism against the extremes of dogmatic skepticism, authoritarianism or rationalism on the one hand and of eclecticism on the other hand itself mitigated against his being accorded the serious attention of his contemporaries. In the end it seemed that death itself entered the lists against him: he died unexpectedly just before he could accept the honor of having at long last been elected a member of the Académie des Sciences Morales. Intimate as well as influential friends of Cournot had long urged him to allow them to present his name as a candidate for this high honor. But here again his great modesty and his aversion to any sort of solicitation, even in a matter as important as this, led him to decline. Finally, his health almost completely undermined, his life having all but run its course, he half-promised his friends that he would undertake to complete the preliminary formalities: but he died suddenly on March 30, 1877, before these preliminary steps could be completed. Thus he missed the opportunity to have his name inscribed upon the official roll which carries the names of so many of France's "Immortals." [2] Yet as M. D. Parodi says, "Among the illustrious thinkers of the nineteenth century, Cournot remains closer to us than any other; ... if not his books themselves, in all their parts, at least his spirit and his manner, his characteristic mien as a philosopher represent almost exactly the method and manner of the distinguished philosopher of our day." [3]

IV

Although essentially median, the philosophy of Cournot is more congenial to the continental tradition of rationalism than it is to the British tradition of empiricism. This congeniality is due in part, no doubt, to the fact that Cournot, like Descartes and Leibniz before him, was first of all a creatively competent mathematician. However, his rationalism was persistently critical and "open," rather than dogmatic and "closed"; it is a rationalism illuminated and made cautious by a careful study of Kant's critical philosophy. But Cournot's critical rationalism is as disposed to a realistic interpretation of the nature of things as Kant's critical philosophy was to an idealistic interpretation. Human knowledge is relative, to be sure, but it is relative to structures which are part of the nature of things, and not projections of the structures of our senses, understanding and reason. However these latter may or may not affect our grasp and knowledge of the connective structures among things, processes and events, they do not determine these structures whether as known or as existent. Cournot is confident, and apparently with good reason—except when excessive

---

[2] See Gabriel Tarde, " L'Accident et le rationnel en histoire," *Revue de métaphysique et de morale*, XIII, 3 (May, 1905), p. 347; and H. L. Moore, "Antoine Augustin Cournot," *op. cit.*, p. 543.

[3] " Le criticisme de Cournot," *op. cit.*, p. 452.

skepticism makes us unwarrantedly dubious about such matters in the wake of the collapse of some equally unwarranted dogmatism —that the "laws" or statements of rational, structural relations among things which give order to our knowledge and which are the implicit condition of successful action and behavior, lead us toward a grasp of objective relations among natural phenomena, relations which exist independently of us. Our approximate grasp of these relations is the measure both of our knowledge of these phenomena and of our ability to use them with any degree of understanding.

Cournot was not insensitive to those aspects of their common philosophic milieu to which Comte reacted with his positive philosophy. Comte was born three years before Cournot and died two years before Cournot set to work to prepare his own *Souvenirs*. In spite of the fact that they were contemporaries, there is little evidence that either was significantly aware of the work of the other.[1] However, we can be reasonably sure of this: Cournot did not agree with either the positivistic or skeptical limitations of Comte's position, nor with its strange culmination in the Religion of Humanity. Elements to which attention is called by positivism, by sensationalism, or by empiricism may well be among those which occasion inquiry, although such elements are not always primary or initiatory and are certainly neither final nor exhaustive. Knowledge can no more be validated by these elements than it can be limited to them however necessary it may be to refer to them for the purpose of verifying or refuting ideas or judgments as true or false.

Cournot insists that structures underlie our knowledge, structures which neither our senses nor any other sort of direct experience not only cannot discover, but for the discovery of which they are to a surprising degree both inadequate and unnecessary. Indeed Cournot argues that experience in any aspect would not only be intuitive but essentially inexplicable unless it were informed by the structure of things, processes, and events which enables us to grasp the "reason of things." These structures, related intrinsically to nature as ordered, are the basis of what we

---

[1] See Gabriel Tarde, "L'Accident et le rationnel en histoire," *op. cit.*, pp. 323–325; and R. Audierne, "Classification des connaissances humaines," *op. cit.*, pp. 509–519.

call the " laws " of nature. While these structures may never be understood absolutely—if, indeed, they are in any significant sense at all even conceivable as absolute—they may be progressively laid bare. For Cournot knowledge would be radically limited to description and to chronological reporting if this were not the case. Thus these " laws " are both metaphysically and epistemologically important; they represent our insights into the structures of nature which account for things " hanging together " in so far as they do " hang together." The data of the senses, and all the empirical happenings which William James so felicitously called " brute facts," and to which any investigator, scientific or otherwise, must ultimately refer if he is to verify or refute as true or false his own or another's hypotheses or rational constructs or the conclusions which may be inferred from them—these data, these " happenings " are dumb as well as brute. They can be given voice and made illuminating and decisive factors in knowledge only in so far as they are amenable to or refuse to be structured in terms of objective orders which give them significant relational locus and meaning. It is only upon the basis of and to the extent that we discover objective structures and " orders," and relations among things, events and processes that we have any basis for making a claim to knowledge.

This notion of objective " order " among phenomena which may account for sensations and other experiences but which in turn cannot be accounted for by the fact that we and others have sensations and experiences—this notion of " order," and the closely related idea of " the reason of things," is one of the fundamental and pervasive elements in the construction of Cournot's philosophy. Both the notion of order and that of the reason of things are related to the nature of things as well as to the condition and validity of our knowledge; indeed, they relate to the latter only because they relate first to the former.

Cournot gives numerous instances in which he makes it clear that our comprehension of phenomena goes beyond any possible experiential information. His observations in this respect could be extensively complemented by reference to the development of our understanding of various natural phenomena which have been the subject of inquiry since Cournot's death and which show rather conclusively our ability to come to know in a surprisingly extensive and adequate way structured relations among phenomena which lie beyond the competence and range of our sense organs. Yet Cournot eschews both the notion of a possible universal and absolute comprehension of nature—even by any superhuman intelligence or reasonable being which the universe may harbor—and the notion that a single all-inclusive warp holds everything together in a single, unbroken and singly concatenated web.

Cournot argues that reality presents a pluralistic network of independent series of events and processes which reveal necessary and determinable relations *within* each series, but not inclusively *among* all series. Therefore, while we can anticipate and predict events *within* a *given* series with assurance, once we have learned the law, that is, the structure of the series, other events which result from the *intersecting* of *diverse* and *independent series* may be anticipated only within such limits as the calculus of probabilities may determine. In the latter case we are limited to probabilities whether we look to the past or to the future: in fact, in his discussion of the nature of history and of how historical knowledge differs from scientific knowledge Cournot argues that we may be less able to recover the past with assurance than we may come to be able to anticipate the future. The past may well have included and undoubtedly did include singular events which have left no trace. To the extent that such events occurred and in proportion to their relative frequency, elements of the past would be left both unknown and unknowable with respect to their antecedents and to their consequences. Our means of becoming aware of and of recording the occurrence of such unique events in the present are both more extensive and more adequate than were those available in the past. Therefore the probability of our being able to previse the possible future ramifications of present unique events— within the limits of probabilities applicable to such events—is greater than our ability to recover evidence of the occurrence and of the nature of singular events once they have passed without trace. In both cases, however, uniqueness and singularity are coincident with the unpredictable and unanalyzable linkage or intersection of two series of events which are external to one another, except at the moment and in the manner of their intersection, so far as the implicit order and structure of each series is concerned.

Such events involve what Cournot calls *l'hasard*; they are

fortuitous, chance, and contingent occurrences. For Cournot this *contingency* is *objective*; it is *metaphysical* in the sense that it is intrinsically involved in the nature of things; it is not *epistemological* as would be the case were the grounds for it to be found to lie in the conditions and limitations which determine the subjective conditions of our knowledge. We are not simply ignorant of the nature of things which would be clearly revealed to a more adequate understanding; contingency among events and processes in nature [2] is so intrinsically a part of the nature of things that

[2] This position reminds us of that of Cournot's successor, Émile Boutroux. In his *De la contingence des lois de la nature* Boutroux raises the question as to whether or not the world may not manifest a certain amount of irreducible contingency. He examines the bearing of this question upon a number of persistent philosophical questions, including that of human freedom. This calls to mind the often unseemly and frantic haste with which numerous contemporary writers have seized upon the Heisenberg principle—which has grown out of attempts during the past seventy-five years to resolve certain problems coincident to the development of modern physics—as a basis for arguing the existence of human freedom, particularly with respect to its moral import.

It certainly seems unlikely that Cournot could have gone along with this interpretation of Heisenberg's work. For one thing, this principle is really one of uncertainty and not of indeterminacy: it says that if we wish to know certain things about physical quantities we must forego our desire to know simultaneously certain other things about the same quantities; it does not assert that the latter are unknowable but simply that they cannot be known along with certain other specific things under certain conditions; nor does it assert that the quantities which cannot be known under specific conditions may not be determinately related to the others which may be. In other words, the Heisenberg principle reflects first of all a limitation of our tools and techniques in the examination of certain questions. Although this limitation may rest upon an objective and metaphysical indeterminacy, the principle in question does not necessarily imply this conclusion—the limitation may involve simply technical or other epistemological inadequacies.

For both Cournot and Boutroux contingency has an unequivocally metaphysical significance however further our knowledge may be limited and the means of acquiring it complicated by epistemological factors. Therefore, while we may be able to reduce the extent of the probable in our knowledge, in so far as the probable is a function of our capacity to know and of what at any given time we can lay claim to as known, the probable cannot be completely eradicated from knowledge for what is known is in part contingent and to that extent unpredictable and unknowable prior to the fact of its becoming actualized.

And suppose the work of Heisenberg and Schrödinger does turn out to rest upon a metaphysical rather than an epistemological indeterminacy. This would at best establish indeterminacy as a part of the nature of things at the level

even superhuman minds would have to make use of and have their knowledge limited by the conditions of the calculus of probabilities.

In sum, the notions of " order " and of " chance " are equally objective in their reference and equally important in our understanding of the nature of things. Order is related to the actual, determinable interdependence among things, processes and events independent of our knowledge or ignorance of such actual interdependences: the category of *order* is the basis for our notion of the " reason " of things as distinguished from our idea of the " cause " of things. Our knowledge of the ordered and structured relations among natural phenomena varies as a large number of other epistemological, psychological, and cultural factors vary. But the *order* which makes possible and conditions our knowledge is not affected by these factors. Orders are real, but order is limited; we cannot deduce from it the notion of a necessary and exhaustively inclusive interdependence and interrelationship among all phenomena. The " brute facts " which smash any such overweening and dogmatic notion of a single and inclusive orderedness also exist.

Chance involves the existence of independent series of events whose coincident intersection—where they have previously been independent—or deviation—where they have previously been linked together—cannot be deduced from the most adequate possible knowledge of the nature of the reason for the linkages within each such independent series. Therefore, probabilities replace complete and theoretical certainty as the measure of our possible

---

of physical processes and events. The contingency from which any conclusions concerning the nature of man as a free and moral agent could be legitimately deduced would have to be a contingency established at the level of human and moral behavior. At least this is the case unless one accepts the implications of a physicalistic monism—in which the principles for all things are found in those of physical processes and events—or of a humanistic or spiritualistic monism —in which the principles of all things are found in those of uniquely human and particularly in those of moral behavior. Cournot would reject both of these alternatives.

Like the later proponents of " emergent evolution," Cournot does not believe that the nature of living and still less of rational beings can be exhaustively deduced from the nature of physical, that is of non-vital and non-rational existence. There is no one least common denominator to which everything— matter, life, mind and those dimensions of the real which we denote as moral or ethical, aesthetic, or religious or spiritual—can be reduced or from which everything could be deduced. Cournot's pluralism is fundamental.

knowledge of the relatedness of phenomena even when, as not infrequently happens, probabilities are so high as to give practical certainty, particularly in concerns of an essentially pragmatic nature. But contingency and chance must not be equated with that which is unexpected, or with that which surprises us. What is unexpected or occasions surprise is a function of epistemological, psychological or cultural factors. Natural objects, processes and events, independent of our experience or knowledge of them, involve both order, i.e., intrinsic relatedness, and chance, i.e., intrinsic unrelatedness. Both order and chance have their foundations in the nature of things; neither stems from the conditions of our awareness and understanding however limited or extensive, incidental or necessary these conditions may be.

V

Before more is said about Cournot the philosopher, let us say something of Cournot the man. He was born on August 28, 1801, in the town of Gray [1] in the Haute Saône, not far from the former ducal city of Dijon. In his *Souvenirs* [2] Cournot tells us his birth

[1] Within recent years the College at Gray has been named after Cournot.

[2] This is the only source of primary biographical information about Cournot which we know to be extant. When Monseigneur Bottinelli set about to publish the *Souvenirs* in 1913, he was given a number of letters and other papers by A. A. Cournot belonging to members of the Cournot family. When in the company of Cournot's grandnephew, Professor Jean Cournot, I visited Monseigneur Bottinelli for the first time, Professor Cournot asked about these materials. We were told that the papers had been passed on to a Belgian priest. In spite of his efforts to do so Monseigneur Bottinelli has not been able to find either the priest or the papers since the end of the Second World War: no one can say whether one or the other or both may not be casualties of that struggle.

In the hope that some clue to the whereabouts of these family papers might be uncovered, should they have survived the war, I asked Professor Jean Wahl, who is presently the editor of the *Revue de métaphysique et de morale*, to insert a note of inquiry in that journal to see if someone might be able to give any information about the whereabouts of the papers. So far this request has brought no response. Since I do not know exactly how many or what kinds of papers were included in this collection, it is impossible to guess how much or how little importance they might have to students working on Cournot. It was evident to me that Professor Jean Cournot thought these papers would be of interest to me in my study.

was hastened by the onset of the Revolution. Probably because of the size of his own family Cournot spent his early years not with his parents but with a grandmother, an uncle and two maiden aunts. He tells us that the influence of these relatives contributed greatly to his development and particularly to the independence of his judgment. The uncle was a Jesuit who remained faithful to his belief throughout the trying anti-clerical years of the Revolution and the period which followed during which, under Napoleon, the form of Catholic institutions was revived primarily to serve the interests of the Emperor. Cournot says that he owes to this uncle everything good that is to be found in him. The older of the two aunts was identified with movements and held to opinions diametrically opposed to those espoused by her brother and other members of the family. Against this background of differences Cournot developed the capacity to reach his own conclusion—and to keep his peace. Here seem to be the roots of his tendency to keep himself aloof from active participation in political matters and to be moderate in such political opinions as he did express.

In 1809 Cournot entered the College at Gray where he continued his studies until he was fifteen. In the four years which followed he studied alone, seeking to gain some knowledge of law [3] and other subjects which interested him, particularly mathematics and natural science. Convinced that a knowledge of mathematics is a necessary condition for an understanding of the sciences, he hoped to go on with his study of mathematics at the École Normale in Paris. He completed his requirements, which consisted partic-

---

If these papers could throw more light on Cournot as a person, they would be interesting without doubt. For while the *Souvenirs* are Cournot's own record, they are written in a form so impersonal and objective that we get from them no vivid picture either of Cournot himself or of his family, of which he really says very little. He used his *Souvenirs* as a medium through which to record his judgments about issues larger than those which affected him or his family alone or even primarily. Furthermore, he unfortunately added nothing to them after their compilation in 1859. Thus we lack any autobiographical statement whatsoever concerning almost a third, and in some ways the most important third, of his life.

[3] It is of passing interest to note that on the basis of this study Cournot won a lawsuit which was the source of considerable income to his family when he was only seventeen years old. In this incident we may well see evidence of Cournot's acumen, of his self-reliance, and of his ability as a student working without benefit of formal instruction.

ularly of some work in mathematics, at the College at Besançon. In August, 1821, he was notified of his admission to the École by Baron Georges Cuvier.[4]

While a student in Paris Cournot acquired knowledge, experience, and friends which stood him in good stead later on. It was inevitable that faculties and students became embroiled in the political controversies of the time. The director of the École Normale at this time was a royalist; most of the students were not. Cournot tells us that he alone, or nearly alone, spoke for a " juste milieu." In 1822 the École Normale was suppressed by the government: some of the students received appointments to other institutions as instructors; others were dismissed and given a pension of fifteen francs a month for a period of twenty months. This latter group, in which Cournot found himself for reasons he never understood, was more or less under the surveillance of the police.

Through the efforts of mutual friends, Cournot was employed in 1823 by Marshal Gouvion-Saint-Cyr in the dual capacity of tutor for the Marshal's young son and as his own personal secretary. In the latter capacity his primary responsibility was to serve the Marshal as a critic and advisor in the editing of his *Mémoires sur les campagnes de l'Armée du Rhin.* Cournot retained this post for ten years and made the most of the considerable freedom and means which this employment gave him. This decade proved to be one of the most fruitful of his whole life. Living as he did in the environs of Paris, he was able to make progress with work of his own choosing, to keep in close touch with his professors and friends in the city, and to obtain through direct acquaintance some insight into the character and the habits of military leaders and other men of affairs.

During these ten years he finished his program of formal

---

[4] In addition to being a philosopher and one of the greatest naturalists of modern times, Cuvier also held many important political posts during his lifetime. In 1808 he was appointed councilor of the Imperial University. During the next several years he presided over commissions whose function it was to organize and visit academies and colleges in Italy, Holland and other countries brought within the French orbit by Napoleon's conquests. Napoleon appointed him Master of Requests in 1813, and Councilor of State in 1814. From 1814 until his death in 1832 he was president of the committee of the interior and rendered important civil services to the state.

studies, having completed the two theses then required for a degree —one in mechanics, the other in astronomy.[5] He also published numerous articles and critical notes. These dealt with a variety of subjects, although most of them concerned matters pertaining to mechanics and mechanical mathematics.[6] Two of these articles, "Application de la théorie des chances à la série des orbites des comètes dans l'espace" which was included as a supplementary chapter in Cournot's translation of Herschel's *Treatise on Astron-*

---

[5] *Mémoire sur le mouvement d'un corps rigide soutenu par un plan fixe* (Paris: Hachette, 1829), and " De la figure des corps célestes," unpublished.

[6] Many of these appeared in the *Bulletin des sciences mathématiques, astronomiques, physiques et chimiques du Baron de Férussac*, Volumes 6 through 16.

Due to the fact that they are not clearly and unambiguously signed there is some question as to the authorship of some of the pieces which have been written by Cournot. In one or two instances Cournot's authorship is established by his use of the signature " C . . . . . t "; however most of the articles and notes are identified simply by the initials A. C., leaving some question as to whether some of them were written by Cournot or by Augustin Cauchy, who was also a regular contributor to this journal and whose initials were the same as Cournot's.

This ambiguity is mentioned by Monseigneur Bottinelli on page 273 of his study *A. Cournot: Métaphysicien de la connaissance,* and on page vii of the excellent study of Cournot, *De l'ordre et du hasard,* by Professor Jean de la Harpe. Published by the University of Neuchâtel in 1936, this book is a comprehensive and valuable study of Cournot's thought. The subtitle of the work is *Le réalisme critique d'Antoine Augustin Cournot.**

After making a study of the disputed items mentioned above, it is my opinion that certainly most, if not all, are by Cournot. My judgment in this matter is based upon three considerations: 1. after carefully checking the indices of the various issues to which both men contributed, I discovered that Cournot always signed his pieces in one or the other of the two ways indicated above, whereas M. Cauchy's are frequently unsigned; 2. in many cases it is possible to establish that Cournot is probably the author on the basis of distinctive differences in the style of the two men; 3. a study of those items which are unambiguously identified by the indices as belonging to each author shows that the disputed pieces deal with topics on which Cournot is more likely to have written than Cauchy. Many of these items are so unimportant, however, that it is a serious question as to whether or not it is worth while to try to settle the matter of their authorship.

* Professor de la Harpe means essentially the same thing by his " critical realism " that I have hoped to suggest by the term " critical rationalism." I chose to avoid the former name because of its use by the group of American philosophers who wrote, under the name of " critical realists," a rejoinder to two other co-operative volumes published in this country early in the twentieth century—*The New Realism* and *Creative Intelligence.*

omy,[7] and "Application de la théorie des chances à la statistique judiciaire et à la probabilité" are of particular interest because they are certainly among the earliest attempts to apply the techniques of mathematical probability, that is of statistics, to the solution of specific problems concerning physical phenomena, on the one hand, and social phenomena, on the other.[8]

VI

Some of Cournot's articles came to the attention of Siméon Poisson, who, following the death of Laplace, became the recognized authority in mathematics at the University of Paris. He was very favorably impressed by what he read and undertook to find a position for Cournot where he could use his obvious talent to best advantage. Following the death of Marshal Saint-Cyr and the publication of his *Mémoires* under the direction of Cournot in 1829, our author's friends insisted that he offer his services to the University. Since he did not know Poisson personally, Cournot assumed with typical modesty that he would be given a minor post in the provinces. Instead, he was at once given a temporary position at the University of Paris until a suitable permanent post could be found for him.

Cournot was never happier than he was during this time (1833-1834); and in no other single year did he enjoy so great a measure of "success" as this term is commonly used. He published his translations of Herschel's *Treatise* and of Kater and Lardner: *Elements of Mechanics* to which he added a twenty-third chapter "De la mesure des forces du travail des machines." It is true these two books were only translations, yet their publication brought Cournot some money and also added to the reputation which his numerous articles had already won for him.

Following this year, Cournot filled a series of increasingly important posts, beginning with his appointment as Professor of

---

[7] This translation was published in Paris by Paulin in 1834. In 1835, a pirated edition of it was published in Brussels by Hauman to which Cournot added, with complete impartiality, a second supplementary chapter, this one entitled "Distribution des orbites cométaires dans l'espace."

[8] Cournot later applied the techniques of mathematical analysis with telling effect in the *Recherches sur les principes mathématiques de la théorie des richesses.*

Analysis and Mechanics at the University of Lyon. This position interested him very much and he was happy to be located once more near his home in Burgundy. Furthermore, he had remained in Paris long enough to carry through to completion the work he had in mind when he accepted the obscure position as Marshal Saint-Cyr's secretary. Yet Cournot had been at Lyon only a year when Poisson, without consulting him, made him head of the Académie at Grenoble. Cournot hesitated for some time before he accepted this post for he was happy at Lyon. He was finally persuaded by his friends to return to Paris to accept the appointment.

Cournot's success at Grenoble raised him still higher in the estimate of his friends and assured him further advancement. In 1836, while Ampère and Matter [1] were making the regular tour of the Inspectors General of the University, the former died at Marseilles. Cournot was named temporarily as Ampère's successor and served as an Inspector General for two years without relinquishing his post at Grenoble. In 1838, Cournot was named titular Inspector General as a result of efforts by Poisson and other friends. This new honor both surprised and disappointed him. Cournot loved the valley of the Rhone. He hoped to stay on at Grenoble at least long enough to carry through to their completion plans which he had conceived there but had been able only partially to put into effect. Therefore, it was with real regret that he again left for Paris to take up his new duties.

During the next few years new proofs of the respect in which he was held followed. His health declining, Poisson could do no more than go through the form of fulfilling his official duties. He asked Cournot to preside in his place at the concourse for the examinations in mathematics. Cournot continued to perform this function for fourteen years and in doing so gained almost the same standing among his colleagues as he would have had had he been a member of the *Conseil Royal*. Poisson died in 1840. When Cournot's *Traité élémentaire de la théorie des fonctions et du calcul* was published the following year, it was dedicated to his great friend and benefactor. This treatise grew out of conferences which Cournot held while he was teaching at Lyon. The time which elapsed between the conception of the work and its publication is

[1] Jacques Matter, 1791–1864: a philosopher and historian who was appointed Inspector General of Libraries in 1835.

typical of the care and perseverance with which Cournot addressed himself to all his work. It was not uncommon for him to spend ten to twenty years or more working on projected volumes before he considered them ready for publication.

The next few years marked the second very fruitful period of his life. In 1842 he published his annotated translation of the *Lettres d'Euler à une princesse d'Allemagne*. His *Exposition de la théorie des chances et des probabilités* which appeared in 1843 was followed in 1847 by his final mathematical study *De l'origine et des limites de la correspondance entre l'algèbre et la géométrie*. Finally, in 1851, his first philosophical work, the *Essai*, was published.

This development is significant for it is tangible evidence of Cournot's awareness that fundamental questions underlie and permeate the sciences. The sciences may serve well the function of bringing these questions to light; they cannot, however, undertake the task of attempting to find answers to them. That task belongs to philosophy, and involves much more than any kind or quantity of positivistic knowledge, techniques, and data which philosophy may press into its service. Cournot's respect for science is great; he is well aware of the invaluable theoretical and practical contributions made to the body of human knowledge as a result of the development of science; he understands, too, how far afield a philosophy may go if it cuts itself loose from those corrections and limits furnished by the sciences which keep any system as close-hauled as possible to well-attested fact. But these and other contributions of science neither constitute philosophy nor do they make it unnecessary.

Here we see the root of the marked difference between Cournot's contribution to philosophy and that of his contemporary Auguste Comte. Comte attributed the difficulties and errors of all earlier philosophies to the fact that they had not rid themselves of theological and metaphysical assumptions and hypotheses: philosophy becomes adequate to its task only as it eschews these primitive elements and moves toward the positive stage in which it casts off all vestigial remains of earlier stages. Not so Cournot: to a limited extent in the *Traité* and more specifically and amply in the *Considérations* Cournot indicates why he considers such a thesis both misleading and erroneous.

One would have thought that the period to which we have just been referring might well have been not only one of the most productive in Cournot's life but also one of the happiest. This was not the case, however. He found the work involved in his office as Inspector General burdensome and not infrequently irksome. Furthermore, his life had been made more complicated by his appointment as Rector of the Académie at Dijon in 1854. He became increasingly disgusted with his position as Inspector General, primarily because of his disappointment over measures the government took with respect to the University under the Second Empire. In 1859 he resigned his position at Dijon to work on his *Souvenirs* and to complete the *Traité de l'enchaînement des idées fondamentales dans les sciences et dans l'histoire*, which was published in 1861 and which he thought was to be his last book. In 1862 he retired from all other public functions and went to Paris to live.

By this time an important group of economists had established without question the importance of his first book on economics. Cournot had been greatly disappointed that this phase of his work had attracted no more attention than had been accorded to it and had apparently given up any idea of further work in this field. However, friends felt that the limited notice accorded the *Recherches* was due largely to the mathematical form in which it was written and urged him to prepare another study in which his ideas would be presented and amplified in a more literary form. Therefore, in 1863 he published his *Principes de la théorie des richesses*. But this book, too, went largely unnoticed. In 1864 Cournot published *Des institutions d'instruction publique en France*; this is a sort of unofficial report to the French people based upon his long and varied experience with the French University system and includes many of his pedagogical views.

VII

Another long period—this one of eight years—elapsed before Cournot published again. In 1872 the *Considérations sur la marche des idées et des évènements dans les temps modernes* appeared. This was followed in 1874 by the last of his specifically philosophical works—*Matérialisme, vitalisme, rationalisme*. Finally, in 1877, the year in which he died, his *Revue sommaire des doctrines éco-*

nomiques was published posthumously. This last book was a third attempt to present his economic theories in a form that might attract some attention. The *Considérations* and the *Matérialisme* represent some of Cournot's more interesting works and are worthy of brief special notice.

The *Considérations* is primarily a philosophical interpretation of history in which Cournot attempts to apply to the unfolding of the modern period concepts developed and organized in the *Essai* and the *Traité*. In this study Cournot attempts to substantiate his claim that history depends upon the independence of certain unique events. These singular occasions are the "brute facts" by which the historian is inescapably bound. For this reason history can never be a rational science after the model of physics and chemistry because it cannot give us the kind of certainty that results from our ability to show a definite rational concatenation among a group of objects, processes, and events: history cannot give us certainty. The historian must be satisfied with the same sorts of analogies, inferences, and non-mathematically determinable probabilities as the philosopher, and for essentially the same reasons. The object of science is to discover uniform rational laws; that of history is the discovery of the implications of singular events which, however important or unimportant they may be, cannot be deduced as a necessary consequent of some specific, structured pattern of events. These historical "facts" are accidents, contingent events which may deeply affect the development of both individuals and groups. But these facts must be recognized and dealt with for what they are—individual, independent occurrences which cannot be reduced to laws. Of equal interest in this contribution by Cournot to the study of the movements of thought in the modern period is his thesis that it takes approximately a century for the unique excellence of each era—whether this forte be in the arts or the sciences—to develop from the early stage of the consideration and tentative collating of individually interesting items and expressions of individual spirits to the discovery of their places in a rationally structured order, to the extent that this can be established at all.

The concept of evolution quickens many aspects of the thought of the nineteenth century where it is suggested in such diverse philosophical principles as Comte's law of the three stages and the triadic synthesis worked out in successive expressions by the German Idealists and finally stood on its head by Karl Marx. But in the same year in which Cournot undertook the preparation of his *Souvenirs*, Charles Darwin published the work that seemed to bring the notion of evolution from the level of a nonverifiable principle of philosophic speculation to that of a verifiable hypothesis within the ken of scientific statement and elaboration.

This change had effects too well known to need repeating here. But one thing we must speak of—the effect of this new theory on the thought of Cournot. Cournot had not, of course, dealt specifically with matters which were affected by Darwin's work. Yet the problem of the relation of living to nonliving forms, of sentient and thinking beings to those that are not was a matter of continuing concern to him. This concern is indicated tangentially as an implication of his persistent interest in the problem of the proper ordering of the fields of knowledge into a hierarchy which would reveal their logical relations and dependencies; it is indicated more directly in the consistent if limited way in which he suggests in both the *Essai* and the *Traité* essential differences between those investigations which have as their subject matter the objects, events, and processes which make possible the development of the sciences and such humanly oriented fields of study as economics, history, psychology and ethics, whose subject matter precludes the possibility of their ever becoming rational sciences.

The *Essai* was published eight years prior to the appearance of Darwin's book which stated his thesis as to how new forms come to be: the *Traité* appeared in 1861, two years after the *Origin of Species* was published. The *Essai* was therefore too early to have any special concern with evolutionary views: the *Traité* was finished before the impact of Darwin's work was fully felt and had as its purpose not the critical examination of a new theory of as yet unknown merit but the systematic restatement of ideas already presented more discursively in the *Essai*. However, it is important to note that the idea of a special vital force which is hardly more than suggested in the *Essai* is made one element of a basic dualism in the *Traité*.

Matérialisme, vitalisme, rationalisme was published in 1875 with the suggestive subtitle " Etudes sur l'emploi des données de la science en philosophie." In this examination of the uses that philosophy may legitimately make, indeed must make, of the data of the sciences Cournot not only lays still more stress on this notion of a vital force, but here this force is presented as being a perfectly natural successor—not consequent—to the mechanical and physical forces which precede it but which neither determine nor necessarily condition it. Despite the title, reason is not the final expression of the vital force. Instinctive elements, elements of feeling interpreted in the light of spontaneous aspirations exist and refuse to be determined and fixed according to rational measure. They overstep reason into a realm of the *transrational* without having undergone that sort of transformation which brings about the idea of the images and the affections of the sensibility. Here, then, is the last stage of Cournot's philosophical development: a reaction against the Darwinian hypothesis, which Cournot felt was essentially materialistic; an anticipation of the vitalism of Bergson and of other philosophies in which the organic features, the gestalt so to speak, of objects, processes, and events in all fields of investigation to which increasingly greater attention has come to be given; and a specific anticipation of the position developed by S. Alexander, C. Lloyd Morgan and others at the close of the nineteenth and the beginning of the twentieth century, and which is commonly known as "emergent evolution."

VIII

Coming back now to a consideration of Cournot's philosophy, we find that even a cursory sketch of the development of his thought makes evident its affinity to rationalism. This affinity exists whether one attends to the presuppositions upon which his thought rests or to the conclusions he reaches after having explored the implications of these presuppositions as they bear upon the investigation of problems in the three major areas of inquiry in which his work was done.

Specifically the clue to Cournot's rationalistic bent is the concept of order. He makes a twofold use of this concept: on the one hand, its referent is those structured relations among things which our minds may approximate with increasing adequacy even though they may not be grasped in any absolute sense; on the other hand, its referent is "the reason of things" in the essen-

tially Leibnizian sense of our ability to know things in the order
through which they give rise to one another. Cournot makes his
respect for Leibniz clear in the *Essai* and elsewhere. He is particularly taken with the bases for, and the implications of, the
latter's principle of sufficient reason, even though he criticizes
it by pointing out that the adjective is redundant since it would
be absurd in any context to refer to the insufficient reason for
anything. In its first meaning *order* is the mark of *objective reason;*
in its second reference *order* is the mark of *subjective reason.* We
have knowledge when our minds conceive the reason of things to be
coincident with their objective concatenation. The major function of philosophy therefore, at least as Cournot sees it, is to examine and to criticize perennially the efforts of subjective reason as it
tries to lay hold of objective reason. This places Cournot squarely
within both the realistic and the critical traditions.

The discovery of the reason of things is not to be confused with
the discovery, first, of bases for the classification either of things
or of the fields of study; secondly, of the conditions of what we call
causal as distinguished from rational order, or; thirdly, of the
nature of logical dependencies among our concepts. The principles
upon which our classificatory, causal, and logical systems rest may
be identical with principles implicit in the objective order to which
the notion of the reason of things refers; but there is no necessary
reason why this must be the case. Finding a principle of classification is something other than finding the reason of things. This
is made evident by the fact that classifications may have pragmatic value and utility whether they be artificial or natural, whereas only the latter could be relevant to our grasp of the reason of
things. In fact, the very distinction between artificial and natural
classifications suggests that there may be something quite limited
and arbitrary in the principle of differentiation in terms of which
a classificatory scheme is to be devised—some of our classifications
being relevant to very limited and to peculiarly human purposes
and interests.

There would be no arbitrary element in any system of classification if it could be demonstrated that it rests unequivocally
upon the reason of things clearly discerned and accurately applied.
We call natural those classifications which come closest to being
based upon objective relations among the items classified, in so far

as we have valid criteria for differentiating items on the basis of such relations. As our knowledge increases and our principles become more adequate, each refining revision of our classification will give successively more adequate approximations of the objective nature of, and relations among, the items under consideration. But since our knowledge is relative and not absolute, we should never assert categorically that any given classification really rests upon and reveals the objective order or reason of the things which have been classified—though once more we may hope, and sometimes with good reason believe, that this is the case. Cournot refers to such classifications as rational or natural because they rest upon principles which may be expressed in the form of rational laws. As a matter of fact, any belief that the nature of things has been discovered is only probable. We shall examine below the factors which condition and determine this probability.

Similar remarks apply to the distinction between rational and logical order. This distinction is analogous to what is commonly understood by the distinction between the "objective" and the "subjective." The "rational order" is objective in the sense indicated in the preceding discussion: it pertains to things in and of themselves; it is the order which is the object of man's quest when he seeks to know, in the sense of what he *discovers,* and not at all in the sense of what he *creates.* On the other hand, "logical order" is an order among propositions which in itself need give no evidence of external relevance or of actual existence. Thus, for an idea to be *valid,* in the sense that it is consistent with a more or less extensive system of interrelated propositions and in the sense that it violates none of the laws of logical discourse, has no necessary bearing whatever on the issue of its *truth* or *falsity* when reference carries beyond the necessary relation among ideas to external or objectively real matters of fact.

Finally, when Cournot speaks of the objective order in terms of which one thing is rationally connected with another, he does not have in mind what is commonly referred to as the "relation of causality." As he uses the term, "cause" is inseparably connected with the idea of action or force; it is the productive agent in any situation in which a change occurs. He means by cause essentially the same thing Aristotle—a philosopher with whom Cournot was well acquainted and whom he read with great respect

—meant by efficient cause. Reason, in its objective reference, is timeless, cause is not; in fact, the latter unconditionally involves the idea of time. The action of processes or forces under the dispensation of order is determinate; the action of causes, properly so called, is the condition for the accidental and fortuitous.

The intrinsic reason for any phenomenon is not to be confused with the conditions or circumstances required for its production: the former has to do with the science of the phenomenon; the latter relates to its history. This is the case since there enter into the production of every natural object, event, or process certain fortuitous elements which cannot be reduced to rational order. Chance events are irrational in the sense that it is impossible to predict them and in the sense that it is impossible for reason to reduce them to laws after they have occurred. Finally, not only is the idea of the reason of things nontemporal, but neither does it imply the idea of a necessary and irreversible sequence.

On the other hand, the notion of cause and effect may be illustrated by a definite succession of the sort known as a linear series. Such a series is, to be sure, one type of order among events, but it is not the only one. Order manifests itself in many ways which are unsuited to this type of representation. In other words, the concept of order or of objective reason is more inclusive than that of cause and effect. Therefore, to equate the two would be to confuse the part with the whole.

From another point of view the distinction between reason and cause is that between law and fact. This is a point of no little importance when we turn our attention to the more empirical and positivistic elements of Cournot's philosophy.

IX

Cournot deeply regretted the schism which began to develop between mathematics and philosophy after the close of the seventeenth century. But he would also regret and would look with disfavor upon certain recent and contemporary developments in which philosophy seems to be equated with mathematical or logical analysis. The formal or rational sciences—of which mathematics is one—take logical, although not temporal, precedence in the systematic organization of human knowledge. Intelligible things considered in and of themselves are logically prior to, surpass, control and give reason to phenomenal and sensible things. The reason for this priority is twofold. On the one hand, mathematicians have a feeling amounting almost to conviction that a " law," that is, a determinable correlation, may be discovered through which nearly any body of data may be related in one way or another with any other, no matter how numerous, disparate and unrelated the data may be. On the other hand, mathematics also has the positive value of serving as a check on empirical studies. So far as it is possible to do so, mathematics helps to keep the empirical sciences free from the excesses into which it is quite as possible for an empiricism as it is for a rationalism to fall, however different the roots and consequences of such excesses may be. Empirical scientists are and must be as interested as are philosophers—although perhaps for somewhat different reasons—in ideas of order as well as in processes of movement and change, in rational explanations as well as, and indeed ultimately more than, in empirical description. No science has any significance for knowledge, if it has any meaning, divorced from the idea of law which carries back to the assumption of some objective order.

Reason and purely rational science are not without their own limitations, however. Mathematics serves to warn us against, and in certain cases at least to enable us to rectify, empirical or "material" fallacies. Furthermore, mathematics has given us insights into theoretical possibilities which are not empirically discovered but which may be empirically explored; it has given us numerous techniques which have made it possible for us to work with problems and with types of data of various sorts which would otherwise be beyond our abilities. But in mathematics and logic demonstration proceeds through the use of reason alone; the possibility that a rational science may be applied with success to practical matters is entirely incidental to the discovery of a concatenated set of necessarily related propositions. Both mathematics and logic may fairly be said to consist of the deductive elaboration of hypotheses or of sets of primitive propositions: conclusions consistently developed within the limits of allowable procedures in the rational sciences may be completely and demonstrably certain within the limits of objective relations of implication. But although mathematical and logical conclusions can be shown to be formally necessary and certain, they are perhaps never wholly verifiable—at least in the sense of their maximum possible extension—within the limits of techniques of empirical investigation. This leads to two possibilities: either there are limitations within the rational sciences themselves, at least so far as their utility in enabling us to get exact information about the real as over against any possible worlds is concerned—including the best of these latter; or these limitations must be admitted and accepted as " brute facts " which suggest the existence of certain nonrational elements in reality as well as in experience, unless we propose to dismiss as irrelevant to and as unimportant for the business of knowing those empirical phenomena which resist exact rational formulation.

Cournot accepts certain implications of each of these two alternatives. For better or for worse we must face the fact that human knowledge is relative and not absolute. This means that in spite of the ideal possibility they can set up and develop, and in spite of other characteristics they have which make them unconditionally necessary elements in the organization of our knowledge, the limitations which affect the rational sciences are just as real as those which affect the empirical science. Under the impulsion of such ideally universal and rational principles as those of the continuity and the uniformity of nature rational sciences err in the direction of assuming, on purely formal grounds, connections among objects, processes, and events where none are now empirically determinable and where, perhaps, there are in fact no connections of the types posited. The contrary difficulty with the empirical sciences is that they fail to discern ordered relations among collections of data until such relations have been suggested as possible through the application or the rational extension of some accepted hypothesis, principle or law, or until some judicious guess or intuition has suggested some verifiable relational matrix among the data concerned. Cournot's insistence upon the necessity of the conjoint use of empirical data and rational principles in the development of knowledge points in the direction of the way in which we may hopefully look for a breaking down of the antinomy between traditional rationalism and traditional empiricism. In this he proposes to set up theoretical bases to exploit as well as explain what had already been done in practice by those whom Alfred North Whitehead identifies as having given us that

"Century of Genius" whose pivot is the year 1642 in which Galileo died and Newton was born.[1]

If we lived in a perfectly rational universe we should never have to worry about establishing the truth of ideas beyond showing that we have made correct use of the principles of formal reasoning: in such a world the exploration of the implications of these principles alone would lead us consistently toward perfected knowledge. But Cournot insists that the data of experience reveal indubitably that the world with which we are dealing contains discontinuity as well as continuity. It is convenient for the human mind to think of all things as interconnected. Furthermore both the practical and theoretical pressures that force us into the quest for knowledge seem almost irresistibly to equate this quest with a quest for certainty. Spinoza's ideal of a completely determined rational system in which the total nature of man, of his world, and of their relation can be exhaustively explored by tracing the logical implications of a set of presumably self-evident truths seems to answer to a common yearning of the human mind. But the ideal of the complete interconnectedness of reality is a transcendental postulate from which we can conclude with certainty nothing as to the actual nature of things. To become effective for knowledge, any such rational principle must be checked against experience and in this process be verified or refuted either in its general or in its particular applications to matters of fact. There is no denying the fact that the history of science shows us some interesting developments related to the application of such rational principles as those of continuity and the uniformity of nature. Yet in even the most rational of the empirical sciences, when we come to deal with actual objects, processes, and events, we find ourselves faced with the problem of being unable to reconcile universal rational principles with the brusque termination or origin of series of events, with the sudden and unexpected appearance and disappearance of objects.

X

This conjunction of the rational and the empirical, of the universal and the unique brings us to the consideration of another distinctive and pervasive characteristic of Cournot's thought—

[1] A. N. Whitehead, *Science and the Modern World*, Chap. 3.

his probabilism.   In spite of their empirical limitations universal principles have demonstrated their pragmatic value to us over and over again.   Indeed we cannot avoid the use of such principles if we are to attempt, let alone succeed in any intellectual endeavor beyond the minimal requirements of an immediate, descriptive reporting of the content of experience.   Yet the occurrence of empirical consequences of the sort implied by such rational principles cannot be regarded as verifying the *truth* of such principles with respect to matters of fact and this for two reasons.   In the first place, every particular instance of a rational concatenation which is detected because of some implication of a universal principle assumes the principle itself; in the second place, specific limited and empirically discovered uniformities and other relations deemed to be verifying instances of a rational order are the logical consequents of the general principle.   The truth of the antecedent principle cannot be established by instances which verify the truth of its logical consequents even though the possible and even probable truth of the general principle may be increased in this way.   How then can it be that general principles whose truth cannot be established have proved to be so useful in bringing together what, prior to the development and application of particular rational concepts, had appeared to be completely unrelated phenomena?   For Cournot the answer to this and related questions is to be found in an examination of the nature and conditions of probability.

Before we proceed to examine Cournot's use of probability theory, however, we must mention briefly one other practical and theoretical barrier to the metaphysical and epistemological application of universal rational principles.   This barrier is implicit in one of the specific objectives of science—the desire to attain the utmost clarity.   To be clearly discerned means to be distinguishable from other things.   In other words, clarity, in this respect, is a function of discontinuity rather than of continuity.   This is neatly pointed out by Cournot with reference to the concept of number.   The very idea of number is inextricably bound up with the notion of an unlimited and continuous series.   Yet within the series the actual succession of separate and distinct numbers, between any two of which an indefinite number of others can be imagined to fall, is one of the clearest examples we have of discontinuity. In fact, number series are so completely discontinuous that being given a certain number we cannot say what the next number in the series is unless we have previously explicitly defined the nature of the series and also exactly what we mean by the " next number."

How, then, may cases in which such rational principles as those of continuity and uniformity lead us to the discovery of actual relations—that is, of relations which reflect the reason of things in the realm of actual existences—be distinguished from those in which an extension of a law in terms simply of its formal and logical implications would misrepresent situations which are as a matter of fact discontinuous? This is both an interesting and a fundamental matter for it involves the criteria used by Cournot in his attempt to work out the rational character not only of his philosophy but of the universe as well, in so far as each is rational. It is here that Cournot touches upon the basic problems of induction which so largely occupied the attention of his English contemporary John Stuart Mill, as they do the attention of many of the most competent logicians of our own time.

Cournot applies the techniques of probability analysis to problems of this sort. The final limiting factor is the improbability, and in some cases—particularly those most amenable to mathematical treatment—the impossibility of our accepting any other theory than that which probability leads us to regard as disclosing the reason for a group of phenomena of whatever sort when these are brought together under the form of a law. For example, we may refer to the case of the law which expresses the relationship between the volume and pressure of gases under "normal conditions," a general principle called Boyle's law by English-speaking peoples and Mariotte's law by the French, after the two men who independently identified the relation in question. It is more probable that the discovered relationship is an intrinsic characteristic of the nature of the behavior of gases independently of the numerous particular sets of observations which are the basis, first, for the formulation of the principle and, later, for the innumerable verifications which have been made of it, than it is that for some entirely unknown and fortuitous reason we should have chanced to stumble upon and measured just those instances which are amenable to rational co-ordination under a regular principle. The latter chance is of the sort that almost any reasonable person would regard as highly improbable even though it must be confessed not to be either physically or logically necessary. The probability that the correlation of pressure and volume indicated by the law under specified conditions is objective—in the dual sense that it gives us some insight both into the *nature* and into the *reason* of things—is greatly enhanced each time theoretical values interpolated between or extrapolated beyond empirically established values are found to have been correct theoretical predictions of subsequently established empirical values, within the limits of allowable experimental error. It is highly improbable that such a correlation is fortuitous. Yet it is equally true that the purely theoretical application of the principle becomes increasingly improbable as we attempt to extrapolate values further and further removed from those which have been verified empirically.

In situations of the sort we have just referred to we are guided more by a feeling for the continuity of forms than by anything else in setting up one principle rather than another. But in such instances the separation of the true from the false does not result from formal logical demonstration; it results rather from judgments based upon the relative probability of the truth of one principle as compared to that of others. As should be expected in these circumstances, the estimate of probabilities varies from one mind to another. Yet in some cases the probability of a given principle is so high that only those who are perversely skeptical will refuse to accept it as indicating the reason of things with a degree of probable truth that amounts to practical, although of course not theoretical, certainty. Judgments of probability in cases of this sort rest upon the preference of reason for the condition of simplicity, generality, and symmetry. All these contribute to our understanding of the reason of things. It is thus that Cournot introduces simplicity as an indication of order. The importance of simplicity in this capacity lies in the fact that on this basis we choose as more probably true those among our hypotheses which either lack or share equally other criteria of order.

Reason is still not entirely self-determinate, however. We have mentioned above the decreasing probability of the truth of such an empirically verified principle as Boyle's law when it is extended beyond confirmed values purely on the theoretical basis of an assumed continuity and uniformity in the nature and behavior of gases. Boyle's law is in many ways a model of simplicity. However, this does not in itself guarantee the validity of its universal application. Simplicity, like other criteria of order and of the reason of things, must be checked against experience. The concept of simplicity proved to be a very useful and suggestive criterion in the case of the contrary Ptolemaic and Copernican hypotheses; but this is not always the case. Descartes' criteria of clearness and distinctness were thought to be eminently simple norms for determining the reasonableness of ideas. Still, they did not keep Descartes from accepting notions which eventuated in the strange and difficult theory of an animal-machine and in the positing of a mind-body dualism which was both practically and theoretically insoluble without appealing to an elaborate system of constructs which at one and the same time became increasingly complicated and improbable. Cournot regards this sort of thing as involving too high a price to be paid for a passing satisfaction of certain rational demands. The difficulty in such cases springs from a failure to join with the development of rational principles the processes of empirical verification. Our reasoning may be led astray at any time on the basis of any rational principle when we seek apodictic certainty outside of those sciences which share the purely formal deductive and rational character of mathematics.

The concept of simplicity derives much of such value as it may have as a criterion of order from the fact that it is intimately connected with another idea which is also of fundamental value in our quest for the reason of things—the idea of symmetry or beauty. Here Cournot suggests a transition from a logic to an esthetic of reason. As might be expected, this led him into a trans-rationalism in which ideas of finality, purpose, and God have a place. Here the basic development of Cournot's philosophy reminds us again of that of Kant.

Cournot felt that he had good reason for moving in this direction even though to some this turn in his thought may seem to introduce an apparent discrepancy. Cournot deemed the idea of a fortuitous concourse of events, of processes undirected by *any* end or purpose insufficient to account for the order and harmony which seem so patently to be an objective characteristic of nature.

This does not mean an abrogation of the objective reality of chance, for Cournot does not think the idea of chance is inconsistent with the idea of purposive organization.   Among other things this transrationalism—this essentially irrational element in Cournot's thought—is symptomatic of the persistent strain of empiricism which permeates all his work.   So long as experience acts as a check upon reason, so long as laws and schemata of one sort and another are shattered by brute facts, so long will irrational elements remain with us.   But just as, on the one hand, nothing in thought or experience justifies the conclusion that we and the nature which produced and which nurtures us is totally and exhaustively rational, so, on the other hand, nothing justifies the conclusion that we and it are totally and exhaustively irrational.   The notion of order is fundamental to our thought and to the possibility of our attaining knowledge; the existence of order seems to be an indisputable part of our experiential contacts with both our inner and our outer worlds.   For Cournot, in the last analysis, the very idea of order implies the ideas of purpose and of beauty.   Therefore, he cannot accept as adequate any account of reality based exclusively and exhaustively upon fortuitous explanations.   We experience and deal rationally with unities, with orders, if not with a single, all-embracing order; we certainly do not experience and deal with an all-embracing disorder.   These unities or orders are each internally structured and related, yet severally independent of and externally related to one another.   Their coincidence or intersection is a matter of chance, not of rational predictability.   Cournot believes that reality cannot be accounted for wholly on the basis of the fortuitous concatenation of such independent elements or structures as may compose it distributively; nor can it be accounted for without chance, without fortuitously related elements, unless we choose to disregard cautions which both reason and experience should point out to us.

There is more to man and more to the universe than can be discerned either by formal processes of thought or by empirical demonstrations if these are used independently of one another. The fact that reason does not carry us the whole way to complete certainty and to absolute knowledge in no way detracts from its real contributions to knowledge.   We continue to drive new sectors into the uncharted expanse of the unknown and these new sectors are progressively incorporated into the body of our previously established knowledge. Furthermore, with the addition of each new increment we also get a more adequate glimpse into the probable nature of the whole of reality. Cournot definitely rejects the principle which is sometimes adopted which identifies the unknown with the unknowable; he also avoids the cosmic error which follows from an assumption perhaps more frequently made to the effect that the unknown is essentially identical with the known. The first of these alternatives implies a completely enervating skepticism, for to posit anything as unknowable dooms in advance as futile and unnecessary any inquiry that might be undertaken to discover its nature; the second of the alternatives implies an unwarranted and dogmatic certainty, for it assumes that we may disregard the actual limits of what may be empirically verified concerning the nature of any aspects of reality which have not been explored with patience and persistence and which may not at any given time be rationally defended in terms of probabilities consistent with what has been empirically determined.

XI

There is much in this aspect of Cournot's thought which is consistent with those various pragmatisms that began to find expression late in the nineteenth century—the pragmaticism of Charles Peirce, the humanism of F. C. S. Schiller, the vitalism of Henri Bergson, the voluntarism of William James, the instrumentalism of John Dewey. But whereas these diverse expressions of a common disposition to deal with truth as something which happens to ideas—and in this these philosophies have much in common with the epistemological implications of John Stuart Mill's utilitarianism—Cournot continually affirms a realistic point of view which none of these pragmatic statements, with the possible exception of that of Peirce, would find acceptable. Thus, while pragmatists may well find much in Cournot with which they can agree and might even welcome him as a forerunner, they would also find much in his work with which they would in the end have to disagree. It seems that Cournot's insights may well turn out to have been the sounder ones.

Here it is interesting to note that the late Morris R. Cohen —one of the great philosophical minds in America during the first half of the twentieth century—has in his *Reason and Nature* and elsewhere worked out the substance of a philosophical position significantly similar at important points to that which Cournot developed during the nineteenth century. Both men recognize the developmental and consequently changing character of the conceptual schemes in terms of which we attempt to give expression to the structural relations we discover in the course of our efforts to comprehend the nature of things; both men see clearly the limitations which affect the human understanding and are consequently perfectly at home among ideas whose truth must be measured in terms of probabilities however great man's disposition to seek certainties; both men find it possible to work with speculative ideas whether in the fundamental categories of metaphysical or epistemological inquiry or in the limited and tentative suggestions in particular disciplines as hypotheses whose truth can be determined only through the several techniques of empirical verification; and both men are aware that the pragmatic utility of such conceptual schemes varies through a wide range from the highly probable or improbable to the highly improbable or probable, as the case may be, in the light of new confirmations of predicted events or of brute facts which refuse to fit into implied structures. In sum, Cournot may well have contributed much more to our understanding of the basic problems and processes which determine whether or not ideas of orders of whatever sorts are or are not true as matters of fact than have most of the utilitarians, pragmatists, positivists and others who since his death have tried to work out the logic implicit in processes of the empirical verification or refutation of these ideas. The reason for this is plain even though it may not be sufficient to win the acquiesence of those who tend to eschew metaphysics, seemingly without being aware that such shunning itself implies a metaphysic.

Among the important pragmatists in the Anglo-American tradition only Peirce explicitly refrains from making the truth of an idea dependent upon " success " in some connotation of that term. And even Peirce limits the application of the inductive processes much more radically than does Cournot, for where the former makes meaning coextensive with possible verification, Cournot's position implies that verification is itself meaningless unless first of all we assume statements about natural phenomena whose meaning, as hypotheses rather than as categorically apodictic assertions, gives insight into the kind of procedures—whether rational or empirical—through which verification or refutation may themselves become possible. Verification and refutation are processes by means of which the *truth* or *falsity* of ideas *may be discovered*; they are *not* means *for making ideas true* or *false*. Were ideas not true prior to our being able to proceed successfully in some course of action which we undertake on the assumption that they are true, every " successful " use of an idea in action would be quite as inexplicable as it is in fact in the pragmatic implications of Henri Bergson's essentially irrational vitalism. Our ability to succeed in actions commensurate with the attainment of desired practical ends is evidence of the fact that under the relativistic limitations incident to the time and the place the idea was true prior to the action, else a program of action projected, conditioned and determined by the idea could not have " succeeded "; ideas may be used as instruments for successful behavior only if they are as a matter of fact true antecedent to their proposed use.[1] Much of the philosophy of the last half of the nineteenth century and of the first half of the twentieth century has grown out of an understandable and probably justifiable reaction to various unmitigated absolutisms of the nineteenth century and to various continuing expressions of overweening rationalisms and empiricism of earlier periods. But certainly a critical rationalism of the sort which Cournot presents and which Morris Cohen has continued to work out in our own time seems more consistent with the fundamental needs, aspirations and possibilities which serve to press and entice men into persistent inquiries as to the nature of things than does the complete metaphysical skepticism implicit in much of the intuitional, positivistic and pragmatic philosophizing of the last seventy-five years.

XII

Coming back to Cournot's introduction of beauty as a criterion of reason, another possible difficulty needs to be mentioned. Is not Cournot reasoning in a circle when he uses beauty as a criterion

[1] Cf. my article, " Truth and the Interest Theory of Value," *The Journal of Philosophy*, XXXII, No. 20 (Sept. 26, 1935).

of reason and then appeals to reason to validate beauty when he points to the reasonableness of a law as the basis for its symmetry and beauty? This might be the case were it not for the fact that order is a basic esthetic category as well as a fundamental idea of reason. Cournot could answer the charge in another way, too. In the *Essai,* in the *Traité,* and in *Matérialisme, vitalisme, rationalisme* he spells out in considerable detail his reasons for regarding order as a self-criticizing concept. Therefore, at least as Cournot sees the matter, he is not caught in any such vicious circle as the above question suggests. Because of this unique nature of the idea of order, we are able to move without interruption and without the intermediation of any other concept from the idea of beauty in this signification to that of the reason of things and vice versa, because these are correlative concepts, each of which rests upon the more fundamental concept of order even though the reference of the idea of the reason of things is more inclusive than that of the idea of beauty.

In the last analysis, therefore, the idea of law when attested to by the beauty or symmetry which its assertion introduces into the structuring of our knowledge turns out to be an expression of the idea of intrinsic form, of an established order in the context within which investigations of possible relations among natural phenomena are being made. If Cournot is correct in affirming that there is a reason of things as well as reason in our constructs, then not the least significant criterion of objective order is the symmetry or beauty of our rational representation of such order even though the latter may be only relative and approximative. Thus with a single stroke Cournot brings together the idea of symmetry or beauty—a subjective validation of reason—and the idea of order —the objective measure of the reason of things—as two aspects of the reasonableness both of ideas and of things which makes possible the joint affirmation of a realistic epistemology and metaphysics.

XIII

In this connection we should now look a little more closely at Cournot's theory of the objective reality of chance. The whole question of the truth of propositions outside the limits of formal, logical validation would not only be meaningless but would in fact never have arisen had it not been for the fact that certain characteristics of nature combined with the acknowledged limitations of the human understanding make complete and universally certain knowledge impossible, no matter how attractive such knowledge may be as an ideal. One factor here is the empirical and descriptive recognition and identification of what we call chance or contingency. It has not been uncommon for thinkers to deny the objective character of this feature of our experience and to support the contrary thesis that what appear to us to be fortuitous elements of experience are the results of our ignorance and would disappear if we but had the means for attaining more adequate knowledge. This is the view of Laplace; it is also the view paradoxically shared by such a thoroughgoing empiricist as David Hume. Cournot disagrees completely with this view. He is convinced that contingency is an aspect of the real quite independent of any limitations found in man's capacities to deal rationally with the real. In fact, contingency is so intrinsically a part of the nature of things that other rational beings whose capacities greatly surpass those of man would also be forced to deal with it as an objective characteristic of reality. Cournot argues that if this were not the case the calculus of probabilities would have no object, unless it were to become simply a " calculus of illusions," as he says in his *L'Exposition de la théorie des chances*.

Cournot's defense of the objectivity of chance must not be thought to imply that chance is some sort of a substantive cause which is itself responsible for the occurrence of certain events. Chance refers to an occurrence involving the rationally unpredictable conjunction of two or more externally related series of events. In general, this notion is not unfamiliar to common sense, for it is in accord with much of our normal experience. We frequently deal with experiences and with series of events which, although we have good reason to believe they are completely independent of one another, nonetheless conspire to produce events which are unpredictable. This view of the nature of chance goes far in explaining Cournot's concern to investigate the problem of the nature and the determination of probabilities, a concern which goes beyond the extent to which he might have considered the problem in the course of his mathematical studies.

However, unfortunately, Cournot treats this idea which is so important to his thought almost entirely through the use of interesting and lucid illustrations rather than by using these examples to illuminate a rational and formal development of his theory. As a result it cannot be denied that the idea lacks the kind of exactness so desirable for the careful exposition, development, and defense of any theory. The illustrations used by Cournot are both sufficiently profuse and apt to indicate in general what he has in mind; yet they are not without certain ambiguities. Before we can determine whether or not reference to objective contingency is proper in the case of any specific event for which we are trying to account, the presence or absence of certain determining conditions must be established. The lack of an adequate and exhaustive discussion of these conditions leaves Cournot's doctrine of the objectivity of chance somewhat unclear. The basis for the criticisms by various writers [1] of Cournot's views on the theory of chance as well as of his closely related work on probability theory is to be found in Cournot's own lack of clarity and precision on this important matter.

A possible, although perhaps an inadequate, defense of Cournot at this point may be suggested by pointing out that he tended so far as possible to avoid metaphysically contentious matters. It is true that he uses the last chapter of the *Essai* to compare his views with those of a select group of philosophers. But in general he is content to present his own views as cogently as he deems it necessary or possible for him to do so without becoming involved in lengthy disputations. He is inclined to deal with contentious questions, and particularly with those which seem to him to involve theoretical mystifications, with the foil of positive assertions of what our experience affirms. So in the present case he is inclined to rest his case after pointing out that while it is true that certain events and conjunctions of events are theoretically possible they in fact never occur. The rationalistic possibility of a com-

[1] See A. Darbon: *Le concept du hasard dans la philosophie de Cournot* (Paris: Felix Alcan, 1911); John M. Keynes: *A Treatise on Probability* (London: Macmillan and Co., Ltd., 1929); Jean de la Harpe: *De l'ordre et du hasard* (Neuchâtel: Secrétariat de l'Université, 1936); F. Mentré: *Cournot et la renaissance du probabilisme au xixᵉ siècle* (Paris: Marcel Rivière, 1908); G. Milhaud: *Etudes sur Cournot* (Paris: Librairie Philosophique J. Vrin, 1927).

pletely interrelated system of objects, events, and processes in reality is insufficient to override the fact that actual objects, events, and processes present themselves to us in such ways as to make it appear evident that each natural phenomenon is not inextricably related to every other. Thus, assuming the universal pervasiveness of gravitational relations—whether these relations be accounted for in terms of forces or in terms of the characteristic curvatures of a spatio-temporal matrix or continuum—it is possible that the tossing of a ball would affect however slightly the course of the most distant planet because it would involve some necessary rearrangement of the structure of the total system. Practically, however, no such modification is observed, nor do those whose business it is to investigate such relations expect any such modification in fact. Whereas the monist might say that this appearance of independence and disjointedness among phenomena may be traced to the fact that we are without sufficiently subtle natural and artificial instruments to detect the completely internal relatedness of things, Cournot would say we do not detect it because we have good reason to believe that in fact there is no such complete interrelatedness. "Tough-minded" and critical rationalist that he was in matters of this sort, he would rest his case heavily on empirical evidences lest he be led into some metaphysical absurdity.

Numerous events are *theoretically possible* which at the same time are *physically impossible*: the measure of this impossibility is that such events never in fact occur. In his instructive consideration of probability in the *Exposition* and the *Essai*, Cournot points to such instances as the following: it is theoretically possible to balance a cone on its point, to impart lateral motion to a sphere in such a way that no rotary motion results, to throw a disk on a parquet floor in such a way that its exact center comes to rest precisely at the point of intersection of two diagonals, and so on. Each of these events is theoretically possible; yet they just do not happen unless by the rarest of chances.

What has just been said does not mean that a fortuitous event is an event without a cause; neither is it fortuitous simply because it is rare or surprising. Cournot reiterates this view beyond any possible misunderstanding of his meaning. The idea of chance is simply the idea of the independence of causal series and, therefore, the idea of the consequent unpredictability of any con-

junction which may occur among such series of events. This unpredictability limits the possible knowledge of both men and gods,
yet none knows for sure where the limits of unpredictability may
be found to lie.

XIV

The objective character of contingency does not mean that
those phenomena in whose occurrence chance plays a part are excluded from scientific consideration. If this were the case we
could develop organized bodies of knowledge only with respect to
those aspects of experience and reality which are amenable to
strictly rational treatment. This would exclude much of the content of the so-called natural sciences as well as of other studies
with respect to which the possibility of rational and scientific study
is at best grudgingly admitted by some and is categorically denied
by others. Even certain phases of mathematics, as an applied
science, would be excluded. Fortuitous events just happen; they
are the brute facts of experience which may be recorded and
described but cannot be predicted as the logical consequents of
general principles; by its very nature reason is limited to the
consideration of generalizations which show logical relations among
facts. Yet contingent events are amenable to the calculus of
probabilities.

How can we distinguish " causes " which are particular and
affect only individual occurrences from the " reason " for events
which may be treated abstractly in the form of a statement of
relations having universal significance and verifiable deducible
consequences? The answer must be looked for in each particular
series of events through the sufficient repetition of the empirical
techniques of observation and experimentation. In this way we
may discover patterns or structures among the events in question
which reveal to us not simply causal relations but the explanatory
reason for these relations. In the latter case we may establish the
existence of certain constantly conjoined characteristics intrinsic to
the phenomena in question. These characteristics may be dealt
with rationally and predictively and not merely on the basis of report and description, or of habit. To the extent that this is true we
may anticipate and explain objects, processes, and events on bases

entirely different from those of custom and habit to which Hume and other radical empiricists have been wont to reduce us. If certain characteristics of the remains of a prehistoric animal enable us to establish the fact that it was a mammal, we can assert with confidence that the animal also had certain other characteristics which are no longer discernible in the remains but which are common to all mammals: in the recent synthetic development of elements which lie outside the groupings of the periodic table, as we had until recently known it, only very small quantities are sometimes available from which the physical and chemical characteristics of the new element may be determined; yet the properties which have been established are accepted by physicists and chemists as those which will be discovered in any future sample of the elements that may be synthesized. Similar illustrations of rational predictability based upon the establishment of constant and uniform characteristics and processes could be drawn from every field of organized knowledge. As is well known, such constant and uniform elements have more than once been the grounds upon which exact and later verified predictions have been made as to the existence and, within limits, the nature of previously unknown phenomena. When the same phenomenon is repeated fortuitous deviations of one sort or another from the norm tend to cancel one another out in such a way that at last the unfortuitous character of the phenomena is established and reveals the law of, that is the reason for, the phenomenon in its general and fundamental aspects. Thus in all cases where repeated observation or experimentation is possible the fortuitous elements of any object, process, or event may gradually be eliminated and those which pertain to the nature of things be revealed. Finally, in dealing with homogeneous phenomena, relations have been discovered of such a sort that we may predict with a high degree of probability, if not with certainty, the characteristics of all of a class of previously unknown phenomena on the basis of a single instance.

Judgments which rest upon such processes as those of which we have just been speaking are necessarily probable. This does not mean that they are all alike. Cournot identifies two sorts of probability which he calls mathematical and philosophical, respectively. The basis for this distinction is the extent to which a quantitatively determinable statement can be made of the relation of the number of chances favorable to the occurrence of a specific phenomenon as compared with the total number of possible chances. Mathematical probability depends upon the possibility of our determining the relative frequency of events; judgments of philosophical probability depend upon our sense of what may be anticipated within the nonquantitatively determinable limits of what we understand of order and of the reason of things. The delineation of legitimate bases for the latter type of judgment is the primary function of critical philosophy and gives meaning to Cournot's statement that the category of order is the basis of all critical thinking.

XV

We have referred above to the way in which Cournot's work anticipates the recent application of statistics to the study of natural phenomena in an increasingly large number of fields of inquiry. This is particularly interesting in the light of a commonly accepted contemporary view that all empirical generalizations and all so-called natural laws are statistical, that is, probable generalizations. For reasons implied in the above discussion we have good reason to believe that Cournot would not accept the radical interpretation of this position by contemporary positivists. But enough has been said of this matter. As far as Cournot is concerned, one of the most interesting applications of this notion of the nature and function of the idea of philosophical as distinguished from mathematical probability is that which led him to his novel basis for the differentiation of science, history, and philosophy. This has been mentioned in various connections in what has been said above. Unfortunately, space does not permit of its further development here. We shall be content simply to mention this differentiation again and to refer the reader to the discussion of it in the *Essai*, the *Traité*, the *Considérations*, and *Matérialisme, vitalisme, rationalisme*. This reading should be both interesting and provocative.

For Cournot the ground upon which the distinction between the sciences and other disciplines rests is a consequence of the discontinuity and resulting indeterminacy of nature. The need for a concept of levels or orders of existence may be traced to the

same root. With this notion come all the problems which inevitably attach to a pluralistic point of view. Specifically we find
that such questions as the dependence or independence of these
proposed levels of existence and the whole question of evolution
as a statement of continuity among forms if not among all levels
of existence demand attention. With this latter question comes another—that of the development of complex forms from more simple
ones. The fundamental issue is this: are the categories which
are adequate for the explanation of the relatively simple phenomena which are found at the " lower " levels of existence adequate
for dealing with the relatively more complex phenomena of the
" higher " levels, or is it incumbent upon us to introduce new fundamental categories and new principles of explanation as we move
upward? These questions are among those explicitly and extensively explored in the development of certain pragmatic positions and by those who have developed the philosophy of
emergent evolution.

Implicit in the growth of science is the tendency to extend
principles as widely as possible, the ideal being that all events,
if possible, and certainly all those of a given sort, may be found
to have their loci within a single continuum. A notable success
in reducing phenomena of a certain order to a structure of rational relations through the discovery and application of particular principles—such as the principles of mechanics and later of the
theory of evolution—seems inevitably to call forth efforts to apply
such principles universally. This tendency is due in part to an
apparently natural human disposition to find the least possible
number of basic principles which will enable us to give a rational
account of all phenomena in all their multifarious aspects and in
part to the persistent belief that an increase in the complexity of
any conceptual scheme, which seems to be required by the pro-
liferation of empirical differentia, is evidence that we have not
yet gotten to the fundamental principles in terms of which truly
rational explanations of the phenomena in question may be attained. Here we return to the regulative idea of simplicity and
symmetry as criteria of order and of the reason of things.

The verification or refutation of any principle as a true statement of the nature of things is ultimately experiential. Experience
reports a universe in which there is independence and diversity

among events as well as dependence and fundamental likeness. This discontinuity is an aspect of experience which cannot be suppressed or evaded. We find " breaks " as we move from one experience to another or from one field to another. This leads us to accept the idea of different and independent orders of being in attempting to find the reason for the diverse phenomena we seek to explain.

One of the most commonly agreed upon indications of discontinuity is found as we move from an effort to explain phenomena in the inorganic world to those in the organic world, and this in spite of almost continuous efforts that have been made to find ubiquitous principles by means of which a bridge between the two orders may be established. There is ample reason to believe that these two realms are related, as evidenced by psychosomatic phenomena; but a variety of evidence seems also to point to the fact that these two realms are independent of one another in ways which make it difficult for us to conceive them as identical in substance as well as in aspect. On the basis of the most adequate knowledge we have at the present time it is not possible to deal with the laws and principles of the physical sciences and those of the biological sciences as coextensive and interchangeable. This is true in spite of the relatively recent development of such mediating disciplines as bio-physics and bio-chemistry. This is typical of the brute facts which give reason to Cournot's insistence upon the need for introducing the idea of a special, vital principle or at least for nonphysical categories in our attempts to account for living forms. He contends for the necessity of maintaining some sort of irreducible independence among the orders of material, vital, and rational phenomena. His empirical affirmation of difference makes it legitimate to surmise that Cournot would have been quite as unable to accept Bergson's use of the *élan vital* as an all-inclusive metaphysical principle as we know he was unable to accept Darwin's theory of evolution. He views the latter theory as an attempt to extend mechanical principles, demonstrably applicable to the physical order, as principles equally applicable to the explanation of living things; he would, in all probability, have found Bergson's use of an *élan vital* an equally unpalatable effort to apply an essentially biological principle ubiquitously as a clue to the nature of all things.

Prior to Darwin the burden of proof had always to be borne

by those who wished to defend the possibility of a completely general mechanical interpretation of all reality including man and his works. With the appearance of the Darwinian hypothesis the tables were turned. Now a thoroughly mechanistic explanation backed by what has been regarded by many as adequate and conclusive empirical support was given for the appearance, disappearance and modification of living forms. The materialistic implications of the new theory finally brought Cournot into the lists and he specifically attacks the implications of the Darwinian theory of evolution both in the *Considérations* and in *Matérialisme, vitalisme, rationalisme.*

Cournot wrote too early to be aware of similar developments in psychology even though certain experimental work out of which much of the later psychology sprang had been begun before his death by such men as Weber, Fechner, and Pavlov. However, had he known of the development of the later physiological and behavioristic psychologies, on the basis of what were regarded as either the positive or negative implications of the work of these and other experimentalists, he would most certainly have opposed these theories concerning the nature of psychical phenomena. His opposition would have rested on grounds analogous to those on the basis of which he rejected the Darwinian theory of evolution. In fact, we may recall that it has already been pointed out that Cournot rejected the idea that psychology might become a science although he held that certain phases of psychological inquiry may be organized scientifically.

Cournot never sought to detract from the successes of those sciences which had made significant rational use of mechanical conceptions in explaining phenomena of a physical order. What he objected to were attempts to extend these concepts to fields in which we are essentially concerned with attempts to reach rational explanations of nonphysical phenomena. Such attempted extensions rest on the propensity of the human mind to find a single set of principles in terms of which the reason of all things may be stated. *Ex nihilo nihil* tends to become *ex uno omnes* as a result of rationalistically motivated drives toward monistic inclusiveness, whether the roots of such a monism be materialistic, organistic or vitalistic, or spiritualistic. Again we affirm Cournot's essential pluralism.

So long as we are dealing with physical objects we need no other categories than those which apply to physical phenomena; when we attempt to apply such categories to other types of phenomena, however, basic difficulties arise. Whatever the future may show, the presently verifiable evidence is that the basic dissimilarity between the physical, the organic and the mental is real and objectively significant. Cournot felt, and probably with good reason, that this will continue to be the case, for there is no little evidence that this discontinuity is rooted in the nature itself of the phenomena in question and not simply in limitations of the means at our disposal for becoming acquainted with and for gaining knowledge of them.

It is in the defense of this theory of radical discontinuity rather than in such principles of indeterminacy that are now identified with the names of Schrödinger and Heisenberg that Cournot lays the ground for the defense of the unique character and significance of the moral and esthetic aspects of human experience. All living things seem to be purposive in a way in which nonliving things are not. Certainly an account of the behavior of human beings and of at least those aspects of the behavior of other living things which are consistent with what we experience in ourselves and seem to observe in other persons must take account of actions which are determined by future ends as well as by conditioned responses which exercise a cumulative influence from the past. We have good reason to believe that this is not true of the nonliving— unless one now be forced to make an exception for those uncanny electronic devices whose development has given rise to the new discipline of cybernetics. For Cournot, as for Bergson, S. Alexander and C. Lloyd Morgan, the thing that differentiates living things is a *vital* principle, an *élan vital*, a *nisus formativus*, a principle of harmonic co-ordination, a creative energy; it is a principle of internal harmony and order which works against external causes of complication, disturbance, disorder, and, finally, disintegration. A similar principle is the key to phenomena above the level of living things. In this Alexander and Morgan would be more inclined to agree with Cournot than would Bergson, since the latter gives to reason only pragmatic utility because he regards it not only as incompetent to explore the metaphysical bases of things but as actually misrepresenting reality. For Bergson we know the objectively real only as we have occasion to deal with it practically or intuitively.   While Cournot would most certainly have rejected any physiological or behavioristic psychology, he would as certainly have found much in gestalt psychology and even in some phases of the work of Freud, Jung and others with which he would have agreed except as, in the case of Freud particularly, all explanations of psychic disturbances of whatever sort are carried back to a single explanatory root, be it sex or some other.

In sum, Cournot repeatedly tries to point out the errors and to protect himself and his readers from the fallacies that beset those who are unaware of or who tend to gloss over a fundamental distinction between two radically different things—the *conditions* under which things occur and the *reasons* for their occurrence. The former have to do with local and temporal particularities; the latter are concerned with timeless and changeless relations.   In the final analysis Cournot's criticism of Darwin's theory of evolution —and by implication of later physiological and behavioristic psychologies—comes down to two points: first, evolution involves certain creative ages; second, it seems to involve a plan of creation which goes back to a rational agent or principle as the initiator of the system and whose action is overtly expressed in the vital principle of organization in living forms.   Once created, organisms may survive and new forms may appear on the basis of what Darwin aptly called natural selection.   But the fortuitous concourse of exclusively physical objects, events, and processes is in itself insufficient to account for the appearance of living forms from a purely physical milieu; the survival of living things implies selection in terms of nonphysical as well of physical factors.

XVI

From the above discussion of the reaction of Cournot to the theory of organic evolution it is apparent that he conceives of the universe itself as purposive and as moving toward some goal.   As a consequence of this view the question arises as to the relation between this affirmed finality and the objective reality of contingency on which so much of his thought depends.   This presents a major problem to anyone who wishes to understand Cournot thoroughly rather than to use various phases of his work as a source of insights which suggest possibly fruitful new approaches in the attack on bothersome but persistent problems. The question is never explicitly discussed and Cournot only vaguely hints at an answer in *L'Exposition de la théorie des chances.* Here he implies rather than states directly that God—who, as we have already pointed out, is not ubiquitous—lays out and takes care of the laws or rational elements of reality and leaves to chance the details of those actual occurrences which we know and refer to empirically as matter of fact.

Considered objectively, we find evidences of the purposive aspect of nature in the beauty and order which relate things harmoniously to one another. In man we find it in the desirability of developing a single integrated personality and in the moral principles which inform our practical behavior with order and direction, and, perchance, with beauty. In every case, and at every level, the "higher" is that which is conducive to conservation and unity; the "lower" is that which tends toward disintegration and disappearance—a notion which is basically and strikingly like that which Empedocles envisioned so long ago in his dramatic opposition of the principles of "love" and "hate."

Once again we have come full circle, for this matter brings us back to those tests of the rationality of the real—in so far as it is rational—which we found earlier were applied to the more limited problem of the reason of things. These evidences depend once more on the presence in things of order, symmetry, and simplicity. And let us remember that in the last analysis these are esthetic categories as much as they are rational. Ultimately they are assumptions not amenable to logical analysis, at least so far as this involves the ability of human beings to reason. Their force rests finally upon their ability or lack of ability to win our acquiescence in terms of their philosophical rather than of their mathematical probability. And there is no calculus of such probabilities through the rational power of which one mind can force another to grant the logical cogency of an argument of this sort.

Cournot admits this. In fact his philosophical statements conclude with a statement of the insufficiency of reason in this limited, logical sense. With this affirmation he asserts the need to turn to some sort of transrationalism, but not of any sort of transrationalism, be it pragmatic, positivistic or vitalistic, which denies to human reason the possibility of exploring and discovering the objective reason of things within the limits imposed by those empirical and relativistic factors which hedge us about. While he would not have gone so far as Bergson in the use of intuition, Cournot was not unaware of or adverse to admitting the intuitional character of certain ideas, particularly in the formal sciences like mathematics. As previously iterated, we are rational beings because our minds are expressions of a rational order, of a universe which is to an indeterminate degree rational. But our rationality is that of the part rather than of the whole. To discover the degree to which objective reality is rational we must have recourse to other sources of insight than human reason.

Finding no more satisfactory rubric which will enable him to continue his quest, Cournot suggests that the possible solution of such fundamental questions leads us to esthetics. In this final word his philosophy reminds us again of the critical philosophy of Kant. But there remains an essential difference between the two men. For Cournot we are rational because we are creatures of and live in a world extensively, although not exhaustively, rational, whereas for Kant the world, at least so far as we can grasp it through reason, is rational because its rational aspects are the projective creation of beings who are rational in part but whose true nature is transrational. For this reason it was possible for the flamboyant romanticism of much of nineteenth-century philosophy to spring from Kant's transcendental ego. These excesses in turn gave rise to others in the form of the excessive skepticisms of pragmatism, positivism, and intuitionism. Cournot's philosophy could not have inspired an orgy of world building as a projection of the self; neither could it sire the skeptical philosophies of the late nineteenth and twentieth centuries which, to use the nice expression of William James, have literally tended to " throw the baby out with the bath " in trying to trim away the exaggerations of romantic and absolute idealism, and of any dogmatic realism, too, for that matter. These later and contemporary skeptical movements seem all too nearly to have cut away the essential substance of philosophy as a reasonable enterprise to be undertaken by reasonable beings. Yet, although they have been in the vanguard of those who denounce philosophy as fruitless and impractical, the representatives of these selfsame movements are often heard to cry out against the growing disinterestedness in philosophy by those not directly concerned with its significance and continued existence!

It is the median character of Cournot's philosophy, to which we have given the name critical rationalism, that makes the reading of him a peculiarly salutary occupation at the present time. His work is a rewarding antidote to many contemporary excesses and inadequacies. He encourages neither the stultifying quest for and ennui of security, nor the meaningless void of a cynical skepticism. We do not know all and we never shall; but we do know some and may know more, for we have found and may continue to find, the reason of some things, which makes it possible for us to go beyond description to explanation in our dealings with an increasingly large number of aspects of reality.

It is my profound hope that making Cournot's *Essai* available to readers in English-speaking countries will draw attention to his stimulating and temperate position. It is my belief that only such a modest yet positive confidence in the capacities of men to reason cogently and effectively about reality, including man, may save us from destructive incompetence. There is ample evidence that the basic problems which persistently concern us—our queries concerning the nature of the universe, concerning the nature of man, and concerning the nature of the relation of man to his world —are, to a surprising degree, demonstrably amenable to rational investigation. If such a philosophy as that which Cournot developed could be extended and made generally and practically effective, it might well serve as a surprisingly adequate antidote to those numerous disintegrative forces which beset our times— times already being referred to as the Age of Suspicion and Anxiety, and as the Age of Unreason—names by which our times may be known by those who come after us. Thus will be recorded our antipodal relation, after such a relatively short time, to that more confident and hopeful period which we know as the Age of Reason.

MERRITT HADDEN MOORE

SELECTED BIBLIOGRAPHY

Cournot's Published Works

Mémoire sur le mouvement d'un corps rigide soutenu par un plan fixe, Paris, 1829.

Recherches sur les principes mathématiques de la théorie des richesses, Paris, 1838. Republished with the complements of Leon Walras, Joseph Bertrand, and Vilfredo Pareto, with an introduction and notes by Georges Lutfalla, Paris, 1938.

Traité élémentaire de la théorie des fonctions et du calcul infinitésimal, 2 vols., Paris, 1841. 2nd ed., 1857.

Exposition de la théorie des chances et des probabilités, Paris, 1843.

De l'origine et des limits de la correspondance entre l'algèbre et la géométrie, Paris, 1847.

Essai sur les fondements de nos connaissances et sur les caractères de la critique philosophique, 2 vols., Paris, 1851. 2nd ed., 1912. Reprinted as the 3rd edition in one volume, 1922.

Traité de l'enchaînement des idées fondamentales dans les sciences et dans l'histoire, 2 vols., Paris, 1861. 2nd ed., 1911. 3rd ed., 1922.

Principes de la théorie des richesses, Paris, 1863.

Des institutions d'instruction publique en France, Paris, 1864.

Considérations sur la marche des idées et des événements dans les temps modernes, 2 vols., Paris, 1872. Re-edited Paris, 1934.

Matérialisme, vitalisme, rationalisme. Etudes sur l'emploi des données de la science en philosophie, Paris, 1875. 2nd ed., 1923.

Revue sommaire des doctrines économiques, Paris, 1877.

Souvenirs, written by Cournot in 1859. Published with an introduction and notes by E. P. Bottinelli, Paris, 1913.

Articles in the *Dictionnaire philosophique* by A. Franck, Paris, 1843 to 1852. 1st ed.

Publication des mémoires militaires de Gouvion-Saint-Cyr, 4 vols., Paris, 1831. Edited with an introduction by Cournot.

Translation from English of *Eléments de mécanique de Kater et de Lardner,* "modified and completed" by A. A. Cournot, Paris, 1834. 1st. ed. Chap. XXIII, "De la mesure des forces du travail des machines" is by Cournot.

Translation from English of *Traité d'astronomie de Herschel,* Paris, 1834. 1st ed. With an appendix on "Application de la théorie des chances à la série des orbites des comètes dans l'espace" by Cournot.

Annotated translation of *Lettres d'Euler à une princesse d'Allemagne,* Paris, 1842.

STUDIES OF COURNOT

Revue de métaphysique et de morale, May, 1905, 13° année, N°. 3. Eleven articles on Cournot in a commemorative issue.

Bottinelli, E. P., *A. Cournot, metaphysicien de la connaissance,* Paris, 1913.

Bouglé, C., *Qu' est-ce que la sociologie?* Paris, 1925.

Darbon, A., *Le concept du hasard dans la philosophie de Cournot,* Paris, 1911.

Harpe, J. de la, *De l'ordre et du hasard; le réalisme critique d'Antoine Augustin Cournot,* Neuchâtel, 1936.

Mentré, F., *Cournot et la renaissance du Probabilisme au XIXe siècle,* Paris, 1908.

————*A. A. Cournot,* Paris, 1909.

————*Pour qu'on lise Cournot?* Paris, 1927.

Renouvier, C., *Essais de critique générale: Premier essai, Traité de logique générale et de logique formelle,* Paris, 1912.

Ruyer, R., *L'humanité de l'avenir d'après Cournot,* Paris, 1930.

Segond, J., *Cournot et la psychologie vitaliste,* Paris, 1911.

AN ESSAY ON THE FOUNDATIONS OF OUR KNOWLEDGE

AUTHOR'S PREFACE

It may seem strange that an author should offer a book on pure philosophy to the people of this country at the present time.[1] It might seem even more strange if the author were to admit, much to his embarrassment, that the book has been in preparation for ten years, although it is of no more than ordinary length; indeed, the first draft of it was sketched some twenty years ago. However, although its subject matter has been dealt with time and again, it pleases me to think that anyone who may take the trouble to read it will find in it enough new points of view to justify my ingenuous perseverance, at least in the eyes of those who love philosophy. I may be mistaken in this. Nevertheless, I may still call attention to the importance of revitalizing the older philosophical doctrines from time to time in two ways: first, by taking account of the progress of our positive knowledge and of the new perspectives opened up by it; and, next, by being able to choose examples better suited to the present state of development in the sciences than those which were available at the time of Descartes, Leibniz, or even d'Alembert. These latter examples still serve as legal tender, so to speak, even though they have become somewhat the worse for wear since philosophers began to neglect the sciences and since the scientists have been only too willing to show their haughty disregard for philosophy. It is also true that in going against the practices of one's own time and in ignoring the fashion prevailing in the schools and in books, one runs the risk of being very poorly received. But, after all, each philosopher works in his own way, and each brings to his philosophical speculations the imprint of his other studies and the turn of mind which they have given him. The theologian, the jurist, the mathematician, the physicist, and the philologist can each be recognized at a glance by the way in which he wears the mantle of philosophy. It would be regrettable in more ways than one if this variety were

---

[1] [This preface was written only two months before the nominal end of the Second Republic on December 2, 1851, and just fourteen months before the establishment of the Second Empire on December 2, 1852.]

to give way to a too monotonous uniformity. This would inevitably be the case, however, if philosophy, in its desire to become disciplined, were to become isolated, fragmented, and thus come to resemble too greatly a profession or career.

It is impossible to write on philosophical matters without touching upon some extremely delicate questions, and without exposing oneself to apparent contradictions or to interpretations which go far beyond the thought of the author. I have tried to set forth more clearly than has previously been done the special reasons for the inevitable imperfection of philosophic language. If I have succeeded in demonstrating this theoretical point to the reader, I shall, by the same token, have prepared him to excuse indulgently and to correct with good will many inaccuracies of expression. As for those who may be motivated by a less generous spirit, I shall do no more than refer them to this quotation from Malebranche: "What one writes is difficult, and sometimes irksome and unpleasant if one's statements are too rigorously exact. When an author contradicts himself only in the mind of those who criticize him and who wish him to contradict himself, he must not take it too much to heart. If by tiresome explanations he were to clarify all the points that malicious or ignorant persons might bring up against him, he would not only write a very wretched book, but his readers would be offended by his answers to imaginary criticisms, or these answers would run the risk of going against a certain sense of fairness on which everyone prides himself." [2]

Just one word more. In perusing a book the purpose of which is to depict the pre-eminent role of reason in the elaboration of human knowledge, one might suppose that the author is a rationalist, to use a term which is commonly employed in modern controversies. In the present case this would be a serious misconception. I am completely convinced of the practical insufficiency of reason; and I should not in the least wish to run the risk of weakening beliefs which I regard as sustaining and as having sustained the moral life of humanity for the sake of certain speculative ideas.

A. COURNOT

Paris, August 28, 1851.

[2] [Author's note on the third chapter of Book I of the *Recherche de la vérité*.]

CHAPTER I

ON KNOWLEDGE IN GENERAL. ON ILLUSION AND ON RELATIVE AND ABSOLUTE REALITY

1. Whatever the object or phenomenon we may wish to study, the thing we know best about it is its form. The basic nature or substance of things is very mysterious and obscure to us. Fortunately, our ignorance of this basic or intrinsic nature of things in no way hinders us from grasping, through reason, all the properties belonging to a form of which we have a clear and well-defined idea. Thus, although our notions of the structure of solid and fluid bodies are very imperfect, and although no one has as yet satisfactorily explained how nature gives rise to the physical types and forms of solidity and fluidity by means of molecular activity, it is sufficient for our purposes that these types are amenable to precise and mathematical definition. Mathematicians have been able to discover in solid and fluid bodies, both at rest and in motion, numerous properties which are logically dependent upon abstract definitions of solidity and fluidity. The study of these properties does not require previous knowledge of the hidden means employed by nature to produce a crystal or a drop of water, means through which liquid and crystalline forms become perceptible to us.

Similarly, although it may yet be a long time before we know the true nature of light, in spite of the progress modern physicists have made in the science of optics, this science has already become a great and important instance of applied mathematics. This has been possible because of the tendency of light to travel in straight lines, to be reflected or refracted, when passing from one medium to another, in such a way that its behavior may be stated in terms of laws which are mathematical, exact, and simple. This part of optics did not change at all when the theory of the emanation of luminous particles gave way to that of vibrations in an ether. Recourse to other explanations has been necessary in order to connect these mathematical laws, on which the form of the phenomenon depends, with later theories concerning the physical structure of light, or the intrinsic nature of the phenomenon.

2. What we have said concerning physical phenomena applies even more to those of the sensible and of the intellectual life. If the physicist is a long way from having an exact understanding of the molecular structure of a drop of water or of a crystal, how may we hope to understand the innermost structure through which nature brings about the mysterious phenomena that we call sensibility, consciousness, and perception? How can we know the essence and internal causes of the act through which a being endowed with intelligence perceives or knows objects which lie outside of it? The most delicate anatomical dissections, as well as the most subtle analyses, have failed to reveal this essence, up to the present time, and will always fail to do so. Therefore, it is necessary to disclaim any knowledge of the mechanism of our faculties, except in so far as their form presents certain characteristics that we can grasp clearly, and the consequences of which may be traced by reason in spite of our ignorance both of their true nature and of the development of the faculties whose operations and relations we wish to study. Logicians, and Kant in particular, have long insisted upon the distinction between the form and the matter of our knowledge. Furthermore, they have shown clearly that the form of things may be made the object of judgments that are certain even though their matter or basic nature remains problematic. This distinction must serve as a point of departure in all logical research. We shall make use of it to deal with a more general and more essential characteristic than those with which logicians have been concerned up to the present time, and from which, it seems to us, many more important consequences remain to be drawn.

3. It is clear that all perception or knowledge implies a perceiving subject and an object perceived, and that it consists in any relation whatever between these two. It follows from this that if the perception or relation changes, the reason for the change must be found in a modification of the perceiving subject, or of the perceived object, or of both.

Just so, when two vibrating strings have been tuned so as to give a definite musical interval which changes after a time, the change is due either to the lowering of the pitch of one, or to the raising of the pitch of the other, or to both being changed at the same time. In the same way, if it is found that a certain number of bushels of wheat equals a given number of days' work at one time and a greater number at a later time, the question will be whether the change is due to a rise in the value of wheat, resulting, for example, from a series of poor crops, a tax on imports, and so on; or, again, whether it is due to a lessening of the value of manual labor resulting from an increase in population, the introduction of new machines, or from other causes; or, finally, whether the change may have resulted from the coincidence of both factors.

4. Suppose that we have a group of strings tuned to give certain musical intervals. And let us suppose further that all but one of the strings continue to give these intervals, a change being noted only when this one string is compared with all the others. It can be considered at least highly probable, if not rigorously demonstrated, that the string in question, being the only one which has *not remained in tune,* is the one which, having undergone a change in tension, has given rise to a new condition in the system.

A similar conclusion will be drawn if we view a table which gives the relative value of different commodities at various times. For example, if, as was supposed in the instance given above, the value of wheat rises in relation to a day's wages while remaining constant as compared to other commodities, it may be concluded that the observed change is due not to an absolute rise in the value of wheat, but to an absolute depreciation in the value of labor. At least, this will be the conclusion whenever it is clear that the relationship between the value of wheat and the values of all the other commodities is not such that one cannot fluctuate without giving rise to proportionate alterations in all those which depend upon it.

5. In the long run, of all the examples that we might take none will suit our purposes better, and none will be more clear and precise, than those which are related to the study of motion.

Let us assume that a point has moved if its position has changed with respect to other points which we regard as fixed. When we observe a system of material points at two distinct times and find the respective positions of the points have not remained constant, we have no alternative but to conclude that some if not all of the points have changed their positions. But if we are unable to relate these to other points of the constancy of which we may be sure, it is at first impossible to conclude anything concerning the motion or lack of motion of any particular point in the system.

However, if all but one of the points in the system have maintained their relative positions, we may regard it as highly probable that only this particular point has moved. At least, this will be the case whenever the other points do not seem to be related in such a way that a displacement of one must bring about a displacement of all the others.

We have just indicated an extreme case in which all but one of the points have maintained their relative positions. Without going into details, we know that among the alternative explanations which might occur to us as a means of explaining changes that have occurred in the system, some are simpler than others. We do not hesitate to regard these simpler explanations as more probable than the others. This probability, the origin and nature of which we do not wish to discuss just yet, may be such that it must be granted by any reasonable person.

If we were not limited to observing the system simply at two distinct times, but were able to observe it through its successive states, certain hypotheses concerning the absolute movements of the points of the system would be preferred to others as a way of explaining their relative motions. So it is that, without making any reference to the notions acquired at a later date concerning the masses of celestial bodies and also the nature of the force which moves them, the theory of Copernicus, as compared with that of Ptolemy, would explain the apparent movements of the planetary system in a manner at once more simple, more satisfying to reason, and consequently more probable.[1]

Finally, there are circumstances which make us positive that relative and apparent movements result from the actual displacement of one particular body and not of another.[2] Thus, the position of an animal may indicate to us unequivocally whether it is

---

[1] "We find, therefore, under this arrangement, a remarkable symmetry of the universe and a certain bond of harmony between the movement and the size of the spheres such as cannot be sought on any other basis." Nicholas Copernicus, *De revolutionibus orbium celestium*, I, Chap. 10.

[2] Newton, *Principia*, Book I (at the end of the preliminary definitions).

at rest or in motion. Thus, too, experiments with pendulums prove the diurnal movement of the earth; phenomena showing the aberration of light prove its annual movement; and, as a result, the Copernican hypothesis takes its place among the truths which are positively demonstrated.

6. Notice, however, that these motions, which we shall provisionally and improperly call absolute, and which we shall seek to use to discover the reason for relative displacements, may themselves be only relative. Numerous examples, as familiar as they are frequently repeated, are available to help us make this fundamental distinction.

If, at two different times, we look at the position of two animals on the deck of a ship on which they have been loaded, we shall find that their relative positions have changed. In the absence of any other criterion for comparison, the posture of one animal as compared with that of the other will be a sufficient basis for judging, without hesitation, that the former has moved while the latter has remained at rest. But this judgment is true only in relation to the system of which the ship and the animals are parts. Were one to take account of the motion of the ship, he might find that the animal that was judged, with reason, to be in motion with respect to it, was at rest with respect to the surface of the earth, and that, in relation to the latter, the other animal had moved. This does not make it any the less reasonable to say that the animal which appeared to be walking has really moved; but the reality of this movement is relative to the system of which the animal is a part.

We have already seen that experiments with pendulums and the phenomena of the aberration of light prove the reality of the diurnal and annual movements of all bodies placed on the surface of the earth. But, perhaps as a result of the motion of the planetary system through space, a certain point of the earth, its center, for example, may at a given time be absolutely at rest, while the center of the sun is in motion. This would in no way invalidate the Copernican hypothesis according to which the sun is thought to be at rest with the earth moving around it. The only thing we would have to recognize is that the validity of the hypothesis is entirely relative to the system of the sun and its encircling planets.

7. In order to follow the analogy as closely as possible with the

problem with which we must deal—the purpose of which is to submit our ideas to a critical examination in order that we may distinguish the true from the false, the illusory from the real—we must consider the question at issue in greater detail but without leaving the order of facts from which our examples are drawn. This must no longer be done by making statements about the real movements of a group of moving bodies on the basis of their relative movements as these are determined by an observer from a position he takes to be stationary. Rather, we must speak of the real motions which may affect either the system of visible bodies, taken together, or the position of the observer himself; and that on the basis of the perception of the apparent movements of the external system, with respect to the position of the observer.

The exactness of this analogy did not escape Kant, who has probed more deeply into the question of the legitimacy of our judgments than has any other philosopher. He himself compares the philosophic reform which he initiated to that brought about in astronomy by Copernicus. The latter explained the phenomena of astronomy by means of the diurnal and annual movements of the earth on which the observer is placed. The former proposed to find in the *forms,* that is, the constitutive laws of the human mind, the explanation of the patterns in terms of which we conceive phenomena of all sorts. According to Kant, we err when we attribute external reality to forms. In a word, using terms which are severely technical but with which we cannot dispense because of usage, Kant accords only *subjective* value to concepts to which most men, including the majority of philosophers, attribute an *objective* reality.

Later on we shall discuss the hypothesis of the German metaphysician to see if it should not be rejected for reasons similar to those which make it necessary for us to admit the important hypothesis of his compatriot, the great astronomer. At this point it is sufficient to recall the analogy of two questions in regard to which reason may pass contrary judgments depending upon the data it has concerning them.

8. By the term "illusion" we shall mean false appearance; that is, anything which is vitiated or distorted because of conditions inherent in the perceiving subject, as a result of which it is limited to a false idea of the object perceived. On the other hand, we

shall use the term "phenomenon" when referring to veridical appearance; that is to say, to anything having all the external reality which we attribute to it naturally. Finally, we will distinguish the phenomenon whose external reality is merely relative from the absolute reality which the mind conceives even though there is no hope of reaching such reality through the means of perception.[3] Some examples will again serve to clarify the meaning of these abstract definitions.

When I am on the deck of a moving ship and see the trees and houses on the shore passing by, I am experiencing a sensory illusion which I immediately recognize for what it is, because I have reason to believe the shore is stationary. On the contrary, my senses do not deceive me when they lead me to believe in the movement of a passenger who is walking near me on the deck. This motion has all the objective reality that my senses lead me to attribute to it. In this case my senses neither alter nor complicate that which it is their function to enable me to perceive and know. Yet this objective reality is only phenomenal or relative, for perhaps the passenger is moving with a speed equal to but in a direction opposite to that of the boat. In this case, he remains at rest in relation to the shore, which I rightly take to be immobile. In any case the real motion of the passenger in relation to the shore or to the surface of the earth is determined by the combination of his motion and that of the boat.

But, by admitting the hypothesis that the passenger remains at rest in relation to the surface of the earth, and consequently absolutely motionless, if it were possible to admit with the ancients the absolute immobility of the earth, it is clear to us that his state of rest results from the coexistence of two contrary motions which offset one another. Yet each motion really exists by itself as an instance of what we call phenomenal or relative reality, in order to distinguish it from the absolute reality conceived by the mind even though it cannot be reached through observation.

The uneven curve which a planet seems to describe when viewed from the earth with the stars being taken as reference points, is an

---

[3] What we call *absolute reality* as over against *relative* or *phenomenal reality*, corresponds to what Kant has called *Dinge an sich selbst*. This is a technical expression which English translators express literally as "things-in-themselves," French translators as *choses-en-soi*.

illusion in which the objective reality is distorted by subjective conditions inherent in the position of the observer. On the other hand, the elliptical orbit that a satellite describes around its planet, apart from certain perturbations whose limits have been determined by astronomers, is not simply an appearance. The description of the satellite's orbit is a phenomenon, or, if one prefers, a fact which has phenomenal reality relative to the system composed of the principal planet and its satellites. More accurately, and relative to the solar system of which the planet and its satellites are both dependent parts, the path of the satellite is a more complex curve, resulting from a combination of the elliptical motion of the satellite around its planet and of the elliptical motion of the planet around the sun. Still more accurately, and considered in relation to the group of stars of which the sun is one, the path of the satellite is the resultant of a combination of the motions of which we have just spoken, plus the still little-understood motion of the solar system. And so it goes; we are not able to reach absolute reality, in the strict sense of that term.

9. In order to clarify still further these preliminary notions, let us compare with the above illustrations others drawn from impressions related particularly to the sense of sight. In dark places fatigued or weak eyes experience impressions similar to those that direct or reflected light produces on healthy eyes under normal circumstances. Flashes of light, dark, or variously colored spots are seen. These same effects can be produced by electricity or by mechanical pressure. Such cases are the basis for the type of visual or optical sensations which physiologists call *subjective,* since they refer to no external object which would be disclosed in the ordinary manner as a consequence of the effect on the retina of rays originating in the object. In the case of hallucinations, ghosts and apparitions are thought to be seen. In this case, it is no longer a morbid or abnormal character of the retina or the optic nerve that vitiates the brain images. Rather, the reverse is true. A diseased and abnormal state of the brain reacts on the nervous system in such a way as to pervert its functions. Similar aberrations of the organs of sensation, which may be regarded as being in some sense normal because of their frequency and quasi-periodicity, give rise to dreams. Quite obviously these experiences are only illusory or false appearances. They are no doubt governed by apparent or

hidden laws which govern our sensibility, but which are not connected with any external reality, or at least are not connected with it in such a way as to produce in us a perception or knowledge of such reality.

When a glowing coal is revolved with sufficient rapidity it produces an impression of a continuous luminous circle. We find in treatises on physics a theory of *accidental* colors, that is, a theory concerning the colors that a white surface assumes along the edges separating it from a colored area, or the colors that the white surface acquires for some moments after we have looked at a colored surface for a sufficient length of time. These, too, are instances of appearances which result from the capacity of the eye to have sensations, and which have no external reality. Certain modifications in the structure of the eye or in the tone of the nerve tissue would enable us to follow the movement of the lighted point; but eyes constructed like ours see a continuous circle. However, these illusions, these false appearances, unlike those of the first category, are neither independent of the presence of external objects, nor bound to the presence of these objects, but they belong to an entirely different set of relations than those which give a representative efficacy to impressions of the same kind under normal circumstances. These illusions, on the contrary, require the presence of external objects, and result from a deviation from the usual laws of representation. This deviation is itself amenable to regular laws and is capable of being defined by means of experience and corrected through reason. In this way perception may be protected from the influence of subjective modifications which alter and distort it.

Our first experience of a rainbow surprises us, and, at first, we think of it as a material object. This is so because we are used to seeing colors spread over the surface of solid bodies which keep their color when moved from one place to another. These are what we call *natural* colors. Our first impulse is to think of the rainbow as having natural colors which occupy a definite place in the sky, and which would have the same appearance to observers at different places, except for the ordinary effects of perspective which we normally take into account. But a rainbow does not have this degree of objective reality or stability. It exists in a certain place only in relation to an observer stationed in another particular place.

Thus, if the observer moves, the rainbow also moves or disappears entirely. Nevertheless, it is not an illusion, for while it is necessary for the observer to take a certain position if the light rays are to produce the image for him, and he would not see it if he were some place else, yet we know perfectly well that the light rays follow their paths independently of the observer and of whether he has his eyes closed or open to receive them. The rainbow is a phenomenon; the position of the observer is the condition of the perception but not of the production of the phenomenon: *ratio cognoscendi, non ratio essendi* [the reason for its being known, not the reason for its being].

10. What has been said of the rainbow may be said about *changing* colors found in certain bodies. The perception of colors changes with the position of the observer with respect to bodies; but not, as is the case with accidental colors, as a result of natural modifications which are characteristic of the sense organ or of the perceiving subject. Bodies actually reflect rays of a certain color in one direction and rays of a different color in another direction. Therefore, we shall say that the idea of a body as having a certain color is not an illusion. This idea has objective and phenomenal reality, although it is relative and not absolute. Contrariwise, we will regard as illusory the way the body appears to a person whose eyes are so diseased that they distort colors, or who, without being aware of it, looks at the body through a colored medium of some sort.

In the case of bodies having an invariable natural color, such as perfectly pure gold, physical characteristics deduced from the color will have the highest kind of *value* in the eyes of naturalists and philosophers. In fact, it will have the highest degree of objective stability. This does not mean that when one says, " Gold is yellow," one conceives that there is something in the metal resembling the sensation that gives us the experience of yellow. For the last two centuries metaphysicians have striven to warn us of such great error. Yet we understand, or at least any person who has reflected a little understands without difficulty, that gold really has the property of reflecting light rays of a certain kind. We distinguish these properties from all others by means of their power to affect the sensibility of the retina in a certain way. Furthermore, thanks to the development of science, we shall be able to distinguish them, if necessary, through other characteristics, such as a certain index of refraction or certain calorific or chemical effects.

Let us assume now that some physicist comes along who scrutinizes the optical properties of bodies more carefully. He may notice that metallic surfaces, even when they are not polished, always reflect more or less abundantly white light which is thrown on them, just as a mirror does. He will also notice that this white light when reflected *specularly* blends, in such a way as to mask its true shade, with light that has penetrated ever so little between the particles of the body. In this case the light is subjected to a peculiar action as a result of which the physical particles tend to absorb rays of one color and reflect those of another, depending upon the nature of the object in question. This is the true basis of the natural color of bodies. Developing this idea, by separating the real color of bodies from another phenomenon which complicates it, that of specular reflection, the physicist will be able to demonstrate that the yellow color of a piece of metal is the result of the fusion of rays of white light reflected specularly and of purple rays which have been subjected to the type of molecular action of which we have spoken. He will also notice that light thrown across a sheet of gold leaf actually has a purple color, and that metallic gold in the form of the impalpable powder obtained in a chemical precipitate is also purple. As a result he will conclude that, contrary to what is ordinarily believed, the characteristic color of gold is purple. Thus he will have taken a further step in the investigation of the reality contained in the phenomena. He will have moved forward another stage in the series whose last term would be absolute reality, whether or not this lies at infinity, and whether or not we can reach it.

But the physicist will not be satisfied to remain at this point; and not simply because he knows he has not as yet grasped the absolute reality that underlies the phenomenal appearance, but because he regards it as not at all impossible to penetrate further into the intrinsic reason or real basis of the whole field of optical phenomena, our first notion of which, reflecting as it does conditions characteristic of our organism, is gained through the sensation of an extended color. By virtue of a law of the human understanding, a law of which we shall have to speak elsewhere, the naturalist will be led to find the reason for all these phenomena in relations of configuration and movement, in the play of certain mechanical forces which are themselves thought of only as the causes of motion. On the basis of this impetus he will formulate hypotheses which, in turn, he will test by means of ingenious experiments. Soon the mathematician will redouble his efforts to reduce sensible nature to a purely intelligible nature in which there are only rectilinear, circular, and undulatory motions controlled by the laws of numbers. But by that very token, and admitting the full success of such tentative hypotheses, assuming that optics will be reduced only to a problem of mechanics, we find ourselves coming again upon an order of more general phenomena, from which we have earlier drawn simple and abstract examples, and as a result of which we have recognized, through these examples themselves, that we do not have the ability to attain absolute reality. Yet it may be within the limits of our capacity to raise ourselves from an order of phenomenal and relative realities to an order of higher realities, and thus to gain gradually an understanding of the fundamental reality of the phenomena.

11. When, in turn, the perceiving subject is considered as the object of knowledge, all the modifications it undergoes, even those to which no external and phenomenal reality corresponds, may be regarded as phenomena, and as such may be observed, studied, and made subject to laws. Thus, hallucinations of the sense of sight will be described and studied by physiologists and psychologists who are interested in the capacity for sensation, as much in its aberrations as in its normal state. The sensation of accidental colors will attract the attention of physiologists and even physicists for the same reason, because of certain very simple and purely physical laws according to which accidental colors appear whenever real colors are contrasted.

12. The distinction between the perceiving subject and the object perceived continues to be admissible even when man observes and knows or seeks to know himself, in his characteristic individuality. This distinction is most evident with respect to the phenomena of our physical nature of which we have sensations. But even in the field of intellectual and moral phenomena man can be an object for self-knowledge. If this were not the case, all knowledge of phenomena of this order would be impossible. It is probable that there is a multitude of moral and intellectual

facts, as there is of physiological facts, which are undetected, which are outside of the realm of knowledge, for the simple reason that it is impossible, so far as they are concerned, to distinguish a subject or faculty [4] which perceives from an object or faculty perceived. Whence comes this power of the inner man to play the part of an object of knowledge to himself, a power that we all sense, which, appearing first only in a very rudimentary form, may be strengthened and developed in the same way as other faculties of life, a power which all languages have indicated by metaphorical expressions? Perhaps this is one of the impenetrable mysteries of human nature. It is certainly one of the questions which modern philosophers have treated most obscurely. But fortunately the solution of this question is not indispensable for the purposes we have in mind. The little we shall have to say about it will naturally find its place in the chapter in which we deal with psychology and with the merit of the methods of scientific investigation used by psychologists.

[4] [Cournot frequently uses the word *faculté* in various parts of the *Essai*. It has been translated literally as "faculty," both when it is used with the general meaning of ability or capacity and when it is used with the connotation given it, in the nineteenth century, by the school of faculty psychologists.]

CHAPTER II

ON THE REASON OF THINGS

13. Animals are not limited simply to experiencing pleasure and pain. Like man they have sense organs; these organs are sometimes more perfect than man's and everything indicates that they are instruments of perception and knowledge. To deny that a dog knows his master, or that an eagle soaring at great heights observes its prey, reveals that one is supporting, in the spirit of sect or system, a paradox against which common sense cries out. Or rather, such a case would strip words of their ordinary meaning and give them others which would be entirely arbitrary and systematic. Animals, children, and idiots perceive and know in their own fashion, although doubtless they do not represent objects to themselves in the same way that a normal adult human being imagines and conceives them, thanks to the concurrence of superior senses and faculties that animals, children, and idiots do not have.

Now, one of these faculties is that of conceiving and seeking out the *reason of things*. Indeed, we regard this as the highest of all our faculties.

Like literary taste or the awareness of beauty, this faculty must be exercised and cultivated if it is to develop. It would be absurd to deny that its development may be hindered by certain defects of organic structure or by unfavorable environmental circumstances, such as those which lead a man to concentrate all his energy on labor or on coarse pleasures. Yet one always finds among men who are reputed to be reasonable some evidence of this tendency to inquire into the reason of things; of this desire to know not simply how things are, but why they are one way rather than another; and, consequently, of this awareness of a relation which is not gained through the senses, this notion of an abstract bond by virtue of which one thing is subordinated to another which determines and explains it.

14. A wide acquaintance with philosophers is not necessary before one becomes aware of the imperfections of philosophic language and understands that in it the same term is often used with

different meanings. The term " reason " is certainly one of those the meaning of which varies greatly from one author to another, and indeed from one passage to another in the work of the same author. Later on we shall examine the question as to whether or not this imperfection in philosophic language is a vice which may be corrected, or whether it is an inconvenience which the very nature of things makes it impossible to correct. For the present, there is reason to conjecture that a defect for which so many distinguished minds have found no remedy is a natural and irremediable one. For the time being, it may also be said that the word " reason " like most of those which refer to the faculty of knowing, such as " idea," " judgment," " truth," " belief," " probability," and many others, has a marked tendency to pass from a subjective to an objective significance, and vice versa, depending upon whether attention is focused upon the subject who knows or upon the object which is known. An ambiguity arises in this way which affects similarly all terms of this sort. Thus the word " judgment " refers at one time to the faculty of the mind, and at another time to the products of this faculty. Similarly, at one time the word " idea " may be taken to mean thought itself, grasped in a given manner; at another time it will refer to the intelligible truth which is the object of thought.[1] The situation is the same with the words λόγος, *ratio*, "reason," which at one time indicate a faculty of a rational being, and at another a relation among things themselves. Thus, it can be said that man's reason (the subjective reason) seeks for and grasps the reason of things (the objective reason). It is natural to admit, at least provisionally and until sufficient examination has been made, that the ambiguity inherent in this class of words and the constant tendency to pass from one meaning

[1] " I have said that I take ' perception ' and ' idea ' to be the same thing. Nevertheless, it must be noted that this thing, although single, has two relations, the one to the mind which it modifies, the other to the thing observed, in so far as it is objectively in the mind. The term ' perception ' refers more directly to the first relation, that of ' idea ' to the latter. Thus, the ' perception ' of a square indicates more directly my mind perceiving a square, and the *idea* of a square indicates more directly the square as it is *objectively* in my mind. This observation is very important for resolving many difficulties which result from the fact, which is not well enough understood, that we are dealing here not with two different entities, but with a single modification of the mind which essentially embraces these two relations." Arnauld, *Des vraies et des fausses idées*, Chap. V.

to another results from our inability to conceive and to explain that relation between subject and object which produces, or rather, which constitutes knowledge. Furthermore, the mind tends to disguise this inability from itself by letting the imagination drift along with I know not what vague and indeterminate creations which have something of the nature of both object and subject. Reid saw well the vanity and danger of this tendency with regard to the theory of ideas.

15. Even when the word " reason " is used in such a way as to designate quite positively a faculty of the human mind and in such a way as to avoid all confusion between the subject and the object of knowledge, it has not as yet acquired a fixed and invariable meaning in the language of the philosophers. Reason is often understood to mean the faculty of reasoning, that is, of linking judgments together, of laying down principles, and of deducing conclusions from them. According to writers of Condillac's school, reason, or that faculty which essentially distinguishes the intellect of man from that of the brutes, consists in the capacity to form general ideas and to fix them by means of symbols. According to Kant, reason is a higher faculty than the understanding, just as the understanding is a higher faculty than the sensibility. Just as the understanding " reduces to a unity," that is to say, systematizes the appearances furnished by the senses by subsuming them under laws, so reason systematizes or reduces to a unity the rules of the understanding by subsuming them under principles. Reason, according to other philosophers who speak with no less authority,[2] is the faculty for knowing absolute and necessary truths,

---

[2] " For indeed a knowledge of the necessary and eternal truths is that which distinguishes us from simple living things and makes us sharers of reason and of the fields of knowledge while it raises us to the knowledge of ourselves and of God. And this is that within us which is called a ' rational soul ' or ' spirit.' To the knowledge of the necessary truths and to abstractions from them should be attributed the admission that we may be raised to reflexive actions by force of which we attain the concept which is called ' self ' and think that the former or the latter is in us. And thence it also arises that we ourselves, thinking upon being, think about matter—now simple, now complex—and about that which is not matter but God himself, while we apprehend that what is subject to limitation with us exists without limitation in Him. And these reflexive actions provide the special material of our ratiocinations." Leibniz, *Opera omnia*, ed. Dutens, II, 24.

such as the idea of God, and of the infinite, the ideas of unlimited space and time, the idea of duty, and others of the same sort. Finally, there are authors, including some of the most recent ones, for whom the term reason is nothing but a general heading, which includes all the faculties related to knowledge as opposed to those related to sensibility on the one hand and to action on the other.

We do not intend to question any of these definitions with minute precision. As conventional and arbitrary definitions, they may all be suited to the exposition of certain systems. The only thing we wish to indicate is that these definitions are arbitrary and systematic, and that they do not bring into sufficient relief the most essential character which distinguishes man as a rational being from other creatures to which common sense accords a certain degree of intelligence, but not reason. Is it not immediately evident that when this distinctive character is made to consist in the perception of necessary and absolute truths, of God, and of infinity, we put ourselves on too high a plane; that we are too far removed from the nature and from what may be regarded as the normal condition of human beings? Consider the case of a child who can scarcely talk, but whose active curiosity leads him to ply his parents and teachers with questions. It will be a long time before he has any idea of infinity, of necessity, and of the absolute; yet already he wants to know the how and the why of the things that fall within the limited field of his intelligence. This fact alone makes him infinitely superior to the most intelligent animals. In spite of his ignorance of all the abstract ideas which control the reason of the adult mind, this childish curiosity may be regarded as the mark, and as the germ, of the faculties which he will some day apply to studies of a higher order, and which will give him superiority over ordinary minds.

But, without stopping to consider what is going on in the mind of a child, it is quite clear that the reason of the adult mind, of the philosopher, and of the scientist is often found to be occupied with matters with respect to which it is possible to avoid the notions of infinity and of the absolute, and where it is even desirable to avoid the introduction of such notions. Doubtless the physicist, the naturalist, the economist, and the politician are not strangers to the speculations of the metaphysicians concerning these sublime and obscure ideas. When the former raise questions which pertain to their own fields, they discover the need of at least the germ of these very ideas in their own thinking. But they also understand perfectly well that it is expedient to ignore them since they have no bearing on the progressive development of the sciences which are the special concern of their studies. Yet everyone agrees that philosophy is found in their writings, although the work of some of them has a more philosophical turn than does that of others. For that matter, the very word "philosophy" appears on the title page of some of their works. Therefore, the philosophic spirit, which is nothing but understanding cultivated by rare insight, may be understood without resorting to such notions as infinity and the absolute. Thus, violence is done both to the nature of things and to the ordinary meaning of words if reason is defined as the faculty whose function consists essentially in grasping the idea of infinity and in perceiving absolute and necessary truths. This is the case even though one of the outstanding functions of man's reason, one of the powers of his intelligence which is absolutely lacking in lower minds, at least from our point of view, lies in just that function.

16. If we contrast the definition we have just been criticizing with that proposed by philosophers for whom the essential character of reason is found in the faculty men have for forming general ideas through the aid of symbols, we shall find that they are deficient because of opposite faults. The one immediately carries us into thin air; the other is insufficient to explain the most simple and commonplace acts of thought. The faculty for recognizing resemblances between things and for expressing them in language by means of classifications and general terms must not be confused with the faculty for grasping relations which signify that things depend upon one another and are constituted in one way rather than another. Through the use of the first faculty, the mind succeeds in putting knowledge in order, in facilitating an inventory of it, or, what amounts to the same thing, in describing more easily how things exist. But it is through the other faculty that the mind grasps the why of things, the explanation for the way they exist, and their mutual dependencies.

The truth of the matter is that the how and the why of things are very closely related in the sense that a good description of a thing usually gives reason a clue to its explanation. More properly, one description is regarded as excellent and as preferable to all others only because it places us immediately in the most favorable position for explaining the subject matter in question, for penetrating as far as possible into an understanding of the relations which govern the course of its development and its organization. Therefore, it is quite plain that arbitrary classifications and general terms may be of assistance not only to attention and to memory, through their function as instruments which are useful in research and description, but that they may also assist in making the perception of the reason of things more prompt and clear. We consider this to be the most essential attribute of human reason. Nevertheless, this is no excuse for confusing faculties which are basically distinct, and which are capable of a very unequal development.

For example, although he would like to know the why of everything, the child of whom we were speaking above does not as yet have, to any great extent, the capacity to abstract and to generalize. Some men who are gifted with very penetrating and very inventive minds, at least within the limits of the special fields to which they apply themselves, are unfamiliar with the forms and procedure of logic, with general terms and abstract classifications. On the other hand, scholars and philosophers who are very much given to generalization and classification, who are very fruitful in the creation of new terms and new forms of procedure for dealing with the genera and classes which they set up, are not those who actually make the most progress in the sciences and in philosophy. Therefore, it is a mistake to think that the truly active principle, the principle of fecundity and of life, with respect to all that pertains to the development of reason and of the philosophic spirit, is found in the faculty which enables us to abstract, to classify, and to generalize.

It is said that the great mathematician Jean Bernoulli, being vexed to see that his contemporary Varignon [3] had apparently appropriated his discoveries under the pretext of giving them a generality that their author had neglected to give them, a thing which could be done without any great additional inventiveness, closed a new memoir by saying maliciously, " Varignon will generalize this for us." On the other hand, one is frequently well-advised to use

[3] [Pierre Varignon (1654–1722), an eminent French mathematician, important because of his work on statics.]

the most general methods, since these are regarded as the ones which are at the same time the most fruitful. Both this general truth and the epigram of Bernoulli should be admitted only with restrictions. Generalizations which are fruitful because they reveal in a single general principle the rationale of a great many particular truths, the connections and common origins of which had not previously been seen, are found in all the sciences, and particularly in mathematics. Such generalizations are the most important of all, and their discovery is the work of genius. There are also sterile generalizations which consist in extending to unimportant cases what inventive persons were satisfied to establish for important cases, leaving the rest to the easily discernible indications of analogy. In such cases, further steps toward abstraction and generalization do not mean an improvement in the explanation of the order of mathematical truths and their relations, for this is not the way the mind proceeds from a subordinate fact to one which goes beyond it and explains it. Once more, the source of the discoveries of genius, of the progress of the sciences, and of the most brilliant manifestations of human reason is, therefore, not to be found in the capacity to generalize.

17. Similar criticisms could be made of all other definitions that have been given of reason as an intellectual faculty or power. But, since the important thing is to fix, as quickly as possible, the meaning of the words one uses, we will content ourselves with saying that when using the word " reason " in the subjective sense, we shall mean to designate primarily the faculty of knowing the reason of things, or, in other words, the faculty which lays hold of the order according to which facts, laws, relations, and objects of our knowledge are connected with, and proceed from one another.[4] By

---

[4] " The relation between reason and order is of the highest sort. Order is discovered in things only through reason and cannot be understood without it. It is the supporter of reason and its true object." Bossuet, *De la connaissance de Dieu et de soi-même*, Chap. I, sect. 8. [Jacques Bénigne Bossuet (1627–1704), a famous and distinguished French theologian, bishop, and eloquent orator. The author of numerous works including *Discourse on Universal History* (1681) and *History of the Variations of the Protestant Churches* (1688). He was the recognized spokesman for and leader of the Roman Catholic Church in France during his lifetime. From 1691 to 1700 he carried on an extensive correspondence with Leibniz concerning a proposed treaty for the union of the Lutheran and Roman churches.]

thus giving a precise meaning to a term in accordance with the needs of the problem under consideration, we may depart so far from any of its commonly accepted meanings that our definition may be criticized as being artificial and arbitrary. This procedure will be more fully justified when, later on in this study, we shall try to show more completely that the faculty thus defined dominates and controls all others; that this is indeed the principle of the intellectual pre-eminence of man, that which makes it legitimate to speak of him as a reasonable being as distinguished from animals, children, and idiots, who also have cognitions, and who even combine them up to a certain point.

18. We must be careful not to confuse the idea that we have of rational interconnection or of the reason of things with the ideas of *cause* and *force,* which are also found in the human mind, but which are discerned by it in another manner. Feelings of muscular tension suggest to man the idea of force which, in association with such ideas of materiality as the senses furnish him, is the basis of the whole system of physical sciences. As to the idea of cause, metaphysicians have written at great length to show how it springs from the internal feeling of activity and of human personality, in order to make clear the process of inference through which man carries over into the external world this idea which makes him conscious of his own characteristic faculties. It is not necessary for us to enter into this delicate question again here; for the idea of the reason of things has a generality of an entirely different sort than has the idea of efficient cause, which in turn is much more general than the idea of force. It seems to be neither indispensable nor possible to assign a psychological origin to the first of these ideas; it is clearly perceived in the highest realm of our intellectual capacities, and the spectacle of nature would not be sufficient to develop it in us if the germ of it were not already present in our make-up. This idea may be awakened but not given through knowledge of our personal activity, and still less through the feeling of muscular effort and through sensations properly so called, that is to say through those gathered through the special organs of the senses.[5]

[5] " Certainly the sufficient reason, in Leibniz's sense of the term, is not the same as the efficient cause. On the contrary, the former is established in its generality, which includes the whole system of our ideas as well as the facts

19. We shall fully justify our assertion if we show that the idea of the reason of things, understood in the general sense which it necessarily implies, is often opposed to the idea of efficient cause, at least in the form in which the human mind deduces this latter from the consciousness of its own activity. When, in playing "heads and tails," a long series of tosses shows an inequality in the chances that one or the other of the sides of the coin will appear, this inequality is attributed to a lack of symmetry or to an irregularity of structure in the coin. The observed fact is the more frequent appearance of one side than the other; the reason for it is the irregularity of structure. But this reason has no resemblance whatsoever to a cause properly so called, that is, to an efficient cause. Yet in ordinary language one does not hesitate to say that the irregularity of structure is the *cause* of the more frequent appearance of one of the sides than the other, or that it *acts* in such a way as to favor the appearance of that side. All the molecules which make up the mass thrown really play only a passive role. Using exact philosophical language, an action, a force, or efficacious power cannot be attributed to the intrinsic structure of the molecular system, to the law according to which the mass of the object is distributed, or to the external form of the body. The appearance of a given side after each toss is the result of active causes, the mode of action of which is variable and irregularly variable from one toss to another. These active causes are fortuitous, and, strictly speaking, the appearance of one side rather than the other at each toss is a result of chance. As we will explain later on, the object of making a great many tosses is to reach a result which is approximately free from the influence of chance or of the fortuitous causes which alone play an active part in each particular toss. Consequently, one may not say that the result obtained in this way has a cause, in the strict sense of that term; yet it has its reason for being or its explanation, which is found in the structure of the coin.

When it is said that a flywheel *acts* in such a way as to regulate

---

of nature, only by excluding the latter, that is, productive causality. . . . The sufficient reason, as its name indicates, is nothing but reason in action, or applied to binding facts together in the natural or legitimate order of succession as illustrated in the connecting of consequences with their principles in the logical order of our ideas and of our conventional symbols." Maine de Biran, *Œuvres philosophiques*, IV, 397.

the movement of a machine, or that it is the *cause* of the regularity of its movements, there is no intention of attributing to the flywheel an inherent power which it does not have. It is well understood that actually the flywheel plays a passive role in the movement of the machine, storing up energy at one moment and at another moment returning it to the parts of the machine in such a way as to offset inequalities in the action of the motive force. It does this in all cases as a result of the inertia of its mass, and not as the result of a peculiar force of its own or of an energy with which it is endowed. This means simply that the regularity of the movements of the machine is a phenomenon the explanation and reason for which is found in the connection between the flywheel and the other parts of the machine.

20. Suppose an engineer were to notice that a river tends to flow away from one of its banks and to be thrown against the other. When he looks for the reason for this phenomenon, he finds it in certain accidental configurations of the river bed. His technical knowledge suggests to him the possibility of building a structure which will correct the flow of the river and keep it from disturbing those who live near the affected shore. If he accomplishes this, it will be said that he has found the cause of the trouble and also its remedy. But, once again, the term " cause " is used improperly, although this usage is in accordance with custom. There is really a series of causes which have successively thrown each molecule of water against the threatened bank. These causes have operated in such a way as to bring the molecules of water together from very divergent places as a result of their describing dissimilar curves in the atmosphere while they were in the form of water vapor or droplets. But all these variations in the way in which the really active forces and causes behave are without effect upon the phenomenon with which we are dealing. This phenomenon is constant, because the reason that determines it is constant, and because this reason is found in a fact or a group of facts which is permanent and independent of the series of active and variable causes which have brought about the concurrence of each individual molecule in the production of the phenomenon at a given moment.

21. We have been drawing our illustrations from mathematical and mechanical facts of the simplest, the most fundamental, and, in some ways, the most common sort. But similar examples will

be found among types of events with which we are much more
familiar. What is spoken of in our day as the philosophy of history
obviously does not consist in an examination of the causes which
have given rise to each historic event according to the whim or
fluctuating feelings of men of action, but in the study of the relations and general laws which make the development of the whole
body of historical facts reasonable when taken together. This is
an abstraction which is drawn from the variable causes which have
been the motivating factors for each particular event. Let us suppose that a certain province has been successively conquered, lost,
and reconquered according to the fortune of battle. The reason
that sooner or later should serve as the basis for the final incorpora-
tion or definite separation of the province is found in the geographical configuration of the country, that is, in the direction in which
its rivers flow, and along which arms of the sea and mountain
ranges lie with respect to it, in the similarity or differences among
the races making up its population, with respect to their dialects,
their mores, their civil and religious institutions, and their commercial interests. Fortuitous causes, such as the energy or weakness, the competence or incompetence of certain persons may lead
to the miscarriage or to the success of a conspiracy. Likewise, a
writer who is interested in anecdotal details will often take pleasure
in bringing into view the insignificance of the causes which have
given rise to an event. But the reason of the philosopher will be
discontented with such explanations, and it will remain uncon-
vinced that it has not found in the defects of the constitution of
the government not the cause, if that term is used correctly, but
the real explanation, the true reason for the catastrophe in which
it has perished.

22. The book to which Montesquieu gave the title *The Spirit of
Laws,* in conformity with the accepted terminology of his day, is
obviously a treatise on the reason of law, or, as we would say today,
on the philosophy of law. What the legal philosopher must do is
to show the reason of a law, of an obligation, of a legal disposition,
or of a custom, and not simply the motives which actually, but
accidentally, have brought about certain legislation or introduced
certain customs. His task is to clarify these motives, to distinguish
among them those that are introduced because of facts or interests
of a particular, variable, and transient character. Until he achieves

this end, reason remains unsatisfied. No one will confuse efforts made to give reason its desired satisfaction with studies the motive behind which is curiosity, and the object of which is to establish historically the causes which have acted upon the mind of a particular prince, upon the plottings of a particular group, or which have won the votes of certain members of a political assembly.

23. If we proceed to considerations of another kind, we shall find a no less striking contrast between the idea of the reason of things and what is correctly called the idea of cause. An organism is a being whose parts are harmoniously related among themselves in such a way that a change in the relations would make it impossible for the creature to subsist or maintain itself. Among the diverse ways in which the existence of such relations can be explained, there is one which supposes that, in the course of time, the concurrence of fortuitous circumstances has given rise to a great number of combinations. Among these forms some will fail to bring together the conditions of conservation and perpetuity; these will have an ephemeral existence. In time, however, chance will give rise to a form having the organic unity upon which the stability and the continued life of both individual and species depend. Although we shall come back to this point later on, let us admit this theoretical conception for a moment. In doing so, it will be seen that the philosophic study of an organism consists in going farther and farther into the understanding of the harmonic relations and the co-ordination of its parts. It is in these that the reason for the existence and the preservation of the organism is found, and not in the causes which have acted fortuitously and blindly in such a way as to produce ephemeral combinations as well as those having the conditions necessary for the persistence of the organism.

Thus, when a naturalist studies the laws of the habitat and of the geographical distribution of plants and animals according to altitude, latitude, and climate, what attracts his attention is not the accidental causes which have carried a viable seed from one place to another, or which have brought about the migration of a certain pair of animals which have had numerous offspring. These causes have no more value in the eyes of a naturalist than those which have, in the course of time, brought about the movement of a great number of animals which have perished without being able to reproduce their species. It is sufficient to understand in a general way that the lapse of time, by increasing the number of fortuitous combinations, has inevitably brought about those species capable of producing the permanent and stable results which our observations discover. Consequently, the purpose of the philosophical naturalist is simply to make clear the adaptation of various parts to one another which makes possible a reasonable account of the acclimatization of the species, and of the final equilibrium between the forces leading to its perpetuation and to its destruction; in a word, the conditions which account for the observed results.

If reason finds it repugnant to be satisfied with such an explanation of all the marvels of the world, and if some things reveal themselves to us in such a way as to indicate the intelligence of an artisan who adapts his means to the ends he desires to attain, it is absolutely essential that the philosopher who wishes to appreciate intelligently the marvels of nature shall have in view the end or purpose of the work, and the state of the structure as a whole. The true reason for the adaptation among the different parts is to be found in this structure, rather than in the secondary causes and the details of the process which the providential reason has used, in the same way in which we use an instrument, a blind force, or a passive agent to carry out plans which our minds conceive. Thus, all naturalists, irrespective of the philosophical school to which they belong, and whether they may or may not be partisans of the doctrine of final causes, as that phrase is commonly used, agree on the basis of one consideration or another to seek for the reason of the principal phenomena of the organism in the very purpose of the organism. It is by means of this regulative idea, this guiding thread, to use Kant's expression, that a more and more profound knowledge of the laws of organic nature is reached.

24. The idea we form of the relation between efficient causes and the effects they produce implies phenomena which succeed one another in a temporal order. But, on the contrary, and in accordance with what we have been saying, the idea of the reason of things, and the consequences to be drawn from it, often implies that an abstraction has been made from the order according to which irregular and accidental phenomena are produced temporally so that we may consider only general results without regard for the influence of accidental causes and the manner in which they follow from one another chronologically, or the conditions of

a final and permanent state which is similarly independent of time. In a word, this abstraction is set up to enable us to reach a theory whose central characteristic is that it is uninfluenced by data which have to do with chronology and history. There are good reasons why the sciences which, like mathematics, deal exclusively with truths which are abstract, permanent, and absolutely independent of time are unable to exhibit among the sort of data with which they deal anything which resembles a relation between two phenomena, one of which is conceived of as the efficient cause of the other. However, anyone who has any knowledge at all of mathematics knows that different demonstrations of the same theorem, all of which are both completely irreproachable so far as the rules of logic are concerned, and rigorously conclusive, may be given. And such a person distinguishes that demonstration which gives the true reason of the theorem demonstrated, that is to say, the demonstration which follows, in the logical connection of propositions, the order according to which corresponding truths are reached, and according to which one is the reason for the other. The mind is unsatisfied until such a demonstration has been found; this is the case not because the mind is incapable of extending our knowledge through the acquisition of a greater number of facts, but because it proves the need for arranging these facts in their natural relations, that is to say, in such a way as to make evident the reason for each particular fact. Consequently, a demonstration is said to be indirect when it inverts the rational order, when the conclusion obtained as a consequence of logical deduction is thought of as including, on the contrary, the reason for truths which serve as its logical premises.

Certain demonstrations given by mathematicians are always criticized for constraining the mind without enlightening it. This is particularly true of those which we call reductions to absurdity. This would not be the case were it not for the fact that such demonstrations do not make evident the reason for the principle demonstrated. The mind, therefore, refuses to consider demonstrations of this sort as primitive and rationally irreducible facts, whose reason the mind is not to seek.

25. It is often said that two facts, or two systems of facts, react upon each other in such a way that each plays the double role of cause and effect in relation to the other. But it is clear that in

such cases the terms " cause " and " effect " are improperly used. The mind necessarily conceives of a chain of causes and effects which follow one another in time—a chain in which each term or link plays the role of effect in relation to antecedent terms and that of cause to succeeding terms—as constituting a series of the type which mathematicians call " linear " because the simplest way to represent it is to imagine a series of points placed one after another. The linear series of causes and effects is not known to double back upon itself. On the contrary, we think of it as proceeding indefinitely, in both directions, however far our observations may be extended. But we have no authority always to attribute the same simplicity to the idea of order and to the relations between things which cannot properly be called causes and effects, but which make one another intelligible, or mutually determine and explain one another.[6] For example, the reason which explains those laws and institutions of a people which are destined to last must be based upon the mores of the groups and developed in accordance with their spirit. On the other hand, within limits, the mores of a people are determined by the laws and institutions which govern them. If disturbing factors have not forcibly produced too great disparity between the laws and mores, they react upon one another in such a way as to tend toward a final and harmonious state in which evidences of the original impulses and of subsequent variations are largely obliterated. When this final state is considered, there is no longer any reason to attribute to one element rather than another a predominant role in their observed adaptation to one another. Similar remarks are relevant to the compatibility which appears between the form of a language and the turn of mind of the people who speak it, and to that between the habits of an animal species, a race, or an individual and the corresponding modifications of its organic characteristics. At

---

[6] " But those causes are incapable of ' dissolution ' which are bound together in turn by mutual bonds and, while one produced the other, are born of themselves in such a way that they are never separated from the loving embraces of their natural association." Macrobius, *In Somnium Scipionis*, Vol. I, Chap. 22.

" All things being caused and causative, abetted and abetting, mediately and immediately, and being held together by a natural and insensible bond which joins together those farthest removed and most unlike one another, I hold it to be impossible to know the parts without knowing the whole, quite as much as it is impossible to know the whole without knowing the parts." Pascal.

different times, one or the other of the terms in the relation will have a preponderant influence, yet not so dominant a one that it is unaffected by reciprocal actions. Between the two extreme cases a multitude of intervening steps can be conceived. Thus, a well-marked subordination of the planets to the sun and of the satellites to their planets results from the constitution of our planetary system. Yet, it might be that bodies making up another system having similar relations of masses and distances would influence one another mutually without giving rise to such a distinct hierarchical relation, or even without any trace of predominance remaining.

We may observe a similar instance of this reciprocity of relations in the field of abstract conceptions, in a form which is incompatible with the notion of cause and effect in its proper sense. Many properties of numbers depend upon laws which govern the theory of order and combination in general. Reciprocally, the science of combinations turns up in a thousand places in pure mathematics and in the properties of numbers. The same objects of thought can occupy different positions in the series of abstractions and generalities. The difference depends upon which of their characteristics are considered. As a result, a tangle of relations arises which is incompatible with the very simple idea of linear development as related to the series of causes and effects. Later on we shall follow out some of the implications of these remarks. At this point we desire only to indicate the principal characteristics which keep us from identifying the idea of the reason of things with the idea of efficient cause; from accepting as explanations of one of these ideas that which would be accepted for the other, keeping in mind the supposition that it is possible to discover how and why these fundamental ideas which govern all its operations exist in the human mind.

26. In fact, as we have already noted, the word "cause" is used intentionally in ordinary speech to designate the reason of things as well as to indicate the cause properly so called.[7] In that it has a certain resemblance to the terminology adopted by the ancient scholastics who, following Aristotle, distinguished four types of causes: the *efficient* cause, to which alone the term cause

[7] "CAUSE, *principle:* that which determines what a thing is, that it shall occur." *Dictionnaire de l'Académie* (edition of 1835).

should be applied, according to the usage of modern metaphysicians; the *material* cause, the *formal* cause, and the *final* cause. Indeed, it is sufficient to refer back to the examples we have been using in order to see the necessity of finding the reason and explanation of things at one time in certain qualities of form, disposition, or internal structure, that is, in the material and formal cause, and at another time in the conditions of organic unity, that is, in the final causes. This use of the term " cause," which as a result of common practice has come to prevail in ordinary discourse, is all that justifies the approximation on which the Aristotelian classification rests. Otherwise it would be childish to say, as most of the scholastics do, that the block of marble from which a statue has been cut is the material cause of the statue. Neither can it be seen any more clearly in what sense it must be said, with them, that the idea conceived in the mind of the artist is the formal, rather than the efficient or final, cause of the creation. In this circumstance as in many others, ordinary language, which is a faithful expression of the suggestions of good sense, may be better than technical definitions. Only when the term " cause " is taken in this broad sense can the adage *Philosophia tota inquirit in causas* ["All philosophy is an inquiry into causes"] be justified; for the reason of things, wherever found, is really the constant goal of philosophic inquiry. The search for the explanation and the reason of things is what characterizes philosophic curiosity, no matter what the order of facts to which it is applied. Over against this is the curiosity of learned men and scholars whose object is to increase the number of known facts, taking more account of their singularity and of the difficulty overcome than of their relative worth in the explanation and the rational co-ordination of the system of our knowledge. Consequently, we find that philosophy will sometimes devote itself to inquiries concerning efficient causes, when it undertakes to explain the great geological phenomena observed at the present time as the result of the lifting of the continents and a consequent displacement of the oceans, for example. At other times, as in the instances we have cited, philosophy will be engaged in an examination of formal and final causes in which it is necessary to account for general, definite, or permanent results which do not follow from the action of accidental and irregular efficient causes. If these cases arouse our interest in no way, or if no trace of the manner in which they act remains, they will remain lost in oblivion. On the other hand, should such cases pique our curiosity and our emotions in a dramatic or moral way, as in cases which affect human beings significantly, the anecdotal memoirs and the scholarly compilations of the antiquarian will become the stuff in which true history finds its sustenance. But in neither case will these be the proper object of philosophic speculations.

27. We cannot hope to indicate here how the idea which we wish to present of that which gives philosophy its essential character is related to and differs from that which was formulated by Leibniz when that great man, the greatest genius by whom the sciences and philosophy are honored, attempted to erect the whole structure of his philosophy on the idea of *sufficient reason*, that is to say, on this axiom: a thing can exist in a given manner only if there is a sufficient reason for it to exist that way rather than another. The elegance, the symmetry, and the profundity of the system raised on this foundation may well be admired, for it is a system that may be regarded as the major work of synthesis in metaphysics. And yet this system has experienced the same fate as all systems, just because even the most profound genius is forbidden to repeat the work of God and to reconstruct the universe *de novo* through the efficacy of a single principle. Moreover, it is not our present purpose to present an exposition or criticism of Leibniz's system, but only to present some observations concerning the statement of and the importance of this axiom which has made him famous. We are interested in these observations to the extent that they contribute to the clarification of our own ideas and prepare the reader for the development of them which follows.

First of all, let us point out that the epithet "sufficient" seems superfluous when applied to the reason of things, for certainly the insufficient reason of a thing could have no meaning. If C exists only as a result of the concurrence of A and B, it would be inaccurate to say that each thing A and B, when considered alone, is an insufficient reason for C. It ought rather to be said that the concurrence of A and B is the reason for the existence of C, or the objective reason or, more simply, the reason for C.

A more important observation may be made on the basis of

the negative form of the axiom. In general, negative propositions
have the advantage of leading to decisive conclusions and formal
demonstrations. It is the laws of exclusion which, by making
necessary the rejection of all but one hypothesis, indirectly establish and remove from controversy the sole hypothesis which remains after all others have been eliminated. But, on the other
hand, it is possible to take advantage of these negative arguments
only under very particular circumstances, and only in cases which
are very simple and comparatively restricted. Thus, in terms of
the demonstration already given (24),[8] which is called reduction
to absurdity, the equality of two quantities is established by proving that one of them can be thought of as neither larger nor smaller
than the other. This is the method of demonstration preferred
by the Greek mathematicians in their scrupulous attachment to
the rigor of logical forms. But, as we proceed in mathematics
from the more simple to the more complex, this method of demonstration becomes more and more inconvenient and impracticable
because of complications which it introduces. The result is that
modern thinkers have been led to substitute other principles for it.
The greatest glory belonging to Leibniz attaches precisely to his
success in systematizing these principles exactly. Without his
work a great many important truths would have remained inaccessible to the human mind. The same thing is true of the
application of the principle of sufficient reason. For example,
consider two forces of equal intensity acting on the same point
but from different directions. Then let us ask ourselves from
what direction it would be necessary to apply a third force to this
point so as to counteract any motion it might tend to have in a
contrary direction, and thus to maintain the equilibrium of the
object. It is clear that the direction of this third force must form
equal angles with the directions of each of the first two forces;
for there is no reason why the object should tend more toward
one direction than toward the other, since the original forces are
posited as being exactly equal. Moreover, the direction of the
third force can be found to lie only in the plane which includes
the direction of the other two, for, since the plane is entirely symmetrical, there is no reason why the direction of the third force

---

[8] [In the body of the work, numbers enclosed in parentheses indicate sections
to which reference is made.]

should be deflected toward one side of it rather than the other. In this case the simplicity of the data and their perfect proportion make possible the undeniable application of the Leibnizian maxim. But this very fact itself enables us to understand that there is something unusual and peculiar in the circumstances which permit us to take advantage of the concept.

28. According to Leibniz, mathematics should be distinguished from metaphysics, since the latter should be established on the basis of the principle of identity while the former should be established on the principle of sufficient reason. But, when this second principle is called upon to establish the validity of a mathematical truth—and there are many instances of this not only in mechanics but also in geometry and algebraic theory—the domain of metaphysics is no more invaded than it would be when any other primary notion or immediate datum of reason is brought forward. The distinctive character of mathematics, as we have clearly explained elsewhere,[9] ought to be distinguished from the character of those subjects that have as their goal the truths which are grasped by reason without the aid of experience but which nonetheless always permit of experimental verification. Thus it is easy to imagine an experiment appropriate for verifying the proposition of mechanics which has just now been established by reasoning. On the other hand, this proposition in the Leibnizian metaphysics: " The created world is the best of all possible worlds " —a proposition offered, whether justifiably or not, as a corollary of the principle of sufficient reason—could not in any way be experimentally verified. This would be true even if we knew exactly what characteristics make one world better than another. The principle of sufficient reason may be relied upon to establish not only mathematical truths but also principles of law, of morality, and even rules of judgment, for, because of this principle, it is evident that judgment is offended by anything that disturbs, without sufficient reason, the symmetry of a rule. For that reason we are not authorized to refer to Leibniz's axiom as a metaphysical principle, in the sense that it would serve to direct the human mind in those investigations which bear on what is referred to as metaphysics as over against the sciences which have as their object

___

[9] Cournot, *De l'origine et des limites de la correspondance entre l'algèbre et la géométrie* (Paris, 1847), Chap. XVI.

either the physical world or the moral nature of man. But this axiom may very well be called a philosophical principle in so far as it presupposes, in the negative form of its statement, the positive idea of the reason of things, which is the source of all philosophy.

On the other hand, it seems evident to us that philosophy cannot be enclosed within the narrow limits of the application of a negative principle such as this Leibnizian axiom, any more than mathematics, esthetics, or ethics can be. Just as there are in the mind, in the absence of any rule or exact formula, faculties for judging the goodness of a moral act and the beauty of a work of art, either absolutely or by comparison with other acts or with other artistic productions, so we have faculties by means of which we grasp analogies, inferences, the connections of things, and the reasons for preference among alternative explanations or rational co-ordinations. In default of demonstrations that the nature of things and the organization of our logical instruments do not require in most instances, there are some valuations, some judgments based on probabilities that often have, for common sense, the same value as logical proof. This is the source of our obligation to study carefully the theory of probability and of probable judgments before examining anything else. We will turn our attention to this matter in the next two chapters.

CHAPTER III

ON CHANCE AND ON MATHEMATICAL
PROBABILITY

29. Just as everything must have its reason, so everything that we call an event must have a cause. Often the cause of an event escapes us or we take something to be its cause which is not. But neither our inability to apply the principle of causation, nor the mistakes into which we fall when we apply it carelessly, have shaken our adherence to this principle as an absolute and necessary law.

We always trace an effect back to its immediate cause; in turn, this cause is conceived of as an effect, and so on indefinitely. The mind cannot conceive nor can observation attain any limit to this progressive series. Turning in the other direction, a present effect becomes, or at least may become, the cause of a subsequent event, and so on to infinity. This indefinite chain of succeeding causes and effects, in which a given event is one link, forms essentially a linear series (25). An infinite number of such series may coexist in time. These may intersect in such a way that any particular event, in the production of which many other events have concurred, may be related as an effect to many distinct series of productive causes. On the other hand, such an event may in turn give rise to many series of effects which, with the exception of the first term, which is common to them all, will remain distinct and absolutely separate from one another. An accurate idea of the crossing and the independence of such chains may be gained through the analogy of human genealogy. Each individual has two sets of ancestors, one through his mother, the other through his father; and in each previous generation the paternal and maternal lines divide into two branches. In turn, each person becomes the stock or the common root of many lines of descendants. Once these lines have appeared, although they have a common source, they no longer intersect in the form of family unions, or do so only accidentally. In the course of time, each family or genealogical group forms marriages with a great many others.   But other groups, propagating themselves at the same time and in much greater numbers, remain perfectly distinct and isolated from one another as far as we can trace them; and if they have a common origin, the proof of this origin rests on other grounds than those of science and of historical documents.

Each human generation gives rise to only a two-fold division in the ascending order.   But it is not difficult to conceive of the possibility of a much greater complexity in regard to questions pertaining to causes and effects in general; indeed, there is nothing to prevent an event from being connected with a great many, or even an infinite number of different causes.   Then the groups of concurrent lines, by means of which the imagination pictures the bonds which connect events causally, would be comparable to bundles of luminous rays which interpenetrate, spread out, and become concentrated in such a way as not to reveal any gaps or breaks of continuity in their texture.

30. Whether there is reason to regard as finite or infinite the number of causes or series of causes which bring about the appearance of a given event, common sense holds that some series are *interdependent* or internally related, that is, they are series which influence one another; and some are *independent* or externally related series, that is, they are series which develop parallel to one another or consecutively, without having the least influence upon one another, or, what amounts to the same thing in the long run, they exert on one another no influence which is made manifest through any appreciable effects.   No one seriously believes that stamping on the earth disturbs a navigator who is sailing to the antipodes or upsets the order among the satellites of Jupiter.   In any case, such a disturbance would be so slight that it would not be made manifest to us through any sensible effect.   As a result, we are perfectly justified in not taking account of it.   It is not impossible that an event occurring in China or Japan may have some influence upon events happening in Paris or in London.   But, in general, it is certain that the program a Parisian lays out for his day will not be influenced in the slightest degree by what is then going on in some city of China in which Europeans have never set foot.   These are like two little worlds in each of which series of causes and effects can be observed developing simulta-

neously which are not connected and which exercise no appreciable influence on one another.

Events brought about by the combination or conjunction of other events which belong to independent series are called *fortuitous* events, or the results of *chance*. Some examples will serve to clarify and establish this fundamental notion.

31. Suppose a citizen of Paris decides he wants to make a trip into the country and boards a train to take him to his destination. If the train is wrecked, and the traveler is a victim, it will be accidentally so, for the causes leading up to the wreck are independent of the presence of this particular person. These causes would have developed in the same way even though, because of unanticipated changes in his affairs or for other reasons, he had decided to go by another route or to take a different train. On the other hand, let us suppose there is some object of curiosity which interests a great many people in the same manner, with the result that there is a great increase in the number of travelers on a particular day and at a particular hour. This might disrupt service on the railroad, and this disruption might be the determining factor or cause should an accident occur during this time. In such a case, series of causes and effects which are originally independent of one another cease to be so, with the result that it becomes necessary to recognize a connecting bond between them.

Again, let us suppose that a man who does not know how to read picks out, one by one, from a jumbled pile of type pieces which are drawn in such an order as to spell the word FRIENDSHIP. This would be a fortuitous event, a result of chance, for there is no relation between the causes directing the fingers of the man successively to such and such pieces of metal and those which have led to their forming one of the most commonly used words in our language.

Or again, suppose two brothers serving in the same corps die in the same battle. When we first think of the bond between them and of the misfortune they have shared, we are struck by the coincidence. Yet, upon reflection, we see that the circumstances might very well be such that the two events may not be independent of one another, and that the unfortunate concurrence cannot be accounted for by chance alone. It is possible that the younger brother may have become a soldier only because of the example of the older. Since they are following the same career, it is not unnatural that they should both find service in the same unit, and thus share the same perils with each other when need arose. If the peril has been equally great for both, it is not surprising that both should have been killed. Causes independent of their common parentage have played a part in this event, but the connection between their status as brothers and their joint misfortune is not fortuitous.

Now let us suppose that they served in different armies, one on the northern frontier, the other at the foot of the Alps. Suppose also that there were battles on the two frontiers on the same day and that the two brothers were killed. There would be reason for regarding this event as a result of chance, for at so great a distance, the operations of the two armies constitute two series of events, the initial order of which might come from a common center, but which would develop later in complete independence of each other, as a result of being adjusted to the local circumstances and conditions. Conditions which would lead to a battle on one front on a given day rather than on another would not be closely related to similar events which would result in a battle on the other front on the same day. If the corps to which the brothers belong respectively have participated in these battles, and if both have been killed, there is nothing in the fact that they are brothers that could have concurred to bring about this twofold event. Thus, when the two great brothers-in-arms, Desaix and Kléber,[1] fell on the same day, indeed almost at the same moment, one on the field of battle at Marengo, the other, at the hand of a fanatic, in the city of Cairo, there certainly was no connection between the maneuvers of the armies on the plains of Piedmont and the causes which, on the same day, led the assassin to attempt his work. Furthermore, there was no connection between these diverse causes and the circumstances of previous campaigns along the Rhine which had led those interested in the glory of our arms to link together the names of Desaix and Kléber. The historian noting this unusual event might very well excite

---

[1] [Louis Charles Antoine Desaix (1768–1800) and Jean Baptiste Kléber (1754–1800), each an outstanding French general credited with important victories in the numerous military campaigns of the eighteenth century.]

the surprise of the reader by saying that he can see in this event only a fortuitous happening, a pure effect of chance.

32. Moreover, the events we have taken as our examples must not be called results of chance because they are rare and surprising. On the contrary, they are rare because chance has led to their occurring rather than many others to which different combinations might have given rise, and they surprise us because they are rare. When a blindfolded person draws a ball from an urn containing a mixture of black and white balls, the drawing of a white one is no more rare or surprising than the drawing of a black one would have been. Either event must be regarded as a result of chance, for there is obviously no connection between the color of the balls and the causes which have led the hand of the person to fall on one ball rather than on another.

It is perfectly true that in ordinary language the term " chance " is used by preference where questions dealing with rare or surprising combinations are involved. If one were to draw four black balls, one after another, from an urn containing both white and black balls, this combination would be said to be the result of a great chance. Perhaps this would not be said had one drawn first two white balls and then two black ones, and still less had the balls alternated with still greater regularity with respect to their colors. Yet in all these hypotheses, the causes which have affected each ball of a different color are completely independent of those which, in each case, have directed the hand of the person doing the drawing. We note the chance which has brought about the death of the two brothers on the same day, but pay no particular attention to the chance which brings about their deaths at an interval of one, three, or six months. Yet there may well be no connection between the causes which brought about the death of the older on one day and of the younger on another and the fact that they were brothers. In the random selection of a series of letters from a pile of unsorted type, that is to say, a pile that is unordered so far as our notions and the habitual uses we make of pieces of types are concerned, no one will pay any attention to the drawing of series of letters which do not call to mind sounds which can be articulated or words used in a known language. But here again, there is no connection at all between the causes which have led the fingers of the person doing the selecting to pick certain pieces of metal rather than others and those causes which have led to a certain letter being stamped on one piece and other letters on other pieces, or the causes which have led us to give a certain representative value to the sounds spelled out by these characters. But this connotation of the word " chance," which is both vague and poorly defined, must be rejected when we are speaking in a more philosophical and hence a more exact language. In order to be clearly understood, it is necessary to settle conclusively upon what is fundamental and categorical in the idea of chance, namely the independence or lack of connection between various series of causes.[2] The word " cause " must be taken in the broad sense, in accordance with ordinary usage, to indicate everything that influences the production of an event, and not simply to indicate causes properly so called, that is, efficient and truly active causes.   Thus, in the game of heads or tails (19), the asymmetry of the structure of the coin tossed will be considered as a cause favoring the appearance of one side rather than the other. This cause operates continuously, being the same for each toss; its influence extends over the whole series of tosses taken both

[2] This idea has been anticipated by St. Thomas and still earlier by Boethius (*De interpretatione,* iii).   According to the latter, " Chance is the unforeseen event arising from causes which ordinarily have some other object. . . . If, in crossing a field, one stumbles upon a treasure, the discovery is truly fortuitous; its occurrence requires that someone had buried the treasure and that another person crossed the field, each with a different intention."

A modern and less well-known author has expressed himself still more clearly on this subject. Undoubtedly he would not have remained unknown had he drawn the conclusions from his argument. He says, " Perhaps someone will ask me if I think chance is an empty term, meaning absolutely nothing, a complete non-entity, etc. . . . I reply, I cannot agree with this. I am persuaded that if what has been said is true, then we utter a falsehood each time we say, as we often do, that chance has brought about such and such a thing, for it is certain that a pure non-entity can make nothing happen, can produce nothing, cause nothing.

" For myself, I am persuaded that chance includes something real and positive, namely, a concurrence of two or more contingent events, each of which has its causes, but such that their concurrence could not have been anticipated by anyone. I am badly deceived if this is not what is meant when the term chance is used." *Traité des jeux de hasard, défendus contre les objections de M. de Joncourt et de quelques autres.* By Jean la Placette, a Protestant minister in Holland. The Hague, 1714, *Duodecimo;* (at the end of the Preface).

singly and independently, and as a whole. Yet each toss is independent of the preceding ones so far as the intensity and the direction of the impulsive forces are concerned. Hence these latter are spoken of as accidental or fortuitous causes.[3]

33. Connected with this idea of chance is another which is of great theoretical and practical importance: This is the notion of "physical impossibility." Here again, we shall use examples as the most satisfactory way to present abstract generalities in such a form that they may be better understood.

It is regarded as a physical impossibility that a heavy cone should be balanced on its point; that the force applied to a sphere should be applied exactly along a line passing through its center and, therefore, in such a way as to give the sphere no rotational motion; that the center of a disk thrown on a parqueted floor should land precisely on the point of the intersection of the diagonals; that an instrument for measuring angles should be exactly centered; that a balance should be perfectly exact; that a measurement of any sort should rigorously conform to the standard taken, and so on. All these physical impossibilities are of the same sort and may be explained with the help of the idea that we have found useful in explaining fortuitous occurrences and the independence of causes.

---

[3] Even among purely abstract conceptions, in which events are produced as a result of rational necessity, and not by efficient causes such as those which give rise to phenomena, the notion of chance or of the independence of causes still has its application. Thus the mathematician Lambert, in the *Mémoires* of the Berlin Academy, decided to observe the succession of numbers in the expression of the ratio of the circumference of a circle to the diameter. He found, as must be the case, that the ten numerals in our decimal system reappear in this series, however far it is carried out, without resulting in any regular order of succession, yet in such a way that the mean value of the numbers differs but little from $3\frac{1}{2}$, no matter how far one may wish to carry out the process. This is exactly as if the figures were introduced successively by a drawing of lots from a jar in which all the numbers were found in equal proportion, rather than through a calculation carried out under determined rules. That amounts to saying that the mathematical formulas from which determinations of the ratio between the circumference and diameter result, within an indefinite approximation, are independent of the construction of our decimal arithmetic, and must, when applied to decimal calculations, lead to a series of numbers having all the characteristics of fortuitous succession, since there is no essential difference between the notion of chance and that of the independence of causes.

In fact, suppose we are interested in finding the center of a circle. The skill of the person seeking the center and the precision of his instruments fix the limits of possible error in this determination. But, within certain other limits which differ from and are more restricted than the former, the draftsman ceases to be guided by his senses and his instruments. No doubt, within this more or less narrowed field, the determination of the center is the result of certain causes, but of blind causes, that is to say, of causes which are completely independent of the geometrical conditions which would serve to determine this center without any error were the operation carried out by someone whose senses and instruments were perfect. There are an infinite number of points upon which the draftsman's instrument may land as a result of the operation of these blind causes; yet there is no reason, so far as the nature of the act itself is concerned, why these causes should set the instrument on one point rather than another. The coincidence of the point of the instrument and the true center is, therefore, an act of exactly the same type as that in which a blindfolded person draws the only white ball from a jar containing only one white ball and a very large number of black ones. Now, such an event is reasonably regarded as being physically impossible in the sense that, while it does not imply a contradiction, it in fact never happens. This does not mean that we must learn through experimentation that the event is impossible. On the contrary, the mind understands *a priori* the reason why the event does not occur and experience enters only to confirm this insight of the mind.

In the same way, when a sphere is struck by a body which is moving through space as a result of causes which are independent of the presence of this sphere in a certain part of space, it is physically impossible, in the sense that it does not happen, that, among the infinite number of directions from which the body might strike the sphere, the determining causes should have given the striking body precisely that direction which passes through the exact center of the sphere. Consequently, the impossibility of the sphere being moved along a straight line without acquiring any rotary motion is admitted. If the force were applied by an intelligent person who desired to obtain this result, but who has organs and senses of limited perfection, it would still be physically impossible for him to do this. For whatever his skill, the direction of the impulsive force would be subordinated to causes independent of his will and of his intellect, within certain limits of error. Rotary motion would result, however slightly off center the direction of the applied force might be. The acknowledged physical impossibility of balancing a heavy cone on its point is explained in the same way, although equilibrium in such a case is mathematically possible. Similar reasons could be given for all the instances cited above.

34. Thus, as we have been saying, a physically impossible event, that is, an event which in fact does not happen and which it would be unreasonable to expect so long as only a finite number of experiments or trials can be performed, or, in other words, so long as one remains within the limits of practical conditions of possible experience, is an event comparable to the removal by a blindfolded person of the white ball from a jar containing only one white ball and an infinite number of black ones. In other words, it is an event which has only one *chance* in its favor as over against an infinite number of chances against it. Now the name " mathematical probability " has been given to the fraction which expresses the ratio between the number of chances favorable to the occurrence of an event and the total number of chances. Consequently, it may be said more briefly, in the accepted language of mathematicians, that a physically impossible event is one whose mathematical probability is infinitely small, being less than any imaginable fraction, no matter how small. It can also be said that a physically certain event is one whose opposite is physically impossible, that is, an event whose mathematical probability deviates from unity by no assignable fraction, no matter how small. Such an event, however, must not be confused with that which brings together absolutely all the combinations or all the chances in its favor, and hence is certain, in the mathematical sense of that term.

On the other hand, it follows from the mathematical theory of combinations that no matter what the mathematical probability of an event A may be in a random case, if the same event is repeated a great many times, the relation between the number of cases which result in the appearance of A and the total number of cases must differ very little from the probability of the event A. Thus, for example, if the odds favoring the appearance of A are two to three, and ten thousand cases are examined, the number of cases in which the event A is found will be approximately two-thirds of ten thousand. If the number of cases could be increased indefinitely, it would be possible to decrease indefinitely, or to reduce to as small a value as might be desired, the probability that the difference of the two relations exceeds a given fraction, however small. In this way we approach more and more closely the cases of physical impossibility mentioned above.

35. In the rigorous language suited to the abstract and absolute truths of mathematics and metaphysics, a thing is either possible or it is not; there are no degrees of possibility or impossibility. But among physical facts and among realities with which the senses deal, in which contrary events may happen, and actually do happen, as a result of the fortuitous combination of certain causes which are variable and independent from one trial to another, along with other unchanging causes or conditions which govern the whole group of experiences, it is natural to regard each event as having a stronger tendency to occur, or as being more possible in fact or physically, in proportion as it is produced more often in a large number of cases. Mathematical probability becomes, then, the limit of *physical possibility*, and the two expressions may be used interchangeably. The advantage of the latter expression is that it clearly indicates the existence of a relation which subsists among things themselves, and does not follow from the manner in which we judge and make estimates, which varies from one individual to another. This is a relation which is maintained in nature and which observation makes evident when a sufficient number of trials are made to offset all effects of fortuitous and irregular causes and to bring to light, on the contrary, the effective role, however small it may be, of the regular, constant causes which are always present in natural phenomena and social events.

36. Therefore, it is not accurate to say, as Hume does, that " Chance is only our ignorance of real causes," or, with Laplace, that " Probability is relative in part to our knowledge, and in part to our ignorance," so that for a superior intelligence, which would be able to disentangle all the causes and to follow all their effects, the science of mathematical probabilities would disappear for lack of an object. To be sure, the term " chance " does not indicate a substantial cause, but an idea. It is the idea of every possible

variation and combination among many systems of causes or facts, each of which unfolds in its own system, independently of others. An intelligent being superior to man would differ from him only in this respect: It would be less frequently mistaken than man is, or even, if one wishes to put it this way, would never be mistaken in the use of this fundamental idea of the reason. It would not be liable to regard series of events which really influence each other as independent, or, contrariwise, to imagine relationships between really independent causes. It would depict with greater surety, or even with rigorous exactness, the part played by chance in the successive development of phenomena. Such a being would be able to indicate *a priori* the results of the concurrence of independent causes with respect to instances in which we are obliged to fall back upon experiment, because of the imperfection of our theories and of our scientific instruments. For example, being given a die having a definite form other than a cube, or one in which the density was not uniform, which would be thrown a great many times by impulsive forces the intensity, direction, and point of application of which are determined for each throw by causes which are independent of those present in subsequent throws, a superior mind would know, as we do not, almost exactly what relation there must be between the number of throws which will result in the appearance of a certain face and the total number of throws. For such a mind this knowledge would have a definite object, whether it knew the forces acting in any given case and could thus calculate the effects of each particular throw, or whether this knowledge and this calculation was beyond its power. In a word, it could go farther than we and could apply better the theory of these mathematical relations, all of which are bound up with the notion of chance, and which come to be, in the order of phenomena, so many natural laws, capable of being established on this basis by statistical experiments or observations.

In this sense, it is true to say, as has been so often repeated, that chance governs the world, or rather that it has a part, and a large part, in governing it. This is not in conflict with the generally accepted idea of a supreme and providential direction. This is true whether providential direction is presumed to bear only on average and general consequences which are established as a result of the laws of chance, or whether the supreme intelligence arranges details and particular facts in such a way as to co-ordinate them with designs which surpass our science and theories.

If we keep within the field of secondary causes and observable facts, which is all that science can attain, the mathematical theory of chance, the development of which would be out of place here, seems to us to be the most extensive application of the theory of numbers, and justifies the ancient saying, *Mundum regunt numeri* ["Numbers govern the world"].[4] In fact, no matter what certain philosophers may have thought, we have no basis for believing that we can give an account of all phenomena simply by means of the ideas of extension, time, and motion, that is, in a word, simply by the notions of continuous magnitudes on which the measurements and calculations of mathematics rest. The acts of living, intelligent, and moral beings are by no means explained, in the present state of our knowledge, and there is good reason to believe that they will never be explained by mechanics and mathematics. Consequently, these acts do not take their place in the field of numbers on the same basis as geometry and mechanics do. Yet they find their place in this field in so far as the notions of combination and of chance, of cause and of fortune are superior to (*supérieures*) geometry and mechanics in the order of abstractions and apply to phenomena in the domain of living things as to those that produce the forces which activate inorganic matter; to the reflective acts of free beings as to the inescapable determinations of appetite and of instinct.

37. Indeed, mathematicians have applied their theory of chance and probability to two very distinct kinds of question: To those of *possibility,* which has an entirely objective value, as has just been explained; and to those of *probability,* in the common sense of that term, which is in fact relative in part to our knowledge and in part to our ignorance. When we say that the mathematical probability of throwing a "double six" while playing backgammon is denoted by the fraction 1/36, we refer to a judgment of possibility. This means that if the dice are perfectly regular and homogeneous, so

---

[4] "Everything in the world is taken to depend on certain plans and a consistent law of change; to such an extent that even in things which are of the nature of cause and chance, we are held to acknowledge a certain necessity, as it were, and, if I may use the expression, a certain fatality." Jacobus Bernoulli, *Ars Conjectandi*, Part IV, (near the end).

that there is no reason in their physical structure for the appearance of one face rather than another, the number of double sixes which turn up in a great number of throws, as a result of impulsive forces the variable direction of which from one throw to another is absolutely independent of the marks on the faces, will obviously be 1/36 of the total number of throws. But we may also have in mind a judgment of simple probability. Then we may ignore whether or not the dice are regular or the way in which structural irregularities act, if they are present, for we have no reason to believe that one face will appear more often than another. Then the appearance of a double six, which can result only from one combination out of 36, will be less probable, at least relative to us, than the combination of " deuce and ace," since the latter may be formed by two combinations, depending upon whether the ace is found on one die or the other. This remains true, even though the latter combination is physically less possible or even impossible. If one player bets on " double six " and another on " deuce and ace," having agreed not to count throws that do not result in one or the other of these points, the odds will be determined in the relation of one to two. This rule will meet the needs of fair play as well as possible, if it is assumed that one knows the dice to have perfect structural regularity. But the same rule would be unfair if the referee knew that the dice were loaded and in what manner.

In general, if, in the imperfect state of our knowledge, we have no reason for believing that one combination occurs more readily than another, although, in reality, these combinations may be events the physical possibilities of which are unequal, and if we understand by the probability of an event the relation between the number of combinations which are favorable to the event and the total number of events which our limited knowledge leads us to put in the same class, this probability will cease to express a relation which exists really and objectively between the things. It will take on a purely subjective character and will be liable to variation from one individual to another according to the extent of his knowledge. It will still have a mathematical value in this sense, that it can, or even must, serve to fix numerically the conditions of a bet or of any other contract depending on chance. Moreover, it will have the practical value of offering a rule of conduct sufficient to guide us, in the absence of any other determinant reason, in cases in which it is necessary to take sides. Thus, we shall act reasonably in making our plans in anticipation of event A rather than of event B, if the probability of A, based, as we have said, on the present status of our knowledge, is higher than that of B, even though the unknown possibility of B may surpass that of A. But in such a case the numerical values of the probabilities of A and B will determine only an order of preference. They will no longer be measures in the true sense of the term. Consequently, such probabilities, although they deservedly attract the attention of the philosopher, who analyzes the motives of our judgments, and of the moralist, who seeks a rule for our actions, must be repudiated as lying outside the applications of a mathematical theory which has as its object quantities that can be rigorously compared to a unit of measure.

38. In regard to fortuitous events whose conditions are not determined by man himself, the causes which give a certain physical possibility to a given event are nearly always unknown both as to their nature and as to the mode of action, or are complicated in such a way that we are able neither to make an exact analysis of them nor to submit their effects to mathematical measurement. Even in games where the whole situation is determined by human agreement and invention, the construction of instruments of chance is subject to certain irregularities which modify the chances in ways which it would be impossible to know *a priori*. Consequently, mathematical probability taken objectively, or conceived as measuring the possibility of things, can in general be determined only by means of experiment. If the number of cases of a certain chance were increased to infinity, its mathematical probability would be determined exactly with a certainty comparable to that of an event whose opposite is physically impossible. When a very large number of cases is dealt with, the probability is still only approximately given, yet it is legitimate to regard it as very unlikely that the real value is completely different from the value inferred from observations. In other words, it very rarely happens that an appreciable error results when the value obtained by means of observation is taken as its true or real value.

Even in cases where the number of instances is not so large, efforts have been made to draw from certain mathematical considerations formulas for evaluating numerically the probability of future events on the basis of observed events. But such formulas indicate nothing more than subjective probabilities which are useful for the most part in determining the odds for a bet. They cannot avoid being erroneous if applied, as they often are wrongly, to the determination of the possibility of events.

39. In actual living, we must constantly make up our minds on the basis of so small a number of experiences that they cannot give us a true indication of the possibility of an event. Thus, it would be impossible to determine exactly the chance of our being mistaken in believing in the occurrence of an event, or in judging that the possibility of this event falls within such and such limits. However, it is clear that if event A has happened more frequently than event B in a given number of cases, however small this greater number of cases may be, this greater frequency will be reason enough, in the absence of any other information, for adjusting our conduct in anticipation of the recurrence of event A rather than of event B. If one considers two fractions, one of which is the relation between the number of instances which contain A and the total number of cases, and the other the relation between those instances in which B occurs and the same total number of experiences, the relative size of the fractions will justify an order of preference for choosing between events on the assumed recurrence of which we govern our conduct. But this ground of preference will not be amenable to measurement by the fractions in question or by other numbers which certain mathematicians have proposed for this purpose. In a word, except in determining the odds for a bet, subjective probability, which is the thing at issue here as it was in the case considered above, will be outside the field of the application of the mathematical theory of chances, the essential purpose of which is to discover measurable quantities and relations which subsist between things independently of the mind that knows them.

We need to recall in this connection the philosophical principles of this theory, because, in all our subsequent study, we shall incessantly have to appeal to judgments based upon probabilities which, although not of the same nature as mathematical probabilities and although not capable of being subjected to calculation, nevertheless apply also to the notion of chance and to that of the independence of causes, as we shall go on to explain.

CHAPTER IV

ON PHILOSOPHICAL PROBABILITY.   ON INDUCTION AND ANALOGY

40. In order to state our ideas more precisely, we shall turn first to some imaginary and abstract but very simple examples. Let us suppose that a variable quantity may assume values expressed by numbers between 1 and 10,000, and that four consecutive observations or readings have given four numbers, such as 25, 100, 400, and 1600.   This is a regular progression in which each of the terms is four times as large as the preceding one.   We shall be very much inclined to believe that such a result is not fortuitous. We shall also be inclined to believe that this result is not comparable to four random drawings from a jar which contains tickets numbered from 1 to 10,000, but that it indicates some regular law in the variation of the quantity measured, corresponding to the successive order of the readings obtained.

Four values obtained by means of observation might give some other mathematical law instead of the progression indicated above. For example, they might form four terms in a progression in which the difference between any two successive terms is constant, as in the series 25, 50, 75, and 100; or four consecutive terms in the series of squared numbers, such as 25, 36, 49, and 64; or they might equally well belong to a series of numbers that are called cubical, triangular, or pyramidic, and so forth.   Furthermore—and it is important that this be noted—algebraists have no difficulty in showing that a mathematical law or even an infinite number of different mathematical laws may be stated each of which binds together the values successively introduced no matter how many there are, and no matter what inequalities the table of consecutive values may present at first sight.

However, if the mathematical law introduced to show the relation between the observed numbers were to become increasingly complex, it would become less and less probable, in the absence of any other indication, that the succession of these numbers is not the result of chance, that is to say, of the concurrence of independent causes, each one of which would have introduced each particular reading. On the other hand, when the simplicity of the law is striking, it is repugnant to us to admit that the particular values exist without any interrelation and that chance has given rise to the relation observed.

41. But what precisely is it that determines the simplicity of a law? How can we compare and arrange under this statement the infinitely diverse laws that the mind is capable of conceiving, and to which, at least in questions pertaining to number, it is possible to give a mathematical expression? A given law may seem more simple than another in certain respects, and less simple when both are considered from some other point of view. A smaller number of terms or of operational signs may be involved in one statement, but on the other hand, these operations may be of a higher order, and so on.

In order to reduce to mathematical probability the probability based on the characteristic of simplicity that an observed law presents, a law which is one among many others which might just as well have been given if the alleged law were nothing but a fact resulting from the fortuitous combination of unrelated causes, certain conditions would have to be fulfilled. First of all, one should be able to set up two categories, one of laws which are regarded as being simple, the other of those to which this characteristic of simplicity is not suited. In the second place, it would be necessary that one should be justified in putting on the same level all those that had been grouped together in the same category; for example, that all laws which are said to be simple should be simple in the same degree. Finally, the number of laws in each category would have to be limited, or better, if the number of laws were absolutely unlimited, it would be necessary that, although they increase indefinitely, their relation should tend toward a finite and assignable limit, as happens in the cases to which the calculus of mathematical probability applies. But none of these suppositions is admissible, and consequently, for a threefold reason, the reduction in question must be regarded as radically impossible.

42. When, in the examination of a series of numerical values which have been obtained in the manner explained above, we must choose from among the infinite number of laws which are capable

of relating them that one which first attracts us because of its simplicity and which is followed by observations which subsequently introduce other values which conform to the same law, the probability that this regular progression of the observed data is not the result of chance is obviously increased in proportion to the number of new instances which fit into the series. Thus it may happen that the law becomes such that it no longer leaves the slightest doubt in regard to this matter as far as any reasonable mind is concerned, and this may happen quite soon. If, on the contrary, the assumed law does not hold for the results of new observations, it will clearly be necessary to abandon it for subsequent observations and to recognize that it does not hold for the series as a whole. But this does not necessarily imply that the regularity of the earlier observations is the result of pure chance, for it is well known that constant and regular causes may act in one part of a series but not in others. Both hypotheses will have their respective probabilities. But, for the reasons already indicated, these probabilities will not be of a type that may be evaluated and compared numerically.

It may also happen that the simple law which impresses us when we are considering a certain set of readings does not apply exactly to the observed values, but to other values which approximate them very closely. Thus, for example, in place of the series 25, 100, 400, and 1600, observation might have given the following: 24, 102, 405, and 1597. The idea would then occur to us that the regular effects of a constant and basic cause are complicated by the effects of other accessory or disturbing causes which may themselves be subject to regular laws, constant for the whole series of observed values, or varying irregularly and fortuitously from one value to another. But the probability that this is the case is evidently bound up with the probability of the existence of a regular law in the more simple case that we have considered above, and it would no more permit a mathematical evaluation than would the former.

43. Leaving the field of abstractions and fictions for the time being, let us turn to the period when Kepler, after a number of attempts to discover a numerical law which would express, on the one hand, the distance of the planets from the sun, and, on the other hand, the length of their revolutions, finally recognized that the latter are proportional to the square root of the cubes of the distances. Here is a mathematical law which is stated in a rather

complicated form and which is applicable only to the six planets then known. At that time there was no basis upon which the reason for this particular relationship might have been foreseen. Kepler discovered it while groping in the dark under the influence of Pythagorean notions which have since been repudiated by better thinkers. This being the case, it is perhaps to the point to ask if this relation was not uncovered by chance, as a result of Kepler's need for finding some neat mathematical principle which would relate numerically quantities which seemed to be only fortuitously connected. His contemporaries among astronomers seem to have placed this interpretation on his work; and in spite of the discovery of the satellites of Jupiter which served to verify for this particular system the law observed in the planetary system as a whole, the third law of Kepler, as it is called, attracted little attention until Newton made his great discovery which showed that this law, together with many other results of observation, may be deduced from the principle of universal gravitation.

Kepler had also been struck by a unique relation exhibited by the table giving the distance from each of the planets to the sun. If the then-known planets, with the exception of Mercury, are arranged in the order of their distances from the sun, this is the way they lie: Venus, Earth, Mars, Jupiter, Saturn. The numerical values respectively of the interval between the orbit of Venus and that of the earth, or the differences of the radii of the two orbits, and the corresponding intervals between the orbits of the other planets will be found to be almost exactly proportional to the simple numbers 1, 2, 12, and 16. This led Kepler to make two conjectures: First, that there was a planet as yet undiscovered lying between Mars and Jupiter—the orbit of which was related to the distances of the orbits of Mars and of Jupiter respectively in proportion as the numbers 4 and 8—so as to permit the replacement of the series just given by the progression 1, 2, 4, 8, and 16, the interval always doubling from one planet to the next; secondly, that there must also exist between Venus and Mercury a planet the intermediate orbit of which approximately offsets the anomaly which excludes the orbit of Mercury from the simplest law that can be stated.

This last conjecture by Kepler has not been verified. But the other received striking corroboration in the tardy discovery of a group of telescopic planets, the number of which is fourteen at the time this is being written, but which seems bound to be further increased. All these move around the sun, some at slightly less, others at slightly greater distances than those which would completely satisfy Kepler's induction. These bodies evidently all have a common origin. They must either be regarded as so many fragments of a planet which exploded, or their relations in celestial space and the similarities of their physical make-up must be explained in some other fashion. But even before the discovery of the telescopic planets, the orbit of the planet Uranus, which at that time was thought to be located at the outermost limit of the planetary system, was seen to corroborate the inference in a singular way, since the distance between its orbit and that of Saturn was very nearly double that between the orbits of Jupiter and Saturn. In order better to fix these ideas in the mind of the reader, we have prepared a table giving the actual observed values as compared to those which would rigorously satisfy the law in question. We have chosen Juno to represent the telescopic planets in this table because of its median position in the group. It must be kept in mind that, in this table, the number 1000 represents the radius of the earth's orbit.

Distance Between the Orbits	Observed Values	Theoretical Values
Venus and Earth	277	277
Earth and Mars	523	554
Mars and Juno	1146	1108
Juno and Jupiter	2533	2216
Jupiter and Saturn	4336	4432
Saturn and Uranus	9644	8864

This comparison shows notable discrepancies. Yet, on the other hand, it must be kept in mind that the planetary orbits are not perfect and concentric circles lying on the same plane, but are rather ellipses whose planes are inclined towards each other, the eccentricities and inclinations of which vary with time in such a way that the errors presented in a table of mean values do not exceed the limits within which the physical distances between the sun and each of the planets always oscillate. Furthermore, it is not intended to give to the formula a rigorous precision which excludes the intervention of disturbing and irregular causes capable of altering the principal result which is due to the action of a constant cause.

The anomaly of the planet Mercury remains. This planet lies closest to the sun. Yet the distance which separates its orbit from that of Venus is somewhat less than that which separates the orbits of Venus and the Earth, while the first distance should be only half the second, according to the law in question. In order to avoid, or rather to mask, this anomaly, a different way of stating the law has been suggested. The distance between Mercury and the sun is expressed by the number 4, then that of Venus is found to have a value approaching 4 plus 3 or 7, that of the Earth 4 plus two times 3 or 10, that of Mars 4 plus four times 3 or 16, and so on, including Uranus. Stated in this more complicated and, for that reason, less probable form, the progression of planetary intervals is called Bode's law, after a German astronomer of the eighteenth century. But this schematism broke down when the planet Neptune was discovered in celestial space far beyond the orbit of Uranus, but at a distance much less than it should have had according to Bode's law, and also much less than Bode at first supposed it to have, since the actual distance between the two orbits does not greatly exceed that of the interval between the orbits of Saturn and Uranus, instead of being twice or nearly twice this interval. Therefore, it must be acknowledged that the first and last terms of the series of known planets are exceptions to the law formulated by Kepler. This is not a sufficient reason for rejecting the described progression as being simply the result of chance so far as the intermediate planets are concerned; for we can easily understand that causes of regular distribution, while not excluding complications due to other disturbing and anomalous causes, may control the median part of the series, while the first and last terms are not influenced by them. Here, then, are probabilities and inferences which the natural scientist cannot ignore. However, they are not of a sort which the mind is forced to accept, and it would be chimerical to attempt to express them numerically.

44. Perhaps some readers will find the theoretical considerations presented in section 40 and following easier to grasp if we refer to some examples taken from geometry. Therefore, let us suppose that ten points have been observed as successive positions of a point

moving on a plane; let us also assume that these points are found to lie on the circumference of a circle. No one would hesitate to admit that this coincidence is not fortuitous, but, on the contrary, clearly indicates that the moving point is fixed in such a way as to describe a circular line on the plane. If these ten points fall slightly to one side or the other of the circumference of an accurately constructed circle, we will attribute the deviations either to observational errors or to disturbing or secondary causes, rather than give up the idea that a regular cause directs the movement of the point.

Instead of tracing the circumference of a circle, the observed points might be plotted along an ellipse, a parabola, or along any one of an infinite number of other curves which may be given mathematical definition. Theoretically we know that an infinite number of mathematically determinable curves may be made to pass through such a group of observed points no matter what the number of them may be. Yet the path actually described by the moving point may coincide with none of these curves, and it may be found to be impossible to plot its path in terms of any regular law.

The probability that these points are spread on the plane as a result of a regular influence will depend, therefore, upon the simplicity of the curve along which they can be made to fall either exactly or within certain allowable limits. Mathematicians are well aware of the fact that every classification of lines according to their simplicity is more or less artificial and arbitrary. In certain respects, a parabola may be regarded as being a simpler curve than a circle; however, the usual definition of a circle seems to be more simple than that of a parabola. Therefore, it is not possible, for reasons already stated, to reduce this probability to a numerical evaluation of the sort which results from the distinction of the chances favorable or unfavorable to the production of a given event.

Thus, when Kepler had found that the movement of the planets could be represented in terms of ellipses having the sun at one focus; and when he then proposed to substitute this geometrical conception for combinations of circular movements in terms of " eccentrics " and " epicycles," which astronomers had used prior to his time—guided as they were by the idea of a certain notion of perfection which was assigned to the circle, and which should correspond to the perfection of heavenly things—his novel hypothesis rested on nothing but the idea of the perfection or simplicity of the ellipse. This brought to light many remarkable properties which should at once have attracted the attention and challenged the discernment of mathematicians as regards the properties of circles. Actually the elliptical orbit served to unify the whole body of astronomical data only approximately, because of the errors which had inevitably affected these observations themselves, as a result of disturbing causes which altered the elliptical movement appreciably.

An oval curve, which differs only slightly from a circle, will differ still less from a properly chosen ellipse. But in order to regard elliptical motion as a law of nature, it was necessary to break away from the idea that nature by preference follows simple laws of the sort which guide our abstract speculations. It was necessary to find in the consideration of mathematical relations reasons for preferring the hypothesis of elliptical movements to that of circular movements on the ground that the former are more simple. Now, a situation of this sort could result only in philosophical inferences of greater or less probability; and there was no way in which these inferences could be stated numerically until after the formulation of the Newtonian hypothesis which, by giving at one and the same time the reason for the elliptical movement and for the perturbations which modified it, put Kepler's discovery and his right to lasting fame beyond any serious question.

45. In general, any scientific theory whatever which is devised to relate a certain number of facts established by means of observation may be compared to the curve drawn according to a mathematical definition by imposing on it the condition that it pass through a certain number of previously given points. The judgment that reason gives as to the intrinsic value of the theory is a probable judgment. The probability of this judgment depends upon two things: First, on the simplicity of the theoretical formulation, and, second, on the number of facts or groups of facts which the theory connects. The same group of facts must include all those which make up a series one after another, or those which already explain one another independently of the theoretical hypothesis. If, as new facts are brought to light by observation, it becomes necessary to complicate the theory proportionally, it becomes less and less probable as a natural law, or as one to which

the mind can attribute objective value. Soon it is nothing more than an artificial scaffolding which finally breaks down when, as a result of excessive complexity, it even loses its utility as an artificial system, its utility being to facilitate the work of the mind and to direct research. If, on the contrary, facts acquired by observation after the hypothesis has been formulated, as well as those facts which served as the basis for its construction, are unified by it, and especially if facts predicted on the basis of the hypothesis are strikingly confirmed by later observations, the probability of the hypothesis would be such as to leave no doubt in any sufficiently enlightened mind. Astronomy furnishes us with the most magnificent example of this in the Newtonian theory of gravitation, which has made it possible to calculate the movement of heavenly bodies with a very minute exactness, which has accounted so far for all their apparent irregularities, which has led to the prediction of the existence of several of these before observation had given evidence of them, and which indicated to observers the area of the heavens in which they should look for previously unobserved celestial bodies.

Nonetheless, this continuous agreement does not lead to the same type of formal demonstration which serves to establish mathematical truths. Yet one is not reduced to the sophistic absurdity of accounting for such agreement on the basis of chance. The observed concurrence yields only a probability, but a probability which is comparable to that of a physically certain event, when this term is used in the sense explained above (34), a probability of the sort which influences the conviction of any fair and unbiased mind. It would be contrary to the nature of things that a physical law could be established in any other way.

46. If we continue the use of geometrical comparison presented in section 44, we find it necessary to distinguish clearly inductions which apply to points lying within the limits of observation from those which lie beyond the limit of the observed data in one direction or the other. Thus, a moving point has been observed in ten positions taken by chance as the subject of a similar number of observations; and it has been determined that these ten points are of a sort found to fall along a geometrical line which is not limited and does not turn back upon itself, as is the case with a circle or an ellipse, but along a line that may be prolonged indefinitely, such

as a parabola or hyperbola. On the basis of these data it will be inferred that, had it been possible to observe them, the intermediate positions would have been so many points falling on the same curve. The reason for this conclusion is that it would be very extraordinary that chance should have led to the selection of just the points which may be connected by so simple a geometrical law, while intermediate points would lie outside it. In any case, observations may be sufficiently multiplied to exclude all reasonable doubt in regard to this matter. Furthermore, it will be inferred with a high probability, or even with a quasi-certainty, that the path of the curve described by the moving point follows the same law and may be prolonged along the same parabola or hyperbola slightly beyond the extreme points given by observation. For how can it be admitted that circumstances which are fortuitous or are completely independent of the movement of the point, and which have led us to begin and to end our observations at one place rather than at another, should have given us as end points just those at which the moving body begins or ceases to follow the simple law which holds for all the intermediate points? However, the farther we go beyond the observed limits the more uncertain the inference becomes. This is so because the laws which control the movement of the point may be modified suddenly, or by imperceptible degrees, or may be complicated by the intervention of disturbing causes which do not affect the intermediate region within which our observation has been concentrated.

Even though the points given by observation did not fall on a curve which is remarkable for the simplicity of its definition, if they are adequately related and connected by a continuous line, it would be very probable that the path of the curve actually described by the moving point would deviate but little in any direction from the path obtained in the manner indicated. The probability that the line goes through these points will vary in proportion as their arrangement indicates a more or less regular path in the movement of the point. For, if the line actually described had any noticeable irregularities in it, how could it be admitted that chance had led to the making of just those observations which conceal these significant irregularities? Nevertheless, it remains infinitely less probable that the actual path of the body has been followed exactly. Hence, the inference, although highly probable, rests only on an approximation. But what are the chances that certain limits of error have not been exceeded? How does it vary with respect to the intervals of the points which have been determined in an exact manner, and with regard to the path indicated by the arrangement of the points as a whole? These are questions to which no mathematical solution can be given, and we must not be afraid to admit it. Consequently, let us repeat, the probability in question, although always connected with the notion of chance or of the independence of causes, is not of the type which may be resolved in an enumeration of chances and in that way be brought within the scope of measurement.

Not only will the exactly determined points be connected by means of a continuous path, as a result of a feeling for the continuity of form which cannot be mathematically and rigorously defined, but the path will also be extended in both directions beyond the limits of observation. This is another case of inference through approximation, to which a probability corresponds which must gradually become weaker in proportion as we go beyond the limits of observation, until a point is reached at which the least careful mind would sense that the inference was becoming, first, very questionable, and finally, completely unfounded.

47. There is no question in physics which is not such as to furnish us with obvious examples of the application of these abstract notions. Suppose some air, originally at ordinary atmospheric pressure, is placed in a container and then subjected to pressures of one, two, three, and so on up to ten atmospheres. It will be found that the volume of the air becomes successively one-half, one-third, one-fourth, and so on up to one-tenth of what it was originally. This process is the basis of an important law the discovery of which is attributed to Mariotte or to Boyle, and that we know under the name of Mariotte's law.[1] Strictly speaking, these ten readings are not a demonstration of this law so far as intermediate pressures such as two and a half atmospheres are con-

---

[1] [Edme Mariotte (1620–1684), a French mathematician and natural philosopher. Mariotte was one of the earliest experimentalists in France, who discovered the law of the elasticity of fluids that is named after him. He was a contemporary of the English natural philosopher, Robert Boyle (1627-1691), who, apparently independently, discovered the same principle which is known in English-speaking countries as Boyle's law.]

cerned. The judgment that leads us to affirm that the law holds
for all values from one to ten atmospheres includes incomparably
more than any experience can include, since it holds for an infinite
number of values while the number of possible experiences is necessarily finite. Now, this inductive judgment is rationally based on
the fact that in the experiment we have just mentioned, the choice
of the points observed, or the values of the pressures for which the
experimental verification has been undertaken, must be considered
a product of chance. This is true because reason recognizes no
possible relation between the causes which, on the one hand, bring
about variations in the mass of a gas in proportion to the pressure
exerted upon it, and the circumstances which, on the other hand,
determine the intensity of gravity at the surface of the earth and
the mass of the atmosphere from which the value for the weight of
the atmosphere or for atmospheric pressure results. In order to
contest the validity of this inference it would be necessary to admit, on the one hand, that the law which relates pressures to volumes has a very simple form for certain values, and becomes complex, for no apparent reason, when applied to intermediate values.
In addition to this, it would be necessary to assume that, among
an infinite number of possible readings, chance had led in several
successive cases to the selection of just those for which the law in
question takes a constant and simple form. This reason will not
admit, and should it be found that ten instances are insufficient,
or that they should be spaced less regularly, only the terms of the
example will have changed. In cases in which inference rests upon
a probability of this kind, it will always be found that reason does
not retain the least doubt, in spite of every sophistical objection.

Now suppose we wish to extend Mariotte's law in either direction beyond the limits of experimentation, for example, to pressures of eleven or twelve atmospheres or, in the opposite direction,
to pressures of nine-tenths or eight-tenths of an atmosphere. This
extension will be made on the basis of another inference, and a very
reasonable one, for it would still be infinitely improbable that
chance should bring experimentation to a halt at precisely the point
where the law ceases to hold. Yet, just as soon as we go at all beyond the limits of experimental evidence, it is not infinitely improbable that the law undergoes no appreciable alteration, although
it may remain very probable, when the distance from the limits of

observation is small, that the law will hold, at least with a very close approximation. In general, the probability of the constancy of the law decreases as the interval between any given term and the last observed term is increased. Yet it is impossible to assign a mathematical relation to the variation of the distance and that of the corresponding probability; it is impossible numerically to evaluate this probability which depends upon the degree of simplicity of the observed law, and on other experimental or theoretical data which bear upon the nature of the phenomenon. In the particular example in question there are numerous grounds for admitting the possibility of serious errors beyond the limits of experimentation, since, even within these limits, Mariotte's law is not absolutely verified by the most delicate and recent observations.

Again, let us consider the case of a series of experiments conducted for the purpose of discovering how the tension of water vapor varies with the temperature of the liquid producing it. Here we find no law which may be stated in a form as simple as Mariotte's. For want of such a formula we must set up a table in which numbers stating temperatures at which the experiments are made are placed alongside numbers which measure the corresponding tensions. For intermediate temperatures, for which no direct experimental results are obtained, we " interpolate," that is to say, we insert between the numbers obtained experimentally others which seem most likely to fit in with the general progression of the observed values. These interpolated numbers can be absolutely exact only through an accident the probability of which is infinitely small. Yet it is extremely probable that they differ very little from the exact values when we consider that neither experience nor theory indicates any causes which would suddenly disturb the interval. Once again, it is possible to extend the table to apply to values a little above and below those observed, with a high probability that there will be little deviation between these values and the true values. But, at a considerable distance beyond these limits, the absence of any simple form makes legitimate inference impossible, and it is not possible to indicate the course of the phenomena even approximately.

48. We do not pretend to have enumerated all the forms to which judgment by induction is susceptible. However, the instances we have given are sufficient for our purposes. And even

though we have intentionally put them in terms having the simplicity and also the dryness of mathematical statements, they enable us to see how we must interpret analogous judgments growing out of other circumstances in which we may be dealing with matters of an altogether different sort than the determination of quantities or the formulation of a law according to which one quantity is related to another. For example, if each improvement of optical instruments had led to the discovery of new details in the analysis of an organic tissue, we should infer without hesitation, but not without doubt, that each portion of organic tissue is, in turn, made up of organic parts, and so on infinitely, and above all that other details of organic structure would be made visible by still more perfect instruments. For, if we are not justified in affirming, on the basis of a large number of terms in a series, that it extends to the infinite, it is infinitely less probable that it should stop just at the point where we reach the limit of our means of observation because of a system of causes entirely independent of those which pertain to the nature of the object perceived.

In all these cases, we see how little basis there is for the assertion of the majority of logicians that inductive judgments rest upon the belief in the stability of the laws of nature, and upon the maxim that the same causes always and everywhere produce the same effects. To begin with, this maxim need not be confused with the hypothesis of the stability of the laws of nature. If the same causes acting under the same circumstances were to produce different effects, this diversity itself would be without a determinant cause or reason. Such a situation is repugnant to a basic law of human thinking. Judgments expressed as consequences of this fundamental law—such as the axiom of mechanics which was used as an example in section 27—are *a priori* judgments and must not be placed among inductive judgments. So far as physical phenomena are concerned, some are governed by laws independent of time, and others develop in time, that is to say, in accordance with laws in the expression of which time is a factor. Thus, because a freely falling body actually does fall to the surface of the earth, we cannot legitimately infer that it would fall in the same way and with the same velocity if the experiment were repeated at any time whatever. For, if the velocity of the rotation of the earth were to increase with time, a period would come when the intensity of the

centrifugal force would first balance that of gravity and then become greater than gravity.   As a matter of fact, we know both
through theory and through experiment that the velocity of the
earth does not allow for such acceleration.   But in such a case,
extrinsic knowledge of this sort is necessary to make valid the inference from an actually observed fact to a future fact.   Conversely,
we should make a great mistake if we were to infer from the fact
that the temperature on the earth has for ages been favorable to
the existence of living things and does not even seem to have undergone any appreciable change during historic time, that it always
has been and always will be suitable to the conditions of plant and
animal life as we know them, or even of any plants and animals.
The judgment on the basis of which we believe in the stability of
certain laws of nature, or on the basis of which we affirm that time
does not enter into their definition, rests either on a theory concerning phenomena, as in the case of the earth's gravitational attraction
which we used as an example, or on an inference analogous to those
presented by other cases cited earlier.   But it must not be said
inversely that induction arises from such a belief.

It is still true to say that we are led to conceive all the laws of
nature, even those into the expression of which time enters, as
emanating from more general laws or permanent laws which are
immutable in time.   But this belongs to an order of higher considerations to which logic and science properly so called may not
attain, and which we may, and even must, exclude from our
present considerations.

49. Judgment by *analogy* is often compared with judgment by
induction, and is not always clearly distinguished from it.   According to Kant,

Induction proceeds from the particular to the general in accordance
with the principle of generalization, namely, that principle which states
that what holds for many things belonging to the same genus will hold
for all other things belonging to the same genus; while analogy draws
conclusions from the partial resemblance of two objects of the same
genus to their complete resemblance. . . . Induction extends empirical
data from the particular to the general with respect to many objects;
analogy, on the contrary, extends the known qualities of an object to a
great number of qualities of the same thing.[2]

[2] *Logic,* Chap. III, Sect. 3, Par. 84.

But there are many sorts of induction which have no relation to the notion of genus and species. This is the case, for example, when the path of a curve is prolonged or completed by means of induction, or when a physical law, like Mariotte's, is extended beyond the terms established through experiment. Even in the particular instance Kant had in mind, no one knows with absolute assurance what difference there is between attributing to a thing inductively what is characteristic of others of its genus, or concluding on the basis of analogy that it possesses the quality found in its congener. Many gases have been successively liquefied in proportion as they have been subjected to increased pressure or to increasingly intense cold. On this basis it will be affirmed by induction that all gases could be liquefied if it were possible to create a sufficient amount of pressure or if the temperature could be lowered sufficiently. Or, better still, this judgment might be regarded as resting on analogy because of the resemblance we notice between the properties of all gases in regard to just those characteristics which depend upon variations of temperature or pressure. From this it may be inferred that there is a reason, implicit in the general nature of substances in the gaseous form, why they become liquefied when the pressure rises above, or the temperature drops below, certain limits, and why, in all probability, in the case of gases not yet liquefied as well as for those that have been, the specific differences of composition can act only when certain limits are approached or passed.

The *Academy* [3] says that to reason from analogy is to found an argument upon the resemblances or the relations which one thing has with another. In order to give complete philosophic accuracy to this definition it is necessary to say, " based on relations or on resemblances to the extent that they indicate relations." In fact, in reasoning from analogy the object which the mind has in view rests solely upon relations and upon the reason for resemblances. These resemblances are of no value when they give no evidence of relations among the order of facts to which an analogy is applicable.

On the basis of analogy, chemists admit the existence of elements that have not as yet been isolated. They even assign the families or groups in which these unknown bodies must be placed.

[8] [The *Academy* here refers to the Academy of Plato and his successors.]

But in doing this they take account only of the analogies presented, in accordance with this type of chemical action, and the compounds into the formation of which the unknown elements are thought to enter. It will be of no importance to them whether these bodies exist in a solid, liquid, or gaseous state at ordinary temperatures or whether they are clear or variously colored. In sum, they do not limit themselves to establishing resemblances, nor do they base the probability of such and such an hypothesis in the field of chemistry on the basis of the number of resemblances. They will take account chiefly of the importance of particular properties, the importance of which has been indicated theoretically or has been established by previous experiments. The same method will be followed, with even better reason, in the study of organic bodies in which the variety of relations, together with a well-marked subordination of characteristics, offers an entirely different field to analogical judgments. In this field especially, analogy furnishes us with those irresistible probabilities that must be likened to physical certainty. No naturalist seeing a member of a previously unknown species of animal suckling its young could fail to be perfectly sure in advance that dissection will reveal a brain, a spinal cord, a liver, a heart, lungs characteristic of arterial and venous circulation, and so forth. A patient study of living organisms makes evident laws whose nature does not vary in the course of innumerable modifications to which they are subject in certain forms of organisms. And although, in most cases, the reason for these laws is beyond our knowledge, we can neither doubt their reality, nor admit that a fortuitous concourse of causes independent of one another produced an illusion of them.

If we consult the etymologist, who is nearly always the best guide, we find that the term "analogy" (ἀναλογία) must be understood to mean more specifically a method through which the mind proceeds, by means of the observation of relations, to the reason for these relations, it being impossible for it to descend from the immediate conception of principles to the explanation of relations that are derived from them and of which they are found to be virtually composed. On the other hand, "induction" (ἐπαγωγή) is more specifically the operation by which the mind, instead of stopping suddenly at the limits of immediate observation, follows along the path suggested by it, prolongs the described line, and yields, for some time, so to speak, to the law of the motion which it had projected, but not in an irrevocable or blind manner. Reason tells itself why it would be wrong to resist, and charges itself to justify fully what originally may have been an instinctive tendency.

50. In all the judgments we have been considering, the mind proceeds not by way of demonstration, as it does when asked to prove a theorem in geometry or to draw the conclusion from premises by means of the rules of formal logic. Therefore, in addition to what is called "apodictic" proof, or formal demonstration, there is another type of certainty which we, in accordance with writers in this field, have often called "physical certainty," in so far as it is applicable to natural events. This type of certainty may also be called "philosophical" or "rational" because it results from a judgment which, in evaluating diverse suppositions and hypotheses, admits some because of the order and connection they introduce into our knowledge and rejects others as irreconcilable with this rational order which human intelligence seeks, in so far as it lies in its power. But while the certainty acquired through logical demonstration is fixed and absolute and admits neither of shadings nor of degrees, this other judgment of reason, which under some conditions produces certainty or firm conviction, in other cases leads only to probabilities which tend to become weaker by indiscernible nuances, and which do not impress all minds in the same way.

For example, as science now stands, certain physical theories are reputed to be more probable than others because they seem to be better adapted to the rational connection of observed facts, because they are more simple or because they bring out more remarkable analogies. But the force of these analogies, of these inductions, does not impress all minds to the same degree, not even those that are most enlightened and most impartial. Reason is impressed by certain probabilities which are nevertheless insufficient to yield complete conviction. These probabilities change with the progress of science. Such a theory, although rejected in the beginning and opposed for a long time, may finally win unanimous assent. But some succeed in this more slowly than others. This proves that something which varies from one mind to another is involved in the rudiments of this type of probability.

In regard to other questions we shall never be able to attain anything but probabilities which are insufficient to bring about complete conviction. The question regarding the habitation of other planets by living and animated beings is of this sort. We are struck by certain analogies between the other planets and the earth. It is repugnant to us to admit that, in the scheme of things, one little globe, lost in the immensity of the celestial space, should be the only one on whose surface the marvels of organic life develop. But we can hardly expect the progress of science to throw any new light on those things which God seems pleased to have put beyond our means of observation. Relatively near us is a planet the dimensions of which are comparable to those of our own and which seems to be placed in such physical conditions that it would be impossible for any such organic beings as those which inhabit the earth to survive on it. The mind will accept or reject the philosophical opinion of the plurality of worlds to the degree that it is struck by the analogies or disparities between other planets and the earth.

Seeing a fragment of bone from an animal of a species which has disappeared, but congeneric forms of which are still found today, a naturalist will conclude with certainty not only that the animal belonged to the class of mammals, but that it had a four-chambered heart, bi-lobed lungs, venal and arterial circulation, and so forth, and, in addition, whether it belonged to the order of carnivorous animals, that is, to the genus of the *cat,* or was a ruminant, belonging to the genus of the *stag.* By means of this very cogent induction he will establish with certainty all the important characteristics of the animal's organic structure, habitat, and mode of life; but he will have to remain satisfied with probabilities in regard to certain particular differences between the extinct species and those of its congeners. He will also have to remain in complete ignorance of other details. In the case of a species the generic type of which has disappeared, and especially of one which is not found among types actually known, the certainty of the inductive judgments will rest only on the most general characteristics; and its probability will become gradually weaker as it turns to the explanation of details and secondary characteristics, but it will not be possible to give a quantitative statement of its continuous deterioration.

51. This variable, subjective probability which occasionally ex-

cludes all doubt and produces certainty *sui generis*, but which at other times appears to be no more than a vacillating hope, we call " philosophical probability " because it is connected with that higher faculty by means of which we take account of the order and the reason of things. The vague awareness of such probabilities exists in all reasonable men; it determines, or at least justifies, those steadfast beliefs which we call " common sense." When it becomes distinct or when it is applied to delicate questions, it is found only in trained intellects and may even be an attribute of genius. It is applicable not only in the search for the laws of physical or animate nature, but also in the investigation of hidden relations which connect the system of abstract and purely intelligible truths (24). The mathematician himself is most often guided in his investigations only by probabilities of this sort, probabilities which enable him to foresee the sought-for truth before it has been possible for him to give demonstrative evidence of it by deduction and to impress it, in this form, on all minds capable of grasping a series of rigorous arguments.

52. Philosophical probability, like mathematical probability, is connected with the notion of chance and the independence of causes. The more simple a law appears to us, the better it seems to satisfy us with respect to the condition of systematically connecting scattered facts, of introducing unity into diversity, the more likely we are to admit that this law has objective reality; that it is not simulated as a result of a concourse of causes, which by acting independently of one another on each isolated fact would have given rise fortuitously to the apparent co-ordination. But, on the other hand, philosophic probability differs essentially from mathematical probability in that it is not reducible to a numerical statement. This is not so because of the present imperfection of our knowledge with respect to the science of numbers, but because of the true nature of this form of probability itself. This is not the place either to enumerate the possible laws resulting from the continuous or discontinuous variation of some numerical element, or to arrange them according to magnitudes in relation to that propriety of form which determines the degree of their simplicity and which, in various degrees, gives unity, symmetry, elegance, and beauty to the theoretical conception of phenomena.

As we have already explained, mathematical probability may be taken in two senses: objectively, as measuring the physical possibility of events and their relative frequency; subjectively, as furnishing a certain measure of our actual knowledge of the causes and circumstances which enter into the production of events. This second meaning has incomparably less importance than the other. Philosophical probability rests, no doubt, on a general and generally true notion of the real nature of things.[4] But its nature is such that, in each application, it changes with the state of our knowledge and according to individual variations that distinguish one person from another.

The idea of unity, of simplicity in the system of natural laws, is a conception of reason which remains unchanged in the passage from one theory to another, whether our positive and empirical knowledge is extended or restricted. But the moment it is understood that in our role as observers we are reduced to recognizing only fragments of the general order, we gradually become aware of our mistakes in the partial applications we make of this regulative idea. When only a few vestiges of a vast building remain, an archeologist who undertakes to reconstruct it may easily be mistaken in his inferences as to the general plan of its construction. He will project a wall passing through a number of fragmentary remains, the alignment of which, in his opinion, cannot possibly be reasonably accounted for as a fortuitous coincidence. Yet, if other remains are exposed, he may be forced to change the original plan of the restoration, and to recognize that the previously observed alignment was a result of chance. This does not follow from the fact that the remains had not always been part of a system and of a regular plan, but from the fact that the details of the plan had been co-ordinated only in view of the observed alignment. These observed remains would be like the extremities of so many chains which were attached to a common link, but which were not immediately reconnected with each other, and which must be considered independent of one another in everything that does not follow necessarily from the bond which connects them to the common link (29).

---

[4] [Original text: . . . *de ce que les choses doivent être;*]

CHAPTER V

ON THE WAY IN WHICH PROBABILITY ENTERS INTO THE CRITIQUE OF OUR IDEAS OF THE HARMONY [1] OF RESULTS AND OF THE FINALITY OF CAUSES

53. The idea of the finality of causes, like that of chance, keeps continuously cropping up again and again, quite as much in ordinary conversation as in the discourses of philosophers and scientists. We feel the close connection between these ideas and are led to make a comparison of them even when we cannot verify this connection rigorously. If one of these ideas remains indecisive or obscure because of faulty definitions, the same reasons are bound to function in such a way as to make the other also obscure and indecisive. If it has been our good fortune to give greater clarity to the idea of chance, to indicate its characteristic traits more clearly, to draw consequences from it which assist somewhat in perfecting the theory, we may hope, without fear of presumption, that we may have the same good fortune in throwing some

[1] [Cournot uses the terms *l'harmonie* and *harmonique* frequently throughout the *Essai* and particularly in the present chapter and in Chapter IX. In most cases these terms have been translated directly by "harmony" and "harmonic" even at the risk of a tendency toward what some may consider a stilted and pedantic form. However, in some cases, synonymous expressions like "organic unity" have been used where it is unlikely that their use will result in ambiguity. Such ambiguity might result in some cases if these synonymous phrases were used, because Cournot uses these terms in a way that reflects the fact that he is sensitive to and is influenced by both the mathematical and organic signification of *l'harmonie* and of words derived from it quite as much as, or indeed perhaps more than, he is of the esthetic connotation of the term. Thus, while he uses the term *l'harmonie* in such a way as to show that he is well aware of its general and specific philosophical meaning, one need not go far beneath the surface of his thought before one discovers that Cournot is also sensitive to the basically mathematical and rational connotation of the word *l'harmonie*, which is also common and useful. This is but one of many evidences, both in the *Essai* and elsewhere, of the persistent manner in which Cournot's earlier and profound acquaintance with mathematics influenced his later economic, social, and historical thought and writings.]

light on questions relative to the harmony of things and on the part that chance plays in final causes, by following the same analysis or an analysis of the same sort. These are questions that excite the restless curiosity of the ignorant as well as of the learned. Mankind cannot remain unconcerned or indifferent to the investigation of them in any stage of its development.

Whenever the production and continued existence of an object requires the agreement or the harmonic concurrence of various causes, that is to say, of a combination unlike another, reason has only *three* ways of accounting for the observed harmony. The first method is by exhausting the fortuitous combinations in the unlimited extent of space and time, during which all unstable combinations will have disappeared without leaving any observable traces, so that our observations reveal and can reveal only those combinations which have brought together fortuitously the conditions which assure duration and persistence. The second way of accounting for the observed harmony is by an intelligent and providential direction which fits the means to a desired end, or which gives to blind and secondary causes the ability to act in the same way as would intelligent forces which would be conscious of their actions and of the ends they set up for themselves. In the third place, we may account for the observed harmony by mutual reactions whose operation would be sufficient to introduce into the final state which we observe a harmony which did not exist originally (24) and which, being the necessary result of blind forces, in itself shows no indication of providential co-ordination or of an end in view.[2] Thus, when the question is to explain the agree-

---

[2] " Let us imagine two clocks or two watches which are in perfect agreement. Three alternative explanations are open to us to explain this agreement: the first consists in the mutual influence of one timepiece on the other; the second, in the attention given them by the person who takes care of them; the third, in their inherent exactness. The first alternative was experimented with by the late M. Huygens, with results that greatly astonished him. He had two large pendulums attached to the same piece of wood. The continuous pulsations of the pendulums were transmitted by similar vibrations in the particles of wood. But these different vibrations did not maintain their identity, and, at least as far as the pendulums were concerned, although there was no direct contact between them, it happened almost miraculously that when their oscillations were disturbed so that they moved irregularly in relation to each other, after a short time, they again moved with the same oscillation, like two strings of uniform

ment between a prediction and the event predicted, it is also true that we can form only three hypotheses: first, out of all the great number of predictions made by chance, only those have been retained with respect to which the play of fortuitous causes has led to confirmation; second, the prediction is the effect of a natural or supernatural knowledge of the causes which must lead to the event; third, the prediction and the event predicted have reacted on one another in such a way that either the statement of the prediction has been modified to agree with the event, or the account of the event has been made to agree with the prediction. The last alternative is illustrated by the situation in which troops lose courage and give up the battle when their spirit has been broken by an oracle which has predicted their defeat.

54. Let us speak first of the explanation in terms of the influence of mutual reactions. This involves going back to what is meant in natural philosophy by the initial state and the final state and indicating by examples how order and regularity tend to be introduced in the passage from the former to the latter. Let us imagine a regularly formed body, such as a sphere, which has originally been heated unevenly and in such a way that the variations of temperature from one part of it to another follow no regular law. If the body is then placed in an environment the temperature of which is uniform and constant but much lower than the mean temperature originally given to the different particles composing the object, the latter will gradually lose its heat. Its mean temperature will decrease and will tend to become that of the surrounding

---

tension." Leibniz, *Premier éclaircissement sur un système nouveau de la nature et de la communication des substances.*

What seemed to be a *kind of marvel* at the time of Huygens and Leibniz is today among those physical phenomena which are best known and most completely explained by mathematical analysis. There is another hypothesis that Leibniz had no need to consider, and hence omitted in the passage quoted. This hypothesis is to the effect that, in the great number of pendulums found in a clock store, and among the great number of fortuitous causes associated with them, chance would finally lead to the appearance of two pendulums having the same period. As to the last two hypotheses, which it is convenient for him to distinguish for his particular purpose, we should relate them to a simple explanatory principle, namely, that of a co-ordination which is intelligent or has some end in view. Like Leibniz, we also have three explanatory hypotheses or principles, but they are not the same ones.

medium.   But at the same time, the heat within the body will tend
to become evenly distributed.   The particles in the center, which
originally would have been heated less than the others, will have a
higher temperature than those near the surface, because, on the
one hand, the former will have dissipated part of their initial heat,
and, on the other hand, the particles near the center are farther
from points through which the body as a whole radiates heat at the
expense of its mean temperature.   After a sufficient length of
time has elapsed, the temperature of the outside layer will be
perceptibly the same as that of the surrounding medium, and, as
we go from the surface to the center of the body, the temperature
will increase so that it will be possible to divide the mass of the
body into spherical and concentric layers in such a way that all
the particles in each layer have a uniform temperature.   Thus
the heat will be distributed in an increasingly regular manner
which finally becomes perfectly regular even though there had been
no trace of regularity in the initial manner in which the heat was
distributed.

In the same way, if we imagine a scattered mass of material
particles, irregularly distributed at random distances from one
another and moving at many different speeds but still under the
influence of forces which draw them toward one another, we shall
find after a sufficient length of time that the particles will come
together to compose a body having a regular form.   The regular
motions of rotation and of translation of this body will be a sort
of mean between the different motions which actuate the diverse
particles in their original sporadic state.   Order will spring of itself
from an originally chaotic condition.

Similarly, to use a final example, if air or water is agitated
irregularly as it flows from the opening of a pipe or of a canal which
has a regular form, the motion will be propagated in such a way
that at a certain distance from the mouth nothing will be perceived
but regular waves whose law or explanation will no longer depend
on the form of the original disturbance.   In all these phenomena,
the order which finally establishes itself—like the constancy of
relations discovered by means of statistics—attests only to the final
preponderance of a regular [3] and permanent influence over anomal-

---

[3] [I have taken the liberty of correcting the text at this point in what seems
to be a typographical error which was not caught in the first edition of the
Essai and has persisted through the third edition. The text at this point reads,

ous and variable causes. This is the consequence of mathematical laws, and all we can do is admire it in the same way in which we admire a theorem of geometry which impresses us because of its simplicity and because of the fruitfulness of its applications.

55. The same thing is true of the harmony finally established among many phenomena or series of phenomena as a result of the influence that one series exercises upon the others or because of their mutual reaction. Thus it is that, in accordance with the curious experiment mentioned above (53, note), if two clocks having beats which are not perfectly synchronized, and the movements of which are not strictly co-ordinated, are fastened to a common support, it is noted after a certain length of time that the transmission of the movements of one clock to the other through the medium of the common support has led to their becoming synchronized and exactly co-ordinated. In general, bodies which can transmit their vibratory motions to one another tend to vibrate in unison, even though originally endowed with vibratory motions having periods which are out of phase and which are of unequal duration, provided that the discordances and inequalities do not exceed certain limits. Our planetary system offers examples of such phenomena on a grand scale. The moon always turns the same face toward the earth because it takes the same length of time to make one rotation on its axis as it does to complete its orbit around the earth. It would be very unusual if the original conditions which had determined these two periods independently of one another had been such as to adjust them so as to produce such an exact concurrence between them spontaneously. But if it is admitted that, in the beginning, the two periods were slightly different, and if it is further admitted—and there is every probability that this was the case—that originally the mass of the moon, like that of the other celestial bodies, was fluid, the force of attraction of the earth has been able to modify the form of its satellite in such a way as to bring the two periodic movements into phase in the long run

---

" Dans tous ces phénomènes, l'ordre qui s'établit en définitive n'atteste (comme la constance des rapports trouvés par la statistique) que la prépondérance finale d'une influence *irrégulière* ou permanente sur les causes anomales et variables." Vol. I, 106, of the first edition; page 78 of the third edition (italics mine). Both the immediate context in this passage and Cournot's general position make it seem highly probable that he means at this point to contrast a *regular* and permanent influence with anomalous and variable causes.]

and to produce the phenomenon we now observe, namely, that one hemisphere of the moon is forever turned away from our view. There are reasons to believe that the satellites of the other planets exhibit the same process, due to the same cause. Thus the velocities with which the satellites of Jupiter circle around their planet also have unique relations among themselves which may be explained in an analogous manner by means of mutual reactions which must tend to adjust the parts of the system harmoniously, subject always to the condition that the parts were originally placed in relations sufficiently like those which their reactions within the system tend to establish, or to re-establish when external causes happen to disturb them.

56. In phenomena of an entirely different order, and of orders which are very dissimilar to one another in that, unlike the preceding examples, they are not amenable to mathematical treatment, similar harmonious structures may be discovered. These, too, follow from mutual influences or reactions, which, however, operate effectively only within certain limits. Consequently the initial state of the system must be regarded as having been, if not precisely in the harmonic condition which establishes itself in the long run, at least in a condition not too much unlike this. An organ acquires greater vigor and becomes more developed as it is exercised, and, by the same token, when it is used more frequently and in more varied ways, it develops qualities suited to its new uses. On the other hand, an organ which ceases to be exercised atrophies and disappears along with the need the animal had of it. We have a celebrated example of this in the eye of burrowing animals such as the mole. In the social order, needs stimulate industry, and new resources develop in such a way that they square harmoniously with new needs. This is seen notably in the equilibrium which establishes itself between the population and the means of subsistence. This equilibrium does not lead us to suppose that the productivity of marriages had been adjusted in advance to the fruitfulness of the soil, and it leads us still less to believe that the latter had been allotted in view of the former. The introduction of a foreign body or an injurious substance into the organic system of an animal irritates the tissues, and as a result of this very irritation nature tries, as we say, to throw off the substances which are harmful to it and the foreign bodies which injure

it. The organism then tends to recover or to re-establish its normal condition which has been temporarily disturbed, provided its injuries have not resulted in lesions or alterations which are too deep-seated. When any disturbance whatever has occurred in the animal or in the social economy, the restorative forces acquire a higher degree of energy for that very reason. Thus, after copious bleeding or long abstinence, the appetite of a convalescent person is whetted and ailments clear up more rapidly in proportion. So it is also that after a war or revolution has decimated the virile population and dissipated the capital of a nation, men tend to multiply and capital to renew itself so rapidly that a few years of peace and of a wise administration are sufficient to remove the trace of past calamities.

57. But besides harmonies of this sort, which are established after the fact and are self-explanatory, there are others which cannot be explained in the same manner, because they occur between diverse events or orders of independent events, which could not react upon each other in such a way as to produce a harmonious relation that did not exist originally, or to re-establish a previously existing harmony that has been accidentally upset. In order to enable us to understand this point better, let us take another example from astronomy. In the theory of the movement of the stars, as in the theory of the movements of any system of bodies whatever, there are two things to consider: first, the forces to which the bodies are submitted so long as they are moving; second, the initial data, that is to say, the positions the bodies had occupied and the velocities with which they had moved during a period which we use, together with certain forms of calculation, as the basis for assigning all the phases through which the system must subsequently pass, or even—with the exception of certain limitations which we shall have to discuss later—to trace back the various phases through which the system must previously have passed. In order that the movements of our solar system may perpetuate themselves with the regularity and harmony which are so striking to us, it is not only necessary that matter be submitted to the permanent action of a force the law of which is very simple, like that of universal gravitation. It is also necessary that the masses of the sun and of the planets, their respective distances, their distances from the stars, and their velocities throughout a given

period should have been so proportioned that these celestial bodies will periodically describe nearly circular and invariable orbits, except for trifling perturbations which alter them now in one manner and now in another, and which are found to be restricted within very narrow limits. This is what we mean by the conditions of stability of the planetary system. Our knowledge being what it is at present, we have no basis for supposing that the phenomenon of this stability is an instance of those which are established or re-established as a result of a trait inherent in the mutual reactions and of the bonds of interconnection within the system. The question at issue here is not whether or not this phenomenon may be an absolutely primitive fact, to explain which we must have recourse to more or less arbitrary hypotheses. We wish only to note clearly that the fact of these initial dispositions of the parts of a material system and the fact of the submission of the parts of the system to the action of permanent forces of this sort are two facts between which reason does not recognize any essential dependence, the one being in no sense whatever the consequence of the other. Thus the concurrence of the two facts, the establishing and maintaining of an order whose harmony impresses us, among an infinite number of other possible arrangements, is not a necessary result and may be attributed only to a fortuitous combination, or to the action of a higher cause which finds the reasons for its acts in the end that it pursues.

58. Let us take another example, more closely related to phenomena properly called organic. The chemical elements of bodies that we have been able to submit to analysis are very numerous, but by no means do they all play the same role in the economy of our terrestial world. Some are abundant, others are rare; a small number of them are found in the most variegated and complex combinations and are, for that reason, adapted to furnish organic nature with its essential materials. Now it is certain that the causes which have determined the proportions and distribution of the various chemically heterogeneous substances in the structure of the earth are, by their nature, independent of those which have brought about the development of organic and living beings. On the other hand, although living things undergoing the influence of physical conditions can, as a result of their marvelous fecundity, adjust themselves to very different physical

conditions through the modification of forms by means of apparent or hidden devices in such a way as to make them compatible with new conditions, it is equally certain that this power of modification has very narrow limits as compared to the range between which physical and external conditions are able to fluctuate. Let other proportions or a different distribution of the chemicals which compose the surface layers of the earth be imagined, and the development of plant and animal life would be impossible for want of the requisite conditions; let the mass of the atmosphere decrease sufficiently, and the entire surface of the earth would be in the same condition as the icy summits of the Alps; let the proportion of silica on the surface of the earth increase, and the continents would all be as sterile as the desert sands are for us; let the proportion of sodium chloride in the ocean increase, or let certain injurious elements be introduced into it, and it would be as devoid of living forms as the Dead Sea. The mass of the atmosphere, to speak of this condition only, must be related to the distance between the earth and the sun, the warmth of which it must retain and concentrate. At the same time, the atmosphere must be related to the way in which the forces act which direct the evolution of living things. Without this relation—and observation itself shows us that this is possible—the conditions requisite to many remarkable phenomena would be lacking. Both reason and experience give us sufficient evidence that in these cases there is a concurrence of independent causes, a harmony which is not necessary, in the mathematical meaning of necessity, and for the explanation of which there are, as we have already said, only two hypotheses: that of fortuitous causes and that of the subordination of all concurrent and blind causes to one which seeks to attain some end or purpose.

59. Among the infinite number of examples of harmony exhibited by organic beings, whether they are considered by themselves or in their relation to external agents, let us take as one of the least complicated that which results in changes in the color of the hair of an animal with changes in climate. We are completely ignorant as to how the climate acts in such a way as to cause the hair of an animal to become thicker when it moves to cold regions, and to get thinner when it goes to warmer regions—indeed, what else can we be in these matters! But, on the basis

of all appearances, the effect of heat and cold upon the sensibility of the animal and the disturbances which may result from it in the internal structure of its organism, have no more effect on the action of temperature in modifying the development of hair and skin, than they intervene in the action that light has on the tegumentary system, to the point of adorning the coats of animals which live in the heat of the tropics with the most vivid colors, and, on the contrary, making the coats of animals or birds living in the polar regions pale and dull. Without doubt, it is not the need of more brilliant ornamentation that gives the hummingbird its metallic brightness; very likely, also, it is not the discomfort of the cold felt by animals that move toward icy regions that stimulates the growth of fleecy and more abundant hair. If this judgment is accepted, we must admit a concurrence, whether fortuitous or pre-established, between the needs of the animal and the action of the surrounding medium upon the development of the tegumentary system. As a matter of fact, it would be rash to affirm absolutely that the effect of the cold upon the sensibility of animals is not the immediate cause of an increased development in the tegumentary system. But we are using this case only as an example and as a hypothetical case, if one wishes to regard it in that light. In any case, the probability of the consequence that we deduce from it obviously will be subordinated to the probability of the hypothesis, the state of our knowledge being what it is.

It has been remarked that the hair of animals frequently takes on a color similar to that of their environment, as if nature had wished to give them a means of protection from natural enemies so that the species might be preserved. Thus, hair which is white in snowy countries takes on a ruddy tint in arable regions, and in the midst of the great African desert approaches to a remarkable degree the color of the sands which form the background of that melancholy landscape. Whether or not the fact is more or less constant, and whether or not it can be explained through physical laws, is not what we have to examine. Rather, we are interested in discovering whether or not the fact that an animal is hunted by its natural enemies and makes efforts to avoid them may always be admitted to contribute to the change of its coloration, so that, if there really is a connection between this and the need for protection, we shall know whether it is to be accounted for by chance, or by a finality which controls the manifestations of a higher cause. It cannot be an instance of those harmonies which are established by themselves by means of influences or reactions which result from the interconnections among different parts of a system.

60. Moreover, the marvels of organic life make it possible for us to find examples which, if not so simple, are at least equally decisive. Let us admit for a moment that the feeling of cold and the resulting discomfort to the animal are sufficient, so far as the operation of organic forces is concerned, to account for the stimulation of the hair cells and for the modifications of size and structure that the hair undergoes. Granted that this may be the case, will it be enough to persuade us that the eye may be formed and improved as a result of the influence of light; that the properties of this physical agent and all the very complicated and amazing apparatus of vision may be brought into accord with one another, in the long run, as the result of an influence comparable to that which ultimately synchronizes the periods of two clocks fastened to a common support? Everyone knows that, if lack of excitation is sufficient to explain the atrophy of the visual apparatus in animals whose mode of life takes them away from light, and that if this deficiency paralyzes the formative force, which tends to result in the most complete development of the structure where circumstances are favorable to nourishment and excitation, it would be impossible to conclude inversely either that light possesses this formative ability, or that the impact from light is a sufficient reason why the operation of the organism tends to construct a device for seeing without a preliminary accord, *parte in qua*, between the physical properties of light and the characteristic laws of the organism.

The fact that the faculty of vision is in general use throughout the animal kingdom must not be made the basis of a misconception. Electricity plays just as considerable a part in the physical world as does light. However, while nearly all animals have eyes of one sort or another, there is nothing more special and more rare than the existence of electrical devices such as those which serve the torpedo and the electric eel as means of defending themselves against their enemies and of attacking their prey. If these fish were placed in surroundings where they were unable to charge their electric " batteries " these organs would no doubt atrophy. This would simply be another application of the general law of organic life which indicates that the development of any unused organ is arrested, or that the organ atrophies after its complete development. But can we conclude from this that the influence of electricity is the force which has created and developed in the torpedo and the electric eel the beginnings of the electrical device? If this were the case, why would not the same influence, which is present universally, tend to produce a similar organ in all aquatic creatures, or at least in all species of the same family or genus, which, besides living in the same element, are so completely similar to the torpedo and the electric eel in all other organic details? It must be concluded from this that the eye has no more been fashioned as a result of the action of light than the " battery " of the torpedo has been by electricity, and that the generating cause of these structures is a formative force which is inherent in animal life and moves toward the realization of a determinate type in each species, and is directed by laws which are appropriate to itself. If the most general characteristics of the organ of sight seem to belong to animality as such while the electric " battery " appears only as a very special and accessory detail in particular types of organisms, and if, moreover, the former serves a very important function and is suited to meet a very general need while the latter satisfies a function which is accessory only to a need that nature has numerous other ways of satisfying, the reason for the difference between them is not to be found in the dissimilarity and unequal importance of the role of the physical agents, to the influence of which animal forms can only passively submit. The reason for the difference must be found in laws which are peculiar to animal nature.

We become increasingly convinced of this autonomy in proportion as we learn more and more about organisms. Then we come to understand that the function of an organ and the use the animal makes of it for the satisfaction of a certain need are not what is most fundamental, best established, and most characteristic of the organ. Although a type which is fundamental and persistent in its general characteristics comes to be modified in a number of ways in its details in passing from one species to another, the organ, the identity of which cannot be mistaken throughout all these successive modifications, often performs very different functions; and reciprocally the same functions are often taken care of by organs which are clearly distinct. In sum, it is not generally possible to define an organ by the function that it performs; the attribution of a given function to a certain organ seems most often to be an accident and not that which essentially characterizes the organ nor that which determines the fundamental relations between it and the total structure of the organism. Now, if the physical world and that of living things, which are governed respectively by laws which are peculiar to them and which have their special reasons, are found in opposition and in conflict, the harmony that is observed between the two for the accomplishment of the functions and the satisfaction of the needs of the living being with respect to everything which exceeds the part that one may reasonably attribute to mutual influences and reactions, can be accounted for only by a fortuitous coincidence or else by the end which governs the determinations of a superior cause and which sets up the general laws of the physical world as well as the special laws of the world of living things.

61. The comparison between these two hypotheses or explanations must now be established. First, we shall deal with the former which rests upon the idea of fortuitous concourse, and of the exhausting of fortuitous combinations in an unlimited space and time. Whatever may be said for this explanation, which is constantly being brought forward and as constantly being opposed, it may not be passed over in silence or dealt with contemptuously, since it is the one most satisfying to reason, so far as certain details are concerned and within certain limits, or even the only one reason can accept. It is clear that an organism in which the whole structure does not concur to conserve the individual is doomed to perish. Likewise, a species can continue to exist only if all the circumstances concur which are required to assure its propagation and perpetuation. We may conclude from this that, among the infinite number of combinations to which the continual play of natural forces has given rise in the unlimited field of extension and duration, all those which have not embodied the conditions of stability have disappeared, leaving only those which found conditions of sufficient stability in the wholly fortuitous harmony of their parts to remain alive. And in fact we see that species and individuals are very unequally divided in their means of resisting the action of destructive forces. For some, the length of life is

shortened; for others, multiplication is restricted. When the destructive forces become more intense or the means of resistance more feeble, the germ cells will fail to develop, individuals will not be born viable, or the species will disappear. Now, observation teaches us that some species have in fact died out, and that every day individuals become feeble and fail to piece together the conditions of viability.

62. It is worth while to carry this point still farther; for these considerations apply not only to organic and living beings, but also to all cosmic phenomena in which indications of order and harmony are found. Our planetary system, which is so remarkable because of the conditions of simplicity and stability which it satisfies, is itself only a speck of dust in the heavenly spaces, one combination among an infinite number of others that nature must have produced. No matter how slight our knowledge of other systems or of other tremendously distant worlds may still be, we already derive from observation reasons for believing that by varying the combinations nature has not bound herself to incorporating the conditions of simplicity and permanence to the same degree. As we have already seen, some very special conditions had to be realized before an atmosphere could be formed around the earth, so proportioned and constituted, as a consequence of the distance between the earth and the sun, that it may exert an influence on solar heat and light peculiarly suited to the development of animal and vegetable life, and, at the same time, furnish a chemical element indispensable to the maintenance of respiration and life. Also, among the heavenly bodies, the one closest to us offers a first class example of a heavenly body placed in altogether contrary conditions by the fortuitous circumstances of its formation. The moon has no atmosphere, and we have every reason for inferring from observations that its surface is doomed to a permanent sterility. It was essential that the solid material forming the surface of the earth should have had a certain chemical composition and that the inequalities of the surface should have assumed certain contours in order to permit such variety and richness in the development of forms and organisms. It was also necessary that, where these conditions were lacking, deserts, barren rocks, and ice-covered areas would be found in which cryptogamia and microscopic animalcules, crowded together by the millions, are

the last and lowest products of a weakened and diminished creative force. In some places wild, torrential, or stagnant waters, sources of destruction and of noxious emanations for all the higher forms of both plant and animal life, are found instead of those brooks, rivers, or lakes, those well-arranged bodies of water whose distribution and regular pattern by contrast bring out even more the disorder and irregularity presented by other portions of the geographic picture. If, in the present state of affairs, the ravaged and sterile areas represent only a small portion of the earth's surface; if the limits of Typhon's empire have receded nearly everywhere before the action of the organizing and life-giving principle; the geological monuments still remain to instruct us that the order has not always been the same, and that eons of time have had to elapse and numerous convulsive disturbances to occur before the order we now observe gradually came into being, and this order, in spite of its relative stability, cannot escape the causes of dissolution, over the course of the ages, any more than can other natural combinations.

63. This is essentially the same type of argument that would have been used by a member of the Greek school of Epicurus or by a medieval thinker, except that it is stated in terms and applied to examples more appropriate to the state of modern science. We shall also ask science to furnish us with examples and inferences, not for the purpose of destroying the argument, for it has its value and its legitimate applications, but to combat its extreme consequences and its exclusive tendencies.

Let us assume that the earth will undergo no more shocks of the type that in past epochs have raised up mountain ranges and brought about dislocations and irregularities on the surface of the continents and on the ocean bottoms. The action of air and water, combined with that of gravity, will tend very slowly, but nonetheless constantly, to loosen rocks and carry their debris to the bottom of valleys and basins. In a word, they will tend to level all that has at various times been raised up, to fill in all low places, and to smooth over the surface of the earth as though the materials of its outer crust had been originally fluid. Even in the present state of things, topographic inequalities of the earth's surface, although enormous as compared to our size and to our meager constructions, are so small in relation to the dimensions of

the earth that astronomers have been able to neglect them in making most of their calculations.   Indeed, they have been so struck
by the similarity of the general form of our planet to that which
the laws of hydrostatics would have given it, according to the
hypothesis of an original fluid state, that they have not hesitated to
regard this hypothesis as demonstrated by the very form of the
earth.   Let us disregard for a moment all other proofs and all other
inferences furnished by the progress of geological observations
which make it impossible for us reasonably to doubt the original
fluid conditions.   The agreement between the spheroid form of the
earth and the laws of hydrostatics may still be explained completely without the assumption of an initial fluid state, and starting with any original form whatever, as a result of the indefinitely
prolonged action of causes which, even today, tend to reduce the
roughness of its surface to a perfectly level plain.   An infinite time
is at our disposal to meet the needs of this theoretic conception,
as it is for the using up of all fortuitous combinations, however
huge the number of elements to be combined, and however
unique the combination to be accounted for may be.   Nevertheless, the time required to reduce a solid body the size of the earth,
no matter what its shape, to a form that the same mass would
spontaneously assume if it were in a fluid condition, and to do this
by wear and tear and by a process of gradual dissolution and disintegration, so immeasurably exceeds the duration of the great geological periods—however great this duration may be in comparison
to what we call historic time and to which we go back by human
tradition—that, in the absence of any other indication, reason does
not hesitate to prefer the original hypothesis of initial fluidity,
which is so natural and so simple, to an explanation which makes
such an excessive claim.   Thus, when we see that in the relief
of ancient geological formations, no matter how far back in time
we may be able to go, nothing gives greater evidence of a different
form than the present appearance of the general direction of strata,
we reject as absolutely improbable the explanation based upon
the slow disintegration of surface layers, without even needing to
refer to inferences drawn from volcanic phenomena and from the
increase of temperature as we go below the surface of the earth,
both of which lead us to admit that, at a rather shallow depth,
the earth is still in a state of igneous fluidity.

But this lapse of time, which reason would reject as an explanation of such a phenomenon as the elliptical form of the earth, is only a moment, in comparison to the period that would be required before one could reasonably admit, on the basis of the rules which guide us in matters of probability, that, after innumerable forms had been created only to be destroyed, other combinations occurred by chance in which all the conditions of harmony needed to assure their stability were found through the evolution of fortuitous combinations alone, and beyond the limits within which mutual reactions suffice to explain the final harmony. Thus it would be by chance, after an incalculable number of combinations, that the eyeball would appear with all its tissues and fluids, the curvatures of its septa, the different densities of the refracting materials of which it is composed and combined in such a way as to correct the aberration of light waves, the diaphragm which dilates or contracts to control the size of the pupil, the pigment which covers the back of it to prevent the trouble that would be caused by internal reflections, the accessory organs which protect the eyeball, the muscles which move it, the branching of the optic nerve in a sensitive network so well suited to the formation of images, and the connection of this nerve with the brain, which is no less especially adapted to the sensation which is transmitted. All that would not prove a pre-established harmony between the physical properties of light and the plan of the organic structure of the animal. It would be only too easy to insist upon the inexhaustible details of this inductive argument. This has been done so often and sometimes so eloquently that we need do no more than indicate here its place in the discussion. It would be even less appropriate to reassess the commonplace topics of the schools on the throwing of dice and on collections of letters and to repeat the fictitious examples which have been repeated over and over again even in antiquity.[4] Modern science has a more satisfying

[4] " You ask why these things come about in this way, and by what means they can be understood. I confess that I do not know, but maintain that you yourself see that they do come to pass. You say they are accidents? Indeed, can they be? Can anything be a matter of chance which has in itself all the marks of truth? Four dice were thrown, and each showed a different number; would the same combinations occur by chance in each of a hundred successive throws? Colors scattered at random on a canvas may form the outlines of a face. Do you think the beauty of the face of Venus of Cos can be achieved by

and more decisive response than those scholastic reasonings. It
has deciphered the records of the ancient world. There it has
seen that in a certain geological period living things did not and
could not have existed on the surface of our planet. Consequently,
the condition of an unlimited time for the evolution of fortuitous
combinations is ruled out completely. We know that races have
succeeded one another, and also that they have very probably been
modified by external conditions. But no more then than now did
nature proceed by means of myriads of rude first drafts before
fortuitously stumbling upon an organic type capable of maintaining itself as an individual and of perpetuating itself as a species.
The existence of a formative force, which itself proceeds from the
conditions of unity and harmony which are characteristic of it, and
which always puts itself in relation with external circumstances
and undergoes their influence, is, henceforth, for all reasonable
persons, not only the probable consequence of an abstract argument but also the indubitable consequence of the data of observation themselves.

64. Most often the three principles or guiding threads that
we have mentioned must be accepted concurrently in order to preserve the important part of each in proportion to our knowledge
and to the value of the inferences that may be drawn from it. A
gardener brings a wild plant under cultivation, places it in new
surroundings, and soon the organic type, submitting to the external influences, adapts itself to these new conditions, and consequently to the needs in view of which its cultivation has been
undertaken. Certain organs disappear or atrophy; other organs,
such as the flowers, fruit, or roots, which are useful or interesting
to men, become larger, more vigorous, and more beautiful. Here
we see the role of reactions and influences which are capable of
leading to a final harmony, and which, in this case, substitute a
new harmonic order, brought about by the efforts of man, for the

---

the random scattering of colors? Suppose a sow imprints a letter 'A' on the
ground with its snout. Can you, on that account, believe she could write the
Andromache of Ennius? Carneades related that a head of the infant Pan appeared on a slab of marble when it was split in a Chian quarry. I admit that
some such figure could have appeared, but not such a one as might have been
fashioned by the artist Scopas. For it is certain that a perfect imitation is
never made by chance." Cicero, *De divinatione*, i. 13.

order which the primordial laws of nature had established, apart from human action and prior to the introduction of this new force into the scheme of things. If the same gardener plants some seedlings at random, among the great number of individual varieties which result fortuitously from the different characteristics of the seeds, combined with the accidental influences of the atmosphere and soil, he will find some which combine the conditions necessary for propagation, in the sense that the person who cultivates them is interested in propagating them in preference to other forms which he allows to die out. In their turn, those that are saved produce a multitude of others among which those are selected which, by fortuitous circumstances, combine to a still higher degree the qualities which made the earlier generations valuable, qualities which thus are strengthened and made more and more pronounced in being transmitted successively from one generation to another. The formation of cultivated races, which are like new types artificially substituted for those found in a natural state, is explained in this way. This example serves to give the idea of the part that chance and the indefinite multiplication of fortuitous combinations play in the establishment of the final order and of the harmonies which are observed in it. But there are limits to this participation of chance, as to the influences arising from cultivation. The most important role in the establishment of the final harmony is always found in the primitively generative and formative force associated with the original type, because of a pre-existing harmony which the art of cultivation may modify noticeably, but can neither supply nor create out of whole cloth.

What we have said of this little fragment of nature when brought under cultivation may be applied equally well to the unrestricted behavior of nature in an uncultivated state, except for the magnitude of the dimensions involved. Without doubt, many forms and species have been created and strengthened by a fortuitous concourse of accidental circumstances, as a result of the diversity of climates and of the long time which has elapsed since the age when living things first appeared. But, as far as we can tell with such knowledge as we have, this explains only the most minute fraction of the varieties of type and of organizations, and we must take account especially of the varieties inherent in the primordial plan of nature in the construction of organic types.

Similarly, in order to explain the final harmony of the individual organs with respect to one another and of the whole organism and the surrounding environment, it is no doubt necessary to recognize the part played by reciprocal influences and reactions which, within certain limits, are sufficient if needed to re-establish a harmony that has been disturbed accidentally. But, principally and above all, attention must be given to the essential harmonies of the primordial plan. If it happens that the paw of a Newfoundland dog gives a rudimentary suggestion of the webbing which is adapted to its aquatic life, if it also happens, according to the remark of Daubenton, that the intestinal tract of the domestic cat is some-what elongated because of the fact that it lives partly upon vege-table food, then these facts make clear two points. They demon-strate to us the unique influences of environing conditions and of acquired habits in the modifying of organic types, but only within narrow limits, in such a way as to adapt these types to new con-ditions of life. They also show us, because of the very narrow-ness of the limits and the organic imperfection of the products, the complete distinction that must be made between such external influences and the formative force which resides in the organic type itself and in the harmonic co-ordination of its parts. Other-wise we might as well compare the callosity that habitual fatigue develops, often too late, but in a definite manner, on just those parts of the skin which need protection, with such specific pro-tective organs as claws or hoofs, which undoubtedly go back to the original adaptation of parts to one another in the specific form.

65. Ordinarily, when the final *consensus* arises out of mutual influences or reactions, the roles of the different parts which tend to form a unified system are not equally important. Usually one part has a predominant influence because of its mass or for some other reason, and it may even subject the other parts to its in-fluence without, in turn, undergoing any appreciable reaction. In other cases, when the mutual influences and interactions are not sufficient to explain the observed *consensus*, and when reason feels obliged to seek in the finality of causes the explanation which it would otherwise fail to attain, it must not assume either that, in general, all the parts of the harmonic system play equally im-portant roles in the order of finality. Where finality is most mani-fest, as in the organic structure of living beings, such a parity of

roles cannot be attributed to all the parts of the organism without going contrary to all the notions science gives us regarding the subordination of organs and of organic characteristics, which have neither the same stability from one type to another nor the same importance when considered simultaneously within the same type. Thus, when we see an elephant it is evident that its massive structure makes necessary the peculiar prehensile organ which we call the trunk, and consequently that there is a remarkable harmony in the organization of this animal between the extraordinary development of the nose which makes it, by exception, an instrument for grasping, and the modifications of size and form in other parts of the body.   It would be ridiculous to suppose that the nose of the elephant had been lengthened as a result of persistent efforts made by its ancestors to obtain by means of their noses objects which would serve them as food.   That exceeds the part played in such instances by mutual reactions; paleontology gives no evidence of this progressive lengthening; the race would have perished long before such an end could have been attained.   So reason is brought to recognize an original harmony, a final cause.   But it is also evident that nature did not make the animal heavy and massive and did not deprive it of means for reaching objects which serve as food directly with its mouth, because it had provided it with a trunk.   On the contrary, it is because the general conditions of size and structure had been given for this form, as a result of higher laws which direct the principal modifications in animal forms and the distribution of species into orders and genera, that nature, getting down to details, modified a secondary organ in such a way as to fit it for the carrying out of a special need imposed by the dominant conditions.   In the order of finality, the general conditions of structure and size are the antecedent term; the exceptional development of the nasal apparatus is the consequent term.   Reason would be offended if the order of these terms were reversed, as they would have to be if one obstinately persisted in seeing in this harmonious adaptation of parts to one another only a fortuitous coincidence.

At other times, the diverse terms of the harmonious relation appear to be on the same level of importance so that, at least in the present state of our knowledge, there is no reason for regarding one as subordinate to another.   It is necessary that carnivorous

animals be sufficiently agile to capture their prey, muscular enough
to force it to earth, and that they have claws and teeth for tearing
and devouring it. But we have no decisive reason for regarding
the characters which lead to the development of teeth as dominating those which result in the formation of claws, or the other way
around. These characters appear to us to be of the same order
and to concur in the same manner, and for the same reason, in
the general harmony of the organism (25).

66. We shall have reason to make analogous remarks, if we
move on from the consideration of the harmony which exists among
the parts of an organism to the study of the harmony which the
relations of one organism with the organisms around it afford us,
or even to that of the harmonies that the arrangement of things
in the physical world manifests on an even grander scale. Thus,
it is not permissible to say offhand that vegetables have been
created to serve as pasture for herbivorous animals, or that the
latter have been formed so that they may live on vegetables. The
development of vegetable life is the prior, dominant fact to which
nature has subordinated the construction of certain types of animals
which are formed in such a way as to find their foodstuffs in the
vegetable kingdom. This is not a proposition that can be demonstrated with logical rigor; but it is a relation that we come to
understand through the sense we have for the reason of things and
through a view of phenomena as a whole. Only bees would
imagine that flowers had been created for their particular use. As
for us who are disinterested spectators, we see clearly that the
flower is a part of a group of organs designed primarily to make
possible the reproduction of the plant and constructed toward that
end, and that, on the contrary, it is the bee whose organism has
been so modified as to enable it to draw nourishing juices from
flowers and to assimilate them into its own substance. It would
be ridiculous to say that an animal had been created to serve as
the host for a parasitic insect; yet it cannot be doubted that the
organization of the insect has been accommodated to the nature of
the tissues and the fluids of the animal upon which it lives. If
we keep this in mind when examining most of the examples that
have customarily been cited to ridicule recourse to final causes, we
shall see that the ridicule is based on an inversion of relations and

misconceives the natural subordination of phenomena to one another. But, just because materials like stone and wood were not
created to serve in the construction of a building, it does not follow
that we are obliged to explain the appropriateness we observe between the material and the ultimate design of the building by
means of blind reactions or by a fortuitous coincidence. Now, so
far as we are able to tell, in the general plan of nature the same
objects must successively be seen, first, as things created by nature
for their own sakes, ingeniously using pre-existent materials for
that purpose; then, as materials that it employs with no less skill
in the construction of other objects. To invert this order continually when it makes itself so clearly evident would offend reason.
Yet this has often been done by those who have sought to consider man as the center and the end of all the marvels of which
he is only the intelligent witness, and of which he still, most frequently, has only an imperfect notion.

67. Natural phenomena, being linked to one another, form a
network all the parts of which are connected with one another,
but neither in the same manner nor to the same degree. The fabric
can be compared neither to a system having an absolute rigidity,
which would be capable of moving only in one piece, so to speak,
nor to a whole, each part of which would be free to move in any
direction with absolute independence. In one case, the bonds
linking the parts together are relaxed, and there is more chance
for the operation of fortuitous factors; in the other case, on the
contrary, they are more closely knit, and the systematic unity is
more strongly indicated. Thus, the pattern of a leaf is perfectly
definite so far as the principal veins are concerned, while, so far
as the ultimate ramifications of these veins and the agglomeration
of cells which fill the space between them and form the parenchyma
of the leaf are concerned, the fortuitous play of accessory conditions gives rise to innumerable modifications and details which
are not at all the same from one individual leaf to another. We
are just as far from a faithful interpretation of nature when we
disregard the systematic co-ordination of fundamental traits in
which it distinctly shows itself as we are when we imagine, without
warrant, bonds of co-ordination and interdependence where there
are collateral series each of which is governed by laws character-

istic of them after their separation from the common trunk, and which subsequently have only accidental relationships and fortuitous connections with one another.

It is an axiom of human reason that nature is governed by general laws, and we go contrary to this axiom when we invoke a providential decree, when we have recourse to a final cause, that is, to a *deus ex machina*, for each particular fact, for each of the innumerable details that the picture of the world presents to us. But we have no basis for saying that nature is governed by a single law; and as long as its laws do not appear to us to be derived from one another or all to be derived from a higher law through a purely logical necessity, indeed, as long as we may conceive them, on the contrary, as having been decreed separately and in an infinite number of ways, all being incompatible with the production of harmonic effects like those we observe, we shall have a basis for seeing in the effect produced the reason for a harmony which the interdependence of concurrent laws or their logical dependence on a supreme law does not take into account. This is the idea we find expressed by the term "final cause." It follows from this that the more we can reduce the number of general laws and independent facts through the development of our positive knowledge, the more we will reduce proportionately the number of fundamental harmonies and distinct applications of the principle of finality. But also, as each fundamental harmony, taken by itself, acquires value and probative force as evidence of the finality of causes and of an intelligent co-ordination, it will become necessary for us to judge the perfection of a system by the simplicity of its principles and by the fruitfulness of its implications. Thus, if we deem it advisable to come back to a single principle which would explain everything, this unique principle, or primordial decree, would be the highest expression of knowledge as well as of supreme power.

Moreover it must be clearly understood that the considerations with which this chapter has dealt do not lead us to such a height. We have in mind only the philosophic interpretation of natural phenomena, in the light of science and of reason, in so far as it does not take us beyond the limits of secondary causes and of observable facts. We have inquired neither how it is possible to have a supreme direction among details which themselves are left to the operation of fortuitous causes, nor how, in a supernatural order toward which it is also natural for man to tend because of religious feelings, chance may be the minister of Providence and the executor of its mysterious secrets with respect to particular facts (36). Still less do we have the temerity to ask what the final end of creation may be. The finality which we find ourselves unable to disregard in the works of nature is, so to speak, an immediate and special finality, a chain of which we are able to follow only scattered fragments. A certain organism is admirably adapted to the performance of a certain function, and the operation of the function is equally well adapted to the needs of the individual and to the preservation of the species; but what end did nature have in view in the creating and propagating of this species? This is what is not indicated to us and what we are unable to attempt to define without making gratuitous assumptions which sometimes are ridiculous and are always unworthy of a careful thinker. So the field of our knowledge is limited in regard to what we would have to know in order to throw out conjectures as to the general disposition of the world without being guilty of a shocking presumption.

68. In general, the mind does not decide between the two explanations referred to—one, the finality of causes; the other, the exhausting of fortuitous combinations—on the basis of a rigorous proof and a formal demonstration. Although common sense is offended, no rule of logic is violated if we attribute to a providential arrangement the most insignificant relation and the one which may most easily be thought of as resulting from fortuitous combinations. This is also true if, conversely, and giving ourselves free rein, we attribute to the operation of fortuitous combinations, by a harmonious concurrence of innumerable circumstances, the most marvelous result in which the intelligent adaptation of means to end shines forth with the greatest splendor. However justified the mind may be in its preference for one solution or the other, depending upon the case at hand, it will run into a sophistical contradiction. This is not a transitory contradiction of the sort found in all scientific truths until they have been definitely established and accepted by science, but a permanent contradiction arising from the fundamental inability of human reason to resolve it by means of a categorical demonstration in the absence of direct observation.

Does this mean that man should and can be indifferent to the choice of the solution to be given to these eternal problems; that he must renounce, in so far as his faculties enable him to do so, taking account of the principles of order and harmony found in the plan of the world, of the part which these diverse principles play, and of the manner in which one is subordinated to another? Shall we conceive a picture of nature in which these considerations would find no place, and in which we should be limited to describing plants, animals, rocks, and mountain ranges without having anything to say as to the relations among them, of the parts to the whole, and of the way in which reason understands these relations? At this point it becomes necessary to make a profound distinction between scientific knowledge based upon the observation of facts and the deduction of their consequences, and speculative philosophy which bears upon the inquiry into the reason of things. All the remaining portions of this book will have as their purpose to set forth again and again this important distinction between science and philosophy, by trying to portray the role of each and to show that neither may be sacrificed without bringing about an abasement of the human intelligence and the destruction of the harmonic unity of its faculties.

69. Now, as it is the nature of philosophic speculation to proceed by inferences and judgments of probability and not by deductions and categorical demonstrations, it must and does happen that probability has innumerable degrees; that reason is sometimes led irresistibly to see, here the consequence of a pre-established harmony, there the result of the indefinite multiplication of fortuitous combinations; while in other cases it remains undecided, tending to favor one explanation or the other depending upon inclinations which vary with intellectual habits, the state of enlightenment, and external impressions.

When it is seen that the sun, which is the center of the planetary movements which it dominates and regulates because of the enormous preponderance of its mass and because of the great distances nature originally placed between the planets, is also the source of the light that illuminates them and of the heat which makes it possible for life to develop on them, we cannot fail to recognize the admirable arrangement which brings together harmonically, in the production of these beautiful phenomena, such natural forces as gravitation, light, and so on. And even though these forces were to be considered as emanating from a single principle, they would be no less characterized, considered as secondary principles, by distinct laws having the same independence from each other as do streams which spring from a common source, but which, after separating, maintain their own course and their own rate of flow and accommodate themselves to the accidental features of the terrain through which they pass (52). On the other hand, although it is permissible for a pensive and poetic imagination to think of the moon as having been created solely to illuminate our nights with its soft light, a more exact mind, aware of the accidental and irregular features in the distribution of satellites among the major planets of our system, could never be brought to invoke the principle of finality to account for a harmony the importance of which is subordinate, and which even only partially fulfills the end that one may wish to assign to it. Still less would anyone whose mind is enlightened by the progress of geological studies admit that, if prehistoric upheavals of the earth's crust have buried masses of incompletely decayed vegetation, this was done so that man might later find in them combustible material which would be needed for the development of his industry, although some persons have ventured to suggest this. The interval between the extreme cases we have used as examples may be filled up with whatever number of degrees may be desired.

70. As for the principle of a final *consensus* through mutual influences or reactions, when the progress of our scientific knowledge makes it possible for us to explain some particular harmony in this manner, this explanation is definitely taken over by science and no dialectic subtlety can weaken it. No doubt the number of particular cases explained in this way is small, but a few examples are sufficient to show that the application of the principle does not exceed absolutely the intellectual powers of man, and that the field of applications may be extended to the degree that our positive knowledge is perfected and extended. If the application of the principle in question requires—as would appear to be the ordinary case—that, up to a certain point, the initial disposition of the parts had had to be related to the final conditions of harmony, it would still be necessary to fall back upon one or the other of the two principles which enable us to account for this original condition.

And for that reason, we shall once more fall into the inevitable ambiguity indicated above.   The remaining part of the explanation, based upon the mutual reactions of various parts of a more or less unified system, retains all the certainty of a scientific demonstration.   Further, there will be some leeway in the basic assumptions as to the initial state—as will be seen by a discussion appropriate to each particular case.   Moreover, there will be reasons for refraining from an appeal to final causes or to the exhausting of an immense number of fortuitous combinations, in order to account completely for the harmony observed in the final state.

71.   In closing, let us comment briefly on the use of the principle of finality as a guiding thread in scientific research.   This usage may consist only of the application of the common adage: " Whoever wills the end, wills the means."   When the end, that is to say, the result, is a given and incontestable fact, it is necessary, for logical reasons, to admit the means, that is to say, the uniting of the circumstances without which this result would not have occurred. This condition imposes a direction on experimental investigations, until we have verified through direct and positively established observation that of which the necessary existence had earlier been established by rational means.   Thus, we are justified in deducing, from our knowledge of the habits of flesh-eating animals, the necessary presence of claws suitable for seizing and tearing their prey, and a type of digestive system suited to a carnivorous way of life, and so on.   In this way (50), it is possible to *reconstruct* extinct species, at least so far as their most basic traits are concerned, from nothing more than fragmentary fossil remains.   Such great advances have been made in this sort of work that a very profound knowledge of the harmonic adaptation of parts in animals has been acquired.   Such work does not imply the complete solution of the philosophical problem as to the origin of and the reasons for the observed harmonies, and does not require us to have taken sides in favor of one or the other of the three principles of explanation among which it is necessary to choose in order to account for them. It is simply a question of concluding logically from a certain fact the conditions without which that fact could not have occurred. In this operation, the mind proceeds with all the assurance and all the demonstrative rigor which pertains to logical deductions.

But still another road is open to the mind.   This consists in its being guided by the presentiment of a perfection and harmony in the works of nature which is far superior to that which our weak intelligence has so far been able to discover. Although this presentiment is not infallible, because the point of view from which we must judge the works of nature allows us only a restricted horizon and because the greatest perfection in regard to details is not always compatible with the simplicity of the design and the generality of the laws, yet it will most often happen, principally when observation has to do with living forms, that the observer, by yielding to this feeling and directing his investigations in accordance with it, will, for that very reason, find himself on the road to discoveries. This indefinable premonition, which must be highly valued even though it does not have the certainty of a logical rule, is like that by which the mathematician is put on the trail of his theorems and the physicist on the track of a physical law, to the degree that it appears to them that the suggested theorem or the law satisfies the conditions of generality and simplicity and symmetry which contribute to the perfection of order in all things and which long practice in the sciences has made familiar to them.

72. The considerations upon which we have entered are pertinent here not simply because the idea of a harmonic order in nature is essentially correlative to the notion of chance and of the independence of causes, and for that reason connected with the theory of philosophical probability, but still more because they have an evident and direct influence upon the judgments we make concerning the reality of our knowledge and the objective value of our ideas in general. In fact, is it not clear that, if there is such a harmony in all the details of creation, particularly in the structure of living things, there must also be a harmonious relation between the system of external causes which act upon us in such a way as to give us knowledge and ideas, and the system of knowledge and ideas which results from it? Ought not that which is particular, accidental, and abnormal in the impressions received and in the results produced as we move from one individual to another, or from one phase of the existence of the same individual to another, be obliterated and disappear, so that finally there will be agreement between the fundamental notions or the rules of the intellect, and the fundamental laws or general phenomena of the external world? On the other hand, if such a *consensus* must necessarily

be finally established, is it not clear that it results from the influence of external causes upon the production of ideas, and not from the influence of our ideas on the constitution of the external world? Those strange metaphysical systems which draw the external world, or at least the order we observe in it, out of the very order of our ideas, are, in the last analysis, only the extreme exaggeration of the error into which one falls in improper applications of one or the other of the two principles of interdependence and finality (65 and 66), when, instead of conceiving that particular facts have adapted themselves or have been adapted to general and dominant facts, one imagines, on the contrary, an adjustment of general and dominant facts in view of or through the influence of particular and subordinate facts.

73. What is true of the adaptation of parts between the intellectual make-up of a rational being and the constitution of the external world is also true of all the other harmonies in nature. We may assume that it does not exceed the power inherent in the influences and reactions of one system on another, just as we may also believe that it would be inexplicable without a pre-established harmony; and finally, the third explanation, through the exhaustion of fortuitous combinations, offers itself here as elsewhere, at least on the basis of a scholastic quibble. But, however we conceive the reason for a certain harmonious adaptation, it is evident that it occurs necessarily only to the degree that it is required for the direction of intelligent beings in their relations with the external world. This is the true basis of the distinction posed by Kant between speculative and practical reason; for it would be contradictory to hold that the ideas of an intelligent being had not been harmoniously related to its needs and to the acts it intends to carry out as a consequence of its ideas and its needs, just as it would be contradictory to hold that an animal whose stomach and intestines are adapted to the digestion of living prey had not received from nature means with which to catch it. If one goes beyond the circle of needs and acts of intelligent beings, which depend entirely upon their relations with the external world, so that one may indulge in speculations concerning what things are in themselves, independently of their relation to intelligent beings, it is incontestable that nothing can be concluded from the action of the general principles which direct the harmony of creation,

any more than Descartes was justified in appealing, in a similar instance, to the principle of the veracity of God.   For, if it is evident that God could not have deceived us in regard to the rules He has imposed on our intelligence for the direction of our actions, what right have we to assume that He must have given us infallible rules for reaching absolute truths, knowledge of which has nothing to do with our being able to work out the destinies that He has laid out for us?   Therefore, it is necessary to resort to other principles for the critical discussion of the value of our ideas where speculative rather than practical questions are involved.   These are the principles that we are going to attempt to indicate, while apologizing for the dryness of the technical explanations.   The question is worth going into, however, whether we believe in the possibility of a solution, or whether we have in view only the comparison of the systems which have occupied the human mind for so long a time.

CHAPTER VI

ON THE APPLICATION OF PROBABILITY TO
THE CRITIQUE OF THE SOURCES
OF OUR KNOWLEDGE

74. In the three preceding chapters we have dealt with probable judgments based upon the state of our knowledge. We have taken it for granted that no one contests the validity of this knowledge and that the discussion refers exclusively to the value of the consequences that may be drawn from it. For example, suppose the question is raised as to the probability that the planets are inhabited. The existence of space and of bodies, and of the planets in particular, will be admitted as undeniable. What astronomers tell us about the dimensions, form, distances, and movements of these bodies will be accepted as beyond doubt. With reference to probabilities, we have intended to discuss only those analogies and inferences which, following the acquisition of knowledge that is regarded as certain, lead us to believe that the planets are inhabited. Now, on the contrary, the problem is to apply to the critical examination of the sources of human knowledge the fundamental ideas of the reason of things, of order, and of chance—that is to say, of the interconnection and independence of causes—and the consequences that can be deduced from them as to the nature of probabilities and of probable judgments. This is the principal object of our inquiries throughout the whole of this book.

All the faculties by means of which we acquire our knowledge are or appear to be subject to error. The senses suffer from illusions, memory is capricious, attention wavers, and mistakes in reasoning and calculation escape us time and again. Therefore, we rightly distrust ourselves and regard as established truths only those which have been examined and are accepted by a large number of competent judges who have reached their conclusions under different sorts of conditions. Skeptics have availed themselves of this rule of good sense in order to deny the possibility of distinguishing the true from the false; other philosophers have concluded on the basis of it that our knowledge, although never absolutely certain, may attain probabilities which come closer and closer to certainty; still others regard unanimous or nearly unanimous assent as the unique and solid basis for certainty. These attitudes are found in the philosophy of every age.

Let us admit that each of the faculties through which we get our knowledge may be likened to a fallible judge or witness. A superior intelligence which would understand the entire scope of these faculties, which could penetrate into the mysterious devices of memory, for example, would be able to determine exactly the chance of error in the operation of each function and in the use of each faculty, for each individual and under certain determined circumstances. It is possible that such an intelligence would be able to recognize that error becomes physically impossible for certain individuals under certain circumstances; for, in the last analysis, we have no basis for affirming absolutely that there is any operation of reason, however simple it may be, which does not carry with it the possibility of error.

An intelligence incapable of drawing such conclusions *a priori,* but possessing an infallible criterion for distinguishing cases in which one of our faculties deceives us from those instances in which it furnishes us with exact information, would be able to determine experimentally from this data itself (38) the chances of error inherent in the use of this faculty, provided it could carry out sufficiently numerous series of experiments and properly set up the conditions under which the experience takes place.

75. Were such an intelligence not in possession of an infallible criterion, observation would lead it to determine numerically the chances of error, which would be unknown *a priori,* provided it is assumed that the chance of truth is greater than the chance of error. This condition must be admitted if we are to admit that the end and the consequences of man's intellectual faculties, when they are functioning normally, is to instruct him rather than to deceive him. If this is the case, erroneous perceptions and judgments may be thought of as the result of accidental disturbances of the faculties and functions. This statement rests upon a theory that cannot be properly developed without the help of mathematics, but the principles of which we wish at least to indicate here, since we must not neglect anything which is so clearly connected with our problem.

In order to fix these ideas through the use of an example, let us suppose that an observer who makes a practice of regularly observing the condition of the sky may be in the habit of forecasting at sunset each day the time at which the sun will set on the following day. If he has kept a record of his forecasts over a sufficiently long period of time, the relation between the number of forecasts contradicted by events and the total number of forecasts would give, without appreciable error and in a purely experimental way, the determination of the chance of error which affects the judgments of the observer under the conditions indicated. There would be no limit to the precision of this experimental determination provided the experiment could be prolonged indefinitely, and if, moreover, the observer were neither to gain nor to lose in perspicacity in the course of his observations. This latter assumption is made for purposes of greater simplicity. If, after an initial series of observations had determined the chance of error with sufficient precision, another series of experiments was begun, always under the same conditions, approximately the same relation would be found to hold between the number of forecasts contradicted by the events and the total number of forecasts. In each series of experiments, the great number of observations made would appreciably tend to compensate for the effects of causes which vary irregularly from one experiment to another. Thus, only the effects of regular and permanent causes or of causes which control the whole series of tests would remain.

76. Now, let us imagine that two observers, who work independently of one another, make their observations simultaneously and record their results. The chance of error will be very different for each of the two, but—so that we may argue from the simplest case —let us suppose that it happens to be the same. Finally, let us also assume that the causes which influence the truth or error of the judgment of one of the observers are completely independent of those which affect the truth or error of the judgment of the other, resting as they do in the moral and physical dispositions found accidentally in each man. A mathematical connection will be found to exist between the number which measures the chance of error for each observer and the relation of the number of cases in which they agree, to the number of cases in which they give contrary judgments. If, for example, each observer is mistaken

once in every five times, or if the chance of error is one-fifth, the two men will agree in their forecasts seventeen out of every twenty-five times; and the tabulation of the records will show approximately this relation of seventeen to twenty-five provided that a sufficiently large number of instances is dealt with, so that the fortuitous irregularities cancel one another out. In this way it will be possible to pass from the first relation to the second, or inversely, by means of a mathematical formula.

However, it is easy to see that in passing from the second value to the first, an ambiguity is found which is not present in the direct passage from the first to the second. If it happens that the two observers agree seventeen times out of twenty-five, when both are mistaken once in five times, it is evident that they must still be in agreement seventeen out of twenty-five times when they are both mistaken in four out of five instances; that is, when the fraction one-fifth represents not the coefficient of error but that of truth. The extreme case in which they would always be in agreement, without in any way communicating with each other, would clearly indicate that what each says is always true or always false. This ambiguity inherent in the nature of the problem must be, and actually is, found in the mathematical formula. But if there are, *a priori*, sufficient reasons for believing that the chances of truth outweigh those of error, that fact alone will remove the ambiguity. For example, if the mathematical formula gives these two possibilities: a chance of error of *one-fifth*, and of truth, *four-fifths*; or a chance of error of *four-fifths*, and of truth *one-fifth*; only the first possibility is admissible, and the second will be rejected.

Thus it is that we conceive the possibility of determining a chance of error empirically, not only through direct observation, as in cases in which we possess a criterion of truth—such as that which would result, in our example, from the comparison of the forecasts with the subsequent events—but indirectly and by means of relations furnished by mathematical calculations whenever a similar criterion does not exist. Thus, when a physician prescribes a treatment for his patient, an infallible criterion as to the truth or error of the doctor's judgment cannot be drawn from the evidence of a single instance; for it is possible that the patient may die even though the prescription may be the best possible under the circumstances; or, on the contrary, the patient may recover in spite of mistakes in the treatment. Let us assume, now, that two physicians are called in separately for consultations on an extensive series of pathological cases. There will be no way of determining directly the chance of an error in the judgment of each of them. But the record of the consultations will show how many times the two doctors have been in agreement and how many times they have come to contradictory conclusions. On the basis of the explanations given above, we are able to conceive how we could proceed to determine these chances indirectly and without ambiguity, if we were justified in believing, as no doubt we are, that the professional study of medicine, without resulting in infallibility, tends more toward truth than toward error, and that, in general, it is better to consult a physician than to resort to throwing dice in case of illness.

77. In inquiries where calculation is possible, and even in all cases where rigorous logic may be applied, it is necessary to begin with hypothetical and abstract cases which serve to lead gradually to other cases which are more complex and more nearly approach applications to real conditions. This has been the method of procedure in the present discussion. Actually the chances of error vary from one person to another, and, in general, even from one judgment to another by the same person. When several people make a judgment about the same fact, the causes of error which operate in the case of one person are not completely independent of those which operate in the case of someone else. If one is dealing with a long series of judgments such as the decisions of a court of law, the theory whose basis we have just indicated may still be applied after all inexact hypotheses have been amended, provided we have a sufficient quantity of experimental data. Then the numerical values found through calculation would no longer indicate the chances of error for a certain person and for a certain type of case, but would reflect the average of all the values the chance of error may have for a great many persons and for a great many kinds of cases. In this way a truly exact theory can be reached about the average and general results of certain judicial institutions. These are the results which engross the legislator and interest persons concerned with the science of social organization. Were this not so, nothing could be concluded from the application of the

principles to any particular case, as many people have believed and still do believe.

No doubt it would be interesting and useful, for the progress of the body of knowledge concerning our intellectual nature, to have a table giving the mean values of the chance of error in perceptions or judgments in fields other than that of legal decisions, just as it has been useful in getting knowledge of man's physical nature to have tables of mortality, of average sizes, of weight, of muscular force, and so on, for different ages and different countries. Furthermore, the theory of average chances must not be neglected even when we see no way of obtaining the statistics which would make the theory applicable; for, in the first place, the theory may stimulate experimentation, just as experimentation often corrects theories; and, in addition, it is good, as Leibniz has said, to have methods for everything that may be discovered through reason even though circumstances may in fact impede the application of the method. But at the same time it must be recognized that the problem that concerns us throughout this work is to evaluate, in each particular case, the motives which lead us either to give, to deny, or to suspend our assent. Now, in this connection the theory of mathematical probability, when properly understood, would not often be of any help; when wrongly understood, it would lead to false conclusions.

78. Taking another example, let us suppose that experience has clearly shown that once in twenty times each of two persons, A and B, is liable to make a mistake in a given type of numerical calculation. From this it will not follow that when B has carefully checked A's figures and found them to be correct, the probability of simultaneous error is one to four hundred times, as could be concluded by comparison with the probability of drawing a black ball twice in a row from a jar containing nineteen times as many white as black balls. In fact, there is reason to believe that, just because it is B's intention to check a result already obtained, his attention is increased, and he is more on guard against the chances of error. Even though B were working in ignorance of the result obtained by A and had no intention of checking it, it would be very extraordinary that, of all the possible mistakes, he should fail to catch exactly those which A had not found, or that he should overlook another mistake which would affect the same figure of the

final result in exactly the same way. Consequently, if the results found by the two calculators were to agree exactly, the probability of error in the common result, based on these notions of combinations and chances, would be much less than one in four hundred. The calculation of this probability would be a very complicated problem. Its solution would depend upon the form of the numerical calculations which had led to the two concordant results, upon the number of figures used, and similar factors. On the contrary, if the errors in calculation are the result of mistakes in method common to the two men or to a mistake in the tables which they use, the probability of a common error in the two results will be much greater than one in four hundred. In other words, more than once in four hundred times the two men would obtain results which, although they agree, would none the less be false.

79. Now let us admit that the result found by the two calculators satisfies some simple law, suggested by a theory already verified for other cases, the confirmation of which is being sought. Under these conditions everyone will agree that it is extremely improbable, or even impossible, that a fortuitous error in calculation would give exactly the result needed to make the result square with the theoretical law. No one will doubt the correctness of the result obtained or ask whether the two calculators were liable to be mistaken once in twenty or once in a hundred times. We have taken for our example a mathematical calculation, that is to say, an intellectual operation of the most mechanical type. But it is clear that a similar discussion could be given with respect to all acts of the mind which tend to make anything known. This is true even though the evaluation of the chances of error, whether *a priori* or *a posteriori,* appears inescapably to involve difficulties which are less easily overcome since they deal with more complex operations than those to which we have been referring, or with those which bring into play the less well-known motivations of our intellectual nature.

The greatest mathematicians have fallen into error, and some propositions accepted as true, even in pure mathematics, have later been rejected as false or inexact. However, it would be very extraordinary, and for that reason alone improbable, that so many geometricians, throughout a period of twenty centuries, would have been mistaken in finding the demonstration of the Pythagorean theorem, as it is given in Euclid, irreproachable. But, when we remember that this theorem can be demonstrated in different ways and that it is co-ordinated with a whole system of perfectly connected propositions, we shall be completely convinced not only that the demonstration conforms to the regulative laws of human thought, but, what is more, that this theorem belongs to an order of truths which subsist independently of the faculties which reveal them to us and of the laws to which the functioning of these faculties are submitted.

80. Similar remarks may be applied to the credibility of witnesses. I have a friend in London, and he informs me that a serious event has happened in that city: a great fire has caused enormous loss and utterly destroyed a section of the city. In telling about this event, my friend adds certain particular details to his account. Soon after I receive this word, one of my friends in Paris, who also has a correspondent in London, shows me a letter in which the same facts are reported together with the same particulars. More to the point, I know that his correspondent and mine do not know one another, have no relation whatever with each other, and consequently could not possibly have collaborated with the intention of deceiving us. In such a case, I should not think I had been misinformed even though both writers may be subject to some sort of hallucinatory experiences once in ten times or once in a thousand times, or may be inclined to fool their friends with malicious jokes. For how could such a bizarre caprice have struck them both on precisely the same day? And, even granting that this might have happened to them, how without any collaboration could the fantasies of their imaginations happen to have contrived the same story, even to particular details? Doubtless such a thing is not mathematically impossible; but it would involve such a prodigious chance that reason could not be led to admit such an explanation when so natural a one is available to it, namely, the actual occurrence of the event described. However, I should suspend my judgment as to certain details in spite of the similarity between the two letters. For everyone knows that during the experience of a great disaster the mind is inclined to exaggerate both to itself and in its accounts to others the extent and the sequence of what has happened. Men love the marvelous and the supernatural. This results in a deviation from the truth, which

has affected, or at least may have affected, the two correspondents in the same way without collaboration.   Ten letters or a hundred letters, received on the same day from different persons who had not compared their stories, would still leave me very suspicious of exaggeration in regard to certain details.   I should wait until the imagination had had time to be quieted and until inquiries of a sufficiently exacting character had been started, before accepting such details as accurate.

In general, if many witnesses are unanimous in reporting an isolated fact, if we know there has been no possible collusion between them, if they have not been under the influence of, and, as it were, in an atmosphere in which the same causes of error or deceit might influence them; if, on the contrary, no possible connection exists between the causes capable of impairing the testimony of each of them independently, then the mathematical theory of chances would be a sufficient reason for our rejecting as extremely unlikely the supposition that they were all mistaken, or that they all wished to deceive us.   But, if the event witnessed is complex, if all the circumstances are completely consistent with one another and with other facts regarded as certain, then another judgment of probability, based upon the idea of order and upon our need of taking account of the rational connection of events, will put the event in question beyond doubt, even though the number of witnesses is not large, or even though they have been exposed to manifestly related causes of error.

This sort of judgment applies even more particularly to historical evidence.   We firmly believe that a person named *Augustus* once lived, not only because of the great number of original writers who have mentioned him, and whose testimonies concerning the principal facts of his life agree among themselves and with the testimony of historical monuments.   Our belief is based even more, and, indeed, principally, on the fact that Augustus is not an isolated person, and that his life makes reasonable a large number of contemporary and later events which would be without foundation and unconnected if so important a link in the chain of historical data were omitted.

To imagine that certain peculiar persons may take pleasure in doubting the Pythagorean theorem and the existence of Augustus in no way weakens our conviction concerning them.   We should not hesitate to conclude from this fact that the ideas of such persons have been deranged in several respects and that they are at variance with the normal conditions which must exist in our faculties if they are to fulfill their functions.

It is for this reason that our belief in certain truths rests neither upon the repetition of identical judgments nor upon unanimous or nearly unanimous assent. It rests primarily upon the perception of a rational order according to which these truths are connected with one another, and upon the persuasion that the causes of error are anomalous causes, affecting each perceiving subject differently, from which it would not be possible for such a co-ordination among the perceived objects to result. In a word, the *critique* of our judgments and perceptions, and of the judgments, the sensations, and the reports of our fellow men, rests principally, and it may even be said essentially, on probabilities of the sort that we have called " philosophic." Actually this is the way in which this critique always functions, whether in solitary meditation, in oral discussion, or in books. Sometimes this critique passes unnoticed, so startling and indisputable are the conclusions to which it must lead. In a great many cases it leads us to probabilities whose values cannot be fixed by numbers or by precise symbols of any sort, which do not strike all minds in the same way, and hence only give rise to endless controversies.

81. Must we stop here? May we not, and should we not still undertake to criticize our faculties, our ideas, and our judgments in this way, when they are considered no longer from the point of view of individuals but from the point of view of the species, and also when general rules and ideas and not simply particular objects and facts are being considered? The grounds for making a decision are the same. The use of our senses, and in general of all the faculties by means of which our knowledge is elaborated and perfected, is guided and controlled by that superior and regulative faculty for which we reserve pre-eminently the name " reason " (17). This is the faculty which grasps the order and reason of things by moving from phenomena to laws, from consequences to principles, and from appearances to reality. Furthermore, this is also the faculty that must tell us whether the opinions and the ideas which we get from the use of all our other faculties, after we have offset all fortuitous causes of illusion and have made corrections for all

individual and accidental anomalies, are true only for men and adapted to the nature, circumstances, and laws characteristic of our species, or whether, on the contrary, these faculties have been given to man so that he may attain in some measure real knowledge of what things are intrinsically and independently of our contact with them.[1]

A man might be limited to seeing things only through a prism or a lens which would alter all visual angles, distort all shapes, and change all relations of size and position. Such a person would be unable to make out any of the laws which govern the material world. He would find only confusion and disorder among phenomena which impress us because of their simplicity and harmony. At least this would be the case unless, through the assistance of other senses, or even through reasoned discussions of experiments performed under favorable circumstances on the sense of sight itself, he could distinguish what part of his perceptions is due either to the form of the device or to the organ through the intermediation of which visual rays reach him.

This hypothesis is not purely a figment of the imagination. We actually do observe the stars through such a medium as has been indicated, namely, the terrestrial atmosphere, which deflects rays of light unequally in proportion to the distance of the star from the zenith and does so in such a way as to change the zenith distances, to alter the apparent distances among the stars themselves, and to disturb the shapes of the groups in which we arrange them. Because of this disturbing cause, which is called " astronomical refraction," the phenomena of diurnal movement appear to lose their harmonious simplicity. The stars no longer describe, with uniform motion, perfect circles around the axis of the earth. But even though we were unable, with our knowledge of the composition of the atmosphere and of the manner in which light is propagated, to assign the physical cause of this illusion and to calculate the effects of atmospheric refraction, we should not hesitate to recognize that the irregularities of the diurnal movements of the stars are purely apparent and due to optical illusions, the real cause

---

[1] " It is a great defect in the senses that they establish the boundaries of things from the point of view of man and not according to the measure of the universe; this can be corrected only by reason and universal philosophy." Bacon, *Novum Organum*, Bk. III, Chap. 40.

of which is the medium in which we are immersed. To become convinced of this, it would be sufficient to observe that these irregularities are more or less noticeable, depending on the condition of the atmosphere, and that they result in greater errors as a star moves toward the horizon, so that at just the moment when it acquires its maximum size from our position, it becomes smaller or disappears for an observer some distance away. Finally, even though this last decisive experiment could not be made, or even though it were impossible for us to compare observations of the same star simultaneously from positions very distant from one another, it would be sufficient for us to notice that our horizon has only an accidental relation to the axis of the diurnal movement, and that the direction of our horizon depends upon our position on the surface of the earth—a circumstance which has nothing to do with the motion of the stars. That would be sufficient, we maintain, to enable us to conclude, with that high probability which compels the acquiescence of reason, that those irregularities, the size of which depends upon the height of stars above the horizon, are related to the circumstances of our observations and not to the nature of the stars; they are only apparent and do not affect the real motions.

82. One cause of optical illusions, comparable to that which results from the ocean of air in which we are immersed, the nature and effects of which astronomers know how to measure so well, may be found—as Bacon suspected in passing—in the composition of the human eye itself, in the structure of the media and parts that go to make it up. After all, it would be difficult if not impossible to verify this directly by means of the sensibility of the retina and of the nerve fibers which connect it with the cerebral nerve center.[2]   If this internal atmosphere—to use a figurative expression—really exists, and even if we had only a few reasons for suspecting its existence, it would also be necessary to doubt the legitimacy of the laws of diurnal motion and to assume hypothetically that the laws which really control the phenomena are complicated by the laws according to which our vision takes place. The whole structure of the astronomical sciences, resting as it does on the laws of diurnal movement, would be shaken to its very

---

[2] For example, the optical illusion, known among astronomers and physicists as " halation " appears to be related to the form of the sensibility of the retina.

foundation. But this would occur to no one, and least of all to an astronomer. The beautiful simplicity of the observed laws is a sufficient guarantee of the absence of any internal cause which would complicate the laws without our being aware of it. It is repugnant to reason to admit that a defect in the structure of the human eye, far from disturbing the order and regularity of external phenomena, would introduce an order, regularity, and simplicity among them that would not be found in them otherwise or would be found in them only in a lesser degree of perfection. Thus we are firmly convinced that observation does not lead us into error; that we really locate the stars in their true *optical positions*, after we have taken account of the deviation caused by the intervention of the atmosphere and by other irregularities arising from the movements of the earth, irregularities which are themselves reducible to regular laws the theory of which may be discovered. The very small anomalies that may still be found in these corrected observations are reasonably accounted for (44) as errors inherent in every operation of measurement made with the senses and with instruments of limited perfection. Should they not compensate for one another with an approximation which becomes proportionately greater as a larger number of observations that they affect is accumulated, they will indicate the existence of a constant cause of error or defect either in the instruments used, or in the sense organs themselves, or in the habits of the observer. A constant tendency to *overestimate* or to *underestimate* slightly, either in the operation of measuring angular quantities itself, or in the process of reading the scale of an instrument, is an example of such a cause. Finally, if the anomalies to which we are referring do not become appreciably less noticeable when compensations occur among measurements made by a large number of observers working under a variety of circumstances, they will actually indicate a constant cause of error, and therefore an imperfection which follows from the nature of human beings themselves. Such an imperfection should not be surprising, since, in general, while meeting the conditions of harmony required for the maintenance of its plan and for the conservation of its work, nature does not seem to impose upon itself the necessity of satisfying them with a mathematical precision. Quite to the contrary, nature seems to have a constant tendency to admit tolerances and deviations of which, moreover, reason takes account—as we have seen in the preceding chapter—by the very explanations that can be given of the order and harmony of the world.

83. The comparison that we have drawn from physics can be duplicated in a slightly different form which has particular advantages. Therefore, let us suppose that, instead of seeing objects directly, we have a mirror in front of us which reflects their images to us. It is by means of such a mirror that Herschel probed the depths of the starry sky, and there are worlds into the heart of which the eye of man has penetrated only in this manner. But Herschel understood perfectly the construction of his telescope, since he himself had invented it. On the other hand, it is possible to imagine a mirror in front of which nature has placed us without apprising us of the fact and without informing us directly of the form it has been pleased to give it. However, if the mirror were curved, the deformity of its images would produce the same effect that would be produced by the introduction of the prism or lens mentioned above. By distorting all the appearances and introducing an obstacle to the arrangement of phenomena according to a simple and regular order, we should be led to suspect the existence of a disturbing cause which affects, not the objects we perceive, but our instruments or organs of perception, and consequently our perceptions themselves and all the ideas connected with them. But, were the mirror a plane one, the order in which all phenomena would be connected with one another would suffice to justify us in concluding that we are placed in conditions favorable for seeing external objects as they are, whether we have a direct intuition of them, or whether they are revealed to us only through the mediation of certain images which, though perhaps weak, are nevertheless faithful in the sense that they retain fully the principal forms and the characteristic traits of the original.

Nevertheless, even in the case of a plane mirror, there would be a very fundamental difference in form between the objects and their images. This difference would be similar to that found between the right hand and the left, or to the difference which the study of anatomy discloses between the arrangement of the internal organs of most men, in whom the winding course of the intestines occurs uniformly, and that found in some persons, in whom the arrangement of the intestines is the same except that it is anomalously inverted. The same sort of inversion would affect at once the movements of celestial bodies, the action of electric currents on magnets, the action of crystals on light, and a multitude of general or particular traits of the world we know. But, for the very reason that it would affect simultaneously the whole and all its details, it would not disturb in any way either the regularity of the whole or the harmony of the parts; it would not make things any more complicated. In such a case we should have no reason for preferring one or the other of two such perfect symmetries and could place no confidence in any inference leading us to believe or not to believe in the hypothesis of a reflection, or of an odd number of reflections, from which the inversion of the geometrical relations would result. If such a question were ever to cease to be problematical, it would be necessary for observations of a different nature, based upon other properties of light, to show us other characteristics by means of which we could distinguish direct rays from reflected rays and those which had been reflected only a few times from those which had been reflected more often. The progress of the science of optics has actually put us in possession of a new criterion for making this distinction. But the recent acquisition of this new criterion serves better to show the insufficiency of another criterion for distinguishing the image from the real object, although this latter criterion suffices even now for deciding that we have before us, if not a real object, an exact image, not a phantasm.

84. To return to our first example, it is for this reason that, after having disentangled the observation of the diurnal movement of the stars from the cause of the trouble and complication which results from the intervention of the atmosphere, we do not doubt that we refer the stars to their true optical positions. We certainly do not believe there is any flaw in the formation of the eye or in the structure of the *sensorium* which falsifies all measurements of angular distances to such a degree that the simplicity of the laws of diurnal motion would be nothing more than the product of a fantastic illusion. Yet, on the other hand, the phenomenon of the diurnal rotation of the celestial sphere maintains the same characteristics of regularity and of mathematical simplicity whether it is explained by the rotation of the entire system of stars, or by a rotation in the contrary direction impressed

on all terrestial bodies. This leads to an ambiguity like the one we mentioned above; to solve it we should need the help of new knowledge about the physical structure of heavenly bodies, knowledge which would furnish reason with other analogies and other inferences. By means of this new knowledge, not only would the question in regard to the direction of motion be solved, but we should also become certain that the scheme of the optical positions of the stars, or what we call the celestial sphere, is only a phenomenon (87), an image *sui generis*, so different from the real object that the image of it would still shine for us for years after the object it indicates had ceased to exist. Nevertheless, although we may have been given the capacity to penetrate much further into the knowledge of the reality from which these phenomenal appearances emanate, it is always correct to say that our constitution does not falsify the phenomena in any way and does not keep us from knowing their real law or from having an exact idea of them entirely independent of the peculiarities of our own organic structure.

85. The senses are not always in the same condition, nor do they always function in the same manner. Nevertheless, and in spite of objections raised by ancient skepticism, there are a number of reasons why neither aberrations of the sensibility among a few individuals in certain unusual conditions, nor those aberrations which occur habitually and periodically in sleep, are sufficient to shake our faith in the ordinary testimony of the senses. This is so because the notions they give us of external objects when we are awake and when nothing interferes with their normal operation agree perfectly with one another. It is so, too, because impressions of different kinds, received through the various senses, may be perfectly connected, systematized, and co-ordinated by the hypothesis of external objects as the understanding conceives them. It is so, too, because memory establishes the identity of ideas which the senses have given us from the obscure period of earliest infancy during which their training is completed, in spite of the variety of painful and agreeable feelings that in each of us has accompanied the perception of the same external objects at different times in our lives. Furthermore, it is so because the identity in the perception of the same objects by all men whose faculties are unimpaired, although not capable of being demonstrated formally,

clearly manifests itself in our dealings with one another, whereas there is no regular connection between the dream of yesterday and that of tomorrow, nor between our dreams and those of other men. Finally, in spite of our meager knowledge concerning the fundamental principle of sensibility and of the operation of psychological functions, it is so because we know enough about them to make out that disturbances of the sensibility in sleep or in other circumstances of animal life result from the suspension or obliteration of certain faculties or from the weakening or injuring of certain organs. *Exceptio firmat regulam* [" the exception confirms the rule "].

Sometimes the senses expose us to illusions that could be called normal, because they are shared universally, and because, far from resulting from an accidental disturbance in the harmonious adaptation of the faculties to one another, they are the constant result of that adaptation itself. Such are the optical illusions as a result of which the sky assumes the appearance of a flattened vault, and the moon seems to us to be much larger when it is at the horizon than when it is near the zenith. Numerous explanations have been offered for these and many other illusions. But even if they remain unexplained, the concurrence of the other senses and the intervention of reason are not slow in rectifying the errors of judgment which may accompany them at first. When two faculties apparently contradict one another, our mind finds no difficulty in deciding between them; it recognizes the pre-eminence of one faculty over the other and does not hesitate to conceive the phenomenon in the only manner which affords a regular and systematic co-ordination of the sort which alone would satisfy the highest laws of reason.

86. Just as nature has constructed the eye for perceiving optical angles without altering them and optical configurations without distorting them, and has evidently done so in order to adapt them to the needs of beings endowed with a sense of sight, so it has fashioned the understanding not to co-ordinate impressions coming from external things according to a form of its own and foreign to objective reality, but to penetrate into this reality always to the degree necessary to fulfill the destiny of man.

Now although, in philosophizing, man cultivates faculties the rudiments of which he derives from nature, it is clear that nature did not create man to philosophize. If one wishes to think of it thus, this activity may be regarded as the destiny of some individuals, but it is surely not the destiny of the race. It is therefore perfectly clear that the acts in which man is most like animals suggest to him instinctively the fundamental perceptions or intuitions which he needs to be guided in the exercises of his animal functions, functions of which animals themselves appear to have at least a vague consciousness. It is also clear that, for the accomplishment of the acts which raise him above animality but which pertain to the fulfillment of the destiny of the species, man has natural beliefs [3] that can be called spontaneous, not because they suddenly appear in the mind, but because they far precede any philosophic or rational control. In this sense it is true to say with Pascal that "nature confounds the skeptics"; but the other member of the antithesis, "reason confounds the dogmatists," cannot be admitted as this austere genius admitted it. Reasoning, but not reason, confounds the dogmatists to the extent that it reduces them to the inability of demonstrating formally the theses of dogmatism. But reason properly speaking, that is, the sense of the reason of things, succeeds, as occasion arises, in legitimating certain natural and instinctive beliefs and in rejecting others among the prejudices or illusions of sense.[4] This sorting of the true and the false in the beliefs or intellectual preferences that we hold to by nature, this critique of the instruments by means of which we come to have knowledge of things, cannot, without contradiction, result from formal demonstrations of the sort used by mathematicians, as skeptics of all ages have pointed out. This sorting out or criticism never results in anything more than judgments based on probabilities. But these probabilities may, in certain cases, acquire such force that they irresistibly gain the acquiescence of reason, although they throw only an indecisive light on other parts of the field of speculation.

87. The system of critical philosophy here indicated is none other than the system of criticism followed in the sciences and

[3] " Nor did nature will that any one be completely deficient with respect to these things." Cicero, *De oratore*, Bk. III, Chap. 50.

[4] " I see all truths with an inner light, that is to say, with reason by means of which I judge the senses, their organs, and their objects." Bossuet, *De la connaissance de Dieu et de soi-même.*

in the conduct of life. We must content ourselves with the highest probabilities in the solution of the problems of philosophy, just as we must be content with them in physics, in history, and in practical affairs. And, just as there are things in history and physics which are beyond doubt although not logically demonstrated, the same may, and must, be true in the field of philosophic speculation. It is necessary to know how to recognize the gradual and continuous weakening of probabilities wherever they are found, whether in philosophy or elsewhere. The pretense of reducing everything to logical demonstration, and even the tendency to look for this sort of proof by preference, can only result in skepticism, as the experience of all ages attests and as the laws of human intelligence indicate *a priori*. The idea of proceeding in philosophy in the same way as the mind proceeds elsewhere is without doubt so simple an idea that no one would be able to view it either as an invention or as a reform. It was also a simple idea that extended to celestial bodies the laws of inertia and of gravity which control the motions of material things on the surface of the earth; but from this simple idea came all the great astronomical discoveries of the seventeenth century. Moreover, the idea that when we think we are guided entirely by probabilities of unequal force is not new. This was the opinion professed by the Greek school known as the "Third Academy"; Cicero was the best interpreter of this school among the Romans, and he remains the best for us. But among the ancients the notion of probability was always vague and confused. In modern times, when the progress of the exact sciences had resulted in the flowering of the theory of mathematical probability, at about the time when philosophy and the exact sciences were beginning to pull apart and each go its own way, it seems that this discovery itself prevented giving to the philosophical doctrine outlined by the Greeks the methodological rigor and the precision without overfussiness which characterizes the modern spirit. We must go much more deeply than we have into the fundamental idea of chance and the independence of causes; we must clearly distinguish the notion of philosophical probability from that of mathematical probability, as the mathematicians understand it; we must see what these ideas have in common and in what they differ, to the point of being essentially irreducible to one another.

It is especially necessary that we distinguish that subordination of our faculties which alone may lead to a control and to a solution of apparent difficulties. Unless this distinction is made, there will be no more philosophical discussion, properly so called; facts claimed to be primitive or irreducible will be multiplied indefinitely, and concerning them we shall appeal ceaselessly to common sense. That would be equivalent to the indefinite multiplication of occult qualities in physics, a procedure which excludes all theoretical organization.

88. There is a common prejudice among enlightened people that man, being able to make judgments only through the aid of his faculties, could not know how to criticize them. But if man has different faculties, and if they are arranged hierarchically and not simply connected, the plausibility of this *a priori* judgment vanishes at once. The explanation given so far, and which we shall continue to give in what follows, makes evident, or we hope will make evident, the fact of this hierarchical arrangement. The senses are only the instruments of reason. And just as man succeeds in assuring himself, by means of his senses, of the causes of error inherent in the instruments he has created, so he is able, under certain circumstances, to make sure of the causes of error which reside in the natural instruments which his reason uses.

Taking a new example, let us suppose that we are interested in measuring a certain magnitude, and that this magnitude must be estimated by sight without the aid of instruments, so that instrumental errors do not complicate those resulting from the imperfections of the senses. We are quite certain, prior to any experience, that such an estimate will be tainted with error, for the chances that anything depending upon the senses and our contact with the physical world would be mathematically exact are infinitely small. In such a case, what we must attempt to discover experimentally is the presence or absence of a constant cause of error, which, combined with other causes of error whose influence varies fortuitously and irregularly from one measurement to another, would tend to make all the measurements too large or too small in such a way as to give an appreciable error in a given direction with respect to the average result, after the effects of variable and fortuitous causes of error had largely compensated for one another. Now let us imagine that all measurements gotten

in this manner are arranged in a table according to their magnitude, either above or below the average depending upon whether
they are larger or smaller than it. If there is no constant cause
that is either organic or inherent in the nature of the mind, or
that results from the action of the environment which tends to
favor the production of a larger or smaller number of errors, the
particular excessive or deficient measurements will be found to be
symmetrically disturbed on one side or the other of the mean value,
from which the true value in such a case will probably not differ
appreciably.   As one gets farther from the average value, in either
direction, the particular values become less frequent and are separated by larger intervals from those which precede and follow them.
This is so because, as the hypothesis would lead us to expect, the
probability of a smaller error should be greater than that of a larger
one.   The particular values will be equally grouped or equally
spread out at the same distances from the average.   If, however,
such a symmetrical distribution is found in the table of particular
values, it will still not be proven, but it will be at least very probable, that in making the measurements in question the eye has
not been influenced by a constant cause of error, and that the
average does not differ appreciably from the true value that was
being measured.   If, on the contrary, the symmetrical distribution
to which we refer does not occur, we shall be certain that the
chances of errors in one direction outweigh those of errors in the
opposite direction, provided we work with sufficiently large numbers.   For example, we shall be sure that a constant cause favors
larger errors.   In this case it will be extremely improbable, if not
rigorously impossible, that the average will be found not to differ
sensibly from the true value.   A simple idea of the mind, a purely
rational conception, will have argued the truth or error of the
sensible perception and of the judgment of comparison or of measurement which follows from it.

89. It is said that man necessarily makes himself the center
of everything, that he necessarily relates all things to himself.
That this may be an instinctive tendency of his sentient nature is
not to be denied.   But the history of the sciences furnishes many
proofs that man has in reason the means of combating and sur-
mounting this tendency and of raising himself above pure *functions
of relation*, as the physiologists appropriately call them.   What is

more consistent with this instinctive tendency than to suppose that the earth is immovable and the center of the motion of the heavenly bodies?   Yet, by means of a series of analogies, inferences, and proofs which address themselves to reason and not to the senses, man has been led to give up this prejudice.   And he has done so in spite of many other obstacles which tend to go contrary to the judgment of his reason.

Reason and science have led naturalists to quite different results.   The gradation that they establish in the series of animal species which people the earth puts man at the head of the series and reduces the other species to a lower rank in proportion as they become increasingly unlike him with respect to their characteristics taken as a whole or with respect to those features which the whole body of our observations forces us to regard as basic and dominant.   Nevertheless, it is perfectly clear to all zoologists that this gradation must not be prejudiced by our position in it.   Such an order is not artificial, because it does not present any of the inconsistencies that an artificial order set up in accordance with man's accidental position in the series would inevitably present.   This is so because the progress and the concordant results of zoology, comparative anatomy, embryology, and paleontology have long since put this matter beyond all doubt.   This is being reaffirmed each day on the basis of new discoveries.

The discovery of the order of natural affinities which thus, through rational inferences, makes us certain of the pre-eminence of our species, has resulted from scientific investigations and from methodical and persevering work.   At first, stimulated by the unique instincts of his sentient nature, man in fact arranges the creatures of the earth in an artificial order based upon the services they render him and upon their advantage to him, or at least—if he wishes to free himself from matters that touch him personally—according to their size, their external form, the length of time it takes them to attain their growth, or the environment in which they live.   In a word, they are grouped on the basis of characteristics to which man is naturally led to attribute a value they do not have basically, and which stands in the way of a more penetrating understanding of the nature of existence, in proportion as the progress of science brings to light less evident facts and permits the reason to grasp more essential relations.

This does not mean that, in the arrangement which it is reasonable for us to regard as the most natural and the most true, there are still no traces of a relative and artificial order, adapted to the manner in which we conceive things rather than to an exact representation of what things are intrinsically and absolutely. We shall examine this matter later, and we shall discover that its cause is a consequence of the way in which some of our faculties develop.   As a result, this application of the principles of criticism in another way will only serve to give the principles a new confirmation.

If man were in contact with rational beings having a nature different from his own; if we knew in fact many different types of rational animals, as we do know a number of species that are very much like us so far as their organic structure and their animal functions are concerned, no one would doubt that there might be many other means of completing the critique of our knowledge and of distinguishing that which belongs in reality to things from that which is imposed by the constitution of the species.   But we lack such terms of comparison, and the distinction between various human races is too slight, so far as consistency in specific distinctions is concerned, to disclose methods sufficiently sure and sufficiently significant for philosophic inference.   However, even though this is true, all critical judgment is still not impossible.   It is undoubtedly very natural to believe in the physical and intellectual pre-eminence of the race to which one belongs.   But this natural prejudice may be confirmed or denied by reason.   Thus, for example, should it happen that the same characteristics which serve to establish the gradation of animal species and the incontestable pre-eminence of human beings over other animal forms were also of a sort that could serve to establish a gradation among the races in the human species, reason would clearly have to admit the superiority of the race which combines the distinctive characteristics in the highest degree, and this would have to be admitted independently of any accident of birth.   The inference to which nature yields in such a case is of absolutely the same sort as that which leads us to extend the path of a curve which brings us to prolong, beyond the discovered end points, a curve whose path is suggested to us by a sufficiently large number of points that have been determined (46).

90. If the order we observe among phenomena were not that which is found in them but the order in which our faculties place them, as Kant maintained, it would be impossible for us to criticize our faculties, and, together with this great logician, we should all fall completely into the most absolute sort of speculative skepticism. But it is not enough to lay down such a hypothesis gratuitously. We must check it against the facts; and we have shown that the facts are inconsistent with it. Unless we exaggerate idealism to the point of admitting that thought creates the external world out of whole cloth—and the criticism of such errors of speculation is not the object of our study—and so long as we give ideas only representative and not productive efficacy, it must be agreed that an order exists in things which is independent of our manner of conceiving them, and that, if there were no agreement between the order of our impressions and the order inherent in the objects represented, only an infinitely improbable chance would be able to bring about a mutual relationship between these two kinds of events in such a way as to produce a simple order or regular connection in the system of representations.[5] It is precisely because this harmony is not perfect and is no more capable of a rigorous precision than the other harmonies of nature (73), that it can present, and does in fact present, partial disorders, omissions, and contradictions in our conceptual system.

The idea of order has this peculiarity and eminence: it contains within itself its justification or its regulation. To know whether our other faculties deceive us or not, we look to see whether the ideas they give us are or are not connected with each other in an order which satisfies reason. But the idea of order can be given to us only by order itself. And if it were possible for this idea to arise in the human mind independently of all manifestations of external order, it could not be maintained against the perpetual evidence of disorder. Simply because we have the faculty of reason, and because this faculty is not condemned to impotence or to being destroyed from the beginning through lack of use, we must believe that the authority it claims is a legitimate authority. The eyes cannot testify to their own validity, nor taste to its

[5] " It is not according to nature that that which has its basis in disorder and in confusion should have the same character as that necessarily derived from a consistent state." Confucius in the *Ta-hio* or the *Grande étude*, Sect. 7.

own; but reason can bear witness in its own behalf, and at the
same time it can testify for or against the eyes or taste, as the case
may be.   Moreover, it would be fanciful and even absurd to seek
a criterion for that faculty which itself criticizes the others, since
that would involve us in an infinite regress.   It is fully evident
that it would then be necessary, without any discussion, to adopt
the most radical skepticism, and to say with the Greek, "We know
only that we know nothing."   But again, unless we have been
completely deluded, we are dealing here with a discussion much
more basic than pedantic subtleties, and we readily renounce any
attempt to convince those who do not themselves admit the authority of reason.

As Jouffroy says:

Just as reason, receiving evidence from the senses, from memory, and
from consciousness, asks itself what the validity of this evidence is and
to what extent it ought to rely on it; so, to the degree that it passes
judgment on these faculties in proportion as it conceives, beyond what
they reveal to it—realities and relations which escape them—it asks itself
what worth its own judgments and conceptions have, and to what extent
its confidence in itself, which is the final and the *supreme basis* of all it
believes, is justified.   Thus reason, which controls everything within us,
controls itself.   This is not merely a supposition, but a fact which observation immediately confirms in us, and which philosophic debates
have served only to carry over into the field of history. . . . But from the
fact that reason raises doubts concerning itself, does it follow that it has
the ability to resolve them?   By no means. . . . About what does reason
doubt?   It doubts principles which make it what it is, principles that
for it are the very criteria of what is reasonable and true.   What means
does it have for resolving this doubt?   It has and can have only these
very principles.   Therefore, it is able to judge these principles only by
themselves.   It tests itself, and if it doubts itself to the point of feeling
the need of being tested, it will be unable to trust itself when it undertakes this examination.   This is so evident that common sense would be
offended by our insisting upon it.   There is in us an ultimate basis for
belief, and it cannot be otherwise; indeed, if we doubt this ultimate
reason, it is evident that this doubt is invincible; otherwise this reason
for believing would not be ultimate.   This is what the Scottish school
means when it maintains that to attempt to prove basic truths implies
a contradiction; for were it possible to prove them they would not be
basic.   For the same reason, it is foolish to wish to demonstrate self-
evident principles; for if they could be demonstrated they would not be
self-evident.   This is repeated by Kant when he maintains we cannot

objectify the subjective, that is to say, make human truth cease to be human, since the reason which discovers it is human.   This impossibility can be expressed in twenty different ways, yet it always remains the same and is always insurmountable.[6]

This quotation, which we have taken the space to transcribe verbatim, contains a mixture of incontestable truth and of false applications which must be separated.   The confusion results entirely from differences in the meanings given to the term " reason," some being more and some being less inclusive.   In the analysis of the faculties and the organs of the understanding, after we have come to understand the part played by the senses, by memory, and by consciousness, the evidences of which admit of examination as is acknowledged by Jouffroy, we find that the mind is governed by certain rules, conceives and judges things according to certain ideas and principles imposed upon it by its own nature, which can come neither from the senses nor from memory nor from consciousness.   For example, it is found that the mind necessarily conceives of an unlimited space and time in the midst of which phenomena occur; that it is led invincibly—as the structure of all languages proves—to attribute the destructible qualities which it perceives to an indestructible substance which it does not perceive, and so forth.   These laws, ideas, and principles that cannot be derived from the senses are, when taken as a whole, what many philosophers call reason (15).   But reason thus conceived is something multiple and complex, the diverse data of which raise in us doubts of fact and of law and must be submitted to the control of a higher principle, just as is the case with the evidences of the senses, memory, and consciousness.   To justify the prerogative of the supreme and regulative principle, it is necessary that this principle should have some feature that distinguishes it from all others.   Now, in the first place, if we examine the question as to what principle it is by means of which reason criticizes the evidence of the senses, memory, and consciousness, upon what principle historical criticism, scientific criticism, the criticism of judicial evidence, and generally all types of criticism rest, we will find that reason does not proceed by invoking the idea of an infinite space or of an indestructible substance, or by any other principle of that

[6] Preface to the translation of the *Works* of Reid, p. clxxxviii.

kind, but that, on the contrary, it always proceeds by making reference to the idea of the order and the reason of things, by rejecting that which would result in contradiction and incoherence, by admitting or by tending to admit that which leads to regular coordination. In the second place, we do not understand at all how such an idea as that of an indestructible substance or of an unlimited time could test itself, or test the idea of the order and reason of things; while we can very well conceive how the latter idea can serve us as a basis for testing the former ideas in so far as we see that the latter brings order or introduces incoherencies and conflicts into the system of our conceptions. To the degree that the idea of order tests itself, it would be a contradiction to suppose this idea to be a prejudice of the human mind, or to be true only, as Jouffroy says, as a human verity, and that we shall none the less find order in nature in proportion as our study is extended.

Thus reason—when this term is taken in a sense which, according to us, is much too inclusive—doubts itself and the principles which constitute it, and not without ground. But regardless of what Jouffroy says, it is not brought to a condition of serious, to say nothing of insurmountable, doubt concerning the single regulatory and supreme principle by means of which it criticizes its several constitutive principles and all the other human faculties, any more than it raises serious doubt about mathematical axioms. However —and this is an entirely different matter—it is the nature of this regulative principle to furnish only probable inferences, but inferences which may sometimes exclude all reasonable doubt; it is by no means its function to furnish rigorous demonstrations like those which can be deduced from mathematical axioms.

It is a far cry from this hierarchical organization to the confusion of the Scottish philosophy, which prides itself upon multiplying, rather than reducing, the number of basic truths, and in which the appeal to common sense—which is such an easy way out—would dispense with testing the evidences of the senses, of memory, and of the imagination—which Jouffroy still submits to the examination of reason—as well as the principles of reason themselves, the testing of which, we see, is not at all impossible. As will be shown in what follows, there is no less difference between the theory we are presenting and that of Kant, who not only maintains that absolute truth cannot validly be inferred from human reason, in which he was perfectly correct, but who also systematically rejects everything that is only probably and not rigorously or formally demonstrable. For that reason he is led to impute to the constitution of the human mind, in spite of the most significant analogies and inferences, everything that we are led, with reason, to regard as pertaining to the nature of the external objects of our perceptions.

CHAPTER VII

ON THE SENSES AS INSTRUMENTS OF KNOWLEDGE; ON IMAGES AND IDEAS

91. The union of two correlative faculties, that of feeling and that of movement, appears to constitute the fundamental and distinctive character of animal nature. When these two faculties begin to manifest themselves clearly, we see that they depend upon an organic structure that is called the nervous system. The ramifications of one branch of this system go to the external covering of the animal and transmit impressions from this covering to the central parts of the system. A much more complex organization of this central part gives evidence that it is the seat of a very intricate elaboration. The ramifications of the other branch of the system transmit from the central parts to the motor organs the excitation which is to provoke movements in them. Certain ramifications of the first branch, by assuming a particular arrangement and some particular, natural aptitudes, and by adapting themselves to certain completely specialized organs, also acquire special functions capable of undergoing modifications in their capacity for sensation which are very different from one another, and which are distinct from those which affect generally the whole of the structure. These special, distinct, and in some ways heterogeneous modifications of the sensibility are what is properly called " sensations " or " sensory " affections. We observe that sensations are more easily distinguished from one another and give rise to clearer perceptions when they originate in senses having a more perfect organization, that is to say, an organization which impresses us because of greater complexity in regard to details, and greater unity and harmony in the whole. The point of dispute among philosophers lies in the answer given to this question: precisely what part do the senses play in the elaboration of human knowledge? That they furnish materials essential to the fabric of our knowledge is a fact which cannot be disputed.

Man has five senses; no more, no less. Animals which are

closely related to man have the same senses and the same number of senses, except for certain anomalies which result from accidental circumstances. It is necessary to descend very low in the order of animals before we come to species in which these sense organs, or certain of them at least, undergo profound modifications, become dull, and finally disappear. We can hardly imagine, among some species, organs of sensation which are essentially different from ours, which would not pertain to normal types of animality, or which would appear only accidentally and accessorily. Does this number—five—therefore, have some secret value connected with the essence of things? Or if nature, in setting up just this number, has used its discretionary powers, so to speak, is there any reason to believe that with one sense more or less the whole body of our knowledge would be thrown into confusion and not simply extended or decreased? Is it only a thoroughly chimerical pretension on our part that we can have even a superficial and limited knowledge of what things are, with means of perception which are obviously contingent and relative and which are without doubt adapted to the needs of our animal nature, but which by no means conform to the presumptuous exigencies of our curiosity? Let us resume the analysis of our sensations from this point of view, an analysis which both philosophers and physiologists have already made, and with respect to which something always remains to be done.

92. Let us begin with some remarks that apply not to the special organs of sensation, or to the senses properly so called, but to the general system of sensibility. The animal receives impressions of heat and cold from all parts of its sensitive covering. Guided by that sensation *sui generis*, man comes to know, not the inner nature, but the presence of an agent which arouses sensations, which affects all organisms by impressing innumerable modifications upon them, which plays an essential part in all physical phenomena, and which spreads and is dissipated according to laws that have been determined by science, the discovery of which has greatly contributed to the extension of our knowledge of nature. A man insensible to the action of heat would be deprived of information which is indispensable for the maintenance of his animal life. This evident fact can be no concern of ours in the question we are now considering. But would the body of his knowledge be seriously altered? This is the point which must claim our attention.

Anyone who knows something of astronomy can imagine how the sky would look from the moon or from Saturn, or from a world set up astronomically in a way different from ours. We are led to pursue with some curiosity the invention of an imaginary astronomy, and we wonder how an intelligent observer, supplied with instruments similar to those we use, but working from a different vantage point, would be able to progress gradually from the intuition of apparent movements, which differ from those we see from the earth, to a knowledge of the real movements such as our science has enabled us to acquire by examining phases of the problem of which we have a perfectly preserved historic record. For our purpose this is not only a matter of curiosity; it is also useful to indicate how our physics would be reconstructed by adapting it to hypotheses which, although indubitably imaginary, involve nothing that implies a contradiction or which is repugnant to reason in any other way.

93. Therefore, let us pretend that variations in the calorific state of bodies affect our senses directly no more than variations of their electrical condition do or than the magnetism of a bar of steel does. It would not require a very careful study of nature to note that the volume of liquids varies constantly, and that these variations are particularly noticeable when bodies are exposed to the sun or when, if they are in the shade, they are brought toward or removed from a fire. In such a situation it would certainly occur to someone to make these variations perceptible by putting some liquid into a rounded vessel at the end of which is a slender tube. This would not yet be a thermometer or an instrument suitable for the measurement of variations in temperature; but it would be an indicating instrument capable of showing the existence of these variations, or what physicists call a " thermoscope." By placing the thermoscope at different distances from incandescent bodies, by placing a screen between such a body and the thermoscope, by introducing different sorts of conditions such as mirrors or focusing lenses, by coating the bowl with various substances, it would be established that the influence emanating from the incandescent object is transmitted in a time too short to be perceived, that its energy varies in accordance with the inverse square of the distance, that is is modified by the condition of the surface of the instrument through which it passes, that this invisible emanation is reflected

and refracted, and that certain media transmit it while others partially dissipate it or prevent its passage completely. It would be noticed particularly that media which are opaque or impervious to light are highly pervious to this other emanation, the laws of which are being studied. Consequently, this emanation would be related to a principle analogous to light which sometimes does and sometimes does not accompany it, which appears to differ from it in many respects, and which follows different laws in certain cases. By pursuing this idea we should arrive, or could arrive, at a theory of radiant heat which in truth would not differ from that to which the results of our most recent researches have led us.

94. It would not be long until we discovered that dark objects, when exposed to rays of the sun or to the emanations of a luminous body for a certain length of time, also act on the thermoscope in the manner indicated, until they have returned gradually to their original state. Thus the idea would be confirmed that the cause of this emanation must, at least provisionally, be distinguished from light even though light accompanies it when it reaches a certain intensity. A natural inference, confirmed by experiments which can easily be conceived, would lead to the admission that all objects, even when they have not been brought near incandescent bodies or exposed to the rays of the sun, have a radiation of the same nature, although less intense. It would follow that this radiation also occurs in the substance of which the thermoscope is made, but that it has no apparent effect when this body loses as much by radiation to surrounding objects as it receives through the radiation of these bodies. In a word, the notion of "temperature" would be acquired and the theory of the "unstable equilibrium" of temperatures would be developed, just as it is taught in our books.

The experiments that would be made to study the laws of the propagation of heat in solid bodies would for the most part be made in the same way and would result in the same mathematical theory.

Finally, it would be noticed that changes in the molecular state of bodies are connected with changes in their thermoscopic state; that, for example, water expands, or its molecules become separated farther and farther from each other, until it becomes a gas; that it contracts, or its molecules come closer together, until it becomes a solid. It would also be noticed that, if the thermoscope were plunged into snow or into boiling water and subjected to the radiation of an incandescent body, it would show no change until all the snow had melted or all the water had evaporated. This last observation would suggest the construction of the thermometer, or of a graduated instrument having a scale which makes it suitable for defining and for measuring temperatures. It would also suggest the making of a calorimeter, that is, an instrument suitable for measuring the effects of this particular type of radiation, this emanation which, unlike light, is neither tangible nor ponderable. In addition, it would be noted that most chemical actions are accompanied either by the emission or by the absorption of this intangible principle. As a result of all this, heat would come to be conceived as a cause whose most general effect is the tendency of the molecules of bodies to separate and to neutralize the action of other forces which tend to bring them together.

95. In a word—for it is well understood that we must omit or abridge the details—under the conditions that we have set up, it would be possible to get exactly the same ideas of the nature of heat that we have of it, except that these ideas would not be associated with the memory of certain sensations, which in this case manifestly contribute nothing to the clarity of the ideas and do not assist the mind in the work of theoretical construction. We would come to know about heat in the same way in which we know about electricity, that is, by means of scientific and not of common knowledge. There would be no common words in any language to designate the experiences of heat and cold; but there would be technical or scientific terms which would even tend to pass into the ordinary language of intelligent persons because of the generality and the significance of the notions they express. This is how it happens that we may say, with assurance of being understood by everyone, although it would have been unintelligible in the time of Louis XIV, that an orator *electrified* his audience. No doubt the historical order of the various discoveries would have been different, and the point of departure and the order of didactic exposition would not be the same. Yet all these accessory circumstances, although of great interest when one considers man in his dual nature as both a sentient and an intelligent being—as is the case, for example, in matters of education and pedagogy—are of no consequence in relation to questions pertaining uniquely to his

intellectual faculties or to ideas which these faculties elaborate as a result of their particular quality, the essence of which does not change—whatever the sensible soil, so to speak, in which they may have been planted.

96. Not only is the capacity of our senses to receive impressions of heat and cold not the essential condition of our knowledge of the nature of heat and its effects; not only does this not contribute to the scientific perfection of this knowledge; it may actually impair it, if reason is not on guard against illusions of which it is the source. The modifications of the nerve fibers with which sensations of heat and cold are connected may be stimulated by organic disturbances as well as by the physical action of heat. A person shivers when he has a fever, even though his surroundings may be warm, and so on. In the absence of organic disturbances, habit dulls, modifies, and weakens sensations that the physical action of heat would lead us to experience. A bath of constant temperature seems warm or cold depending upon whether we have left a warmer or a cooler room. We find that a cave in which the temperature does not vary appreciably with the seasons seems warm in winter and cool in summer. Thus, in all books on physics, after the author has spoken briefly of the effect of heat upon our sense organs, he hastens to point out, on the one hand, that we must not form our conclusions on the basis of this impression, and to make known, on the other hand, the construction of instruments the exact indications of which, being independent of the condition of our senses, must guide the observer without regard to misleading suggestions of the sensibility, at least within certain limits of precision.

97. Therefore, it is not without reason that, from the very beginning of philosophy, speculative minds have spoken out against the errors of the senses and have insisted upon the necessity of disencumbering sensible perception from whatever is variable, relative, and inherent in our organic nature, in order to reach the idea or the true knowledge of things. This doctrine has been overworked. It has often been very badly attacked and badly defended. It has been connected with venturesome systems or mystical visions with which it has, however, nothing in common. Above all, the type of demonstration or refutation of which it admits is generally misrepresented. Instead of taking, for purposes of analysis, common knowledge—knowledge which has re-

mained in a rudimentary state, so to speak—preference must be given to scientific knowledge, that is to say, to knowledge which is organized, developed, and perfected. Natural philosophers know very well that in the rudimentary state all types, all organic threads are or seem to be confused, and that in order to know their distinctive characteristics fully, it is preferable to study them in the organic forms which are the most perfect. Animal forms and plant forms, which are so clearly distinct in the higher species, become confused in proportion as one goes down the scale to lower species. Therefore, if science is the organic perfecting of knowledge, there are good reasons to believe that in seeking to discover how the senses contribute to the organization of science, we shall be in a better position to grasp fully the essential part the senses play in the elaboration of knowledge, even in its elementary or rudimentary state.

98. Before leaving the example that has suggested these general reflections to us, we must not fail to draw the attention of the reader for a moment to the principles by means of which we succeed in finding, with respect to temperature and to quantities of heat, the fixed terms of comparison that the organization of our sentient nature is unable to furnish us. If we construct thermometers containing different liquids, such as water, alcohol, and mercury, it will be found that these instruments, which are really artificial senses conceived to give us precise indications of the temperature of the materials with which they are put in contact, do not move with perfect agreement, and at first we shall have no basis for knowing which of the readings should be preferred. If, however, it is noted that all these thermometers are in essential agreement so long as the liquids in them are some distance from the temperatures at which they boil and at which they become solid; and that the errors of each particular thermometer are greater as the temperature of the liquid used comes closer and closer to either of these two points, it will be realized that the errors are due to disturbing causes which follow from the specific nature of each liquid, and which cease to have any appreciable effect over the intermediate portion of the scale in which all the thermometers are found to be essentially in agreement. Later on, someone will conceive the idea of replacing the liquids with gases, that is to say with fluids in which we have strong reasons for believing that the molecular structure has reached

a higher degree of simplicity and regularity than is found in liquids. Then it will be discovered that these gas thermometers agree not only with each other at all temperatures, but also with liquid thermometers throughout that portion of their scale in which the disturbing causes connected with their specific structure no longer exert a noticeable influence.   When these two steps have been taken, we shall be convinced that gas thermometers are the best regulative instruments for use in standardizing the others and for fixing the degrees of the temperature scale absolutely.   This is a probable judgment exactly analogous to that by which we make statements concerning the relative and absolute movements of a system of bodies (5), and the motives for choosing among the evidences of different artificial senses are of exactly the same nature as those for choosing among the evidences of the various senses and faculties with which nature has endowed us (85).

Now let us pass on to the measurement of the quantities of heat released or absorbed by a body when its physical state is changed, when it enters into chemical combination with other bodies, when its temperature changes, and so forth.   These quantities are neither tangible nor ponderable; they escape the ordinary processes of measurement by means of the senses of sight and touch, and a rational conception must make up for the defect of the senses.   If two quantities of heat, A and B, have been used to raise the temperature of two liters of water, one from 10 degrees to 50, and the other from 10 degrees to 90, we are not justified in saying that B is twice as large a quantity as A; for it may well be that a liquid mass already heated by 40 degrees, and consequently one in which the molecular structure is already modified, requires more or less heat to raise its temperature another 40 degrees.   The conclusion would become very much more probable if the two quantities A and B were used, one to raise the temperature of two liters of water from 10 degrees to 50 degrees, and the other to raise the temperature of four liters of water from 10 degrees to 50 degrees.   The conclusion would be still more likely if the quantity A has served to melt one kilogram of ice, and the quantity B to melt two kilograms.   This follows from the fact that we should find it difficult to think that the simple juxtaposition of two pieces of ice or of two quantities of water had influenced the quantity of heat necessary to melt each piece of ice or to bring each amount of water from a temperature of 10 degrees to one of 50 degrees. But what each experiment taken by itself indicates with a high degree of probability may no longer be reasonably doubted on the basis of the concurrence of the two experiments whose results reinforce one another. For, seeing the disparity of the effects produced, it is inconceivable that an exact proportion would have resulted if the quantities of heat which produce them were not also in the same proportion. By multiplying experiments and agreements of this nature, the conclusion we have drawn will be established beyond any doubt. In this way, through the concurrence between the senses which observe and reason which interprets, we are able to pass, without supposition, the limits of sensible observation, and to reach, without falling into a vicious circle, to use the common figure of speech, that *quid inconcussum* [" constant thing "] which we need in order to evaluate the structure of the theory.

99. Now let us go back to the problem we have raised for discussion. We have shown that the abolition of one faculty of the general sensibility, such as that of perceiving impressions of heat and cold, leads neither to the curtailment nor to the modification of the system of our ideas. We now propose to examine the characteristic influence on our ideas of each of the special organs of sense. We will begin with the one having the simplest form, and in which—in the opinion of all physiologists—the general sensibility has received the least refined improvements, namely, the organ of taste. The importance of this organ for one of the principal functions of the life of the animal is clear enough; but in proportion to the degree that this importance is great, so—by a sort of compensation of which there are numerous examples in nature—the utility of this organ is slight and even nil so far as matters of knowledge are concerned. The sensation of taste is the result of the chemical action which molecules dissolved in a liquid exert on the neural papillae of the organ of taste. This organ is a chemical reagent which is sometimes endowed with an exquisite delicacy, which is able, by the perception of characteristic tastes, to detect in a mixture of various things the presence of particles which would escape the balances and the reagents of the laboratory. But the perception of taste throws no light on the nature of chemical or molecular action. It is an experience of the sentient

subject which gives no image and implies no knowledge about the object sensed. To learn through the sense of taste that sea salt has, as it is said, a pure taste and that ferrous sulphate has an astringent taste is to learn that each of these two salts has its own peculiar effect on the taste organ. But this tells us nothing concerning the nature of sea salt or of ferrous sulphate. In the same way, a gouty pain tells us that there is something in the humors or tissues of our organs capable of producing a painful sensation, without, for that reason, advancing our knowledge of the structure of the tissues, of the composition of the humors, or of the nature of the morbid condition. Therefore, taste contributes to our knowledge only indirectly and as a reagent. That is to say, after we have recognized that a certain substance gives us a clearly recognized, and, as we say, characteristic sensation of taste, the taste later enables us to recognize the presence of the substance in a mixture in which it is blended and in which we should otherwise not be able to distinguish it, either because the quantity is too small or for some other reason. Like any other reagent, the sensation of taste may also inform us, in certain cases, not about the specific nature of the substance, but about the class of substances to which it belongs, and, consequently, about the characteristic properties which it shares with other substances of the same kind. Thus, when a substance causes us to experience an acid taste, we know that it is capable of uniting chemically with salifiable bases, and also that, if the product of this reaction is decomposed by means of an electric current, the same substances, liberated once more, will go to the positive pole of a voltaic cell, and so forth. We know all these things because experiments have taught us that the property of uniting with salifiable bases and that of moving to the positive pole of a cell are constantly associated with the property or quality of exciting in us the sensation of acid taste. However, from that we shall not gain a better understanding either of the chemical characteristics by which acids and bases are contrasted, or of the connection that exists between the chemical compositions of acids and the property they possess which cause us to experience an acid taste. Even though we were to come to know precisely the nature of the chemical action of acid substances upon nervous tissue, this would not alter in the least our insurmountable ignorance as to how that chemical action engenders one sensation rather than another. This ignorance springs precisely from the fact that the sensation of taste has, in itself, no representative value and throws no light on the causes which produce it. As a reagent, the organ of taste is itself of little use to us in the development of our scientific knowledge. Surely, no chemist would imagine that Scheele or Lavoisier would have failed to achieve any of their memorable discoveries even though they had been absolutely deprived of the sense of taste. The supposition that the sense of taste might become a reagent which would be used as habitually by chemists as litmus paper is for recognizing the presence of acids would lead no one to believe that the possibility of acquiring our actual knowledge of chemical phenomena springs from the accidental fact of the sensitivity of the organ of taste to certain chemical actions, any more than that it might be thought to follow from the very particular and very accidental fact of the presence, in the pith of certain plants, of a coloring matter which is extremely sensitive to the action of acids. Furthermore, what concerns us here are the essential conditions of knowledge or of insuperable ignorance, and not the accidental circumstances which may facilitate or hinder our inquiries, or tend to hinder them, or impress on them a preference for a certain direction.

100. The sense of smell is much superior to the sense of taste so far as complexity of organic structure is concerned. It is at once more special and more highly perfected, for perfection of organization is always indicated by a greater specialization of function. The sensory nerves involved lose their tactile sensitivity at the moment they cease to be in immediate connection with the motor nerves. The sense of smell is distinguished from that of taste by this double characteristic and has a certain resemblance to those of hearing and of sight. Although it is far from equaling the perfection of these two superior senses, it is manifestly designed, as they are, to enable the animal to perceive objects at a distance; and, in certain species, it acquires a degree of subtlety that, in conjunction with the faculties of touch and of locomotion, enables it to provide for the needs of the animal as well as, and perhaps better than, the senses of sight and of hearing do. But, on the other hand, there are evident connections between the sense of smell and that of taste. These connections may be either anatomical, that is to say, they may pertain to the structure and the

disposition of the organs, or they may be physiological, that is to say, they may have to do with the analogy and correspondence of functions. For this reason, we may suspect among certain animals, and notably among certain ruminants, the existence of an organ adapted to the search for their food, acting as an intermediate sense or establishing the passage from one sense to the other. Both are directly connected with nutrition and develop in a way that parallels what we call instinct rather than in a way that parallels the intelligence of the animal. Both taste and smell are adapted to the perception of molecular actions, or to actions which have had their source in material particles which exist in an extremely finely divided state, whether this division is the result of chemical or of mechanical action. Finally, both of these senses, and particularly the sense of smell, must be considered, so far as their relation to the acquisition of knowledge is concerned, as being extremely sensitive detectors, but detectors which nevertheless lack the capacity for revealing to us the nature of the causes which produce the reactions. An odor, like a taste, is an affection of the sensing subject, which yields no image and which by itself neither implies nor determines any knowledge of the object sensed. Condillac spoke advisedly when he said of his imaginary statue, which he thought of as having only a sense of smell, *it would feel rose color*, provided that our language, suggested by a structure and by habits of an entirely different sort, is successfully adapted to represent the obscure phenomena which would be produced in this hypothetical state. It is incontestable that the sense of odor alone would give such a creature no idea of the external world and that, as men are normally constituted, this sense adds nothing to the theoretic or scientific knowledge of the external world. It furnishes the physicist with some further examples of the extreme divisibility of matter; sometimes it serves—as we have said in regard to taste—as a reagent for chemists. But the development of science would not be hindered if the sense were done away with. Had nature not endowed man with it, the only result would have been some disturbance in the operation of his animal functions, but, all other things remaining equal, neither his knowledge nor the nature of his mind would be affected in the least.

101. However admirable the structure of the eye may appear to us, there are good reasons to believe that the sense of hearing

involves an even more complicated apparatus and an even greater organic perfection, and occupies the highest rank in the series of the sense organs. Without going into the explanations that modern anatomists give on this subject—which is not within our province —we can observe (89) that the sense of sight is less perfect in man than it is in some species which are very unlike him and undoubtedly occupy a lower place in the scale of animals; while the apparatus for hearing attains its perfection in man, in whom it is, in connection with the faculty of producing articulate sounds in such a way as to influence the formation of language, the organic condition for the development of all our intellectual faculties. However, since the influence of language on the development of thought should be studied by itself and within the context of considerations of another sort, we shall disregard at this point that indirect influence of the sense of hearing upon the development of intelligence. We shall assume that man has a sign language or a written language, or an entirely different instrument analogous to speech and capable of the same development; then, always proceeding by means of a process of successive limitations, we shall deal with the sense of hearing before we deal with that of sight; on the ground that, under the conditions stated above, this should result in less profound modifications of the organization of our knowledge.

In fact, although physics has two major divisions, optics and acoustics, the names of which are sufficient to indicate their dependence upon our two highest senses, that dependence is not as rigorous for one as for the other. Sound is produced by vibrations which particles of bodies set up when their condition of equilibrium is disturbed. These vibrations are very rapid, yet they are not so rapid but that they may be measured, indirectly and by means of mathematical calculation, without the aid of hearing. This is made possible because of the connections that theoretical developments have brought to light between the length of vibrations and other phenomena which can be measured. Indeed they may even be measured directly with the aid of certain ingenious instruments for the invention of which we are indebted to modern physicists. Even though deprived of his sense of hearing, man would still be able to discover, through the senses of sight and of touch, vibratory motions impressed on very small particles of matter; and if his

intellectual faculties were weakened to no greater degree than those
of the well-informed deaf-mute, he would reach, through identical
mental operations, the same theoretical conception of the causes
of these motions, and the same mathematical formulas which are
the highest expression of them.   As a matter of fact, when dealing with vibrations having frequencies which lie beyond certain
limits, the physicist and the mathematician are in exactly the same
position as the deaf-mute. The ear is insensitive to vibrations which
are too slow or too rapid, that is, to sounds which are too low or too
high.   But this does not prevent the physicist from including them
all in the same theory and the mathematician from dealing with
them all in the same equation without regard to the limits of the
range of sensibility which, probably because of organic factors, is
likely to vary from one individual to another and from one species
to another.

No doubt, in the case of sounds which the ear cannot hear, the
physicist finds himself deprived, not only of a sensitive detector
which serves in the same way as do odors and tastes, but also of
an instrument of measurement which often acquires a marvelous
precision in persons whose ears, as a result either of natural dispositions or of acquired habits, perceive musical intervals with
great accuracy. We readily admit that while it would not be
difficult for a deaf-mute to teach acoustics—in the same way that
the blind Saunderson [1] taught optics—it would be difficult for him
to be as brilliant an experimentalist in this field as a Chladni [2] or
Savart.[3] It is also undoubtedly true that, although a mathematical

[1] [Nicholas Saunderson (1682–1739), an English scholar and mathematician
who was totally blind after the first year of his life. In spite of this handicap,
he became an able student of the classics and of the exact sciences. He was
appointed to fill the Lucasian Chair of Mathematics at Cambridge University
in 1711. His writings were primarily in the field of mathematics, one of his
books being *On Fluxions*. He was a friend of Newton and other speculative and
natural philosophers of his time.]

[2] [Ernst Florens Friedrich Chladni (1756–1824), a German philosopher, inventor, and mathematician. He did significant work in the development of
the theory of acoustics and in perfecting the theory of sound.]

[3] [Félix Savart (1791–1841), a French scholar, physician, and physicist. He
gave up his career in medicine to devote his time to the problem of acoustics.
Following Ampère's death, Savart succeeded to his chair in the Collège de
France.]

formula represents virtually all the details of phenomena, there are certain consequences which would be overlooked if they were not brought to our attention by sensible experience. It would be difficult to read these consequences in the formula unless the results of experience had forced us to reflect upon them for some time. But again, we are concerned with the essential conditions of knowledge; and from this point of view everything that is actually found to be included in the statement of a law, everything that may be deduced from it by the power of reason, is supposed to be given us by the knowledge of the law itself. It is not a question of what any particular man may attain, depending upon the extent of his individual abilities; it is a question of the results which human reason is capable of reaching, and ought to reach, if no accidental obstacle arises to arrest its indefinite progress.

The sensation of an isolated sound can no more give us the idea of the cause which has produced it than can a sensation of taste or odor, even though we might have every reason for believing that the physical modification of the nerve fibers with which the sensation of sound is connected consists in a vibratory motion which is strikingly analogous to the external phenomenon which produces it. In fact, the vibrations of the nerve fiber, like those of a sounding body, follow one another so rapidly that we are unable to have any knowledge of their distinctness or of their succession. But when the ear is struck by two or more sounds which form a definite musical interval—one of which is an octave or a fifth above the other, for example—a simple relation is established between the different vibratory motions in the nerve fiber which is similar to those in the vibratory motions of different sounding bodies. Although consciousness has no faculty for counting or for distinguishing the vibrations one by one, it is on the contrary fully capable of perceiving the regularity of the periods to which the vibrations of the nerve fiber are subjected. This is why, in the matter of comparing two musical tones, the ear not only reacts to them but is also an instrument of measurement. This is how we explain the pleasure the ear finds in harmonic consonances and its aversion to dissonances, whereas we have no idea at all of the physical cause of the attraction or of the repugnance that we feel for a taste or a smell. As we move up in the scale of the senses, we find that sensations begin to acquire a representative value and cease to be simple affections, cease to be incapable of telling us anything about the nature of the causes that produce them.

102. We have just pointed out that the sense of hearing contributes in two ways to the growth of our knowledge and to the development of our ideas: first, directly through the perception of sounds and other phenomena with which acoustics is concerned, and secondly, in an indirect and more general way, by influencing the construction of tools of language by the help of which we form and communicate our thoughts of all kinds. Similarly, the sense of sight must be studied from two points of view; first, in so far as it gives us directly, through a sensation *sui generis*, the perception of light, of colors, and of all the phenomena dealt with theoretically in the study of optics; and, secondly, in so far as it contributes indirectly to making known to us the universality of phenomena of the physical world by putting at our disposal the light which illuminates everything. The action of bodies on light makes manifest to us their existence, their forms, their dimensions, their movements, and the modifications they undergo as a result of their reciprocal actions. Of these two functions of the sense of sight— the one direct and special, the other indirect and general—which constitute vision properly so called, the latter must be regarded as primary. For, although light itself may be a very worthy object of study, and although the eye would still have to be regarded as a very precious organ even though it were to serve only to reveal to us the existence and some of the properties of so important a natural agent as this, it is clear enough that this is not its natural end, and that nature has given eyes to us, as to other animals, so that we may see the objects disclosed by light, and not at all to provide us with the satisfaction of coming to understand to a greater or lesser degree the nature of light and of the laws which control optical phenomena. Now it must be kept in mind that the act of seeing depends essentially neither upon the inner nature of the essential character of luminosity, nor upon the special manner in which light acts upon the nerve fibers, or upon the special form of sensation which is the immediate result of this kind of action. The retina might become insensible to the rays of the solar spectrum which now produce in it different sensations of color, and it might receive through actually invisible rays—such as are known to exist both above and below the limits of the visible spectrum

—sensations of which we have, at present, no idea.   But this should not alter the essential conditions of the visibility of bodies, namely, the indefinite radiation of light in straight lines and in all directions, and the reflection and dispersion of rays when they pass from one medium to another.   Every radiation is subject to these mathematical laws, and although in other ways physically unlike light, each can, like light, be detected by making use of an instrument designed for perceiving objects at a distance, and may be the intermediary for this *perception at a distance* entirely independently of the sensation *sui generis* which is associated with it, and which results—without our knowing how—as much from the inner nature of different rays of the luminous spectrum as it does from the special structure of the nervous tissues of the retina and of the optic nerve.

The assumptions we have made for purposes of our analysis are not purely gratuitous.   There are certain organic anomalies which suffice to suggest the idea of them.   The eyes of some persons are naturally, or become accidentally, insensible to certain colors. Cases may be cited in which the distinction between colors seems to have completely disappeared, and in which images of illuminated objects continue to be seen like figures in an engraving or a picture done in shades of gray.   Persons in whom vision operates in this anomalous way may be compared to those in whom the ability to distinguish tastes has been destroyed, although when eating they still perceive impressions of hot and cold and other tactile sensations.   What is fundamental in the function and in the sensation which accompanies it still persists even after the suppression or the dulling of this special and accessory sensibility that nature uses, in this circumstance as in many others, for stimulating a sensing subject, or for embellishing the object sensed, in such a way as to attain more completely or more surely the end in view of which the whole organism functions.

When we put colored glasses in front of our eyes, or when we illuminate objects with light from which certain rays found in sunlight have been artificially eliminated, we place ourselves voluntarily in conditions analogous to those in which the persons of whom we were speaking above find themselves placed by infirmity or illness.   Nevertheless, vision would operate as in ordinary conditions in such a way as to give us the same ideas of the distance, the forms, and the dimensions of bodies, and in general of all the phenomena of the physical world, with the single exception of the coloration of bodies and images. Our theories of mechanics, of astronomy, of general physics, of chemistry, and of physiology would be absolutely the same, even though nature had included in the spread of the solar spectrum which is visible to us one ray more or less, or even if, without modifying the sensitivity of our sense organ, it had changed the nature of light by substituting for our sun one of those stars which appear red or blue to us, or the light from which, without offering very conspicuous differences, is none the less found by prismatic analysis to have a composition unlike that of solar light.

We are indebted to the physicist Brewster [4] for an ingenious theory according to which the graduated colors of the solar spectrum would be accounted for by the superimposition of the three spectra, red, yellow, and blue, in each of which the light has the same color throughout but varies in intensity from one point to another, with the result that the points at which each color attains its *maximum* intensity would not coincide. At one point one color and at another point another color would be most pronounced in the spectrum formed by the superimposition and the blending of the three elementary spectra. According to this theory, which there is no reason for us to discuss but which we may cite as a hypothetical example, the human eye would be sensitive not to one light but to three distinct lights, three types of radiations or effluences adapted, so to speak, to the needs of vision, among other radiations which are of no value to vision but which produce other perfectly definite physical, chemical, or physiological effects. Were we to take account of things in this way, we should understand still better to what extent the phenomenon of the distinction of colors, the gradation of which would be reversed by a simple dis-

---

[4] [Sir David Brewster (1781–1868), a Scottish experimental philosopher and inventor noted particularly for his work in the field of optics. He was awarded the Copley medal of the Royal Society of London for his essay " On the Polarization of Light by Reflection." He invented the kaleidoscope and an improved method for lighting lighthouses. He was one of the founders of the *Edinburgh Philosophical Journal* and of the *Edinburgh Journal of Science*. He was the author of a *Treatise on Optics*. In 1868 he was elected a foreign associate of the Institut de France to fill the place left vacant by the death of Berzelius.]

placement of the points corresponding to the *maximum* intensity of each elementary shade, is accessory and accidental in the act of seeing. From that it is but a step to the suppression of one or the other of these three visible emanations or to the substitution of one which is actually invisible for one of those now visible.

Certainly, we are not foolish enough to believe that it is possible to produce hypotheses and to imagine plans which could be substituted for the plan of nature. No doubt there are good reasons why our sensations and their causes are what they are down to the smallest detail. We are interested only in something which is both possible and permissible for reason, the distinguishing of the essential and fundamental conditions of a phenomenon from the conditions which are accessory and on which the perfecting of it depends. It must also be recognized that in this matter the essential conditions are mathematical and not physical, that they are conditions of form and not of stuff or matter (1).

103. What would happen if our eyes were to cease to be sensitive to visible rays which now produce in them the sensation of a certain determinant color, or if the limits of the visible spectrum were more restricted? Evidently the situation would then be exactly what it now is in the case of rays which are invisible, the existence of which we are able to establish only through the action they exert on magnetic needles, on thermometers, and on chemical reagents in regard to which we are quite able to establish laws of reflection, of refraction, and of polarization that are identical with or analogous to those which hold in the case of visible rays. Thus, fundamentally, the sensations of color are like those of sounds, of odors, and of tastes; they could be abolished without producing necessarily any gaps in the body of our knowledge. Should light, taken as a whole, that is to say, the whole system of visible rays, lose its special action on the retina, and thus become an invisible emanation, we should still be able to attain not only knowledge of the external world and of objects at a distance. What is more, we should even be able to discover the existence and the characteristic properties of the light which has been rendered invisible, provided the retina were to become sensitive to some other emanation, governed by the same laws of radiation, that would consequently satisfy the mathematical conditions of vision and of awareness of objects at a distance. Indeed, for rays actually visible, the eye

is much more sensitive and—what has still quite a different scientific importance—is an even more precise instrument of measurement than are the thermometer, the magnetic needle, or chemical preparations. Consequently, in the imaginary hypotheses which we presented in order to facilitate our analysis, it would be no more difficult to create the theory of invisible light than it would be to create the theory of heat without the clue of sensations of heat and cold, or chemical theories without the aid of organs of taste and smell, or even the theory of the vibrations of bodies without the aid of the sense of hearing. But, once again, we must remind ourselves that, in this whole analysis, we are interested in the question of the essential conditions of knowledge, of those the lack of which is a cause of insuperable ignorance. We are not concerned with the accessory circumstances which facilitate the progress of learning and develop its natural potentialities in such a way that scientific theories develop from them. Any advance in physics would be singularly hampered if we possessed neither a solid and transparent substance like glass nor a metal like mercury which is liquid at ordinary temperatures. The point we are making is simply that it would be a mistake, in a critique of the sources of human knowledge, to assign a fundamental role to these specific and very particular properties which give an exceedingly great importance to such substances as glass and mercury when it comes to the skillful conducting of experiments.

Moreover, sensations of color are in all respects comparable to those of heat and cold, and to sensations of tastes, of odors, and of sounds. They often result from a disturbance within the nervous system that is not provoked by any external excitation or from irritations produced by electricity or mechanical injuries. In a word, they may result from causes other than those which produce the same sensations in the normal and the habitual state. We have no idea of the relations that must exist between the specific nature of each ray of light and the special sensation of color of which the former is the determining or the provocative cause. The sensation of color, like that of taste, has in itself no representative efficacy; the former no more reveals to us the specific character of rays of light than the latter indicates to us the molecular structure of palatable substances.

The scale of colors in the solar spectrum has sometimes been compared to the scale of musical tones, and the harmony or the contrast of certain colors to musical consonances and dissonances. But these comparisons are very inexact, especially from the point of view of the analysis we are making, because they would tend to establish a parallelism between two senses of which one, that of hearing, is very superior to the other in this respect.   In fact, we have found (101) one reason why the ear perceives the numerical relations of tones or their musical intervals in the very manner in which the branches of the auditory nerve come to vibrate in unison with the vibrations of sounding bodies and of the medium that surrounds them.   This does not mean that the ear is able to count the vibrations, for they are so rapid that hundreds of them occur in the short interval of a second.   What the ear does apprehend or count in its own way, because of the exact correspondence between the vibrations of the auditory nerve and the vibrations of the sounding body, are very simple relations between these large numbers which escape direct perception, the one being two or three or four times that of the other, for example.   Consequently the ear is not only a focal point for agreeable and disagreeable sensations; it is an instrument for the immediate perception of musical intervals, the sensation itself having a representative value which still results from a formal characteristic.   This is made known by the periodic recurrence of the same impressions, and not by the nature of the impression produced.   Thus, the perception of a musical interval remains the same whatever the absolute pitch of the tones compared, or whatever the other accessory qualities which modify the sensation in so far as it is purely affective.   The sense of hearing is quite notable because of its power of being able in this way to separate from the basis or from the *stuff* of sensation an abstract mathematical relation which—as language itself abundantly indicates, and as the history of philosophy testifies—has become the pattern for our highest conceptions of the order and harmony of existence.   The sense of sight possesses no such power with respect to the perception of colors.   Certain combinations of colors delight or offend it, just as at a concert certain combinations of voices delight or offend, or just as certain combinations of flavors please or displease the sense of taste.   But although it is possible to give physical reasons for what we call the harmony or the contrast of colors, the sensation of the harmony of colors, unlike that of the musical interval, is not the perception of a mathematical relation which remains the same when the terms of the relation, that is to say, the associated colors, happen to change. By admitting the doubtful fact that the accident of color may be related to the rapidity of the vibrations of the ether, it remains certain that the one is in no way the image of the other; and that, consequently, the eye cannot become aware of the vibrations of the ether, the rapidity of which is out of all proportion to that of the vibratory motions of ponderable bodies, but, moreover, that it is incapable of grasping the harmonic intervals or the simple relations between these numbers which are so enormous that the imagination is overwhelmed by them.

104. We must now return to the study of the general function of the sense of sight, namely, vision properly so called, which we have recognized to be fundamentally independent of the specific distinction of rays and colors. We are immediately struck by this circumstance, that to the degree that we are ignorant of the relations between the sensations of tastes, of odors, and of colors, and the causes which make them what they are specifically, we are also ignorant of the correlation between the thing perceived and the nature of the perceiving organ; in cases of the perception of a colored surface, not in so far as it is colored, but insofar as it is extended. The retina is a "sensitive tablet" (*tableau sentant*); this phrase makes comment unnecessary. It is simply the case of applying to the sense of sight the remarks that we applied earlier to that of hearing. We discern nothing which is capable of connecting the sensation of a certain timbre with a certain kind of random vibratory motion among the particles of a resonant body, any more than we discern what connects the sensations of yellow or green with the action of certain rays of the spectrum, or a certain taste with the chemical action of the molecules of a certain substance. Therefore, such sensations are affective and not representative. But we do find, in the very form of motion in the auditory nerve, a reason, deduced from the correspondence and the synchronization of the vibrations, why the ear may be capable of immediate representation and consequently of direct perception of the numerical relations or intervals between tones. And in the same type of vibration in the nerves of the retina we find a still more immediate, more apparent reason why the eye perceives

geometrical relations and relations of position and of size among objects from which the light rays emanate, except, of course, for changes of perspective, which we learn to correct by subsequent further training of the sense of sight. Psychologists have said much about this matter, with which we are not particularly concerned at the moment. In the one case as in the other, the representative power results from the fact that the phenomenon of sensation is the translation or the image of the external phenomenon, not with respect to its matter or stuff, but with respect to its form, which is the sole basis of the representation.

Even though we were able to conceive of the successive suppression of the senses of taste, of smell, and of hearing, and even of the abolition of the distinction between colors, without having essentially modified the body of our knowledge, and without thereby necessarily having condemned the rudiments of any of our scientific theories to destruction or to failure, it is evident that the suppression of the sense of sight would actually hinder nearly all scientific development, by making the acquisition of a great body of information absolutely impossible. But it must be carefully noted that if this were the case, the body of our knowledge would be mutilated rather than disorganized or vitally altered. It would be like a tree whose small branches had been cut off, with the result that its ornamental value is lost, but which still had its trunk and its main branches. Inversely, if one were to restore the sense of vision to a blind man who had previously been taught, as one of a group of blind persons, by means of the sense of touch alone, his knowledge would be extended, developed, and reoriented, but he would not find it necessary to reconstruct the framework which supports it according to a new plan. One sense adds something to the other; ideas are not supplanted by other ideas, but new ideas are added, or rather adapt themselves to ideas already acquired. It is to be fully understood that we are speaking only of those ideas to which sensible perception, illuminated by reason, necessarily leads, and which are clarified by reason; we are not speaking of those ideas which originate in doubtful inductions or which the imagination has created out of whole cloth by exceeding the limits of legitimate observation and inference.

Instead of assuming the sudden and complete abolition of vision and of perception at a distance, let us suppose that vision is progressively curtailed, or that the ability to perceive objects at a distance is progressively reduced. The effect of this increasing myopia would clearly be to introduce a restraint in the operation of the animal functions, to deprive the myopic person of the enjoyment of certain sights, and to exclude him from certain kinds of studies and discoveries. But it would not go so far as to upset the whole system of his ideas, and he would conceive of things exactly as do other persons in whom the sense of sight has maintained its natural range. Inversely, when, as a result of the discovery of various optical instruments, the natural range of human vision is extended, the subject matter of our studies and the basis of our knowledge are undoubtedly increased, but the form of our knowledge or the essential structure of our ideas is not changed any more than it is changed when we discover any new physical instrument that augments our means of observation. All that would have happened, as now happens with every discovery of this sort, is that our observations, by being extended, would have made it possible to indicate other analogies and to understand other inferences, and consequently to modify our scientific theories in so far as they have been premature and hypothetical.

105. Psychologists have raised the question whether or not the sense of sight, without the concurrence of that of touch, would give us the notion of extension in two or three dimensions, or even whether the sensations it would provide in this isolated condition would be sufficient for the sensing subject to conceive the idea of an external world and to learn to distinguish itself from it. Experience is obviously incapable of answering this question, and it may be regarded as a pure fantasy, since the hypothesis to which it is related is contrary not only to the structure of our species, but to the fundamental plan of all animal life. What would actually happen if a sense designed to make possible the perception of objects at a distance were not associated with the faculty of moving toward some objects and away from others, and, consequently, with the consciousness of the muscular effort which produces movement and the train of tactile sensations which accompany the use of this faculty? The general harmony of nature would be impaired, giving rise to a monstrous combination, completely devoid of any stability. Given a fundamental instrument and a function, we could easily conceive of accessory improvements being

successively added to it or taken away from it; but it would be absurd to assume the accessory improvement by restricting the fundamental part. Consequently, and even assuming that reason can give a solution which is not arbitrary to the question at issue, we should be unable to draw from this solution any legitimate inference, or any proof for or against the foundations of our knowledge, since all legitimate inferences must be drawn from the consideration of the general order of nature (81) and not from a hypothesis which places one in direct contradiction to this general order. Nevertheless, we believe that if we were to amuse ourselves by imagining an intelligent animal which is deprived of all means of locomotion and of tactile sensation, but which is provided with an organ of sight similar to the eyes, attached to flexible and retractable appendages such as nature has given to certain inferior species, and with which the creature would be able to make explorations in all directions, and with consciousness of this control, then we should have to regard such a creature as being capable of acquiring the notion of the *externality* of things. It would be necessary to assume that it would gradually discover, through changes of perspective, which of its impressions had resulted from the displacement of perceived objects over which it has no control and which had come from voluntary movements of its visual organ. Thus, even though it were unable to attain a clear perception of three-dimensional space, it would at least have gained an idea of a continuum having not less than two dimensions or a sort of rounded surface upon which it would project all images, without taking account of the distance they might be from it. This idea is very much like that which children and childlike persons have of the heavenly sphere. But let us spend no more time on a discussion infected so inevitably with vagueness and arbitrariness.[5]

106. It is better to proceed with our analysis and to turn finally to an examination of the sense of touch. This is not, like the other four, a special sense and does not reside in a distinct organ, but must be regarded as a general instrument of sensibility, as though the animal itself were coming into contact with the external world

[5] See the *Traité des sensations* by Condillac; the article by Reid entitled, " Géométrie des visibles," II, 486ff., of the French translation of his *Works;* but especially the *Manuel de physiologie* by J. Muller, II, 270ff., of the French translation by Jourdan.

through the whole of its sentient covering. Here the sensitivity characteristic of nervous tissue is not heightened for the perception of the most delicate impressions. Rather, it is dulled and weakened by protective organs. This accounts for the innumerable differences in tactile impressions, depending on the varieties of shape, of structure, and of functions of the organs of the animal and of the coverings which protect them.

Passive touch, or the vague sensation of contact, has, with reason, been distinguished from active contact or touch properly so called. In this respect, too, the sense of touch is distinguished from the other senses. This distinction does not mean that there are any absolutely passive sensations. That action and reaction are inseparable is just as true in physiological phenomena as in every other; the optic nerve or the olfactory nerve could not be affected by light or by scented particles unless there were a reaction by the nerve fiber. The fact of reaction serves to make the phenomenon of attention and of the persistence of the reproduction of emotions and images comprehensible, although still quite imperfectly so. In the same way, all sensations may stimulate, and in general are intended to stimulate, movements which reveal the clearest manifestation of the activity of the animal. But, although the movements which follow a sensation of taste or of smell ordinarily contribute nothing, or contribute little, either to strengthening the sensation or to making it more clear, the ordinary effect of tactile sensations is to stimulate movements by which these sensations are repeated, understood, and localized, until they have attained the degree of clarity that the animal seeks by performing these movements. The sense of touch thus becomes a surer instrument of perception because it is more susceptible to voluntary direction; and it is precisely this prerogative of the inferior character of tactile sensibility, in the anatomical and physiological order, which explains why this sensibility does not devolve exclusively upon organs having an entirely special perfection. Moreover, all animals are provided with organs which are not so much special as they are unique, by which the acuteness and the activity of the sense of touch are heightened, like the hands of human beings, the trunk of elephants, the whiskers of cats, and the antennae of insects, for example. In this respect there are great differences among the most closely related species, because the modifications affect accessory structures having neither the fixity nor the value characteristically reserved for deep-seated and fundamental traits of the organism, although they may have the greatest influence on the habits of the species, and although they may be found to be quite evidently related to the environment in which these species must live, with their diverse instincts of hunting or of propagation. These same accessory modifications are not of such a nature that they change either the fundamental conditions of knowledge or the essential characteristics of perception. No doubt, there is good reason why all naturalists from the time of Aristotle have seen in the remarkable structure of the human hand one of the causes of the pre-eminence of our species, one which has effectively concurred not only in the development of man's industry, and consequently of his actual power over nature, but also in the development of his intellectual and moral faculties, the superiority of which gives this actual power the consecration of right. Everything in the economy of nature is joined together and marvelously co-ordinated: it gives at one and the same time the intellectual superiority and the mechanical instruments which intelligence must use. Nevertheless, while recognizing this providential harmony, it is always necessary to distinguish the basic from the accessory, the essential from the accidental. Because it would be difficult, and perhaps practically impossible, to develop a reputation for ability in physics, in chemistry, and in anatomy without having received from nature a certain manual dexterity, it must not be concluded that manual skill is what essentially makes the great physicist or the great anatomist. No one would ever take it into his head to deny that the ideas of the external world formed by an unfortunate person who from birth had been deprived of the use of his hands would be entirely in conformity with those of other men.

107. We have already recognized that the sensations of heat and cold, which are part of the tactile impressions, could be abolished without the body of our knowledge being altered thereby; and that consequently they do not contribute directly and essentially to knowledge. From this it may also be concluded a priori that in themselves they have no representative value. Moreover, the same thing must be said about all those affections of the tactile sensibility which remain obscure and confused among

most men, but which, in certain blind persons, are said to acquire
a surprising delicacy and clarity.   One person is able to distinguish
by touch between the polished surfaces of glass, wood, marble, or
metal, even when all these substances have the same temperature
as that of the hand; another person will go farther and will distinguish the polish of oak from that of beech, that of porphyry
from that of statuary marble, that of steel from that of copper.
Yet all these sensations will lack any representative value and
will give no notion of the varieties of molecular structure which
must probably be related to them as their cause.   The same thing
will be true of these diverse tactile sensations that is true of the
sensations of taste, of smell, and of color; they could be entirely
abolished, and in fact they are practically nonexistent among most
men, without decreasing in the least our knowledge of external
nature.   The only difference would be that we should no longer
have at our disposal a sensitive instrument which would become
valuable, when others are wanting, for indicating what bodies we
are dealing with, assuming that we had acquired in some other
way knowledge as to the nature, the constitution, and the properties of these bodies, knowledge which it would be impossible
to derive from the sensations in question.

Some German physiologists, Weber and Valentin, among others,
have presented the ingenious idea of measuring with precision the
degree of acuteness of tactile sensation in different areas of the skin.
In the area being explored the skin is touched with a pair of
dividers, the arms of which are unevenly separated—the points
having been covered with cork so as to avoid injuries or impressions that are too sharp.   The distance between the points
is noted at just the moment when the impressions they make
begin to be distinguished from one another.   Values are obtained
in this way which vary greatly from one area to another, and from
them tables may be set up to make comparisons easier.[6]   Pursuing this same idea, let us imagine that we have marked off an
area on the surface of the skin within which tactile sensations
acquire their *maximum* acuteness and clarity.   This area will have
a great likeness to the retina, although it would without doubt
be very far from being able to attain the amazing delicacy of the

[6] The reader will find one of these tables and accompanying observations in
the *Manuel de physiologie* by Muller, I, 606 of the French edition of 1845.

retina. If pads of silk, or wool, or other substances were substituted for those of cork, the tactile impressions would be altered
as the visual impressions are when one color is substituted for
another. But, although the sensation will be changed with respect to its matter or substance, and in a way which has no
influence upon perception or knowledge, the essential form of the
sensation, in which its representative capacity is to be found, will
be found to remain constant for each organ and, moreover, the
same for one organ as for the other. Instead of two, it is possible
to conceive of a larger number of points acting simultaneously
upon as many sensitive spots, which are, furthermore, variously
spaced and assume variable patterns. Tactile sensations circumscribed in this way will be capable of engendering the image of a
one- or two-dimensional space, under the same circumstances in
which this image would result from the impression produced on
the retina by visual rays (105). They would also give rise to
illusions comparable to those affecting the sense of sight, and of
which one may get an idea from the well-known example of the
ball that feels like two when it is rolled between two crossed fingers.

Now, without spending more time on these arbitrary suppositions, let us restore the sense of touch to all the organs by means
of which an animal acts upon external bodies and upon his own.
Let us give the animal the internal awareness of the muscular effort
which determines the movements of the organs. Let us further
take it for granted that the organs yield to the incitement of instinct and to the impulsion of the will, by adapting themselves in
their articulation to resistant bodies, patterning themselves after
them by simultaneous or by successive contacts to which the
faculty of recollection lends a quasi-simultaneity. The notion of
space in all its clarity will emerge from this series of complex
sensations through a judgment of pure form, and, moreover, whatever the nature *sui generis* of the tactile sensations of heat and
cold, smoothness or roughness may be. This would be true even
though these sensations were replaced by others of which we have
no idea at present, and even though the knowledge of the movement produced would result in a sensation of a different sort than
that which accompanies the contraction and the exertions of the
muscular fibers in us. In fact, the resistance which objects set
up against the use of muscular force adds to the image of space and

to the notion of " externality " by suggesting to our mind the ideas of solidity, of materiality, of mass, of inertia, and so on, which we use in imagining and explaining the diverse phenomena of the physical world. Here again, visual sensations would reveal no more to us than those which come through taste, smell, or hearing. But all the tactile sensations of heat and cold, of smoothness and roughness, and so on, would give us no more if they were not accompanied or followed by the use of muscular force, under the influence of that branch of the nervous system which controls voluntary movements, a branch that is now known—through the discoveries of modern physiology—to be clearly distinct from that which is intended to receive and to transmit the various sensory impressions: the tactile sensations of heat, of cold, of smoothness, and of roughness quite as much as the sensations of colors, of sounds, of odors, and of tastes. This important physiological discovery is marvelously useful in the philosophical analysis of sensations, since it forces us to distinguish, in the complex situation that the physiologists have designated by the name of active touch, that which is truly a sensation from that which is connected with the use of an active function belonging to an organic structure perfectly distinct from that of the sensory nerves, although it is necessarily connected with this structure. Now, for a moment let us concern ourselves only with the sensory impressions and with the notions or representations which they are able to give us by themselves. We shall examine later on the representative value of the physical notions which are, for our understanding, the less immediate products of the consciousness we have of the functions of the nervous motor system and of the exertion of muscular force.

108. So, after all, of the five senses with which nature has endowed man and the higher animals, all of which surely have a great, although unequal importance in the order of the functions of animal life, there are really only two which for men may be essentially instruments of knowledge. As instruments of knowledge, these two senses are in some ways identical; they are homogeneous, or they furnish homogeneous images and knowledge; namely, the representation of space and the knowledge of the relations of magnitude and of geometrical configuration. In each of these two senses the representative property is connected with the form

and is independent of the matter of the sensation, with *ratione formae* and not *ratione materiae*. The other senses, or the functions of these two senses with respect to which they must be considered heterogeneous, contribute to the increase of our knowledge only in an indirect and accessory manner, by furnishing reactions; that is to say, by furnishing means of recognizing the presence of agents concerning the nature and structure of which we know only what sensations endowed with a representative capacity have taught us. The fundamental sense for knowledge—active touch— is not connected with a special structure with which it has pleased nature to furnish certain privileged species, and of which individuals may be accidentally deprived without being deprived of the means of self-preservation and of dealing with the external world. In its essential characteristics, if not in particular and individual perfections, touch consists of that which is most fundamental to the general character of animality. The consequence that must naturally be deduced from this is that other senses, or an increase in the perfection of the senses we now possess, should assist in the progress of knowledge, just as the discovery of a new reagent or of a new instrument would. These probably would put us on the track of phenomena the existence of which we do not now suspect, and which perhaps we shall never suspect. A change in the number or in the acuteness of our senses would not alter the formal conditions of the representation and the knowledge of phenomena. Such senses would modify our actual theories in so far as they are conjectural and go beyond observation, but not in regard to that which is only the pure co-ordination of observed facts. We do not intend to offer this either as a demonstration or even as an inference, like those to which reason cannot refrain from yielding, but as a very probable inference which there are good reasons to accept, not from motives which are connected with routine or blind habit, but from those which spring from a reasoned analysis of observable facts. Therefore, when we speak of certain nervous states in which the organic conditions of sensibility seem to be all confused, in which the operation of certain senses would be suspended while others were abnormally livened, in which even senses unknown in normal states are brought into play, so to speak, we shall prudently avoid denying phenomena of which we can say nothing other than that their explanation surpasses our understanding.   But we shall not hesitate to reject those which imply the upsetting of the essential conditions of our knowledge and which are none other than the fundamental laws of living creatures and themselves necessarily related to the laws of the world in the midst of which animated beings live and act.

109. If we trim away from a sensation or from the memory of a sensation everything which in itself has no representative value, everything that does not contribute to knowledge or that contributes to it only indirectly as a reagent or as a sensitive detector—as has been explained, everything that could not occur in it without necessarily halting the ordinary development or the scientific progress of knowledge—what will remain will be the " idea," or the true knowledge, of the object.   If, on the contrary, we take a complex sensation, the idea with its accessories, or rather, the idea with its sensory support, we shall have what may be called, in opposition, the " image " of the object.

Were we to consult only etymology, " idea " and " image " would be equivalent terms—the one derived from the Greek, the other from the Latin.   But because the former has only recently become a part of our common language, and because it was for a long time limited to the philosophical vocabulary, usage, by varying the meanings of one term or the other, has always associated the former with intellectual functions of a higher order.

From the point of view of etymology, the terms " idea " and " image " would refer to nothing more than impressions received from the sense of sight and to recollections of these impressions. The picture formed on the retina when the eye is open to the action of light would be carried, so to speak, into the brain or into the mind.   No doubt this is a very crude way of accounting for the perception of visible objects and of their representations in the mind.   But there is no other which presents itself naturally, and in turning to this metaphor, man has only obeyed the law which requires him to fix all purely intellectual notions by means of symbols or by sensory comparisons.

If the word " image " and its derivatives had never been changed from this original acceptation, the faculty of imagination would be inseparable from that of perceiving colors; the blind would have no images; and the only images in the writings of the poets would be those which apply to the eyes or which consist in the reproduction of impressions produced through visible objects. Writers have not observed this distinction, and usage has not drawn a rigorous line of demarcation between the meaning of the terms "idea" and "image" in either familiar or literary style. Generally speaking, the latter is used to designate perceptions which come more immediately from the senses, the former to designate those which have required a more active co-operation of the forces characteristic of the mind. The features of a person who is very dear to me and the sound of his voice are images presented to my thought; but I treasure the idea of his good nature. In speaking of a weak person who has been in danger, the expression is commonly used that he looks like the image of death; and of a fervent Christian, that the idea of death is the object of his habitual meditation.

110. In so far as we need a more precise language, we may have it without difficulty if we make use of the distinction established above and of the analysis which has provided and motivated this perfectly clear distinction. Thus all ideas of the form and the dimensions of bodies will be the same for the blind as for all persons having normal vision, although the former certainly imagines objects differently from the latter and independently of accidents of light, shadow, and color. Habit does not permit us to lay our representations of these entirely aside, even when our attention is concentrated upon qualities or properties which are independent of any accidents of light. Those who are born blind and those who have normal vision, when thinking of the same geometrical demonstration, will construct in imagination the same figure, and will have identical ideas of it, but not the same image; and because both think by means of intellects which are organized in almost the same way, they will both need images, but not necessarily the same images, to think of the same idea.

The idea of objects that a person with normal vision will have is more complete than that acquired by a blind person if left to himself, because the former will have the idea of the property inherent in these bodies as a result of the reflection of light rays which are distinguished from other rays by certain inherent characteristics, such as that of having a certain index of refraction, or that of exerting a certain chemical action. But the sensation of color, although entering into the images normal persons have of bodies, will not enter into the formation of the idea of the body, although it may have suggested one of the elements of the idea.

A deaf-mute who has had sufficient instruction will have the same idea of acoustical phenomena that we have; a man whose organism would make him insensible to heat and cold, as has been explained, would have all the ideas that we have of the nature of heat and of its effects. But the image that such a person would have of heat, like the image of the phenomena of acoustics for the deaf-mute, would imply only movements, changes of distance and of configurations; they would be devoid of the train of sensible impressions which, however weak, the words " heat " and " sound " by themselves arouse in us.

111. There is an analysis which separates objects, and an analysis which distinguishes them without isolating them.   Thus, in experiments with the refracting prism, rays of different colors which had always been connected with each other are found to be broken up unevenly and separated throughout the remaining portion of their paths.   This is an example of an analysis which separates and isolates.   But suppose, as Brewster ingeniously posited in the construction of the hypothesis already noted (102), that rays of different colors had the same index of refraction.   We should then have no way of isolating them in such a way as to describe their different paths.   Nevertheless, if a certain medium has the property of reducing the intensity of rays of certain colors—if not entirely, then at least in a proportion which increases with the density of the medium traversed—we should still be able to separate the beams which would have the same path and to refine the medium progressively with respect to one or the other.   We could do this by drawing a conclusion from data furnished by observation, through a legitimate inference, the principle of which has been set forth above, provided it were possible totally to exclude the beam which may be only partially and gradually extinguished.

An analysis of this second type is applicable to the distinction between the pure idea and the train of sensible impressions which necessarily accompany it.   This is true because of a law inherent in the nature of the human mind; because the mind is not a pure intelligence, but one which functions through the assistance of organic structures; and because the intellectual life of man is closely associated with an animal nature from which it gets what is needed to nourish it and strengthen it.   Even though we may not completely free the idea, we may at least refine it successively, and gradually weaken the sensible impression or the image which remains connected with it in the operations of thought; and we may recognize clearly that neither the essential characteristics of the idea nor the results of the operations of thought depend either on the kind or on the intensity of the image or sensible impression. Nature itself, by gradually weakening certain sensible impressions through the mere effect of habit, undertakes to provide this analysis which must then be completed by a judgment of reason which has been expressed in the adage, as true as it is forceful, *summum principium remotissimum a sensibus* [" the highest principle is farthest removed from the senses "].

112. Because of the necessity for giving substance to the idea through the use of sensible images, the need arises for conventional symbols, which play so important a part in the development of the human mind, and concerning which we will have some important observations to make later on. We are already in a position to point out that the sensible impression of the sound of the articulate voice or of the letters of the written word becomes duller through habit, and consequently takes away from the idea a lesser part of attention in proportion as spoken or written language becomes more familiar to us. However, the idea can never be wholly freed from the need of support by sensible impression, even when language serves us only as a means of conversing with ourselves and as needed for our solitary meditations.

By imagination, we ordinarily mean an essentially active and creative faculty, an aptitude for grasping with ardor, and for expressing with energy, through images drawn from sensible nature, the emotions of the soul and the inspirations of the heart. But beneath this poetic faculty is another less brilliant one, which also consists of the ability to associate sensible images, as used by the mind, with ideas which are often of the driest sort and the least able to excite our enthusiasm and to arouse our passions. This faculty is found very unequally among men, depending upon whether it is perfected by exercise or weakened by inaction, and, quite probably, also because of certain individual variations in organic structure. Certain persons can easily and distinctly think of a regular polygon of six, seven, or eight sides; but no one is able to imagine a polygon having a thousand sides, and to think of such a figure it is necessary to use artificial symbols. Nevertheless, the properties of the latter polygon are as fully known to the

mathematician and the idea that can be formed of it is as clear
as are those of a hexagon or a square. We can picture to ourselves the motion of a body, provided the motion is neither too
slow nor too rapid; but we can form no image of the vibration
of a taut string which makes five hundred oscillations a second,
although we have just as exact an idea or understanding of that
motion as we should if it were moving a hundred times more
slowly, and might, therefore, be gotten hold of by that faculty of
the imagination of which we are speaking.   Certain marvelous
aptitudes, such as that for doing certain very complicated calculations very rapidly in the head, must be related to the singular
development of this faculty.   This ability has nothing in common
with mathematical genius—as some persons, including some who
should know better, are tempted to believe about such prodigies;
for that genius occupies itself with ideas, and discovers new relations among them.   Nor has that ability anything in common
with the talent which makes it easy to follow and to co-ordinate
the discoveries of the genius in the field of ideas, even though the
ability to imagine may aid the genius or the talent, just as a good
memory would, without our being justified on that account in
saying that a good memory is the cause either of talent or of genius.
We shall completely ignore the organic causes of a finer memory
or of a greater ability to retain and construct images of things.
But we may understand them, even though we are probably still
far from knowing the organic causes of the superiority of the genius
working on the ideas, supposing this superiority to be attributable
to organic modifications.

The question of knowing whether the animal or the young child
has ideas brings us back to the question whether any knowledge
of external objects, and of the qualities pertaining to these objects,
is connected, among animals and children, with the images and
the impressions of the sensibility. Since we have no doubt that
animals and children may have at least the beginnings of knowledge, we shall admit without hesitation that they have ideas.
But these ideas will be incomparably less refined and less clearly
distinguished from sensible impressions than are those of the adult
man, and, especially among animals, they will be incapable of that
indefinite perfection, of that continual progress the glorious
privilege of which God has reserved for man.   On the other hand,

since we do not mean by " idea " the capacity of knowing, but an actual knowledge, we are not concerned with any question about latent ideas, or innate ideas, since no one can doubt that the first traces of knowledge and of the intellectual life appear only after a previous development of the functions of animal life, and when the sensibility has already been awakened by a mass of diverse impressions both general and restricted.   Nevertheless, all these questions are connected; for it is a law of animate nature, once the most curious and now the best established, that during their development the higher species pass through phases which if not identical with are at least very similar to those at which the lower species finally stop.   Consequently, since there are species which are limited to the most obscure sensory impressions even after their complete development, this in itself gives good reason for presuming that there must be, even for the highest species, transitional phases in which the sensible impressions are obscure and confused to the same degree.   But we shall content ourselves with indicating this relation here and shall come back to it later on when we glance at the whole of psychology and at the connections between the faculties of intellectual life and the faculties of animal life.

CHAPTER VIII

ON OUR NOTION CONCERNING BODIES, AND ON OUR IDEAS OF MATTER AND OF FORCE. ON THE DIFFERENT CATEGORIES OF PHYSICAL PHENOMENA AND THEIR SUBORDINATION

113. Scholastic philosophers and, since the discoveries made in the field of experimental physics, even modern metaphysicians have insisted at great length on the distinction between the *primary qualities* of bodies and their *secondary qualities*, meaning by the former the qualities of extension, impenetrability, mobility, and inertia, and by the secondary qualities those which produce on our senses impressions of taste, odor, color, heat, cold, and so forth. We propose to submit this classification, which has the sanction of long usage, to a new and more exact criticism. But first we must point out that if we mean by primary qualities those which we are completely unable to explain by the help of other properties, and which in this sense constitute so many primitive and irreducible facts, then, so far as we are concerned, none should figure more prominently among the primary qualities of bodies than those which philosophers have been in the habit of calling secondary. In fact, we have already recognized that sensations of tastes, odors, and so on, are so many modifications of our sensibility which have no representative value whatever, which by themselves would be unable to give us the notion of bodies and of the existence of the external world, and which imply no knowledge of the reason why they are invariably found to belong to one species rather than to another. Consequently, the capacity that objects have of producing certain sensations in us is necessarily an inexplicable one, the connection of which with other known characteristics we are unable to demonstrate, even though experience has taught us that it is constantly associated with other characteristics. Thus, from the fact that a body has caused us to experience an acid taste, we are clearly able to conclude, on the basis of earlier

171

experience (99), that it must also have the property of combining with salty bases and that, when the product of this reaction is decomposed by means of an electric current, the substance in question will move toward the positive pole of a battery. Nevertheless, we remain completely ignorant why chemical compounds which are characterized by this double property act upon our organs of taste in such a way as to produce an acid rather than a bitter, acrid, or astringent taste. The same chemical compounds also turn litmus paper red. Although at the present time no one is able to explain exactly why this is the case, it is possible that the day will come when someone will explain why it is that when litmus paper is brought into contact with acids, it gives off the least refrangible rays of light, rather than others which occupy a different position in the spread of the solar spectrum. What will never be explained is why these rays which are least capable of being refracted make us experience the sensation of red rather than that of blue or of yellow; in a word, the reason for this is to be found in the relation between the index of refraction of the color and the nature of sensation. Not only do we not now know the cause of such a connection, but the nature of things precludes the possibility that we may be able to know it, and we can affirm that we shall never know it. In this sense, therefore, and relative to us, the properties of visible bodies which have been termed secondary qualities are just those which most deserve to be called primitive and irreducible facts.

On the other hand, if, so long as explanations and proof are lacking, analogies and inferences are taken account of, there is reason to believe that the different specific qualities by means of which all, or at least certain, bodies act upon our organisms are far from being so many fundamental qualities of these bodies from which all others would be derived. They are most often not even immediately connected with the truly fundamental qualities, but are, on the contrary, separated from them by a considerable number of intermediate links in the chain of causes and effects, of principles and consequences. The bark of cinchona, or the quinine which is extracted from it, has the property of causing in us a very bitter taste; at the same time it has what, for us, is the more unique and the more interesting property of breaking up fever and forestalling periodic returns of it. These are accidental, inexplicable, or unexplained characteristics, but not primary characteristics, of this substance, at least not in the sense that a person would attempt to find in them the reason and basis of the other characteristics. We scarcely notice such specific qualities unless we are motivated by some wholly private interest, or unless it is a matter of discovering those plants having similar properties which are beneficial or harmful to other animals, but not to man. It is on the basis of the nature of a thing itself, and not on its relation to other things on which it may accidentally act, that we must derive the classification of its different properties according to their intrinsic importance and their real subordination. Thus no one will attribute to the two specific properties we have just cited the same intrinsic value that will be attributed to the chemical nature deduced from the property quinine has of combining with acids in the same way as a salifiable base. Therefore, leaving aside all specific qualities which relate to a mysterious action on our organisms, among which must be placed those which produce different affections of our sensibility, we shall distinguish among the other qualities of bodies, not primary qualities and secondary qualities, but fundamental or primordial qualities and derived or subordinate qualities, which may in their turn be conceived as being distributed hierarchically, to the degree that their characteristic value grows weaker and as they are a less immediate result of the fundamental properties.

114. Even when we are completely without any understanding how to explain the qualities of bodies or how to connect them with other properties which would be the primary cause of them, we are amply justified in regarding them as not constituting fundamental and absolutely irreducible qualities so long as we see that they fail to persist and that they may disappear and reappear, depending upon the circumstances in which the body is placed and upon the modifications which it undergoes. For the same reason, a certain property will be regarded as being much more closely related to what is fundamental and essential in the nature of the body or as being farther removed from it, in proportion as it shows more persistence or more instability. For example, the substance that modern chemists know as carbon and which we find in two such radically different forms, namely, the diamond and free carbon produced as a by-product of the combustion of vegetable or animal

matter, possesses great stability in both forms.   That is to say, if it is not absolutely infusible and non-volatile, it is at least very difficult to volatilize, and above all to melt, even under the action of very intense heat.   Here we have a persistent quality and one that should therefore be regarded as being more closely connected with the fundamental properties of carbon than are the translucence or durability of diamonds, with which the opacity and friability of carbon are contrasted in such a strange way.   Finally, this same character of stability or infusibility disappears in chemical compounds in which carbon is one of the elements, and so cannot have as fundamental a value as other properties of bodies in general, or of carbon in particular, which remain unchanged throughout a whole series of chemical reactions in which the same particles of carbon may be successively involved.

Mineralogy offers us the most varied examples of this gradation of characteristics.   Thus, limestone, or, to use its scientific name, calcium carbonate, is found in a great variety of forms—earthy, compact, fibrous, crystalline, laminated, saccharine, granular. These are evidently related to entirely accidental circumstances of formation; they can serve as a basis for characterizing clearly neither this substance to the exclusion of others, nor even the specimens of this substance in which they change from one structure to another.   On the contrary, if a study is made of the crystalline forms of limestone which are prodigiously numerous and which characterize so many well defined varieties of the same mineral species, one will be discovered, namely Iceland spar, of which the others may be regarded as so many derivations or secondary modifications; for that reason it may be called the primitive or fundamental form.   Therefore, the property of assuming crystalline forms which can be reduced to this fundamental type has an altogether different sort of importance for the mineralogical species of calcium carbonate than the characteristics deduced from the fibrous, laminated and other structures have for the varieties and for the specimens.   Yet this crystalline type has itself only an inferior value in comparison to those which can be drawn from the chemical composition of the substance. This is the case since, in addition to the great number of crystalline varieties of calcium carbonate which can be reduced to Iceland spar, there is another basic form, aragonite, the crystalline structure of which is essentially different although its chemical composition is exactly the same. Thus, in this case as in all analogous cases of *dimorphism*, the chemical element or molecule persists while the crystallographic element or molecule is destroyed and replaced by another. However we conceive of this subordination of characteristics, it cannot be doubted that, in a great number of cases, the characteristics that are associated with chemical composition surpass in importance, in an equally large number of cases, the characteristics connected with the crystalline type. Crystallographically there are two species of limestone, the distinctive characteristics of which are fundamental with respect to those which distinguish separate varieties from the point of view of their regular or confused crystallization; and these two forms are based in a single chemical type the characteristics of which have a still more fundamental value.

The subordination of characteristics as general and particular is one thing; their ranking as fundamental and secondary is something else again. Without doubt, and in qualities which are equally persistent, we are reasonably led to regard as more fundamental the quality common to a larger number of objects, and with much stronger reason, that quality which, without exception, would be found in all bodies. But, if such a quality persists in a certain type of object and endures throughout all the modifications to which members of this species may be subjected, we must regard it as more fundamental than that which is common to a much larger number of specifically different bodies, even though it is less persistent in each of them in particular. Thus, although the property of being solid at ordinary temperatures, or of being liquid or gaseous at the same temperatures, are properties each of which is common to a great many substances, such a property cannot be thought of as having the importance or the characteristic value of those properties which pertain only to one species or to a few species of substances, but which are indestructible in them and which resist all the causes under the influence of which the substances change state, by passing from a solid to a liquid, and so on.

It is through considerations and inferences of this sort that in zoology and in botany the different characteristics of organic beings are assigned varying degrees of importance or of value by deducing all the elements of this classification from the evidence of observation, from the careful comparison of observed facts, and from the weight of inferences or analogies. So it is evident that in matters of this sort we are unable to affirm anything a priori as to the relations of subordination between facts whose inception is hidden from us by such a thick veil!

115. The problem is not the same when we are dealing with properties of objects that are called inert, with regard to which we find ourselves in the presence of a belief which is native to the human mind, a prejudice common both to philosophers and to the common people. This prejudice consists in admitting a priori the existence of certain fundamental properties or qualities, common to all bodies, as constituting their essence, and by necessarily involving the reason or the explanation of all secondary qualities, whether we can or cannot give this explanation in the present state of our knowledge. It is for this natural belief or philosophic prejudice that we must try to account. It is necessary for us to examine its legitimacy, by taking advantage for that purpose of all the information for which we are indebted to the progress of experimentation and the perfection of the sciences.

Extension and impenetrability have customarily been placed at the head of the list of primary or fundamental qualities. But from the first, the common notion of impenetrability, like that which we gain from touching a solid body and from the feeling of resistance it gives to our muscular effort, corresponds to a very complex phenomenon, of which the highest mathematics have not yet been able to give us a really satisfactory account up to the present time, in spite of its unsparing production of hypotheses. This phenomenon of impenetrability is that of the very composition of the solid body, by means of atoms or molecules that exist at a distance from one another. If the solidity is no longer attributed to the bodies themselves or to the molecular aggregates, but to the ultimate molecules which may be the constituent elements of bodies, a hypothetical conception is introduced in order to satisfy a propensity of the imagination. Experience can neither deny nor confirm this hypothesis which really plays no part in the explanation of phenomena. The pretended primary quality will turn out to be only an imaginary quality, and, from our point of view, will certainly be a gratuitous assumption.

In fact, it must be noted that in the hypotheses to which

modern physicists are led, namely, that atoms which exist at a distance from one another, and even at distances which—although imperceptible to us because of their extreme minuteness—are very large in proportion to the size of the atoms or other elementary bodies whatever they may be, there is no obligation to conceive these atoms as little hard or solid bodies, rather than as small masses which are soft, flexible, or liquid. In bodies which are large enough to be sensed, solidity and rigidity, like flexibility, softness, or liquidity, are so many very derivative and very complex phenomena. We try to explain them better to ourselves by means of hypotheses concerning the law of the forces which maintain the elementary particles at a distance, and concerning the extension of their sphere of activity, with respect to the number of molecules included in this sphere and to the distances which separate them. But whether these explanations are satisfactory or not, it is incontestable that they prejudge nothing and can prejudge nothing concerning the condition of hardness or softness, the solidity or fluidity of the elemental molecule. Our preference for thinking of them as hard rather than as soft, our tendency to imagine the atom or the ultimate particle as a *hypermicroscopic* solid rather than as a fluid mass of the same degree of smallness, are only prejudices resulting from our training, prejudices which follow from our habits and from the conditions of our animal life. We should have other prejudices and other tendencies, if nature, while allowing us the same degree of intelligence, had so arranged things that our adult life would be spent immersed in a liquid environment, which was one of the conditions of foetal life, without habitual contact with bodies which, it is true, do not have that absolute hardness or ideal solidity which we attribute without reason to the elemental particles of matter, but which would nonetheless approach more nearly the condition of perfect solidity or rigidity than any other ideal state.

But, it will be said, impenetrability is not rigidity, and a body is no less impenetrable simply because it is liquid, in the sense that while the mass is penetrated by the separation of the parts, the parts themselves are not. Without doubt these atoms which can never come into contact may still less penetrate each other. And precisely because of that, reason has no basis for admitting, as far as these elementary particles are concerned, a so-called essential or fundamental quality which, on the contrary, would be a useless and insignificant quality and never would nor could come into play. The impenetrability of atomic particles is nothing more or less than their mobility and their effective displacement through the repellent action which other particles exert at a distance. That being the case, impenetrability must not be made into a primary quality, distinct from mobility. Otherwise it may be said that we have asserted without foundation the existence of a distinct quality which never manifests itself and which never plays any part in the explanation of phenomena.

116. The same thing must be said of extension, considered not as the space occupied by bodies but as a quality of bodies. No doubt objects which we experience by means of our senses give us the idea of a bit of continuous, formed, and limited extension. But this is only a false appearance or an illusion. Just as the whitish color and the continuous appearance of the Milky Way are resolved by a powerful telescope into an agglomeration of distinct luminous points of absolutely imperceptible dimensions, so conclusive experiments resolve the phantasy of an extended, continuous, and formed body into a system of atoms or other infinitesimally small particles. It is true, the laws of our imagination oblige us to attribute a form and dimensions to these particles, although there is no rational basis for this, since all explanations that can be given for phenomena, whether physical, chemical, or of other types, are independent of the hypotheses that may be proposed concerning absolute or relative forms and dimensions, and atoms and elementary particles. Both experience and reason seem to indicate that these particles are centers from which attractive and repulsive forces emanate; but no observation makes it possible to learn, or even to assume, whether they may have the form of spheres, ellipses, pyramids, cubes, or whether they take any other curved or polyhedral form. It would seem at first sight, and it was believed for some time, that the laws of crystallography pointed to the polyhedral form of the elementary particles. However, as these laws came to be better understood and more adequately interpreted, every conclusion of this sort has been found to be without weight and to be contrary to the inferences of careful scientists. Thus, to recall only one fact that has been previously mentioned (114), the dimorphism of certain substances

obliges us to admit that crystalline form is not an invariable and inherent property of the ultimate molecules of bodies, but the more remote result and consequent of a mode of grouping among the molecules, the form of which—if indeed they have form— remains as completely undetermined for us as it was prior to the study of the phenomena of crystallization. Indeed, this consequence, to which the development of our study leads us, can be and ought to be foreseen; for it is repugnant to reason to admit that, with the organs and faculties with which nature has endowed us, we should be able to attain the essence of things and the ultimate and absolute reality (8 and 10). Yet we should be able to reach this through the atomistic system, if we were able to determine the form of the ultimate elements, of the indestructible atoms, the existence of which would explain all physical phenomena, while in itself it can be nothing less than an immediate decree of the creative will. Reason would find it no less difficult to admit that a primitive and inexplicable decree had given preference to one polyhedral form over another, had chosen a certain number of degrees and minutes rather than others for the inclination of two sides or of two edges, had given to the sides of polyhedrons, to the radii of spheres, to the axes of ellipsoids, some one fraction of millionths of a millimeter rather than some other; as if, in fact, largeness and smallness were not entirely relative and that there might be intrinsically some reason why atoms and the system of atoms had been constructed to one scale rather than to another. We do not wish to insist at this point on this completely Leibnizian argument, which rests upon considerations which we shall develop a little later in connection with the critique of the idea of space.

Is this to say that we must substitute for the common hypothesis of atoms having definite although extremely small dimensions, and having definite although unknown forms, another hypothesis about the constitution of bodies of the sort that has been proposed by Leibniz himself and by many other philosophers called " dynamists? " Not at all, since this would reproduce under another form the pretension that we think is untenable, namely, that of penetrating into the essence of things and of establishing their primary natures. On the contrary, we shall admit that the atomistic theory has a necessary use; that we could not get along without it in the language of the sciences, because our imagination must rest upon something, and because that something, as a result of the facts we have analyzed in dealing with sensations, can only be an extended and formed atom or corpuscle. But reason intervenes to abstract the idea, or that which constitutes the object of true knowledge, from the image that serves to sustain it, the necessary intervention of which is only a consequence of the laws of our own nature. The atomistic hypothesis is one of a number of hypotheses the use of which, as happens so frequently in the sciences, cannot be criticized so long as we do not commit the error of mistaking the material which enters into scientific construction for what is only external scaffolding; and so long as it is fully recognized that these hypothetical conceptions are not introduced as ideas but as images, and introduced because the human mind finds it necessary to graft ideas onto images (112).

117. It is time that this discussion lead us to speak of the properties of bodies which must really be considered fundamental. And first, in order to distinguish more clearly positive facts, the indubitable results of observation, from hypothetical conceptions that may often be confused with them, let us briefly recall the facts in the order in which they are established by experience. On the one hand, observation teaches us that the form, appearance, and condition of bodies may change, that they may be broken up and scattered, but not destroyed. Thus, if all the new products which may be formed, if all the integral particles of bodies which have apparently disappeared are carefully collected, a sensitive balance will indicate this very important fact—the total weight remains exactly the same, without increase or decrease. On the other hand, this result of observation squares well with a law of our understanding which leads us to conceive something absolute and persistent in everything that is revealed to us through relative and variable qualities. In short, more delicate observations and a more advanced theory show us that this constancy of weight in bodies, in spite of changes of state, is connected with a more general law in virtue of which the parts of bodies, taken in their totality, offer the same resistance to the action of moving forces or require the same expenditure of force to attain the same velocity, whatever the appearance and the manner in which the particles are grouped together and whatever the nature of the

force expended to produce their motion. Finally, in order to show that it is necessary to exert twice or three times as much force, or to repeat the application of the same force two or three times in order to impress on body A the same velocity found in body B, we say that the *mass* of A is two or three times that of B; so that all results of experience which enter into the question are expressed by saying, on the one hand, that the weight of a body is proportionate to its mass, and, on the other hand, that the mass of a body is invariable, fundamental, and remains constant throughout all modifications that the body is capable of undergoing, and is, moreover, a measurable quantity, *sui generis*, in regard to which one body may be compared to another body, but which, in the same body or in the collection of parts of which a body is composed, cannot in any way be increased or decreased.

This indicates to us the meaning that must be attributed, in the language of physics, to the term "matter." However, when this term is used in other contexts, it has or has had some very different meanings in ordinary language and in philosophical terminology. Those objects which we call bodies and which fall immediately under the senses may be destroyed, so far as their individuality is concerned, by the separation of their parts. That which persists after a body has been destroyed or changed, by remaining invariable in the collection of parts, is what we call matter. Matter is the ground, the unknown and imperceptible *substratum* of which mass—which falls within the scope of observation—is for us the constant and characteristic attribute. The nature of our mind is such that we find ourselves forced to conceive a *substratum* or an imperceptible basis as a foundation for all the qualities that the mind grasps, and we are similarly forced to suit language to the necessary form of our conceptions. If, in addition to such properties as mass, which is common to all bodies which are ponderable and the constituent parts of which are indestructible, there are others by means of which these bodies or the elements of these bodies differ radically from one another, so that the qualities to which these specific differences relate must be thought of as being primitive and irreducible just as weight and mass are, the idea of matter will also imply that of a ground in which these different qualities inhere. It will also be necessary to admit bodies that differ from one another not only as a result of different arrangements among the parts of their homogeneous matter, but also in being composed of diverse or heterogeneous materials.[1] This is what experience is able to teach us concerning the composition and the essence of ponderable bodies. Anything that the imagination is able to add to the representation of this essence is of no value in the eyes of reason. If, for the reasons of which we have spoken, we are led to picture to ourselves the bodies which we know by means of the senses as being made up of other bodies which elude the senses—bodies which are perfectly impenetrable, rigid, and indestructible corpuscles or atoms, of invariable forms and dimensions—this conception remains purely hypothetical. We do not know whether or not the masses of these elements are or are not proportional to their volumes, are or are not dependent upon their forms, or upon other qualities of which we have no idea.

118. If we are insuperably ignorant of what the essence of tangible and ponderable matter is, there is all the more reason for our having no real knowledge of those intangible, uncoercible, and imponderable principles to which we relate the marvelous phenomena of light, electricity, and heat, in which we should see not simple accidents of ponderable bodies but rather, in all likelihood, the manifestations of something which would still exist, even after the destruction of all ponderable bodies. It is one of the laws of our minds to have recourse to the same representation for the one as for the other. Thus, when a physicist undertakes to bring to light the laws of the distribution of electricity on the surface of a conductor or the laws of the distribution of magnetism in a magnetic bar, it is easy for him to imagine a fluid, or many fluids, which are spread out in layers of variable thickness or density. But he clearly understands that these fluids have only a hypothetical existence, that at bottom we have no idea of them, and that they are not objects of real knowledge, either for laymen or for

---

[1] " At that moment, for the first time, there was established the heterogeneity of substances and the nature of forces *which do not manifest themselves by means of motion,* and which also introduced the principle of *combination* and *mixture,* alongside that of excellence of *form* as this was understood by Pythagoras and Plato. Everything we know about matter rests upon these differences of form and of mixture. They are the abstractions under which we believe ourselves able to embrace the totality and the movement of the world by means of measurement and analysis." A. von Humboldt, *Cosmos,* II, 268.

scientists. The physicist is perfectly aware that these hypotheses are used theoretically only as a scaffolding or as auxiliary constructs to assist us in conceiving and formulating laws which govern phenomena the real cause of which absolutely escapes us. Moreover, and notwithstanding this identity of images, everything points to a profound contrast between the properties of ponderable matter and those of imponderable principles. Not only can the latter never be detected by means of balances, as their name indicates, but also they seem not to share at all in the inertness of matter, since they offer no appreciable resistance to ponderable bodies, and since their accumulation or their dispersion gives rise to no observable increase or decrease in the mass. While the mass of a ponderable body is something essentially definite and limited and at the same time something absolutely indestructible, it seems that it should be possible to draw electricity out of or add it to a body indefinitely, provided that one draws out or adds equal quantities of electricity of opposite polarity at the same time. It appears that we should be able to imagine without contradiction that electricity or heat are completely destroyed or created through the effect of chemical or molecular actions. In a word, everything that is the real basis of the idea of matter, with respect to ponderable bodies, either appears to be contrary to our experience, or at least has as yet not been established by experience with respect to the so-called imponderable fluids.

119. Let us come back to the idea of "force" which we have already seen to be necessarily correlated with the ideas of mass and matter. In a great many circumstances, bodies are manifestly inert; that is to say, they are moved only through the action of external and apparent forces. In other cases it seems that bodies, even when deprived of any principle of life, can change their places with respect to one another, or are set in motion by an internal movement; finally, the capacity for spontaneous movement seems to characterize living things. But that capacity changes with external circumstances and with the internal nature of bodies; while that which persists in the elements of bodies or in what we call matter is "inertia," that is to say, that property which requires that, if motion is to occur, a certain force must be expended which is proportional to the mass set in motion when its velocity is constant, and proportional to the impressed velocity when the mass

remains the same.   This explains why we are justified in saying that matter is inert, without making any prejudgment about the inertia or the inactivity of complex things which we call bodies. And on this basis nothing is more natural and more in conformity with the observed subordination between phenomena than to conceive of a force which, by acting upon the matter of which a body is formed, impresses activity and movement upon it even in those cases in which we are unaware of the action of an external and apparent force.

Experience shows us that the inertia of matter consists not only in the fact that matter remains at rest when no force or cause of motion is applied to it, but that it persists in a state of motion and continues to move uniformly and in a straight line when no external force or object checks its motion or changes either its speed or its direction.   Consequently, it is said that the inertia of matter consists in indifference to rest and to motion, so that what we call "mobility" of bodies should not be regarded as a special quality, but only as a consequence of the principle of inertia.

120. The idea of force ordinarily originates in the internal sensations that result from our power as mechanical agents, and from the effort or the muscular tension which is the organic condition of the exercise of this power.   We extend this idea by supposing that something analogous resides in all agents capable of producing the same mechanical effects, and use the expression "the force of steam," "the force of a stream of water," or "the force of the wind."   By a process of abstraction familiar to mathematicians, all the physical qualities which make these various entities so different from one another are put aside.   Account is taken only of the direction in which the forces tend to move the bodies, and the speed they tend to impress on them; two forces are equal when they tend to impress equal velocities on a given mass, no matter what the nature of the agent and the physical conditions of the action that it exerts.   We do not need to investigate further the origin and basis of the idea of force in order to verify by experience and to establish by reason the general principles of mechanics and to deduce remote conclusions from them by means of mathematical calculations.   But natural philosophy does not stop there.   In fact, it is very evident that the energy of a gas or vapor and, even more, the tension of a muscle are derived and complex phenomena which

we need to explain through simpler facts, rather than being able to furnish the primordial type which serves to explain other phenomena. The same thing is true of forces that is true of bodies. In both cases, that which immediately affects our sensibility and that which is the immediate object of our perceptions is not the fundamental and primitive reality but a complicated product which we must endeavor to submit to analysis in order to grasp the principles and the basis of it, if this is possible.

121. The Cartesian school wished to rule out the idea of force by likening it to the occult qualities of the ancient Scholastics. The basis of the Cartesian doctrine consisted in wishing to explain everything by means of corpuscles, some of which are coarser and larger, and others of which are smaller and finer, and which, in their movements, necessarily displace one another because of their impenetrability; as if the impenetrability and the movement of a circumscribed portion of the extended were not also occult and inexplicable qualities of which we are able to form a true or false idea only by means of a complex and unexplained phenomenon, namely, the composition of solid bodies which we sense. Indeed, if we admit, on the one hand, solid and impenetrable molecules and, on the other hand, forces by means of which these molecules act upon one another at a distance without the intermediation of material links formed by other contiguous and impenetrable particles, two hypotheses are set up in place of one, we confess to two mysteries instead of one, and the number of mysteries or of primitive and irreducible facts must not be increased unnecessarily. But it is clear, as we have already seen, that once action at a distance is admitted the extension, form, and impenetrability of atoms or elementary particles no longer apply in any way to the explanation of phenomena and serve rather to assist the imagination; so that, in reality, in Newtonian physics, which is accepted by all contemporary schools since Cartesian physics has long since ceased to be fashionable, only a single hypothetical principle enters into all scholarly explanations, whether that principle is the notion of force or of action at a distance, or the notion of the communication of movement through contact, as a result of the impenetrability of contiguous molecules.

122. It is only through proof, that is to say, through the actual application of a principle to the rigorous and mathematical concatenation of natural facts, that we are able to judge the value of a

principle. Newton has had the distinction of having submitted the idea of force or of action at a distance to such proof, and in a most decisive manner. Whatever may be said of it, he formulated a hypothesis, and a most audacious one, by imagining in all particles of ponderable matter a force of which the gravitation of terrestrial bodies is only one manifestation, and which shows that those particles which are separated from one another nevertheless all continue to act upon one another. But thanks to the genius of this illustrious man and of his successors, there emerges from this hypothesis the most complete explanation of the grandest and most beautiful phenomena in the universe. The simplicity of the law of the gradual diminution of force in proportion as distance is increased, the mathematical form that may be given to it, all concur to lead us to regard it as a fundamental law of nature; and this applies equally to all other forces which play a role in the explanation of physical phenomena and which follow the same law as Newtonian gravitation.

But we no longer find this characteristic in the theories of modern physicists in so far as the problem is, for them, to explain the internal and molecular structure of the phenomena that bodies present to us by always taking as one's starting point the idea of action at a distance and among separated particles. In that case they imagine forces whose sphere of action extends only to distances which are imperceptible to us, but which nevertheless include an almost infinite number of molecules. Here are (115) the two new postulates without which it would become impossible to give any tentative explanations of molecular motions by means of the principles of rational mechanics, that is to say, by means of the notions of mass and of action at a distance, combined with the theorems of geometry. However, this is very far from saying that by means of these same postulates the work of mathematicians and physicists has resulted in a theory which, like that of universal gravitation, explains all observed phenomena and often anticipates the results of further observation.

123. If among molecular phenomena special attention is given to those which make up the subject matter of chemistry, it will be seen that chemical theories are completely independent of every hypothesis by means of which an explanation of these phenomena could be given by mechanics. The progress of mechanics has not

contributed to the advances in chemistry and the progress of chemistry has in no wise reacted upon mechanics. It would not even be difficult to show that chemical phenomena are, in every aspect, incompatible with an explanation which has its source in the conceptions of rational mechanics and geometry. Attraction or repulsion between molecules at a distance can only produce effects which are governed by the law of continuity; chemical affinity, on the other hand, gives rise only to sudden associations and dissociations and to combinations in definite proportions. How would this sharp distinction between different molecular states be accounted for if chemical actions, differing only because of distances, underwent only infinitely small alterations when the distances themselves varied only by infinitely small amounts? Likewise, if the separated elementary atoms were to differ from one another only with respect to their dimensions and forms, or if the groups which constitute complex chemical molecules were to differ only because of the number and configuration of the elementary atoms which are maintained at a distance from one another in the interior of each group, we do not see how it would be possible to explain the essential distinction between chemical radicals and compounds, and to account for all the action of the affinities which produce the compositions and decompositions in which the chemist is interested. Differences of mass are no more capable of explaining all these phenomena than are differences of shape and of distance, since variations in mass are also subject to the law of continuity. In addition, the theory of chemical equivalents manifests a most remarkable contrast between the mass that is dealt with in mechanics, which is measured by the weight and the inertia of bodies, and what is called chemical mass, which is measured by the capacity for saturation. Therefore, from every point of view, we are led, together with M. von Humboldt (117, note), to the conclusion that chemical phenomena are inexplicable by means of the principles of mechanics alone; and that the notions of affinity and of elective attraction, upon which the explanations of chemists rest, are irreducible notions to be inscribed upon the catalogue of primary ideas that reason admits without explaining, and that it is forced to admit if it is to account for the concatenation of the observed facts.

124. Thus, on the one hand, we have the idea of a certain sub-

ordination among different categories in which the phenomena of nature arrange themselves, and among the scientific theories brought forward to explain the facts in each category. On the other hand, we understand that in passing from one category to another, it is possible to present solutions of continuity which hold not only in the present imperfect state of our knowledge and methods, but indeed in the necessary intervention of new principles needed for subsequent explanations, and because of the radical impossibility of following the thread of deductions from one category to another without the aid of these new principles or postulates, and to some extent without change of key or rubric. Nothing would be more useful for the wholesome critique of human knowledge than an exact table of these different keys or rubrics. Beginning with Aristotle, logicians have tried many times to construct an inventory of the fundamental ideas or categories under which all our ideas may be arranged. But a taste for an artificial symmetry or for a too formalistic abstraction has prevented them, at least up to the present time, either from agreeing on the drawing up of the list of categories, or from using it as a basis for the progress of our sciences and methods, for knowledge of the organization of the human mind or of its relations to external nature. On the contrary, now that the sciences have undergone so many developments unknown to the ancients, we determine *a posteriori* and through observation itself what the primary and irreducible ideas or conceptions are to which we must constantly have recourse for the apprehension and explanation of natural phenomena, and which, therefore, must be imposed upon us, either by the nature of things themselves or by the conditions inherent in our intellectual make-up.

Furthermore, it is less important to distinguish clearly the truly distinct categories than to have a correct idea of their hierarchical subordination. Now, on the basis of what precedes, it seems that the course of nature consists in passing from more general, more simple, more fundamental, and more permanent phenomena, to more particular, more complex, and more changeable phenomena. In the scientific study of the laws of nature first place is given to the general properties of matter, the fundamental laws of mechanics and those of universal gravitation. On these laws, and on some others, which although less well known probably have no less compass and generality, depend the great cosmic phenomena, and as it were the framework of the universe of the fundamental characteristics of the plan of creation. The constitution of astronomical systems, the nearly geometrical regularity of the movements and forms of the stars—all those sublime phenomena which impress us equally by their simplicity and by their grandeur, and which have always excited the highest admiration of men, as much in ages of poetic ignorance as in periods in which the rigor of scientific methods, the dryness of calculations and formulae seem no longer to leave any place for emotions of the spirit and for the pageantry of images—all must be ascribed to these laws.

Other phenomena tend to subordinate themselves to them, as the details and embellishments of a design are subordinated to the general traits which characterize the movements and attitudes, as specific and individual varieties are subordinated to the pre-eminent characteristics which denote the type of a genus or of a class. We call these phenomena molecular because we have no way of accounting for them except by imagining matter in a state of extreme division which far exceeds the limits of sensible perception, and by comparing what goes on among the ultimate particles with the actions we observe among bodies the dimensions and forms of which enable us to experience them. But, whatever may be said for this hypothesis, whatever explanation we may wish to give for the internal and chemical differences of bodies, we are dealing here with a cause of specific distinction which, on being added to the general properties of matter from which the great cosmic phenomena result, gives rise to phenomena of another order. The latter are more complicated, more particular, and less stable; and in this very complication, nature moves by degrees, so that, as the complexity of chemical combinations increases, they become less permanent and stable, and the phenomena to which they give rise are more particular, changeable, and, so to speak, ephemeral.

However, let us hasten to add that these inferences which still depend only upon the observation of phenomena of the physical and material world, abstracted from the marvels of organization and all manifestations of vital activity, would be insufficient to set forth clearly the hierarchical subordination upon which our attention is just now fixed. We cannot grasp the law of a series by understanding its first terms, and, above all, we cannot be sure of the law we trust we have grasped.   It is necessary, therefore, to pass on to the examination of phenomena of another order, phenomena which are more varied, richer, and better able to furnish fruitful comparisons.   Let us see if they can be classified, or if they fall into classes by themselves, in conformity with the principle of subordination which has already been presented to us by the comparison of the phenomena of inorganic nature.

CHAPTER IX

ON LIFE AND ON THE SERIES OF PHENOMENA
WHICH ARE SUBORDINATED TO
VITAL ACTIONS

125. By following the course indicated in the preceding chapter we are led to the consideration of the simplest phenomena that we find in the world of living things, which nevertheless far exceed the most complex phenomena of corpuscular physics in the degree of their complication. In order to explain living things it is necessary to take account of the basic characteristics of matter; it is necessary to know how to apply the mechanics of liquids and solids; above all, it is necessary to bring in chemical actions, and even the choice that nature has made of a small number of chemical elements possessing unusual properties, which almost exclusively make up the materials of the organic kingdom. This last consideration alone is sufficient indication that we must seek in chemistry the most immediate conditions for the development of organic forces. On the other hand, if the chemist regards as chimerical the attempt to reduce the explanation of the phenomena he studies and of the laws he establishes to a problem of ordinary mechanics, the physiologist regards as still more chimerical the attempt to explain, in terms of the concurrence of chemical and mechanical laws alone, even one of the simplest phenomena of organic life, such as the formation of a cell or of a drop of blood, or, among more complex phenomena which nevertheless depend most immediately upon the play of chemical actions, such processes as the digestion of food or the assimilation of nourishing fluids. It would be still more difficult to overcome the reluctance of reason to admit that the solution of the enigma of creation may be found in the formulae of mathematics or chemistry. With the appearance of organic and living things, a type of phenomenon comes into being which adapts itself to the great laws of the material universe, and which presupposes their unceasing concurrence, but the understanding and scientific ex-

planation of which evidently requires the express or tacit admission
of forces or principles over and above those which suffice to explain
more general and more permanent phenomena.

126. The same progression is observed if we turn our attention
to more detailed matters. All organic beings, whether animal or
vegetable, have certain common qualities, certain analogous functions. So it seems, as Linnaeus indicated in his aphoristic style,
animals differ from plants only in the insertion of one life upon
another. This idea goes back to Aristotle and was given an illuminating development by Bichat [1] through the sharp contrast
he so clearly set up between organic life on the one hand, which is
common to both plants and animals, is always active and never
suspended, is of the highest importance from first to last, and is
always obscure and without consciousness of itself; and animal
life on the other hand, which is essentially irregular and periodic,
appears much later and comes to an end much sooner, and is
gradually perfected with the system of organs appropriate to it in
the various species of the animal world. In a word—in conformity
with the law we are indicating—phenomena which are marked by
the latter are of a higher and less stable form than the phenomena
of organic life which serve as a basis for them. This is not the
place to examine the detailed objections into which the theory of
Bichat runs. The assumption which underlies his concept has
found its place in science and has become the basis for teaching;
and the objections to it only prove the difficulty or the impossibility of submitting the phases through which phenomena of
nature pass in their progressive evolutionary development to our
rigorous distinctions of categories. Morever, it is clear that even
though all the phenomena of the organic structure of plants were
to be successfully explained by means of physics and chemistry,
and that even though all aspects of animal life may be likened to
the organic life of plants, we should still not have the explanation

---

[1] [Marie-François Xavier Bichat (1771-1802), an important French physiologist and anatomist. He developed and contributed new and significant ideas
in connection with the study of histology and on the distinction between organic
and animal functions. These ideas appeared in his *Treatise on the Membranes*.
He also published *Researches on Life and Death*, his monumental study on
General Anatomy Applied to Physiology and Medicine, and two volumes of an
uncompleted work on *Descriptive Anatomy*, to which three more volumes were
added by his students Buisson and Roux.]

of a single phenomenon of animal life, such as sensation, pleasure, and appetite. In passing from one order of phenomena to another, a gap is always found which we try vainly to disguise by means of the artifices of language or to conceal through the ambiguous use of terms.

127. Does not the contrast between organic life and animal life, which Bichat so skillfully pointed out, have the greatest possible resemblance to the contrast between the body and the mind,[2] between the animal life common to man and the lower species, while differing in its forms and in its degrees of perfection, and the intellectual and moral life which we find in man alone but which, it may be said, is shared by all men although it is also subject to infinite diversity in its manifestations, depending upon the aptitudes and levels of culture of individuals and races? Have not all the great portrayers of human nature, all those who have studied it for a practical purpose, and consequently without being preoccupied with metaphysical systems and with the subtleties of the schools, vividly expressed that final contrast which is attested by man's consciousness, that inner sentiment found in persons of the most ordinary sort, persons who are least inclined to affectations or to mystical enthusiasms? Do not both these men, or rather both these distinct ways of living—although they share for the time being in both the organic and animal modes of life—follow different courses? Do they not have their distinct periods of infancy, youth, virility, and decline? Are not the principles and the tendencies of the one higher, and those of the other more fundamental and more settled in regard to these characteristics? Although Bichat's theory was introduced as a result of the progress of scientific observation, it seems that metaphysics, being more critical, cannot be satisfied with a distinction which the common man perceives, and upon which morals and religion have been based since the beginning of civilization. Especially in modern times, the exclusive importance that Descartes attached to the metaphysical notion of substance (and only in this is he an excessive disciple of Aristotle) his explanation based upon the distinction of two substances—the essence of one of which would consist in extension and of the other in thought—has made it a

2 " I see another law in my members opposing the law of my mind." St. Paul, Romans 7: 23. See in Buffon the article entitled " Homo duplex."

matter of habit to consider the distinction between the sensitive mind and the rational mind as a prejudice unworthy of rigorous logicians. Yet this is a distinction with which the ancients were very familiar; [3] it was proclaimed by the earliest teachers of Christianity; [4] it was conserved in the Scholasticism of the Middle Ages,[5] and it was upheld by Bossuet himself,[6] however much he

[3] " The soul of man is divided into three faculties: sense, understanding, and feeling. Sense and feeling are possessed by other animals: but understanding by man alone." Diogenes Laertius, viii. 30 (Pythagoras).

" Pythagoras first, thereafter Plato, divided the soul into two parts, to one they assign a share of reason, none to the other: the part which has a share of reason they make the seat of peacefulness, that is, a placid and quiet nature; the other part is moved by storms, now of anger, again of desire, contrary and hostile to reason." Cicero, *Tusculan Disputations*, iv. 5.

[4] " Let us see where the point of contact is between the outer man [*l'homme extérieur*] and the inner man [*l'homme intérieur*]. All those aspects of existence that we have in common with the brutes belong to the former. In fact, it is not only the body that must be called the outer man, but also that part of life which sustains the organism. When images of objects retained in memory are recalled by it, this is still an act which pertains to the outer man. Animals, too, can receive impressions of external objects through the senses, retain them in memory, and choose among objects those useful to them and avoid those that give them displeasure. But notice that these sensations are retained not simply as immediate sensations. They are purposefully committed to memory; and when they begin to grow weak as a result of being neglected, they are impressed on the mind again by reminiscence and reflection. The result is that while at first memory furnished material for thought, it is re-vitalized by thought. This process finally gives rise to an imaginary view of objects as a result of a process of collecting and comparing what had been scattered here and there, and distinguishes the probable from the true in this collection, not with respect to spiritual but with respect to physical things. This proof, and all others like it, although made with respect to sensible objects and carried out through the mediation of the senses, cannot be carried out without the aid of reason and belongs only to man. Yet the work of reason is higher than this; it is to conceive physical objects on the basis of ideal and eternal laws." St. Augustine, *Traité de la Trinité* (a fragment translated by M. Villemain in the *Tableau de l'éloquence chrétienne au IVe siècle*).

[5] " The soul has three powers or faculties: those of living, sensing, and judging. The soul uses one of these in plants, and two in animals; in man it uses all three of them; it exercises reflection and judgment as well as life and sensibility; it is what we call rationality or reason." Abélard, *Dialectique*. See the work by M. de Remusat, entitled *Abélard*, I, 462.

[6] See especially the *Traité de la connaissance de Dieu et de soi-même*, Chap. V, sect. 13. Moreover, it is Bossuet who has said elsewhere, " We see the great

may have been inclined toward Cartesianism, along with the great minds of his century. This distinction is simply that between animal and intellectual life, after all transcendental hypotheses concerning the essence of causes have been put aside, so as to take account of what is given by the observation of phenomena.

128. Nowhere, however, has Descartes' metaphysics survived as a principle of the scientific interpretation of nature. The idea of force, which was disposed of by the Cartesian school and which Leibniz brought back to an honorable position, furnished Newton with the remarkable explanation of the greatest phenomena in the universe. Taking their cue from Newton, mathematicians, physicists, and chemists all employ the idea of force or of action at a distance, in one form or another. Physiologists proclaim the necessity of admitting vital and organic forces for the explanation of phenomena exhibited by organic and living beings. Good sense should reject any idea which sees in the intelligent and moral man no more than a machine, plant, or animal having a more complex structure, although it should certainly study the mechanical and chemical phenomena in man, an organic mode of life which serves to sustain animal life, and an animal life onto which the intellectual and moral life are grafted. The absurdity in such a case is an example of confusing or identifying any term whatsoever in a hierarchical progression with those which precede and those which follow it. We should have neither more nor less success in deducing an idea or a rational conception from sensation than we have in explaining the seed of a tree or the egg of a bird by means of the action of chemical processes or in accounting for the sensation of color from the way in which the optic nerve is stimulated. Instead of the unique mystery of the connection of matter and mind—or, following Descartes, between extension and thought—we must admit a succession of mysteries, since all our scientific explanations presuppose the successive intervention and the harmonious concurrence of forces the essence of which is inscrutable and the irreducibility of which is for us a reflection of the irreducibility of the phenomena which emanate from them;

---

work which has its beginning, is continued, and is completed. In his designs, God always moves forward; he goes from matter to life, from life to intelligence, from intelligence to the soul, and he stops only when he has created man, that is to say, the being who knows Him."

so that there always are, in the field of human knowledge, illuminated areas which are separated by dark intervals, such as the eye observes in the solar spectrum when glasses having a sufficiently strong power of magnification are worn.

Finally—and this is the point we wish to insist upon here—the manifestations of these mysterious and irreducible forces appear to us to be subordinated to one another. The hierarchical law is evident. We invariably find some more particular, some more complex phenomena which, by their increasing particularity and complexity, imply the notion of a higher degree of perfection, merge into more general, more simple, and more constant phenomena, and which, by their generality and relative fixity, seem to us to partake of a still higher degree of substantial reality. Because of the natural disposition of the mind, this is the basis of a tendency to evaluate the importance of a certain order of phenomena either in terms of the degree of elevation and perfection or in terms of the degree of generality and fixity. These two contrary inclinations are the thing that is truly characteristic in the antagonism between the tendencies of the spiritualists and those of the materialists, tendencies that may be noted even among those who profess to ignore utterly the problem of what the essence of matter and mind are, and who do not subordinate the study of the laws of nature to ontological systems concerning matters which are beyond all means of our knowing.

129. In the study of living things, one general question looms up above all others: must the vital functions be regarded as the resultant and the effect of organic structure, or is the structure itself rather the resultant and the effect of vital and plastic forces? Here we find ourselves moving hopelessly in a circle, for it is just as impossible to conceive that organic forms precede life, as it is to think that life precedes organic forms, if not in time, at least in efficacy. There is no way of conceiving of life prior to organisms, for what would be the *substratum* of vital and plastic forces as long as no organism existed? On the other hand, it is unreasonable and contrary to all observation to admit that the interrelation of the parts of an organism is what produces life; for we distinguish clearly the vital properties of tissues from their mechanical, physical, or chemical properties which continue after life has been extinguished. In the same way, we distinguish between the con-

dition of a mature but unfertilized germ cell and the condition of this cell after it has been vivified by fertilization. Moreover, even the simplest organic element such as a globule or a cell already gives evidence of a structural plan, of a co-ordination among its parts which cannot be accounted for in terms of the concurrence of physical forces acting from molecule to molecule as is the case with those which we have admitted for the explanation of the form of inorganic bodies. Even if we were to suppose that the formation of the elements of which we are speaking had been related to a form of crystallization *sui generis,* we should be checked at each step in the passage to more complex formations. We should find that we had made no advance in explaining, for example, how the rudiments of the various organs are associated and co-ordinated by proceeding to fall in with one another in the formation of an embryo through epigenesis, or how a crayfish can regenerate an amputated member having the same form and the same parts as the original. We sense, more than we understand, that in such a case the plastic force and vital energy, far from awaiting the formation of organs before acting and far from being the resultant and the consequent of an arrangement of parts introduced by the concurrence of inorganic forces, on the contrary, control and determine the formation of the organism. Nevertheless, as it develops, the organism never ceases to govern and to modify the manifestations of the vital and plastic energy. Thus, in organic and living beings, organic structure and life play simultaneously the roles of cause and effect through a reciprocity of relations which has analogues neither in the order of purely physical phenomena nor in the series of acts which are subject to the influence of a voluntary or reflective determination. It follows from this that we can form no idea or conception of the principle underlying these mysterious phenomena, either through information which is furnished by the senses, or through that furnished by consciousness. The vital fluids that have sometimes been imagined—in imitation of the imponderable fluids admitted in physics—do not even have the advantage in this case of hiding our ignorance to some degree. Careful thinkers seem now to be in agreement in avoiding the superfluity and the abuse of these fantastic creations.

130. The term vital or plastic " forces " which prevails generally,

even though it does not give the mind an idea which can be clearly
defined, at least has this to be said for it: it expresses clearly one
of the most marvelous and most certain properties of the unknown
principle of life and of structure, that of going through different
phases of intensity and energy. The force leading to the reproduction of organs which are destroyed, among inferior species
where this process is observed, is weakened and exhausted by its
action, just as it is weakened and exhausted in higher forms by
the reproduction of simple tissues, if the process of regeneration is
repeated too often.[7] When animals like the earthworm show the
power of regenerating a whole animal from every segment obtained by simple dissection, it is noticed that if the regenerated
sections are subjected to repeated amputation, the animals successively produced become simplified and lowered in terms of the
scale of animal forms. Other species present the still more unusual
phenomenon of a single impregnation which is sufficient for many
successive generations. Nevertheless, the generative ability is
finally exhausted, and it is not transmitted from one generation to
another without being impaired. For an analogous reason, if it
rarely happens that the hybrids which result from the crossbreed-
ing of species are fertile, it happens much more infrequently that
the offspring of such unions are themselves fertile, and still more
rarely that fertility is transmitted to the offspring of the second
generation. If the tendency of the germ cells to reproduce individual varieties shows itself in the series of successive generations, even after interruptions or intervals, the repetition of such
intervals tends to weaken it and, finally, to exhaust it. What we
are saying about productive or regenerative forces applies to all
vital forces, or to all the diverse manifestations of the same force,
which are responsible for the development, repair, and conservation of the organism. The organic force in the embryo, in the
foetus, and then in the young during the whole period of growth,
is seen to be endowed with an energy which becomes weaker in
proportion as the main features of the organism are more fully
established and approach their definitive form. In its turn, the re-
productive force reaches its greatest energy, only to go through
analogous phases of decrease. Finally, the force which tends to
conserve the organs, which vies with the incessant action of the

---

[7] Serres, *Principes d'organogénie*, p. 142.

general forces of nature, and which momentarily brings within its sphere of action the material elements assimilated by the organism and later thrown off by it, is, as every one knows, used up and weakened by its own action until the final traces of it have disappeared.

131. The phenomena of living nature differ essentially from the phenomena of the inorganic world by the bonds of interdependence which unite harmoniously all the vital actions, all the parts of the organism, and all the phases of its development. Using Kant's expression, the cause of the mode of existence of each part of a living body is contained in the whole, while in dead or inert masses, each part carries this mode in itself. It is quite true that, according to the manner in which we conceive physical phenomena and the forces which produce them, the reason for the movement of each particle resides in the action that other physical particles exert upon it. It is on this basis that we explain the principle of inertia in matter. But we admit nonetheless—and in this our hypothesis is confirmed both by observation and by calculation—that the actions exercised on one molecule by another are perfectly independent of each other. There are as many distinct and independent actions as there are binary combinations among the particles. The effects of the actions or the forces are added to, neutralized by, adjusted to, or combined with one another, according to mathematical laws. But the forces themselves change neither in import nor in energy because of the conflict or the union which is established between them. In the organism, on the contrary, the action of each elementary organ or rudiment of an organ is visibly directed toward the realization of a certain function which can be conceived only in terms of the relations between the elementary organ and the organism as a whole. Similarly, the structure of each part is not independent of the way adjacent parts are constructed, as is the case in gaseous or liquid masses, or even in crystals, but rather is manifestly related to the structure of the whole. What is said of co-ordination in space may be said, with still more reason, of co-ordination in time. The organization of the embryo and of the foetus is suited not only to the functions it actually performs but also to those it must fulfill after subsequent development. The instinct which leads the bird to collect the material for a nest is related to func-

tions which will later be fulfilled in the propagation of its species. The instinct of the frugal animal is related to the situation in which it will find itself when the period of hibernation arrives. And we might go on to mention other illustrations of the same nature.

It is apparent from the above that a remarkable contrast is found between characteristics and methods as we go from sciences dealing with the properties of inorganic bodies to those which treat of the nature of living things. We reach order and unity in physics by breaking up or analyzing complex phenomena. The further the analytical reduction is carried out, the more clearly we see the phenomena connected according to a systematic and regular order. On the other hand, living things tend toward the perfection of harmony and unity or individuality by the complication of the organism at the same time that it tends to stabilize the direction of its development and its characteristics. Thus, among chemical phenomena, we find that combinations are distinguished from one another by characteristics which become all the more clear, or by properties which become all the more energetic or contrasting, as they become less complex. On the other hand, we find that those beings which are highest in the scale of organic structure are most easy to study and to classify to the degree that their organic systems are more distinct and have more fully determined functions and when at the same time the unity which co-ordinated them manifests itself more clearly. No one would begin the teaching of botany with the study of algae and lichens or the teaching of zoology with the study of sponges and polyps. On the contrary, we sense the appropriateness of studying first a type in which the organism, whether animal or plant, has reached its highest degree of complexity as well as perfection, so that we may later relate the inferior organisms to this type by taking account of the successive degradations and by indicating at each step the simplification of the organs, the decentralization of functions, and the obliteration of distinctive traits, until we come to the most rudimentary functions, and to beings that there is reason to regard as the first rough draft of the creative power (97).

Indeed, if the question is no longer one of describing and classifying beings, organs, and functions, but of apprehending analogies and transitions, of supplying, by means of philosophic inference, the breaks in continuity without which no system of classification would be applicable to the sequential order of organisms, of developments, and of metamorphoses, the direction in which investigation moves will necessarily be reversed. For example, in order to bring to light the analogies between plants and animals, it is necessary to begin with the inferior types of each series, among which the differentiating characteristics of the two series are still vague and undetermined. It will usually be advisable to go back to the embryonic form, observing in it the still undefined traits and the fleeting transformations. From this point of departure the resemblances and analogies are bound to outweigh the differences, just as later on, after the organisms have gone through all the phases of their development, and after the species has been formed in a definite way and in conformity with the conditions of final harmony, the characteristic differences will outweigh the original resemblances and analogies. As a result of this, science properly so called, that is to say, the methodological knowledge of precise, controlled, rigorously established facts which are capable of theoretical co-ordination, will rely principally upon the study of beings which have reached the peak (*summum*) of their organic development and complication. On the other hand, the philosophy of nature, based as it is upon the perception of continuous transitions and modifications, upon the appraising of analogies and similarities which admit neither of exact measurement nor of determination, will come to be related principally to the observation of simple organisms and those which have deteriorated into an elementary state. In a word, in the physical sciences, in chemistry for example, increase in complexity tends to fill in gaps, to show analogies, to do away with breaks in continuity, and to favor philosophical inference by reducing in just that way the importance of differentiating characteristics which serve as the basis of scientific determination and classification. In the natural sciences the opposite of this occurs through the increase of organic complexity. This is a fact of prime importance which clearly marks the passage from one order of phenomena to another, and the basic reason of which is found in the very essence of the organic structure, which is nothing but a tendency toward unity through the co-ordination of the parts.

132. The harmonic concurrence of forces, organs, and functions

in living beings must not be confused with the general harmony of nature. Although we admire an order and a plan in the arrangement of cosmic phenomena which leads us to recognize in it the work of an ordering intelligence, science as such, which has no means of probing the mystery of first causes, is by no means obliged to attribute any interdependent connections among the forces of nature which act as secondary causes in the production of general phenomena, any more than it has to accept them among the natural forces that man utilizes in a machine or a factory, even though his intelligence has enabled him to adjust the parts and to combine the forces in such a way as to make them concur in accomplishing a given end. In such a case, the force inherent in each part of the system follows its own law no less than it would even though the other parts of the system did not exist. For example, we understand that the planets would continue to gravitate toward the sun and to turn regularly around this star even if it were to cease to be a source of light and heat for them, just exactly as they do in the present arrangement of things in which the regularity of their movements appears to be so fully adapted to the manner in which they are influenced by the sun's rays. Likewise, although a manifest harmony may be found between the organic structure of herbivorous animals and that of plants which are destined to serve them as food, it cannot be thought that the forces which actively concur in the germination and development of the plants have an influence, in a similarly active way and as plastic or efficient causes, on the organic structure of the animal or vice versa. But when we consider the animal itself, as an individual and distinct being, it is, on the contrary, impossible for us not to recognize an interrelation between the plastic forces which bring about the formation of the heart in one place and that of the lungs and brain in another; between the vital actions which fashion the tissues and bodily fluids, and those which must ultimately excite the tissues and employ the secreted fluids in the animal as a whole; between the acts which provide for the accomplishment of a function and those by which the function is carried out. It is, then, no longer a question of a harmony which may be attributed to a providential co-ordination or to a fortuitous combination, but to an immediate, active, and determinative influence, which rests on secondary causes, and

which is exercised by forces to whose energy we immediately relate the production of the phenomena which we witness.   This
definite connection, or this marvelous *consensus* between vital
forces and actions, is what makes it legitimate to call them plastic
or selective forces in matters having to do with the phenomena of
organic life, and instinctive actions when we are considering more
particularly the relational functions, or in general all those which
pertain to animal life.   But consciousness brings to light only our
voluntary and reflective resolutions, while the senses and the
imagination show us only mechanical effects.   Therefore, we find
we are absolutely unable to have even an imperfect notion or
image of the operations of an active principle of which we know
nothing except that it acts with inevitable necessity, without
knowledge and without freedom, revealing itself by means of
products which are much superior to anything that a mechanism
could bring about, and superior even to those that the intellect of
man is able to attain through reflective arrangements.[8]

[8] " The organizing force which, obedient to an eternal law, produces and
animates the members necessary to the existence of the whole does not reside in
any organ; it is revealed through nutrition, even among acephalous monsters,
until the moment of birth; it modifies the already existent nervous system, as
well as all the other organs, in the insect larva which is metamorphosed, so
that many ganglia of the spinal cord disappear and others become wholly
destroyed; it is the reason why, during the metamorphosis of a frog, the spinal
cord is contracted in proportion as the tail loses its structure and the nerves of
the extremities develop.   Proceeding harmoniously and without consciousness,
the activity also enters into the phenomena of instinct.   Cuvier put it very well
when he said that instinct is a kind of dream or vision which always follows
the animals, and that they seem to have in their sensorium images or sensations
which are innate and constant, which lead them to act in the same way that
ordinary or accidental sensations commonly do.   But what excites this dream,
this vision, may be nothing but the organizing force acting according to rational
laws.   This force exists in the germ prior to all organs, in such a way that in the
adult it no longer appears to be connected with any organ.   On the contrary,
consciousness, which does not give rise to any organic product and which forms
only ideas, is a late resultant of the development itself, and is connected with
an organ on which its soundness depends, while the primary impulse of every
harmonic organization continues to act even among monsters having no brain.
Consciousness, together with the nervous system, is lacking in vegetables.
Nevertheless, there is in them an organizing force which acts in accordance
with the prototype of each plant species.   Therefore, it is impossible to regard
the organizing force as analogous to consciousness, and its blind, necessary act-

133. As a result of improperly comparing the principle of the
harmonic unity of living things, that is to say, the very principle
of life, to the principle, whatever it may be, of the general harmony of nature, philosophers have from the most ancient times
compared the world in all its immensity to a living being (μέγα
ζῶον), while physicians and physiologists have taken pleasure in
referring to man as a microcosm (μικρόκοσμος), a term which could
be applied equally well to any other animal as to man. But such a
comparison does nothing more than indicate a complete misconception of the profound distinction between mechanisms and
organisms, between inanimate nature and animate nature. The
world is not a gigantic animal, but a great machine, each part of
which obeys its own law and the force with which it is individually
endowed. So the reason for the harmonious concourse of the parts
must be sought for elsewhere than in the essence of these forces
themselves and in their productive capacity. Likewise, the animal
is not simply a little world, that is to say, a little machine included
in the larger one, but a being which contains its own principle of
unity and harmonic activity, awaiting only suitable external
stimuli and a favorable disposition of the surrounding environment
before it unfolds.

Nevertheless, it must be fully recognized that the bonds of
organic unity and interdependence show themselves to be more or
less limited or extended, depending on the particular case. In this
respect a plant is not comparable to an animal, nor a lower animal
to a higher one which nature has endowed with a more complex
and more perfect organization. Among the most perfect animals
themselves, there are some organs or systems of organs whose sensitivity is more lively, and others which carry out their role in the
total organism with more individual independence. In compound
animals and double monsters, the organisms interpenetrate in such
a way as to confuse the ideas that normal and ordinary cases suggest to us as to the independence of organic beings, and the interconnection of their constituent parts.

---

ivity should in no way be compared to the formation of ideas. Our ideas of
anything organic are nothing but simple images of which we are conscious,
whereas the organic force, the primary cause of organic beings, is a creative
force which imposes harmonic changes upon matter." J. Müller, *Manuel de
physiologie,* in the Prolegomena. From the French translation by Jourdan.

134. Finally, the contrast between purely physical phenomena and those that living beings reveal to us follows from the manner in which we conceive physical forces. They are assumed to be inherent in material particles as their permanent and indestructible *substratum*. The property of vital and plastic forces to which reason tells us the harmonic unity of organic beings must be related, in conformity with the type of each species and with the natural disposition for hereditarily transmissible variations, cannot be conceived of as adhering in a fixed and immutable manner to any material *substratum,* whether simple or compound. Thus, all the difficulties and all the mysteries with which philosophers have been preoccupied, especially with regard to those which have human consciousness for their setting and which give rise to voluntary and reflective acts, become visible on the threshold of physiology. It is not only for phenomena of this order, which are the highest of all, but for all functions of life, that the harmonic unity and formative energy which are always closely connected with organic dispositions and physical stimuli may not, after the manner of physical forces, be considered attached to a material *substratum,* whether simple or complex, to a molecule or to a system of molecules. This necessarily results in a lack of coherence in the systematic ordering of our conceptions and an interruption in their theoretical concatenation when we pass from the description or explanation of physical phenomena to the description or explanation of phenomena which are produced in the womb of living nature.

That is why, in natural history, it is so impossible to conceive the first appearance of living beings, and the formation of the first organism which has not come from some pre-existing organism, as, for example, we conceive without any difficulty the formation of crystals and the first manifestation of chemical phenomena arising from the gradual concentration of nebulous matter scattered in celestial space. In fact, just as soon as the forces to which we attribute the power of producing physical phenomena are considered inherent in the ultimate particles of matter, as though this were their *substratum,* we have no difficulty in admitting that they reside in it in a permanent manner—whether or not circumstances permit them to produce sensible effects. In this situation it is unnecessary to have recourse to an intervention of creative power to endow the material particles with forces or properties of this kind, even at the moment these forces first come into play. In other words, the origin or beginning of chemical phenomena is in no way mysterious to us, although the essence of chemical forces, like the essences of all things, is necessarily hidden from all our investigations.   On the contrary, a veil of mystery cloaks and must necessarily cloak not only the origin of living forms in general, but the origin of each living species and the causes of the differences between species according to time and place.   On the one hand, observation leaves no doubt that these species have not always existed; on the other hand, the data of observation are no less incompatible with our admitting a spontaneous development, a creation out of nothing, and the producing of animals and plants by means other than those of ordinary generation.   Thus we see that the scientists who are the least inclined to have recourse to supernatural explanations would not think of using the word "creation" to designate the formation of minerals and rocks, of strata and veins of ore, of deposits of coal and columns of basalt. They know that in the production of all these objects—even when actual circumstances preclude the production of them at the present time—we have no difficulty recognizing the action of physical forces which are still actually inherent in matter.   However, these scientists do use the terms "animal" or "vegetable creation" to denote the aggregate of species characteristic of a region or of a geological period.   They do not mean by this to make an appeal to a supernatural intervention, but simply to note that we find it equally impossible to admit, on the one hand, the endless duration of the order of phenomena that the whole gamut of living things presents to us, and to conceive, on the other hand, of the natural beginning of these phenomena.   The matter at issue here is not a problem of metaphysics, such as that of knowing whether or not the world is eternal, whether matter is created or uncreated, whether order is the result of Providence or chance; the issue is that of a purely physical or biological question which bears upon facts falling within the compass of the world we perceive and of periods of time of which we may, and in fact do, have extant monuments.   Here is a real gap in the system of our knowledge, a gap that reason feels the need of closing, but which it is unable to close precisely because it is not possible for us to reconcile our ideas about matter and those concerning the manner in which vital forces act by giving them a material *substratum,* and by relating them to the forces which produce the most general phenomena of the sensible world.

135. Now, what value must be attributed to the idea of a *substratum* or " substance " which introduced the incoherence we have indicated?   Whatever one may say of it, this abstract and general idea, which is the first of the categories of Aristotle, the cornerstone of many systems and the basis of all that we call " ontology," has no prerogative which puts it beyond critical examination.   It also asks to be judged by its works, that is to say, by the order and connection it introduces into the system of our knowledge, or through the difficulties it produces therein and the conflicts to which it gives rise.   This idea of substance comes originally from the knowledge we have of our identity as *persons* in spite of the continual changes that age, experience of life, and accidents of all sorts contribute to our ideas, our feelings, and our judgments, and to the judgments that others pass on to us.   Therefore, the idea comes naturally from the nature of the human mind, as the structure of language proves.   But when we use this idea, which is entirely devoid of sensible elements, to connect phenomena which can be perceived, reason could doubt the legitimacy of its application beyond ourselves if experience did not indicate that there is in fact something in bodies that persists in spite of all changes of form, of molecular condition, of chemical composition, and of organization (117).   These evidences from experience are sufficient to establish that the idea of substance, in the application that we make of it to bodies and to ponderable matter, is not simply a logical abstraction, a fiction of our minds; on the contrary, the reason and the basis for it are found in the essence of bodies.   This is true even though we may be condemned to remain in perpetual ignorance as to what this essence is, and even though those formed and extended bits of matter which we have found it agreeable to imagine, or rather that we find it necessary to imagine as a *substratum* for material phenomena and for the forces which produce them, are only purely hypothetical and may even be contradicted by all the evidence of reason (116).

When we extend this idea of material substance or *substratum* by analogy to agents that we call imponderable, our experience is deficient, at least up to the present time.   We observe phenomena, we discern laws, but nothing assures us that a systematization of these phenomena and their laws which would be found to imply the idea of substance is anything but an artificial systematization.

Experience might always have left us just as ignorant in regard to ponderable bodies as it leaves us in regard to imponderable agents. To tell the truth, we were as ignorant in the one case as in the other of the true basis of the idea of substance as long as physics remained in swaddling clothes, and as long as we had no means of establishing—notwithstanding certain crude and deceptive appearances—that there is no real loss of substance, that is to say, of mass and of weight, in the innumerable transformations which we can watch matter undergo. This does not stand in the way of observing the sequential order and the interconnection of phenomena in regard to ponderable bodies,[9] as we still do for phenomena attributed to imponderable agents, and it would be a great mistake to say that without the idea of substance the spectacle of nature becomes nothing but a phantasmagoria. On this basis, those parts of physics which have to do with light, electricity, and heat would still exhibit to the mind no definite limits and no actual reality, and would have to be regarded as erudite phantasmagoria; while the phantasmagoria ($\phi\alpha\nu\tau\alpha\sigma\iota\alpha$) is found, on the contrary, only in that artificial portion of our theories in which the imagination, transcending the limits of experience, creates fictions that reason accepts provisionally, but only in the guise of scaffolding and of auxiliary symbols (116).

If the rigorous procedures of experimentation, which are a product of the genius of modern thinkers, had contradicted the application of the notion of substance, even as related to ponderable bodies, if it had been fully established that there are certain circumstances in which mass and weight are destroyed, as there are in which vital force is destroyed, the circumstances under which this destruction takes place would have to be defined. But ponderable bodies would not, for that reason, have ceased to present us with the spectacle of well-ordered phenomena, *phaenomena bene ordinata*, as Leibniz puts it. Only we should have a better

---

[9] " If bodies were mere phenomena there would be no deception of the senses for that reason. The senses tell us nothing about metaphysical matters. The truth of the senses consists in this, that phenomena agree among themselves; nor are we deceived by events if we follow them with reasons rightly founded on experiment." Leibniz, ed. Dutens II, 519.

and more decisive argument for condemning the hypothesis of those formed and extended atoms that our reason already has so many reasons for rejecting, but that our imagination is still unable to give up.

136. The difficulty that we should experience in making this supposition, in conceiving physical forces which do not adhere to a material *substratum,* that is to say, by definition, which do not adhere to a physical atom or to a system of such atoms which have form and extension, is exactly what we experience in conceiving vital or plastic forces. This is so because matter and form seem to change roles in the passage from the phenomena of the inorganic world to those of animate nature; the persistence—to a certain extent—of the form or type takes the place of the persistence of mass and of weight; and the variability or even—within certain limits again—the neutrality of the material follows the variability or neutrality of the forms. It is to escape this difficulty that hypotheses have been formulated in the different epochs of science which today have been judiciously examined and are definitely condemned, such as those of spontaneous generation, of special creation, and others like them. We have not entered into detailed explanations of such hypotheses, which lie within the field of the anatomist and the physiologist rather than that of the logician and metaphysician. It is necessary to confess this difficulty, and even to recognize that it is insurmountable, since it arises from a contradiction between certain laws of nature and certain tendencies of the human mind. But it must not be further exaggerated. As we have already shown, ordinary physics is itself not free from analogous difficulties and contradictions. And if in certain cases the metaphysical notion of substance becomes a source of insoluble contradictions, would reason not have the right to condemn the farfetched applications of it to some type of phenomena that one might wish to make, simply because reason is completely aware that this notion has its root in the human mind and that it directs the organization of human language? We saw above other examples of contradictions that spring from the same cause, and with respect to which we felt that it was necessary to draw the same consequence, however harsh this may appear to certain minds, however abstruse or however trifling it may seem to others.

CHAPTER X

ON IDEAS OF SPACE AND TIME

137. We believe that in Chapters seven and eight we have demonstrated that the senses contribute directly to a knowledge of the external world only in so far as they give us the representation of extension. We have further seen that this representative power is linked with the form of sensation because it is through the form alone that there is homogeneity between sensations and the external causes of the impression produced. But we do not conceive the extended only as a property of material aggregates or of bodies which fall under the senses. We also conceive of extension as the place occupied by bodies, as the *space* in which bodies move and in which all the phenomena of the external world occur. This idea is such, or seems to us to be such, that it would still have an object even though bodies should cease to exist, even though the phenomena for which space is the stage should cease to be produced. Likewise, our sensations have the essential quality of duration; the memory of them persists or lasts even after the organs of sense have ceased to be affected by impressions from external objects. The phenomena of the external world, of which sensations give us knowledge, themselves have a duration; and from the notion of the duration of phenomena we go on to the idea of the *time* in which these phenomena are arranged and brought about. This idea is such, or seems to us to be such, that it would still have an object even though the phenomena of the external world should be hidden from our knowledge or cease to be produced. It also seems that this object would not be destroyed by our own destruction, by the suppression of that sequential order of the affections and of the internal phenomena which endure and which follow one another in us.

The two fundamental ideas of space and time are, therefore, not merely elements of the knowledge of the external world. They surpass this knowledge; and it is from this point of view that we consider them here. They manifest themselves to intelligence through objects which have a character of necessity that other ideas by means of which we conceive the external world do not have. On the basis of these ideas of space and time, thought of as having this character of necessity which imposes itself upon the human mind, rest the sciences which are capable of *a priori* construction and which borrow nothing from experience, which are independent of the consideration of the phenomena of the external world, and in the study of which images derived from the external world would serve only to aid the work of the mind (110) without having any effect upon the body of the doctrine.

Is this characteristic of necessity apparent or real? Does it pertain to the nature of things or to the nature of the human mind? Are the ideas of space and time nothing but ways in which the mind sees things, nothing but laws of its constitution? Or, on the contrary, do these ideas have a representative and objective value; and, if this is the case, what do they represent? There is no philosophical system to which the answers to these questions would not serve as a sort of a key. There is no philosophical question of any great interest which does not terminate at some point in questions to which these fundamental ideas give rise. We do not as yet wish to examine whether these questions can be resolved or whether they transcend the power of reason; whether the discussions of philosophers have clarified or confused them. What, as we see it, should be a clear result and the fruit of these discussions is the perfect analogy, the exact symmetry that all these questions present in regard to space and time; and present in such a way that the solution given or accepted for one of these fundamental ideas may, for that very reason, be given or accepted for the other, in all schools and in all systems.

138. Thus when Newton and Clarke admit the representative value of the idea of time in all its fullness, they are led to admit it for the idea of space. According to them, neither space nor time can be conceived as substances; they are the attributes of a substance. And, because the ideas of space and time are invested with characteristics of necessity and infinity, they are attributes of necessary and infinite Being. Time is the eternity of God; space is His immensity. The rigorous spiritualism, the religious faith of these great men may have vainly protested against the intention of giving extension and parts to the divine substance; the force of analogy carried them away.

Also, when Leibniz submits the two ideas of space and time to critical examination, based upon his principle of sufficient reason, the result of the test is the same for one as for the other. Neither space nor time can have an absolute existence. This is no more possible when they are dealt with as attributes of divine substance than when they are regarded as created substances. For, all parts of space being perfectly similar, there would be no reason why the world, thought of as finite, should have occupied a certain portion of infinite space rather than any other. And if the world is infinite, the whole universe could always be conceived as changing its position in absolute space while the parts of the system continued to maintain their relative positions, so that there would never be any reason why each element of it should have occupied a certain absolute place rather than another (116). Similarly, all the parts of time being perfectly similar, there would be no reason why the duration of the universe, if this duration is finite, should have corresponded to any given portion of absolute time rather than to another. And, if the world has neither beginning nor end, it would always be possible to conceive a displacement of the whole series of phenomena in absolute time which would not disturb their relative dates. Thus there would be no reason why each phenomenon should have been produced at a certain instant rather than at some other. Therefore, neither space nor time can have an absolute existence; space is nothing but the order of coexistent phenomena; time is only the order of successive phenomena. Do away with phenomena, and the idea of space, like that of time, no longer has an object.

Therefore, as a third and last example, when Kant, taking the converse of Newton's theory, refuses to allow any objective value to the idea of space, he does the same for the idea of time. Space and time are, for him, only forms of human sensibility, subjective conditions for the intuition of phenomena. Neither the idea of space nor the idea of time corresponds to the order of things as coexistent or sequential. They correspond to the order according to which the representations of things must be arranged in order to become objects of our intuition.

One should read the celebrated correspondence between Leibniz and Clarke. This may be taken as a model of dialectic. In it one follows in detail the analogy of which we are only delineating the principal traits. Let us affirm once more that the question is not one of championing a side in these metaphysical systems; it is not even one of knowing whether the pretension of having a system or a side to take in such matters is chimerical or not. Our problem is simply that of establishing an analogy, a correlation which should follow from the nature of things and not from our systems, since it shows itself in the most contrary systems.

139. The analogy in question is even more remarkable since it is not drawn from matters dealing with the psychological origin of the ideas of space and time and the nature of the sensory images which help us to conceive them. This contrast clearly proves that we have the capacity to raise ourselves above laws which follow from our own nature and from the organic conditions of thought, in order to apprehend relations which subsist among objects themselves and which follow from their intrinsic nature (88). Psychologically—and because of a peculiar quality inherent in the formation of our senses, as has been explained—for us extension is the object of an immediate intuition, of a direct representation; the device of allusions and symbols is necessary so that duration may become the object of our intuition. We imagine extension by means of the concurrence of sensible images which are naturally associated with it (110), but we are able to imagine duration only by attributing to extension a property representing duration. We *align*, so to speak, the successive phenomena, until we finally get an image and through that an idea of their positional order in time. This work of the mind manifests itself in the form of language: *antea* and *postea* which refer to order in time, are derived from *ante* and *post* which are more immediately related to spatial order. And, it is also generally true, for the perception of ideas of which sensibility does not furnish us immediate images, that we are obliged to combine images which have representative efficacy only indirectly and at second-hand, so to speak, through the aid of analogies that reason grasps between things that are otherwise heterogeneous. This is the basis for the establishing of symbols and the principle of human perfectibility.

Animals seem to us to have only a very obscure perception of relations of time, of duration, and of all that goes along with them.[1]

---

[1] " Wherefore the present time holds first place, because it is common to all animals. The past is as important for those which are endowed with memory. The future is, in truth, for fewer; in fact, for those only to whom the favor of prudence has been given." Scalig., *De caus. ling. lat.*, c. 113.

This seems to be true even of animals which most closely resemble man. In fact, it may be said that the senses have been given to man and to animals simply for the purpose of directing them in space. On the contrary, the practical use of reason is, for the most part, to guide man in time, to co-ordinate his acts in view of accomplished facts and of future circumstances. Since this superior destiny and these higher faculties have been denied to animals, the clear perception of time becomes superfluous for them. And since, even among the animals that are most like man, the faculty for perceiving time has remained and has had to remain in a rudimentary state, it has not been able, even in man himself, to attain the representative clarity which is characteristic of the idea of space. For in everything connected with the development of vital powers we observe that nature propagates variety without losing sight of a common plan in the whole series of living things, developing in one species what is only rudimentary in another or in all others rather than creating out of whole cloth that of which not even the germ exists elsewhere.

As for the perception of space, the innumerable species of animals evidently have it in very different degrees, depending upon their position in the scale of animal life. We always notice, in so far as induction enables us to deal with this problem, that the degree of this perception is perfectly suited to the sort of movements the animal in question ought to make as a result of its perceptions. Or rather, as has been explained (107), it is the act of movement itself which originally gives the animal a perception of space consistent with the functions it must satisfy. Animals have a more or less obscure and imperfect perception of space, but not a false one. Their perceptions are, to an appropriate degree, correct and conformable to external reality, and in addition are suited to the nature of acts which they must effectively carry out in space as a result of their perception.

140. The measurement of extension or of geometrical magnitudes is carried out by means of superposition, that is, by the most direct, the most sensible, and in some ways the most imperfect method of measuring. The measurement of duration is indirect and rests essentially upon a rational idea. We judge that the same phenomenon must be produced in the same interval of time, when all the circumstances remain the same in each recur-

rence of the phenomenon (48). Thus if nature reproduces, or if we are able to reproduce artificially, the same phenomenon in conditions which are perfectly alike, we believe with reason that we possess a temporal standard, or a chronometric unit, and we use it to measure the duration of all other phenomena. It is in this way that time or duration can be measured with a water clock [2] by taking as the unit of duration the time required for the liquid or the fine sand with which the instrument is filled to flow from one section of it to another and by setting up the principle, which is certain *a priori*, that the duration of the flow must be identical, if neither the mass of the substance with which it is filled, nor the vessel, nor the opening, nor any other purely physical conditions of the phenomenon have undergone any change. Yet we do not understand in a completely satisfying manner, either theoretically or experientially, the laws which govern the duration and the phases of the flow. The phenomenon whose duration is taken as the standard for time is, ordinarily, a periodic movement which is as uniform as possible, in order that the aliquot parts of the period may correspond to equal portions of the duration. But it is a mistake to regard the phenomenon of motion as the essential condition of the measurement of time. Yet this has often been done. The unit of time might be the time that a body—of clearly defined substance, form, and dimensions—requires to pass from one temperature to another in a medium the temperature and other physical conditions of which may be exactly defined. The duration of a physiological, or even of a psychological phenomenon, could furnish a standard of time, if it were capable of being reproduced indefinitely under perfectly identical circumstances, and in such a way that the repetition would not modify the conditions of the phenomenon.

Perhaps it will be objected that, if reason posits *a priori* the general maxim that the duration of the same phenomenon must be the same in absolutely identical circumstances, we have no means of establishing this perfect identity with certainty in individual cases. But here again a judgment of reason intervenes, based upon probabilities which go so far as to exclude doubt. When the first astronomers compared the diurnal movement of the sun with that of the stars, they were in a position to raise the

[2] [or hourglass.]

question strictly speaking as to whether it was the duration of the solar day which remained constant and that of the sidereal day which varies, or the other way around.   But before long many inferences led them to take the duration of the sidereal day as the fixed term, and when subsequently the whole theory of astronomical movements was seen to be interrelated according to regular laws by the hypothesis of this constant duration, while the hypothesis of the constancy of the solar day would have been the source of trouble and disorder with respect to these movements, there could no longer remain any doubt concerning the fundamental hypothesis of the invariability of the sidereal day.   And even long before the laws of mechanics had given the immediate reason for the invariability of the rotational movement of the earth, the period of which coincides with that of the sidereal day, all chronometers had to be regulated according to the movements of the stars, as all thermometers are regulated according to air temperature (98), and for an analogous reason.

Thus, from every point of view, the measurement of time requires the intervention of rational principles; it follows from the notion of the reason and the order of things.   On the other hand, the direct measurement of extension falls immediately under the senses.   This is a circumstance worthy of attention and one which squares with the remark already made, that the understanding of time can only be confused and rudimentary so long as the faculty of perceiving the order and reason of things does not exist or exists only in rudimentary form.

141.   In so far as they furnish the materials of a science, and of a science which can be constructed independently of experience (28), the ideas of space and time are still very dissimilar.   Space has three dimensions, time has only one.   Three dimensions—or three " co-ordinates," as the mathematicians call them—are necessary to fix the position of a point which can be moved in space in any way; two are necessary if the point is restricted to remaining on a surface, for example on a plane or a sphere; only one is necessary if the point can move only along a determined line. Thus the points of a path are fixed when the distances from some point along the path are fixed, such as the point of departure or the *zero point* of the path.   A point is fixed on the surface of the ocean when the latitude and the longitude are given.   But, if the

object is above or below the surface, a third co-ordinate must be set up if we are to know its height above sea level or its depth below this level. On the contrary, in order to set the date of a phenomenon or its position in time, it is sufficient, as it is in order to fix the position of a point on a line, to assign a single dimension, namely, the elapsed time or the time which should elapse, between a given moment of time or an era taken as a reference point, and the instant when the phenomenon occurs. The infinite variety of relations of dimension, configuration, position, and order which are the subject matter of geometry are derived from the idea of space. The idea of time, on the other hand, taken in its extreme simplicity, cannot furnish the material for a theory which merits the name of a science, since it is not connected with the abstract conceptions of geometry or with other notions suggested by the experimental study of the physical world.

142. Having indicated the dissimilarities between them, let us turn to the analogies between these two ideas (138), and let us see whether our methods of critical philosophy have absolutely no hold on those abstruse questions which metaphysicians have raised concerning the great and fundamental ideas of space and time. Are not these ideas in fact, as Kant saw, only laws of the human mind, forms which must somehow shape the more particular ideas the material of which the sensibility furnishes to the understanding, rules by means of which the experience becomes possible which imparts knowledge to us of the existence of external objects? Is the fact of giving an objective value for the ideas of space and time, over and above that which follows from the nature of the human mind, an illusion of the same sort as that which makes us shift to trees on the shore, the movement of the ship upon which we are traveling, and to the system of the stars, the movement of the earth from which we observe them (7)?

But, if this were so, by what extraordinary chance could the phenomena which lie within the province of our understanding be connected in accordance with simple laws which imply the objective existence of time and space? For example, Newton's law, which makes astronomical phenomena so reasonable, implies the existence of time, space, and geometrical relations outside of the human mind. How can we admit that astronomical phenomena, which are so manifestly independent of the laws or the forms of human intelligence, would be so simply and regularly co-ordinated in a system, which would, however, point to nothing outside of the mind, because the keystone of the system would be an intellectual, human fact, inappropriately carried over into the world in which astronomical phenomena occur? What is said of astronomical phenomena may be said of all those which science has reduced to regular and simple laws, and which appear, as a result of that simplicity itself, to follow very closely from the primordial laws that are hidden from us.

Moreover, we are not, for that reason, limited to insisting on such inferences, however cogent they may be. We are able to penetrate and we have actually penetrated more deeply into the nature of the act which enables us to understand space. The analysis of our intellectual faculties has enabled us, so to speak, to put our finger upon the correlation on which nature is founded and the processes it uses to give, not only to man, but also to the other animal species, the representation and perception of space, in proportion to their needs. The daring negation of Kant is found to be refuted in advance by that very analysis which makes evident to us the reason for the representative value of sensible impressions, as these pertain to the configuration and the geometrical relations of the objects from which these impressions emanate. There is no question of the fact that this does not result from categorical demonstrations, and we know that the system of the great German metaphysician regards as valueless everything that is not established by means of a logical demonstration. But reason refuses to follow this path which necessarily terminates either in pure idealism or in the most absolute kind of skepticism, as everyone is aware.

143. Were we to go farther and follow the ontologists in their controversies as to the nature of space and time, we should undoubtedly run up against insoluble contradictions. Equally insurmountable difficulties are involved in regarding space and time either as substances or as attributes of substance. Nevertheless, it is clearly necessary in the ontological hierarchy that objects of our knowledge be grouped either among substances or among attributes of substance. But analogous contradictions do not necessarily testify against the objective value of the ideas of space and time. They are explained just as well and even better if it is

admitted that ontological philosophy proceeds from an abstract principle when it undertakes to classify all the objects of knowledge as substances and attributes. This applies as much to the remarks that have already been made (135 and 136) as it does to those which follow. Nothing in the human mind is clearer than the ideas of space and time and all that is connected with them; nothing in it is more obscure than the notion of substance and everything that we might wish to deduce from it. Accurate criticism requires that we do not judge what is clear by what is obscure; on the contrary, it is necessary that ideas which are inherently clear [*claires par elles-mêmes*] throw their light on the obscure regions of the field of knowledge, and help us to drive out the chimeras.

The ideas of space and time are so inherently clear that they necessarily escape all definition. When Leibniz defined space as the order of coexistent things and time as the order of successive existences, it is perfectly clear that his definitions presupposed the idea of definite objects, and that we should apprehend nothing as to their nature if we did not have the idea of them prior to our definition. But, nevertheless, these definitions have philosophic meaning, since they indicate that the idea of order, in proportion to its generality, dominates the ideas of time and space, not only in the scale of logical abstractions but even more in that of rational conceptions. As a result, we find in the theory of order in general the reason for a great number of properties and relations that mathematicians had at first studied individually under the comparatively less abstract and the more sensible forms of space and time.

144. In having recourse to these definitions, Leibniz intended to express a still more important thought: namely, that if the ideas of space and time are not fantastic illusions or forms of our understanding; if, on the contrary, they have an objective and external reality, then this external reality must not be taken in an absolute sense but in a phenomenal and relative sense—if we may make use of a terminology which is not precisely that of Leibniz, but which we believe to be preferable and the meaning and value of which we have undertaken to fix from the very beginning of this book (8 and following). It is to these terms that we bring back the basis of the dispute between Leibniz and Clarke, although they

themselves did not put it in dry logical form, because, in their discussions, they were especially interested in the questions of natural theology which were involved.

Leibniz supported his theory by *a priori* arguments based upon the principle of sufficient reason (138), the probative force of which we, for our part, admit. But may there not be, in addition to these arguments, legitimate inferences which could corroborate these *a priori* arguments? There are in fact such instances, and of different sorts. In the first place, the two fundamental principles of dynamics, the principle of the inertia of matter (119) and that of the proportionality of force to velocity, are both results of experience and could only be derived from experience in so far as we profess to affirm something as to the absolute or relative value of the ideas of space and time. But both principles of dynamics are also necessary consequences of the Leibnizian theory which attributes only a phenomenal and relative value to the ideas of space and time.[3]   Now if a law of nature has need of empirical proof,

---

[3] Let us think of a system which includes all bodies capable of exerting appreciable actions on one another. And, unless matter is indifferent to both rest and motion, there will be an essential and observable difference between the state of the system when the bodies are absolutely fixed and the state of the system when the particles which make it up are animated by a common translational motion, as a result of which they move with the same velocity in parallel straight lines with no change in their relative positions and, consequently, in the actions that they exert on one another. Experience proves the contrary. But it is also true that we cannot refuse to acknowledge this fact prior to any experience, since we admit, with Leibniz, that the idea of space is only an idea of relation and that reason can conceive only of relative motions and states of rest.

The same considerations apply to the principle of the proportionality of forces and velocities. Let us imagine, for greater simplicity, that the material particles A,B,C, . . . n, which are posited as having equal mass, may be subjected to the action of equal forces F, so that they move with equal velocity in parallel lines, and that in addition another force F^1 acts in the same direction upon the single particle A. It follows necessarily that this force F^1 should impress on A a relative velocity, absolutely independent of the common motion of the system produced by the action of the force F on all the particles A,B,C, . . . n of which it is composed. Therefore, if the particle A is considered in a state of isolation and as being acted upon by the forces F and F^1, the effect of the two forces would necessarily be simply and purely additive, each force producing its effect as though the other did not exist, the total velocity of the particle being the sum of the velocities that each force would impress on it when acting by itself. Consequently, a double force, that is to say, the co-

without prejudicing anything so far as the value of a philosophic principle is concerned; and if, on the other hand, the law is a necessary consequence of that principle itself, then, inversely, the experience which verifies the law will be reputed to give, *a posteriori*, the confirmation of the principle or at least will serve as a powerful inference in favor of it.

We could again bring up the remark already made (116) as to the hypothesis of formed and extended atoms, namely, that if the ideas of space and time had real objects and absolute reality, our understanding would have been able to reach primitive and absolute reality through its own skill by these principles alone. However, we have very good reasons for believing that this would prove to be very improbable, although it is not impossible nor can it be proved that it is impossible. But we prefer to insist upon considerations of another kind, to which we more usually have recourse in inquiries of this sort.

---

incidence of two forces capable separately of impressing equal velocities, will impress a velocity twice as large; a triple force, a velocity three times as large; and so on. In sum, velocities increase in proportion to the forces which produce them.

Mathematicians and physicists of the contemporary school, by admitting the principle of the proportionality of forces to velocities as one of the basic principles of the science of motion, have generally admitted it as a datum of experience and as an observed fact. Others have thought it to be only a definition; still others, an ordinary mathematical theorem, capable of being demonstrated like any other theorem, but in doing so they have fallen into paralogisms into which others have refused to follow them. The principle in question, like that of inertia, with which it is, in fact, identical, may actually be only an empirical datum, while the object of the idea of space is regarded as something absolute; and this holds so long as the distinction between absolute motions and relative motions is regarded as something absolute, and not as a distinction which is only relative. If, on the contrary, we agree with Leibniz that there can be nothing absolute in the ideas of space and of motion, the principle of the proportionality of forces to velocities no longer requires the intervention of experience. It is no longer a mathematical theorem or a purely logical definition; it is a philosophical axiom. See Laplace, *Exposition du système du monde*, Book 3, Chap. 2; Poisson, *Traité de mécanique*, second edition, Vol. I, Sect. 116. The preliminary statement and the first part of the *Traité de dynamique* by d'Alembert may also be consulted. It is curious to see how d'Alembert, who thought of himself as being far removed from Leibniz as a philosopher, uses demonstrations which are nothing but a continual application of the basic principle of the Leibnizian doctrine in order to establish what he calls the necessity of the laws of motion.

Let us admit that the mind has a tendency—as no doubt it has—to attribute absolute reality to what we conceive under the ideas of space and time. Let us also admit that this tendency may be deceptive. Very likely there will be incoherencies in the system of our ideas resulting from a lack of harmony between the nature of the objects of thought and the manner in which we think of them. Reciprocally, if incoherencies and obstacles are manifest in the system of our ideas because of the attribution of an absolute reality to ideas of space and time, it will be necessary to conclude from this, and with a probability of the same order, that these ideas do not have the objective value the human mind would like to accord them, because of a condition in its organization as a thinking subject.

145. Now, such oppositions and such conflicts exist with respect to the ideas of space and time, and give rise to what Kant has described as the " antinomies of pure reason " in the section which, to us, is the most remarkable part of his critical philosophy.[4] It is equally repugnant to think of the world as limited in space and as having a beginning and an end in time, as it is to think of it as having neither limits, nor beginning, nor end: this is the first antinomy. It is also equally repugnant to think either of a limit to the divisibility of matter, or to think of it as infinitely divisible: this is the second antinomy. Both the *thesis* and the *antithesis* may be proved equally well and yet they contradict each other. Kant develops along the same lines two other antinomies of which we have no need to speak here.

These antinomies or, to use less technical language, these contradictions, are real. It is not necessary to appeal to arguments based on the scholastic doctrine of universals to be struck by their force. We need only to survey the works of philosophers to be caught up in the interminable debate concerning them. But, as we see it, they are not contradictions of reason, but of the human mind, which is something quite different. For, if reason cannot contradict itself without losing its unity and its regulative authority, it is perfectly comprehensible that in the complex organization of the human mind taken as a whole, there may be elements capable of contradicting each other; that the under-

---

[4] *Critique de la raison pure*, " *la dialectique transcendentale*," Book II, Chap. 1.

standing, as a result of the way in which it elaborates and relates materials furnished to it by the senses, may have its own illusions, like those found in sensations and in certain spontaneous judgments which are continually associated with sensations and which reason disavows (85).

This is the only plausible motive that could have made Kant wish to refuse any objective value to the ideas of space and time and to consider them only as forms of the sensibility. But in this he goes too far, for it is sufficient to admit with Leibniz that space and time are phenomena which have objectively only a relative and not an absolute reality in order to eliminate the contradictions, in which the point of departure of the two contrary theses is the attribution of absolute objective value to the idea of space and to that of time.

The human mind is organized to perceive relations in space and in duration which actually go beyond it and are independent of it. Thus it penetrates into reality, but into a reality which is relative and phenomenal, the knowledge of which is sufficient for man's needs and for his role in the world. When the mind makes an attempt to go beyond this and to build relative reality into absolute reality, it no doubt yields to a tendency which is native to it. But it is deceived by this tendency, and reason warns of this by exhibiting its bottomless abysses and insoluble contradictions.

146. The two Kantian antinomies are not the only ones which the mind runs into when it has pretensions of attaining the essence of things or absolute reality in the twofold conception of space and time. The *plenum* of the Cartesians is untenable in the present stage of physics, and action at a distance across the void, in the way the Newtonians conceived it, is absolutely incomprehensible. The latter hypothesis is one which the force of scientific habits has made familiar to us, but which ought to offend our reason no less than it did that of Leibniz, Bernoulli, and Huygens, if it involves thinking of the void or of space as something primitive and absolute which subsists independently of the phenomena of the material world, and not, rather, as a relation between phenomena, the basis and essential principle of which absolutely escapes our means of perception and understanding.

But what is such a contradiction in the conception that man can have of the physical world, compared to the contradictions in the conception he has of himself, of the action of his organs on the mind and of his mind on the organs, and, generally speaking, of all the phenomena of organic, animal, and intellectual life, of which some are characteristic of him, while others are shared with such a prodigious variety of inferior beings by characteristics common to all living things?   It is repugnant to think of intelligence and thought, of vital or plastic forces being diffused throughout an extended substance, whether this be large or limited in extent, in a system of particles which are separated from each other as much as in a continuous whole.   It is repugnant to conceive of these forces as being inherent either in a single particle or in an aggregate of determined particles, or to imagine them as being carried from one particle to another, from one material group to another, in proportion to the renewal of the material of the organism.   It is even repugnant to assign a place in space to the principle of life and of thought; to fix—as the mathematicians would say—the co-ordinates of a point in space in which the mind would have its seat, and from which it would act on the organs, after having experienced and perceived modifications of the organism.   Reason runs into contradiction on all sides, if the extended is conceived as something absolute, and if space is something necessary, primitive, and immutable.   But, instead of contradictions in our body of knowledge, there are rather facts which surpass our possible attainments, if bodies and space are merely phenomena, the external reality of which we are clearly able to perceive, but not the absolute ground and basic essence.

There are needs in man's nature which would not be satisfied, faculties which would seem fruitless and deceptive, if the end of everything for him were animal life.   On the other hand, it is repugnant to put the accomplishments of the superior destiny of man in space and time beyond the sphere of organic phenomena and sensible facts.   We do not pretend that reason, if left to itself, may be capable of fathoming these mysteries.   We only say that in the presence of such mysteries and for the conciliation of instinctive or acquired beliefs which seem to contradict one another, reason finds new motives for admitting that the forms of space and time, always thought of as inherent in phenomena and not in the structure of the human mind, still have only phenomenal value themselves.

Let us be on guard against suggesting that the philosophic probability of this solution may be a probability of the same order as those which make legitimate, in the eyes of reason, the belief of common sense in the objective existence of bodies and in that of the external world, in so far as it is found in space and time. Let those for whom such inferences are without value abandon the field of philosophic speculation; they clearly have the right to do so. For those to whom such a desertion would be distasteful, it is necessary to accept the inferences as they present themselves, as far as possible without deceiving oneself, and especially without wishing to cause others to be deceived.

CHAPTER XI

ON DIFFERENT KINDS OF ABSTRACTIONS AND ENTITIES. ON MATHEMATICAL IDEAS. ON IDEAS OF GENUS AND SPECIES

147. We have already indicated in a general way how knowledge or ideas break away from sense impression. In this action of the mind on the materials furnished to it by the senses, there is a series of analyses and syntheses, of decompositions and recompositions. This action is comparable to that which occurs in the elaboration of materials that animals borrow from the external world in order to extract the essential material from them and to build up out of this material others which are used immediately in the development and repair of their organs. The comparison is all the more admissible because, in the one case as in the other, the question is not simply one of isolating juxtaposed parts or of bringing isolated parts together. On the contrary, it is necessary to believe that in both cases the products of the combination acquire properties which are found neither wholly nor partially in the isolated elements. These properties are acquired through the elaboration of primitive material which is carried out under the influence of a vital principle. Inversely, the dissociation of the elements permits the free manifestation of properties which are neutralized or rendered latent in compound substances.

The decomposition or analysis to which the intellect submits the materials of the sensibility is called " abstraction "; and although all our ideas of things, even of those falling immediately under the senses, may be abstracted or separated from the sensible impressions which accompany them (109 and following), the term " abstract ideas " is applied particularly to those which we reach through a final abstraction or separation to which we submit the ideas of sensible objects. On the other hand, the act of composition or synthesis by which thought co-ordinates the materials furnished through the sensibility, by introducing into them the principle of unity and of systematic connection, leads to the conception of *entities*. These are often called abstract ideas to distinguish them from the images of sensible objects, but it is still necessary to distinguish the ideas obtained through this latter process from those obtained by abstraction properly so called. The formation of abstract ideas and entities is not reserved to philosophers and scientists. The activity which produces them begins as soon as the human mind begins to act and is made manifest in the organization of languages, whatever the cultural level of those who speak them. In this chapter, our purpose should be to use the rules of criticism, whose applications we are trying to point out throughout this study, in order to distinguish between that part which arises from the constitution of the object of thought and the part which springs from the regulative laws of the mind, both in the formation of abstract ideas properly so called and in the conception of those purely intelligible types that we do not believe should be called entities. We make this statement well aware of the fact that during a certain period philosophers may have abused both the word and the thing, and that at another time both the thing and the word may have been subjected to an unwarranted disrepute.

148. First, let it be noted that a false, or at least a very imperfect notion of abstraction is obtained, if we see in it only a mental process which isolates the essential qualities of an object so as to study these by themselves and thus to attain knowledge of the object more easily. This is what abstraction means in ordinary logic; and in this sense, abstract ideas and abstract sciences would be purely artificial products of the understanding. But this is true only of certain abstract ideas and sciences. In addition to this there is another abstraction—the same one which gives us the ideas of extension and duration, and of space and time free from any sensible image—an abstraction which enables us to distinguish by means of thought elements which are independent of one another, although sensation confuses them. These are abstract ideas which correspond to general facts, to superior laws to which are subordinated all the particular properties through the mediation of which external objects can be perceived by us. And the sciences which have such ideas as their object, which include the system of such laws and relations, must not be passed over as artificial, conventional, and arbitrary sciences.

As an example that brings out the distinction we wish to establish, let us consider a solid body moving in space. We shall select at random a point on the object and consider the movement of the body as the resultant of the combination of two other motions: the one, a motion as a result of which all the points of the mass move in a direction common to the point in question; the other, a motion as a result of which the body would turn in a certain manner around this point, to which we shall have attributed an ideal fixity. The breaking up of the real motion into these two fictitious motions, the one a translational motion common to all the points of the mass and the other a rotational motion peculiar to a single point in the mass, may be done in an infinite number of different ways, depending upon our arbitrary selection of a given point in the mass as the center of the relative movement. This ideal resolution of the real movement of the body into two others will in turn give rise to further resolutions which, like the original resolution, will be, or may be adapted to the manner in which we conceive phenomena, which will furnish images adapted to facilitate the description and the study of them, but which, in general, will be arbitrary and not based on the nature of the phenomenon itself.

Now suppose we consider the case of the movement of a solid body which is affected by the force of gravity alone and which meets with no resistance from the surrounding medium through which it moves. There is a point in this body, known as a " center of gravity," which has this particular quality, namely, that if it is taken as the center of the rotational motion imagined in the above illustration, each of the two motions of translation and of rotation, being independent of each other, acts separately as though the other did not exist. Consequently, the abstraction which distinguishes or which isolates these two motions at that moment ceases to be an artificial and purely logical abstraction. The basis and reason for it lie in the nature of the phenomenon and give us a conception or a faithful representation of it.

149. When, with a view to studying the conditions of equilibrium and of the movements of solid and liquid bodies more easily, we imagine perfectly rigid solids and fluids deprived of all viscosity, that is to say, of any tendency for their parts to adhere to each other, we disregard some of the natural properties that these

bodies possess. In order to simplify the problems we wish to solve and in order to adapt them to our methods of calculation, we form an ideational construct of bodies, the prototype of which is not found in nature and is perhaps not capable of realization. In reality, and by a happy circumstance, solid and fluid bodies of the sort nature offers us do not differ so greatly from the fictitious conditions of absolute rigidity and fluidity that we may not regard the theoretical results obtained through the use of these conditions as already representing within allowable limits of approximation the laws of certain natural phenomena. This is the basis for the utility of hypotheses or of abstract conceptions which are artificially substituted for the natural types of solid and liquid bodies.

When a study is made of the laws according to which wealth is produced, distributed, and consumed, it is discovered that they may be theoretically established in a simple enough manner, if we abstract them from certain accessory circumstances which complicate them and whose effects, as a result of that complication, can be only vaguely estimated. Consequently, it will be admitted that wealth or commercial paper can circulate without the slightest restraint, can pass immediately from one hand to another, can be realized, negotiated, or exchanged for other papers or for coins, and can be exchanged for values of other kinds at the pleasure of the person who holds them and at the current exchange rate and market price. Similarly, the perfect equalization of price under the influence of free competition will be admitted. None of these assumptions is completely true, yet they come close to being true when they are applied to goods preferred by commercial speculators, and in countries and at times in which commercial organization is more advanced.

Such abstractions, by means of which the mind separates facts which are naturally associated and dependent on one another are what we propose to call " artificial abstractions," or, if it is preferred, " logical abstractions." These are abstractions with which we might dispense, if the human mind were capable of embracing at one and the same time all the causes which have an influence on the production of a phenomenon, and of taking account of all the effects which result from their association and their mutual reactions.

150. There are other abstractions brought about by the nature

of things, by the mode of existence of objects of knowledge, and in no wise by the constitution of the mind or because of the point of view from which the mind envisages them. Surely the abstractions upon which the system of pure mathematics rests, for example, the ideas of number, distance, angle, line, surface, and volume, are of this sort. It is in the very nature of things that certain phenomena result from the configuration of bodies, and are not dependent on the physical qualities of the stuff of which the bodies are formed (1). Thus, when our mind abstracts the physical qualities of matter, for the purpose of studying separately the geometrical properties or those of configuration, we must simply conform to the order according to which, in nature, certain relations are established, certain phenomena develop along with but independent of other phenomena. With still better reason, if the progress we make in the philosophical interpretation of nature tended more and more to give us means of explaining all physical phenomena by means of geometrical relations the laws of which we have been able to study carefully, if physics tended to be resolved into mathematics, it would conform to the nature of things and not simply to the nature of the human mind, to isolate by means of thought a system not only of general, but also of fundamental facts. This is a system of relations which dominates the others and contains the objective reason for them to such an extent that we might be able to see in mathematics the thoughts of God, to speak of God as the eternal mathematician. Such abstractions, which are independent of human thought and superior to its phenomena, must not be confused with artificial abstractions which the mind devises for its own convenience, and we shall call them " rational abstractions."

151. The same sort of motives which lead us to attribute an objective value to mathematical abstractions make us believe in the existence of an external world or tend to make us attribute an objective value to the fundamental ideas of space and time. If the notion of a straight line or of distance were nothing but a fiction of the mind, an idea created artificially, by what chance could the forces of nature, the force of gravity, for example, vary with distance in accordance with simple laws, or—as the mathematicians say—be *functions* of distance so that variation in distance is necessarily conceived as the cause or the reason for vari-

ation in force? What would give rise to this harmonious agreement between the general laws of nature, with respect to which we are simply intelligent witnesses, and an idea which is what it is because of the constitution of our understanding, and which would have value only as a human invention and as a product of our personal activity?

152. But, if the belief in the existence of the external world is and must be a natural belief for carrying out the destiny of man, if nature is charged with combating the skeptics on this ground (86), it is by no means concerned, nor should it be concerned, with combating a purely speculative skepticism, which consists in regarding all abstract ideas as nothing but the work of the mind or as arbitrary creations of the understanding. The latter interests philosophy, but does not interest either practical life or science proper. Our knowledge of physics or of mathematics would be neither more nor less adequate, whether we considered mathematical conceptions as mental fictions having no objective reality, but finding a useful application in the analysis of physical phenomena, or whether, on the contrary, we think of mathematical truths as having an objective value independent of the mind that conceives them, and as containing the reason for the physical appearances which are adapted to the modes of our sensibility. Science is indifferent to this transposition of order, and there is nothing which can serve to demonstrate logically that the order a, b must be admitted to the exclusion of the order b, a. But that which has no direct influence on technical applications and on the progress of positive science is precisely that which is of the greatest importance for philosophic order to introduce among the objects of knowledge, and to illuminate by the light of reason the connections and the relations among scientific facts which have been definitely established. Neither the true character of the mathematical sciences nor the role they play in the system of human knowledge will be accounted for until we appreciate the importance of these questions of order and until we have resolved them on the basis of philosophic inferences, analogies, and probabilities.

The farther we go in our examination, the more we shall discover motives for attaching great importance to the doctrinal distinction between logical abstraction and rational abstraction.

For, if all abstractions are artificial creations of the mind, it would be very simple for the mind to arrange these products of its characteristic faculties to suit its own fancy and the convenience of its nature. But if, on the contrary, there are abstract ideas the prototype of which may exist independently of the human mind, since the mind is able to work with abstract ideas, whatever their origin, only by applying certain perceptible symbols to them (112), certain disagreements will be found between the nature of the symbols which are used for this purpose and the nature of the ideas which are recalled to the mind by these symbols. These disagreements are capable of interfering either with the accurate perception of thought, or with the correct expression by means of language of the bonds and the relations which must be discerned between the prototypes of such ideas.

153. Because the fundamental ideas of mathematics are not artificial products of the understanding, it does not follow that all parts of mathematical doctrine are free from artificial conceptions which pertain less to the nature of things than to the organization of our methods. Thus, the application that we make of numbers in measuring or in expressing continuous magnitudes is without any doubt an artifice of our own minds, and does not belong essentially to the nature of these qualities. In this sense, it may be said that numbers do not exist in nature. Yet, when our thought moves to the abstract idea of number, we are fully aware that this idea is not an arbitrary fiction or an artificial creation of the mind, set up to facilitate our researches, as the idea of perfectly rigid and fluid bodies would be. When we study the properties of numbers, we have reason to believe (36) that we are studying certain general relations among things, certain laws or general conditions of phenomena. This does not necessarily imply either that all properties of numbers play a part in the explanation of phenomena, or, *a fortiori*, that the supreme reason of all circumstances which give rise to phenomena is to be found in the properties of numbers, in accordance with that enigmatical teaching which was handed down from Pythagoras to Kepler, and which had its birth in remote antiquity only to disappear with the introduction of modern science.

It ordinarily happens that after the nature of things has furnished the prototype for an abstraction, the abstract idea thus formed suggests, in its turn, further abstractions and systematic generalizations which are no more than fictions of the mind (16). From this it follows that ideas that we call new, because they throw new light on the objects of our knowledge, have their periods of fruitfulness and their periods of sterility and exhaustion. If these ideas are fruitful, it is so because, far from being created out of whole cloth by the genius who first lays hold of them, they are ordinarily only a felicitous expression of a relation found to exist among things. If their fruitfulness is not unlimited, like the number of artificial combinations into which the mind is able to introduce them, it is so because nature does not subject itself to the logical rules which preside over the systematic co-ordination of our ideas. From this in turn it follows that the general fault of systems is, as we say, that they are too exclusive or include only a portion of the true relations of things and turn away from them entirely in their ultimate conclusions or in their excessive prolongation.

154. The application of the fundamental laws of mathematics to the scientific interpretation of nature does something more than show us that they are not artificial creations of the mind. It is to be noted that many of these ideas, in spite of their high degree of generality and abstraction, are only particular forms and, as it were, concrete species of ideas still more abstract and general, to which we should be able to raise ourselves by other means than those of mathematical abstraction, and by the contemplation of other phenomena than those to which calculation and mathematics are applicable. The ideas of combination, order, symmetry, equality, inclusion, exclusion, and so on, do not invest only geometric or algebraic forms; and certain properties of figures or numbers which belong to a type of order of that kind, and to such a form of combination or symmetry, have their cause or their *raison d'être* in a realm of abstraction superior to geometry and arithmetic (143). For example, the notion of inclusion, or the relation of the containing to the contained, is recognized in logic, in which it serves as the basis of the theory of the syllogism; although the manner according to which the general idea contains the particular idea may be very different from the way in which one quantity or one space includes another quantity or space. The idea of force or of active power is clearly more general than that

of a moving or a mechanical force; and perhaps a day will come when, still in conformity with Leibniz's suggestions, we shall be tempted to make a sketch of that higher dynamics the rules of which we have glimpsed only confusedly up to the present time, but which includes those of mathematical and mechanical dynamics, or at least those parts of the latter which relate to conditions not exclusively characteristic of mechanical phenomena, such, for example, as those which are connected with the special properties and exclusive characteristics of the ideas of space, time, and motion.   The two proverbs, "Every action involves a reaction" and "One can rest only upon that which resists," which are accepted alike by physics, medicine, morals, and politics, are simply so many ways of expressing certain rules of that dynamics which has come to be called higher because it governs the moral as well as the physical world, and serves to make reasonable the most delicate phenomena of organisms as well as the motions of inert bodies.[1]

155. In order to prove that the ideas which are the foundation of the structure of pure mathematics have their prototypes in the nature of things and not in fictions of our minds, we have drawn our inferences from observed correlations between the abstract truths of mathematics and the laws of natural phenomena; the

---

[1] In a letter addressed to King James I, in 1603, concerning a project for the union of Scotland with England, Bacon had this to say about a similar proverb, *Vis unita fortior.*   "When Heracleitus, nicknamed 'the Obscure,' published a certain book which is no longer in existence, some persons took it to be a dissertation on nature, others thought it a treatise on politics.   I am not surprised that this was so, for there is great agreement and resemblance between the laws of nature and those of good politics.   The former are simply the order followed in the government of the world, the latter, the order followed in the government of States.   Thus the kings of Persia were very solemnly initiated into a science which was very much respected at that time, but which today is completely degenerate and the name of which is seldom more than taken amiss. In fact, the *magic* of Persia, this occult science of their kings, was applied to politics on the basis of observations drawn from the world of nature.   The fundamental laws of the natural order were posited as the model for the government of the State."   Therefore, we must not be scandalized when, in one of the last chapters of this work, we find *magic* listed in Bacon's encyclopedic table. Moreover, he says again (*Advancement of Learning*, Bk. III, Chap. 5), "Magic was accepted among the Persians as a lofty wisdom and knowledge of the universal harmony of things."   See M. Bouillet's edition of the *Philosophical Works of Bacon*, I, 522.

former contain the explanation or the reason for the latter. But we should put aside these inferences to consider the system of mathematics by itself, independent of any application to the scientific interpretation of nature, in order to get at the organization of that system, and to find once more sufficient reasons for rejecting the very presumptuous or very timid opinion according to which the human mind may operate only upon the products of its own imagination and, as Vico has said, may demonstrate mathematical truths because it made them. Had this opinion been established, nothing would be easier than to divide the field of pure mathematics into regular and clearly defined compartments, or, in other words, to submit the system of mathematical sciences to a classification of the same sort as those which please us because of their regularity and their symmetry, when we are dealing with those ideas that the mind creates out of whole cloth, and may arrange to suit its convenience (152), without being embarrassed by the obligation to reproduce an external prototype. But on the contrary—and this circumstance is well worth noting—mathematics, the exact science *par excellence,* is one of those in which there is most vagueness and indecision in the classification of the various parts, in which most terms which express the principal divisions are taken now in a larger sense and again in a less inclusive sense. Whether they are taken one way or the other depends on the context of the discourse in which they are used, and on the characteristic views of each author, and without our being able to fix their meaning clearly and rigorously in a common language in whichever sense they are taken. This indicates a complication and an intricacy of relations ill-suited to the logical procedures of definition, division, and classification. Nothing shows more clearly that the object of mathematics exists outside the human mind, and independent of the laws which govern our intellect.

156. It is the function of general philosophy to fix the relative position of mathematics in the general system of our knowledge and to estimate the value of the elemental notions which serve as the basis for that vast scientific structure. But if we enter into the details of the arrangement of its parts to each other and of its internal organization, we see analogous questions springing up to which the same means of criticism are applicable, which have a

special interest to mathematicians because their studies provide the clearest understanding of them, and which, to a large extent, make up what may be called the philosophy of mathematics. It goes without saying that these questions of detail do not lie within the scope of the present volume. We have developed them in other works [2] to which we refer any readers who are curious about speculations of this kind. Here we must limit ourselves to the most succinct remarks.

In the forefront of philosophical questions, in mathematics as elsewhere, are those which bear upon the representative value of ideas, and in which it is a question of distinguishing, as Bertrand of Geneva expressed it, what pertains to things themselves —rational abstractions—from that which pertains only to the manner in which we may choose to envision them—artificial or purely logical abstraction. Is algebra only a conventional language, or is it really a science, the developments of which, while no doubt connected with the use of an arbitrary and conventional primitive notation, nevertheless bring together a body of general facts and of abstract or purely intellectual relations that the mind discovers and disentangles with more or less skill and success, but which it is so far from having created that it must engage in many tentative efforts and make many verifications of these relations before we can have, so to speak, any confidence in its discoveries? Is every calculation of negative, imaginary, or infinitesimal values nothing but the result of rules set up by arbitrary convention? Or are all these pretended conventions only the necessary expression of relations which, because of their ideal and purely intellectual nature, the mind is, of course, obliged to represent by symbols whose form is arbitrary, but which the mind does not invent according to the whims of its fancy or because of the bare necessity of its own nature, and which it is restricted in getting hold of, in so far as these relations are presented to it by the nature of things, as a consequence of the faculty of generalizing and abstracting with which it has been endowed? This is the issue that divides mathematicians into schools; this is the basis of the philosophy of

[2] Cournot, *De l'origine et des limites de la correspondance entre l'algèbre et la géométrie*. Particularly the sixteenth and the final chapters. Also, *Traité élémentaire de la théorie des fonctions et du calcul infinitésimal*, particularly Chap. 4 of Book I.

mathematics, as of all philosophy. It must be noted that all this does not affect the positive and purely scientific part of the doctrine. All mathematicians will apply the same rules of calculation to symbols referring to negative, imaginary, and infinitesimal values, and will derive the same formulas, no matter what their philosophic opinion as to the origin and the interpretation of these symbols may be. But that which is of no interest to the doctrine from the point of view of positive rules and practical applications is precisely what comprises the underlying reason for the concatenation and for the relations among the various parts of the doctrine.

It is no more possible to demonstrate logically that certain ideas are not pure fictions of the mind than it is to demonstrate logically the existence of bodies (151). Yet this double impossibility no more arrests the progress of positive mathematics than it does that of positive physics. But there is this difference between these two cases: the faith in the existence of bodies is a part of our natural make-up, while we must become familiar, through the cultivation of the sciences, with the meaning and the value of the high abstractions used in them. This is the meaning of the common saying attributed to d'Alembert, " Go forward and faith will follow you "; not a blind, mechanical faith, an unreflective product of habit, but an acquiescence of the mind based upon the simultaneous perception of an ensemble of relations which can only successively make an impression on the attention of the disciple, and from which follows a group of inferences to which reason must yield, in the absence of a logical demonstration that the nature of things renders impossible.

157. We now move on to consider that other category of abstract ideas to which the mind is raised by the method of synthesis, in order to bring together in a systematic unity the variable appearances of things which are immediate objects of its intuition. These are the ideas or concepts to which we give the name "entities "; and in order not to offend any readers too greatly by using a term which recalls so much scholastic want of learning, it will be well to select first the most obvious examples and to show how entities intervene with regard to the conception of phenomena which fall most immediately under our senses.

If we disturb a point on the surface of a liquid mass, a *wave* is created, the propagation of which the eyes can follow as it spreads out in all directions from the center of the disturbance. This wave has a characteristic velocity of propagation which must not be confused with the velocity of each of the fluid particles which successively rise and fall slightly, above or below the level which they maintain in a condition of rest. The movements of these material particles to and fro are very small and can be measured only with difficulty; while the wave always travels in the same direction, through great distances, with a velocity of which we are perfectly aware without instruments and which we can measure in the most exact manner with the aid of suitable instruments. If many points at some distance from one another on the surface simultaneously become centers of undulatory motions we shall see many systems of waves which meet and cross each other without losing their identity. All this clearly gives us authority to conceive of the idea of the wave as an object of observation and study, which has its own particular form and its own characteristics and particular attributes, such as a measurable velocity of propagation. Yet actually, the wave is nothing but an entity; the material or substantial reality belongs to the molecules which become successively the loci of oscillatory motions. The idea of the wave is the systematic conception of the way in which these motions follow one another and are connected. But this idea does not have an arbitrary origin. It is immediately suggested to us by sensible perception; it enters as an element into the rational explanation of all the phenomena which result from the propagation of undulatory motions. It is an entity of the sort we call natural or rational, as opposed to entities which are artificial or logical.

158. Suppose a naturalist or an engineer takes the Rhone river as the object of his study; suppose he gives us the history of this river, of its meanderings, of its floods, of slow or sudden changes which occur in the behavior of its waters, of the properties which distinguish them, and of species of animals which live in them. Would anyone believe that a person would go to so much trouble in connection with something that is, after all, nothing but an entity, a symbol, a *flatus vocis?* The history of each drop of water would have to be given; each drop would have to be followed through the atmosphere, in the sea and in all the different ocean currents in which chance had destined it to be caught by turns.

The drop of water is the object endowed with substantial reality. If, on the one hand, the Rhone is considered as a collection of drops of water, it is an object which changes ceaselessly. On the other hand, the Rhone is an object having no reality, if, in order to preserve its real unity, we regard it as an object which persists, after all the drops of water which compose it at one moment of its existence have been replaced by others.[3]

In the example we have cited the objections raised would be ridiculous and would probably occur to no one. Such a monograph as that suggested concerning the Rhone would be read with the same curiosity and scientific interest as would one written about that strange wind that blows across the same country and which is known as the " mistral." We do not take poets seriously when they personify rivers and winds. But, notwithstanding the subtleties of dialectic, we no longer consider rivers and winds as abstractions having no more basis than that given them by a symbol, or a fleeting sound. Winds and rivers are objects of common knowledge as well as of scientific theories; such objects can be neither poetic images nor simple logical symbols.

Actually a river like the Rhone, an ocean current like the Gulf Stream, or a wind which behaves in a characteristic way like the mistral belong to that category of entities the notion of which results either from the perception of a form which remains the same in spite of changes of material, or from a form the variations of which are independent of changes in the matter found in them. Or it may be the perception either of a systematic bond which persists whatever the individual objects accidentally involved in part of the system, or of a bond which is modified by causes which

[3] The example that we have taken furnished the ancients with one of their familiar comparisons. " It is a matter of knowing," says Aristotle, in Book III, Chapter 4 of the *Politics*, " whether the State continues to be the same so long as the same number of inhabitants is maintained, in spite of the death of some and the birth of others, like rivers and fountains in which the water flows without ceasing in order to make room for the water that takes its place." In the Middle Ages another metaphysician, John of Salisbury, said, " Thus, the species of things remain the same in ephemeral individuals, as in running water the moving current remains a river, for we say that it is the same river, whence comes this saying by Seneca: ' We both do and do not step into the same river twice.' " Plutarch mentions the same comparison, attributing it to Heracleitus: " According to Heracleitus, it is not possible to enter the same river twice." *On the Word EI over the Gate of Apollo's Temple at Delphi*, 18.

are independent of those which impose modifications on the individual objects (20). These are entities, but they are rational entities, which do not arise because of the way in which we conceive and imagine things, and which, on the contrary, have their basis in the nature of things, on the same grounds as the idea of substance, which is itself only an entity (135).

159. In addition to these entities there are some which are manifestly artificial. Thus, for example, in physical geography we are interested not only in rivers but in what we call " river basins." Some modern authors have set forth in all their minutiae the systematic distribution of the earth into basins of various sorts, on the basis of the water courses which drain them. Now, among these basins, some are clearly determined by the formation of the terrain and by all their physical characteristics; others, and these are much more numerous, are only artificial conceptions of geographers and arbitrary lines of demarcation between territories that nothing divides naturally.

It may be, as has been maintained by certain medical schools, that those who describe diseases have abused the idea of entities; that in grouping certain morbid conditions under the name " fever " or under any other name, they have created an artificial systematization. This systematization has dangerous consequences if it leads medical men to lose sight of the alteration of the organs which are the seat of the illness, for the purpose of attacking the fever as though it were an enemy which must be bound fast and struck down. But suppose that there is, on the contrary, a morbid condition, such as cholera or smallpox, the symptoms, the behavior, the periods of attack, and of waxing and waning, are fully identified, whether we consider their action among individuals or in large numbers of cases. We should no more misuse abstraction in setting up such morbid conditions as entities by writing a monograph on cholera or yellow fever, than we should if we were to write a monograph about a river or a wind. For, by hypothesis, there is a general progression and behavior that marks cholera which will not be modified or which will only undergo changes of a secondary nature, depending upon the susceptibility of populations or of individuals accidentally exposed to its attack, just as the course and the behavior of the mistral does not appreciably depend upon accidental circumstances which have brought a certain bit of air into the region in which this wind prevails.

In general, philosophic criticism of the sciences, in which entities are constantly making their appearance under common or technical names, like the criticism of ordinary or elementary knowledge, as this is expressed by the forms of common language, will consist, so far as this is possible, in making the distinction between artificial entities, which are nothing but logical symbols, and entities based on the nature and reason of things. They are veritable " brain children," to use an expression which is common, but which has a true and profound meaning, when rightly understood. To the degree that progress in observation and in the development of scientific theories suggests to the mind the conception of entities of higher and higher orders, the comparison of observed facts and inferences which are suggested by them should furnish reason with criteria of judgment on the basis of which it will decide to what extent these entities are pure logical fictions, to what extent they have their basis in nature, and to what extent they clearly indicate the purely intelligible causes of the phenomena we experience through sensation.

160. There is one category of entities or abstract ideas which deserves particular attention, and with which the logicians of antiquity and of the Middle Ages particularly concerned themselves. This is the category of " universals," as the Scholastics said, or that which includes the ideas of classes, genera, and species hierarchically arranged according to their degree of generality, species being subordinated to genera as individuals are to species, and so on. Now, the distinction between artificial or logical abstraction and natural or rational abstraction is nowhere more evident than in this category of abstract ideas.

What is properly called classification is an operation of the mind which artificially groups objects in which any common characteristic is found, and gives to the general group thus formed a tag or generic name. This operation is carried out for the convenience of research or of nomenclature, to assist the memory, with respect to the needs of teaching, or for any other purpose which is related to human ends. Using the same procedure, these artificial groups may be further divided into subordinate groups, or grouped together to form collections and, to some extent, unities of a higher order. This is the nature of classification as seen from the point of view of pure logic. Bibliographies, which every person modifies to suit his own convenience in making a catalogue of his own library, may be cited as examples of artificial classifications.

But, on the other hand, nature offers us specific types in the innumerable species of living things and even of inanimate objects, which surely are neither artificial nor arbitrary, and which the human mind has not invented for its own convenience, and the ideal existence of which it understands very well, even when it finds it difficult to define it. This is analogous to the situation when we take the evidence of our senses as a basis for believing in the existence of a physical object before we have seen it sufficiently close at hand to distinguish its outline, and especially before we have been able to give an account of its structure. These specific prototypes are the principal objects of the scientific knowledge of nature, for the reason that in these species or natural groups, the stable characteristics which are the basis of the specific or generic association dominate and greatly surpass in importance the accidental or particular characteristics which distinguish the subordinate individuals or species from one another. In short, just as there are degrees in this domination and superiority of characteristics in relation to one another, it must and does happen that some genera appear to us to be more natural than others, and that the classifications to which in all cases we are obliged to have recourse, as a result of the needs of our study, most often offer a mixture of natural and artificial abstractions without which it would be difficult or even impossible to demarcate clearly the transition from one to another. A physical example, in which the term " group " is taken in its material acceptance, will perhaps better prepare us for understanding those abstract relations between the groups that we think of under the names of " genus " and " species."

161. We are aware that astronomers have grouped the stars into constellations, either on the basis of ancient mythological traditions, or by imitation, or for the purpose, whether well or poorly conceived, of practical convenience, in dealing with the portion of the starry sphere unknown to the ancients. These are manifestly artificial groups in which individual objects are associated not according to their true relations of size, distance, or

physical qualities, but because, from our point of view, they happen to fall along the prolongation of visible rays which are slightly inclined toward one another. Now suppose that certain very small areas of the celestial sphere, areas which are completely isolated and distinct and in which myriads of stars of the same magnitude (or rather minuteness) appear to be accumulated, are observed by means of a telescope, as Herschel has done. We shall not hesitate to admit that these stars form so many natural groups or particular systems although we have little or no information as to the nature and origin of their systematic relations. No one will be tempted to attribute this apparent accumulation to an optical illusion or to an unusual chance which would have brought together in this way the visible rays which extend from our eyes to all these bodies. This is the case even though actually the stars in such a group may be distributed in celestial space at distances comparable to those which separate stars which belong to different groups. This is mathematically possible, but not physically admissible. Once we are convinced that we are dealing with a real grouping of stars in celestial space, and not simply an apparent grouping on the celestial sphere, we shall again resist the idea that this proximity may be due to a chance of another sort. We shall rather believe that mutual bonds of interrelatedness exist among the stars of a given group. For example, we shall believe that the stars in group A would not be so close together if the causes which have determined the position of each were not dependent upon one another, rather than being dependent upon causes which have brought about the distribution of stars in group B, or in the other groups.

After a telescopic study of the skies has given this notion of a *cluster of stars*, and of a non-fortuitous accumulation or of *natural constellations*, we shall be able to recognize—again as Herschel has done—that the most brilliant heavenly bodies, which give the appearance of being irregularly scattered across the celestial sphere, very probably form, together with our sun, one of these groups or one of the natural constellations. Here we have something whose richness and immensity satisfies and exceeds the imagination of poets, but which, in turn, is swallowed up by another immensity that is revealed by the scientific study of the universe.

Now notice—and this is a very essential point—that the mind

has no difficulty conceiving of an indefinite number of nuances between completely irregular and fortuitous dissemination, which would allow nothing but purely artificial groups to be set up, and accumulation in clearly marked, completely isolated groups which are very distant from one another.   The latter, which cannot be considered as fortuitous, but imply, on the contrary, the existence of a connecting link between the causes under the influence of which each individual has taken its place, give us the idea of perfectly natural systems.   There are innumerable nuances between these extreme states, because in them the connecting links gradually become restrained or relaxed, and because the part played by accidental and fortuitous causes may be combined in variable proportions with the part influenced by constant and jointly responsible causes.   If, therefore, we are forced by the nature of our methods to establish boundaries and groups everywhere, we shall be able to direct this activity in such a way as to bring us as close as possible to the conditions of a natural distribution.   But some of the groups will be less natural than others, and the expression of natural relations will be found to be inevitably complicated by the presence of artificial links, factors introduced to meet methodological exigencies.

162. It is not necessary to have a large number of individual objects in order to make natural groups stand out.   For the discovery of some new planets, it was sufficient to suggest the idea of the distribution of the planets into three groups or classes: one perfectly marked-out median group or class comprises the telescopic planets that are comparable with respect to their physical qualities, their small masses, and the approximate equality of the long axes of their orbits; a so-called inner group composed of the earth and the three other planets, whose dimensions and the period of whose diurnal rotation are comparable to those of the earth; and, finally, an outer group which now numbers four, one of which, the farthest from us, is still almost unknown, but the other three of which resemble each other very much with respect to their mass, their speed of rotation, and their attendant satellites.   Similarly, as the number of chemical radicals has been increased, we have come to see very natural groups standing out among them, although including few individuals, such as the group which includes the radicals of potassium and sodium, or, again, those which include

chlorine and its analogues. Still other radicals remain isolated
or can be related to one another only by means of characteristics arbitrarily selected according to some artificial system of
classification.

163. Completely analogous remarks may be made about the
naturalist's generic types and classifications. A genus is natural
when the species included in it resemble one another in many ways,
and by comparison differ so much from species belonging to genera
most closely related to them that this relation to the one and separation from the other cannot reasonably be accounted for on the
basis of the fortuitous play of causes which would make the types
of organization vary irregularly from one species to another. There
must have been connecting links between the causes, whatever
they may be, which have formed the species of a given genus;
or we believe rather that these causes may be broken down into
two groups: a group of dominant causes which is the same for
all species in the genus, and which determines the generic type;
and a group of causes which is subordinate to the former, but
which varies from one species to another, and which causes the
specific differences.

If, in its turn, the genus is considered as a species of a higher
genus to which we give the name " class," so as to avoid confusion,
everything that we have said of genera and species may be said
of classes and genera.[4] In that case the class and the genus will
be similarly derived from nature, if, as a result of the comparison
of species, we must conceive the aggregate of causes which have
determined the nature of each species as breaking up into three
hierarchically arranged groups. First is a group of causes to which
all others are subordinate, and which, being constant for each
genus, and consequently for all the species of each genus, has
determined the whole group of fundamental characteristics which

---

[4] " The universal creative intelligence has the same relation to the production
of natural things that our intelligence has to the conceptions of genus and
species." Giordano Bruno, *Dialoghi de la causa, principio e uno*, 1584.

" Different organisms are thus united by a superior bond which rests in the
cause of their creation, and which has distributed them into classes, orders,
families, genera, and species. Genera exist only in species independent of one
another, and not as an organism which engenders these species." J. Müller,
Manuel de physiologie, Book VI, Sect. I, Chap. 1.

make up the class. Second are groups of causes which are subordinated to the preceding and constant for all species of the same genus, but which vary from one genus to another, and which, in connection with the preceding, give rise to the generic types. Finally, there are causes of a still more inferior sort, which, being subordinate to the preceding, as we have said, form the specific types.

In the regular system of classification to which we submit things for the symmetry and convenience of our methods, the genus may be natural and the class artificial, or the other way around. There is no more natural class in the animal kingdom than that of birds. Yet in spite of this fact, or perhaps because of it, there is in the class of birds more than one genus about which naturalists do not agree and which we may strongly suspect of being artificial. A genus is artificial when the distribution of the varieties of forms among the species comprising it includes nothing that may not reasonably be attributed to the action of fortuitous causes which vary irregularly from one species to another. In that case, a term is omitted in the series of steps that we have indicated; and the fundamental causes which determine the prototype of the class—that of birds, for example—come to be subordinated without intermediary to causes which vary without restraint from one species to another, and which produce the specific differences.

164. There may be, and ordinarily are, a greater number of steps than we have indicated. Moreover, the conception of these steps is itself only an imperfect image, and we observe in the subordination and the intricacy of natural causes innumerable nuances which our nomenclatures and classifications are unable to express. From this, there arises an inevitable mixture of rational abstractions, which have their prototypes or their bases in the nature of things, and artificial or purely logical abstractions, which we use as instruments, but which, like so many direct objects of knowledge and study, lack that theoretical excellence which excites and sustains distinguished minds. The work of most eminent naturalists attempts to untangle these artificial abstractions, which are introduced into the natural sciences so that they may be studied more conveniently, from rational abstractions through which our mind grasps and expresses the dominant traits in the

plan of nature.   The philosophy of the natural sciences consists principally of a critique of this sort.   The difficulty which stands in the way of complete success in this undertaking stems from the continuity of nature as much as from the infinite variety of modifying causes of which only an imperfect expression can be found in the symbols of language, as will be explained later.

One of the most remarkable characteristics of the scientific work accomplished in nearly a century has been this tendency to extend artificial classifications further and further, so that we can accommodate classifications more and more to the expression of the natural relations among the objects classified, even at the expense of practical convenience.   This movement which marks the work on classification should certainly be made evident first in botany and zoology, in which the objects to be classified are so numerous and involve such complex organizations, and, consequently, are capable of being compared in so many ways.   But it has gradually pervaded all the branches of human knowledge. We have cited above examples taken from astronomy and chemistry.   Other instances might have been taken from linguistics, that very recent science which is very deserving of interest and which seeks to bring out the natural affinities and the common roots of dialects.   These are invaluable evidences of genealogy and of the connections of the human races for the periods concerning which history or monuments are silent.

165. In the philosophic schools of the Middle Ages, in a period when skepticism could be confined by religious faith and could no more have been exhibited with respect to the basis of the ultimate data of knowledge and sensible experience than to the bases of morality, dialectical skill was expended on problems relating to the objective consistency of abstract ideas, rational conceptions, and logical fictions.   From this situation developed famous controversies and innumerable sects that have been grouped under three principal headings: "realism," "nominalism," and "conceptualism."   This tripartite division lacked clearly marked lines of distinction and on the whole it indicates nothing more than the existence of two extreme parties and a middle party, susceptible of further fractionization, as is always the case in these long disputes which separate men and which cease only when the parties to them become exhausted.   Certainly we have no wish to do what

has been done by so many others and reopen this sterile and knotty subject. Nor do we wish to go through this labyrinth of subtleties and ambiguities again at the risk of misleading our readers with respect to it. Yet it is well to point out the origin and the point of departure of these schools and to evaluate the principle involved, evaluating it on the basis of its consequences, the trouble it gave rise to, and the interminable contradictions it raised.

The origin of all these disputes lies in the very foundation of the peripatetic doctrine and in the role that Aristotle allotted to the idea of substance when he placed it at the head of his list of categories and subordinated all others to it. Substance, according to this doctrine, is reality or being *par excellence,* and all other categories share in reality only in so far as they represent the affections or the modes of being of a substance. On the one hand, substance is the apex in the scale of classifications, or of "metaphysical degrees": birds are animal, animal is body, body is substance. Now, if the two extreme terms of the hierarchical series of genera and species, and of classes or metaphysical degrees, that is to say, the individual and substances, are things to which reality and the fullness of being cannot be refused, we may reasonably conclude from this that reality subsists in the intermediary degrees, and that the difference between the one and the other, or that which must be added to the one in order to bring about the other, is a reality.[5] Thus, *corporeality* is added to substance to create bodies, *animality* is added to corporeality and substance to form the animal, and so on until the individual is reached which brings together in itself the constitutive essences of the species and of the higher genera, together with the accidents which characterize it as an individual. Such is the basis of peripatetic realism, and it is principally on this ideational foundation that the controversies among men of letters during the Middle Ages revolved. Listen to what M. Cousin said about this matter:

The principle of the realist school is the distinction in each thing of a general element and of a particular element. Now, the two equally false extremes are really these two hypotheses: either the distinction of the

[5] For Aristotle, however, no universal is a substance: Οὐδεν τῶν καθόλου ὑπαρχοντων οὐσία ἐστι. *Metaphysics,* vii. 13, 1038 b 35. To us, the contradiction appears insoluble.

general element and of the particular element leads to their separation, or their non-separation leads to the abolition of their difference, and the truth is that these two elements are at once distinct and inseparably united.   All reality is dual. . . . The self is essentially distinct from each of its acts, even from each of its faculties, although it cannot be separated from them.   The human genus sustains the same relationship with the individuals which compose it; but they do not make it what it is; it, on the contrary, makes them what they are.   Humanity is essentially one and complete in itself and yet at the same time in each of us.   Humanity exists only in individuals and through individuals, but in return individuals exist, resemble one another, and form a genus, only as a result of the unity of humanity which is in each of them.   Here, then, is the answer we propose to give to the problem presented by Porphyry, " Does the genus exist apart or as a perceptible thing? "   Distinct, yes; separated, no; separable, perhaps; but with that we leave the limits of this world and of actual reality.[6]

166. If the human genus maintains the same relation to the individuals who compose it that the self maintains to each of its faculties or to each of its acts; in other words, if we attribute to humanity or to the human genus the substantial reality that we attribute to the self or to a human person; and if this substantial reality which constitutes the genus is found again at one and the same time in all the individuals of the genus which are distinct, although inseparably related to a particular element; then there is in truth a mystery in all this that is as impenetrable to human reason as are the most profound mysteries of theology.   The obscurity becomes still more profound if we notice that, apparently, substantial reality does not belong to this special element, since we have compared it to the faculties or acts of the self; while it must have substantial reality on the same basis as the general element, if it is to reappear on that basis in the sub-genera or hierarchically inferior species.   But the contradictions disappear and the veil of mystery is torn, without having to go beyond the limits of this world and of human science, if, instead of a hierarchy of substances and essences, we see in our generic terms only the expression of a subordination of causes and of phenomena.   In proportion as this subordination is more or less marked, the genus is more or less natural.   It ceases to be natural when the resemblances on which we establish it, although real

---

[6] Cousin, *Ouvrages inédit, Raillard*, Introduction, p. cxxxvi; πότερον χωριστά γένη) ἢ ἐν τοῖς αἰσθητοῖς. [Porphyry: *Isagoge*, Chap. I].

enough, may be explained by means of chance, that is to say, by means of the concurrence of causes not connected with and subordinated to one another. Thus, for every organic species, and for the human species in particular, there is an evident subordination of the causes which result in the individual varieties to the causes which determine the general and specific and hereditarily transmissible traits. There is a no less manifest subordination of the conditions of existence for the individual to the conditions of the existence and the perpetuity of the species. Therefore, the human species, as naturalists say, or the human genus, to use a term more familiar to philosophers and moralists, constitutes a natural genus. In other words, a human nature exists, and these words are not empty sounds, nor do they represent a pure conception of the mind. Similarly, the class of birds and the still more general class of vertebrates are natural; for, as a result of the knowledge we have gained in zoology, we are led to consider as dominant the characteristics of these classes which, taken together, make a sort of prototype or *form* in conformity with which nature has subsequently and secondarily proceeded—by means which up to the present time remain unknown to us—in the operation of varying genera and species within limits fixed by the dominant conditions. Consequently, the causes, whatever they may be, to which it is necessary to impute the determination of the dominant and constitutive characteristics of a class must be regarded as essential and dominant with respect to the equally unknown or very imperfectly known causes which have introduced the diversity of the species.

167. It is not necessary to believe that the Scholastics absolutely ignored the distinction between natural and artificial genera. On the contrary, they more than once indicated that they intended to apply their theories of metaphysical degrees " only to those things which, having a natural substance, proceed from divine action; for example, to animals, metals, and trees, but not to armies, tribunals, nobles, and so on." [7] But preoccupation with substances and with substantive distinctions always entered their minds to obscure a glimpse quite removed from that degree of clarity through which, in modern times, it has led to a profound study of the organization of existing things. The conclusion to

[7] Rémusat, *Abélard,* Bk. I, 432.

be drawn from this chapter in the history of the human mind is that everything becomes clear when we take as our guiding thread, in the philosophic interpretation of nature, the idea of the reason of things, of the connection of causes, and of the rational subordination of phenomena, that sovereign and regulative idea of human reason.   Everything is obscured and made difficult when we take as the regulative and dominant idea that of substance, which has only a subjective basis, or the objective value of which is confined within limits which the genius of Aristotle ignored, and of which the doctors of the Middle Ages had not the slightest notion (117 and 135).   Nevertheless, it must be recognized that the instrument of language is established in conformity with the idea of substance and is suggested by the consciousness we have of our personality and of our *ego*—to use the language of modern metaphysicians.   The order of the categories proposed by Aristotle conforms to the genius of language and to that which we may call the order of grammatical categories.   From this a veritable contradiction arises (136 and 143), a real opposition between the conditions of the structure of the organ of thought and of the nature of the objects of thought.   This contradiction has tormented philosophers during the centuries in which men should have been all the more preoccupied with logical forms, when the science of things was less advanced, and for the solution of which it would be necessary to know how to disengage oneself from the influence of logical forms and the mechanism of language, without thereby going beyond either the limits of this world and of actual reality or the truly essential conditions of human knowledge.

CHAPTER XII

ON MORAL AND ESTHETIC IDEAS

168. In the general harmony of the world, nothing is more striking than the agreement we observe in all the degrees of animal nature between the arrangement of the organs and of the faculties through which animals receive impressions from outside, and the whole arrangement of the faculties and organs by means of which animals react on the external world so that each may fulfill its proper destiny. The two systems move along parallel to one another, develop, become perfected, and break down together. Alongside the nervous system which carries sensations is the nervous system which transmits the directions of the will; with the more perfect sense organs are associated more powerful and delicate organs of locomotion or of grasping; following more obscure or more distinct perceptions, actions become more indecisive or better determined (91 and 131).

Thus, analogy should be sufficient to allow the presumption that since, in matters pertaining to knowledge, man has faculties which are much superior to those of animals, he is for that reason called to a higher destiny and should perform acts of a higher sort. If this superiority in the order of knowledge were to go so far as to enable man to conceive absolute and necessary truths, that ability in itself should be enough to suggest the intervention of a principle which may have this character of necessity and absolute rigor, so far as the regulation of his acts is concerned. This suggestion is undoubtedly only a supposition based on a rational inference of the sort which would impress an intelligent being who, although not a man and without having direct knowledge of the law which governs the action of men, would observe man as we observe animal species, completely enough to foretell all the relations between humanity and the rest of creation.

Therefore, it is very plain that the philosophical study of man consists of two essential parts which are distinct although connected, and that a philosophical theory of our acts and their control corresponds to every philosophical theory of knowledge. It is also clearly evident that this dual nature of man, this complex of intellectual and rational faculties and of instinctive and animal faculties, this life of relations and this ability to rise through the relative to the concept of the absolute, plays analogous roles in logic and in morals; the one being, so to speak, the counterpart or symmetrical reproduction of the other.

A development of knowledge to which no parallel development of the active faculties of man corresponded would, as far as we are able to judge this matter naturally, be an anomaly, an irregularity, and a disturbance of the general plan of creation. Thus, when organs designed by nature for the propagation of a plant have been transformed into beautiful ornaments, into resplendent but sterile petals through careful work and the artifices of cultivation, no matter how much charm they may have for the senses, reason sees in them only a monstrosity rather than a perfection.

When we consider man as a product of social forces, we must no longer expect to find among individuals that natural and correct balance between knowledge and acts, that parallel development of intellectual and active faculties of which we have been speaking. The division of labor, the distribution of roles among various members of the human family does not permit it. And, independently of social necessities, the abuse man makes of his liberty would be sufficient to disturb this balance. It is in the social body that we must look for and that we may find, at least approximately, the correlation and parallelism that nature has achieved with a surer hand among individuals, necessarily confused by progress and abuse, instead of in the species that it has not destined for a social life.

169. In this book, we have proposed to present a sketch of a critique of knowledge. We have not proposed to seek in the human heart, in the analysis of the tendencies and of the needs of human nature for the rules of private morals, of justice, and of politics. No doubt man may find in his conscience reasons for admitting or rejecting certain theories, depending upon whether or not they appear to him to lead to practical consequences that the heart honestly approves or disavows. This is a criterion like any other, and perhaps the best of all; but it is not what we wish to concern

ourselves with in the present study. On the contrary, we think of moral ideas, from whatever source they come, as objects of knowledge for the understanding. The philosophic question we raise is that of discovering whether or not there is any reason to regard moral ideas simply as human facts springing from the unique nature of the species, or whether, on the contrary, they must be related to an order of facts, laws, and conditions which dominate the laws and conditions of humanity. This is another instance of the problem which has occupied us previously, and the principle by which it may be solved must still be the same.

170. Suppose, for example, we are dealing with the problem of evaluating with perfect philosophical detachment a moral system in which the search for pleasure and the avoidance of pain are considered to be the end and rule of human actions. It would not be difficult to see that such a system is out of harmony not only with certain elements of human nature, but also with what we have discovered of the general plan of creation. We see all around us that nature has introduced pleasure and pain as means and not as purpose, as devices for obtaining certain results and not as final ends. Pleasure and pain are connected with certain impressions from external agents and with certain functions of animal life, precisely in the amount required to conserve individuals and species. Every analogy would be made worthless if man, having acquired faculties superior to those of the animals, had not acquired them for other ends than for that which is not even an end in the order of animal functions and faculties. And the dissonance would not be reduced, if we were to replace the desire for present pleasure or the repugnance for imminent pain with a sort of arithmetical balance of pleasures and pains which must follow one another in the course of the life of the individual, as a result of such a basis of determination. This would be the case even though, in order to establish this balance, as many individual lives or as many successive generations as might be wished were gathered into a single group.

171. Especially in morality, skeptics have had good sport by comparing the opinions, maxims, and practices of one people, sect, caste, or period with those in vogue among other sects, nations, or other generations of men. " A midday nap decides the truth . . . .

Right has its eras . . . . A pretty sort of justice, that bounded by a river or a mountain . . . . Truth on this side of the Pyrenees, error on the other . . . ." And to this formidable objection, epitomized in this way by Pascal in his energetic sentences, the dogmatists have been able to reply only by citing the interests and passions of men, which obscure their judgments in regard to things that touch on practical affairs, leaving them with their habitual clarity only when they deal with speculative truths. But we fail to see why it would be more difficult to agree that in matters which pertain to the moral faculties of man, as in what we properly call spirit, genius, and character, the varieties of races or even the varieties of individuals have a freer field of action than in those which pertain to the organization of the faculties by means of which we acquire knowledge of physical objects and their relations. Every species has certain more constant, more specific characteristics, just as it has others which exhibit by preference individual differences or racial variations. And, in the same way, when we compare many species of the same genus, we recognize that the genus is established naturally only by the persistence of certain characteristics which are more fundamental than those variations which differentiate the species. It is consistent with every analogy to say that the faculties which exist fundamentally among animals most closely related to man, as in man himself, although developed very unequally, have more specific persistence in human beings. On the contrary, faculties which are exclusively characteristic of one species—and consequently less fundamental for envisioning the concatenation of species and the general plan of nature—lend themselves more easily to individual differences and to racial variations, or to varieties which result from the prolonged action of the same external influences, according to circumstances in a given country or period. The task of the moral philosopher is to distinguish within the subject matter of moral ideas, in so far as this is possible, that which is specifically fundamental, which pertains essentially to human nature, and which is restricted only in morbid cases and monstrosities, from that which is left to individual differences or to variations of the sort that we have noted above. But his task is not limited to this. In accordance with rational inferences, he may, and indeed he must, distinguish among ideas and beliefs of different origins, those which have their basis, their reason, their prototype in laws of a higher order as compared with those which have given man his specific nature, and on the same basis distinguish what is morally good or bad in individual differences.

172. This pertains particularly to what is called *honor*, the rules of which, even though they are often tyrannical, have as their sanction not remorse, but disgrace, and are not to be confused with those revelations of conscience the authority of which we respect even when we happen to break them. We shall nowhere find this subject better analyzed than in the book in which a very sensitive author has envisioned from every point of view the consequences of the great social transformations of which we have been the witnesses:

It would seem (says M. de Tocqueville) that men employ two very distinct methods in the judgment which they pass upon the actions of their fellow men; at one time they judge them by those simple notions of right and wrong which are diffused all over the world; at another they appraise them by a few very special rules which belong exclusively to some particular age and country. It often happens that these two standards differ; they sometimes conflict, but they are never either entirely identified or entirely annulled by each other. Honor at the periods of its greatest power sways the will more than the belief of men; and even while they yield without hesitation and without a murmur to its dictates, they feel notwithstanding, by a dim but mighty instinct, the existence of a more general, more ancient, and more holy law, which they sometimes disobey, although they do not cease to acknowledge it. Some actions have been held to be at the same time virtuous and dishonorable; a refusal to fight a duel is an instance.[1]

The author goes on to show, with a great deal of sagacity, how the ideas of honor which are characteristic of certain countries, professions, or classes are determined by the needs or exigencies which rest in the very nature of the country, profession, or class. As a result, these ideas become more singular and have more influence in proportion as they correspond to more particular needs and are experienced by a smaller number of men, and, on the contrary,

---

[1] *Democracy in America*, Volume II, Book III, Chap. 18, [p. 230, of the Henry Reeve text as revised by Francis Bowen and further corrected and edited by Phillips Bradley, published by Alfred A. Knopf, Inc., New York, 1945. Reprinted with the permission of the publisher.]

they tend to disappear as social classes break down and populations intermingle. From this the author finally concludes:

If it were allowable to suppose that all the races of mankind should be commingled and that all the nations of earth should have the same interests, the same wants, undistinguished from each other by any characteristic peculiarities, no conventional value whatever would then be attached to men's actions; they would all be regarded in the same light; the general necessities of mankind, revealed by conscience to every man, would become the common standard. The simple and general notions of right and wrong only would then be recognized in the world, to which, by a natural and necessary tie, the idea of censure or approbation would be attached.[2]

But in this extreme supposition, and on the basis of the author's principles themselves, it would still not be permissible to consider rules of universal morality, appropriate to the general needs of humanity, as a sort of resultant or mean between particular standards of honor or morality which are suited to certain groups of men and are adapted to their special needs. For the simple coalescing of needs cannot change abruptly the character of the moral law, make it more *sacred,* impose it by *belief* as well as *will,* give it as a sanction remorse, on the one hand, and the satisfaction of conscience, on the other. This very remarkable characteristic of universal morality, by which it stands in contrast with the rules of honor which pertain to a class or profession, could not be due only to the fact that the general needs of humanity outweigh the needs of a class or profession and are not the product of conventional institutions. It must be due above all to the fact that as a result of their generality the notions *of the just and of the unjust* overshadow the idea of humanity itself; and they must also be due to the fact that we understand that these notions would still govern societies of rational and intelligent beings other than men, beings having neither the same organs nor needs that we have. We believe in such principles just as we believe that there are principles in our human logic which would direct minds served by other senses than ours, using different symbols, or in which truth might be attained without the mediation of sense impressions, and which would not need the assistance of symbols in order to transmit it.

[2] *Ibid.,* p. 242.

If, in the very core of humanity, there is one indelible distinction which does not have its roots in conventional institutions, and which exercises an influence of prime importance on everything connected with customs and on what we call honor, it is surely the distinction between the sexes. Now, although Christianity has worked to raise the moral dignity of women to that of men, and to impose on man, through his conscience, duties no less severe than those imposed on woman as a consequence of the natural conditions of her sex, the world—to use the language of the Christian pulpit—has persisted in a morality at once more relaxed and tyrannical, full of rigor for one sex and of indulgence for the other. Such are characteristics which we assigned a moment ago to that honor which finds its reason in the needs of society, in the conditions of existence of a class or of a particular group. But while public morals always yield to the necessity of these natural conditions, the innermost voice of reason, or the moral sense, even when devoid of religious faith, would protest against the injustices of the mores. This protest indicates that, above the laws of the physical organism and the consequences which follow from them, we conceive of a reciprocity of rights and obligations among moral persons who are bound by mutual obligation and who are capable of raising themselves in the same degree to the ideas of right and duty, notwithstanding all the native physical dissimilarities that nature has put between them.

173. No one would contest the fact of the successive appearance and development of a certain number of moral ideas, as a result of the culture of societies and of individuals, under the influence of religious and civil institutions and of individual education. But it seems that this fact, which is so natural and so constant, has been correctly interpreted neither by minds having skeptical tendencies, nor by those who may have or who may assume the mission of refuting them. The former have thought they were justified in concluding that moral principles have no basis outside of or above social institutions; the latter have wished, by means of subtle distinctions, to maintain intact the proof drawn from a pretended unanimous consent of people during every epoch of the life of humanity. Yet, why should the idea of a moral progress in individuals and societies offend reason and universal order any more than the idea of progress in the sciences, in philos-

ophy, and in the arts? If, for any reason, the objective value of moral ideas were denied, it would be necessary to call in question the objective value of all scientific truths, which are not the heritage of all minds, and which manifest themselves only to certain finer intellects with the assistance of a large number of instruments and helps of every kind, and are met only in the midst of very cultivated societies. On the contrary, must we not deduce an argument in favor of the objective value of moral ideas from this, if it happens that, in spite of very different initial conditions and under the influence of races, climates, and institutions which differ considerably, moral ideas which have been refined by culture tend more and more to approach the same type, far removed from their original distinctions, and, under the same physical influences, come to be more and more strengthened and pronounced?

The mere fact that the system of moral ideas tends toward uniformity, among peoples whose social culture improves under the influence of different circumstances, would be good reason to admit that this system progressively casts off everything that is based on accessory and variable causes, in order to retain longer only that which may be attributed to the very basis of humanity itself and to the moral constitution of our species, to its natural bents and to its permanent needs. But if, in addition, new needs are introduced in the course of this progressive improvement, it must be presumed that such ideas, of which humanity has not always been in possession although its needs may have been the same, are not true only in a human and relative sense. They must be presumed to depend on the general order of things that we are not always capable of discovering, but that always impresses us as soon as it is shown to us. In a word, they must be part of a body of higher truths. Although it can be argued against them that they have remained unknown to less refined men and to barbaric peoples, because they have been perceived only in the course of the improvement of civilization and manners, their newness itself, that is to say the newness of their revelation, would be the best evidence of the eminent rank that they occupy among the principles man discovers but does not create. Otherwise, how would it happen that a genius, whatever power he may have had, would have been able to impose imperishable beliefs on future generations? In endowing some privileged individuals with the most brilliant faculties of genius, nature has, after all, produced only an accidental and passing phenomenon. If, instead of discovering one of the great laws of nature, Newton had envisioned only an ingenious but imaginative system, it could be said that a day would come when his name would be forgotten. But it will never perish from man's memory, because it is connected with the discovery of an eternal truth. It is a law of the moral order as it is of the physical order that the traces of initial and accidental circumstances are effaced in the long run under the influence of causes which always act in the same way and in the same manner. And even though the traces of these initial circumstances were never to disappear entirely, or require for their disappearance periods which, up to the present time, the span of recorded history has been too brief to encompass, we should realize that in their gradual and temporal weakening they cannot be part of the conditions of a normal and definite state. Thus, moral ideas would have even greater value for statesmen and for political historians, but they would become, so to speak, matters of indifference for philosophers whose thought aspires to set aside accidental and variable facts in order to penetrate better the inner disposition of the permanent laws of nature. On the contrary, if an idea or a moral belief is not weakened by being handed down by tradition, or if it persists or recurs, whether complicated by variable accessories or freed from them, in all periods of human history and among peoples who are most unlike one another with respect to the forms of civilization, then such an idea or belief will have to be regarded as having its root in the nature of the species. This is the case even when, as a result of the absence of traditional transmission, the idea or belief happens not to develop in the individual or to develop in him only under very exceptional circumstances which, in one sense, come back to the general plan of nature and to the conditions of definite and permanent order. We may count on this since everything that happens as a fortuitous and singular case is nevertheless destined to happen sooner or later, after the play of fortuitous causes has finally introduced, in the midst of a mass of combinations which leave no trace, the particular combination which incorporates the principle of its perpetuity. Who can say that among the species which today are the most stable so far as their physical characteristics are concerned, there are not some whose origin resulted from individual peculiarities, which, far from disappearing with the individuals in which they first appeared, have found the circumstances by means of which they have been able to reproduce and to become thoroughly established through their descendants. In passing, it may be noted that the same remark must not be lost sight of when the question of the natural or supernatural origin of language is raised. It may be that most men were formed in such a way that, living by themselves and under ordinary conditions of primitive life, they would not have invented the art of speech. But it is sufficient that some individual persons who had a more fortunate structure, and who were placed in more favorable circumstances, were capable of beginning the rough skeleton of a language in such a form that this rudimentary tongue could then be perfected and propagated by all the members of the species. And in this sense, it would still be true to say that the gift of language belongs naturally to the species, or is naturally a part of the structure of the species.

174. All we have been saying about moral ideas applies, with some slight changes, to that other category of abstract ideas which is concerned with beauty and taste in the arts, the theory of such ideas being a preferred study in modern times, and at the present time is usually given the name " esthetics." It is still less our purpose to develop here a system of esthetics than to present a theory of morals. Nevertheless, it is within the scope of our plan to make it understood that, in esthetics as in morality, the purpose of critical philosophy is essentially to bring about the separation between modifications left to individual or racial varieties, to accidental and transient influences, and the basic characteristics pertaining to the normal and specific constitution. After this separation has been effected, the further goal of critical philosophy is to seek to discover whether the ideas which pertain to the normal state and to the specific constitution do not have their objective prototype or their *raison d'être* in the very nature of the external objects which suggest them to us, or in more general laws than those which have impressed on mankind its particular organization and structure. Finally, for all this work, critical philosophy can only set forth rational inferences, analogies, and probabilities of the same sort as those to which we have been ceaselessly calling attention up to this point.

Does an object delight us because it is beautiful in itself, and because we receive from nature the gift of perceiving this quality of external things and of delighting in it; or do we rather call a thing beautiful because it delights us, because we have no other basis for the idea of beauty than the very pleasure which the object causes in us as a result of fixed laws of our particular nature, or because of accidental modifications which it has undergone? This is how the problem which we have run into everywhere presents itself to us in esthetics. This problem consists in distinguishing the role of the feeling or perceiving subject and that of the object seen or felt in the act which brings them into relation with one another and from which a feeling or perception results. In esthetics as elsewhere, there must be extreme cases in which the solution of the problem, in one way or the other, leaves no trace of doubt in a perceptive mind, in spite of the fact that this result may be reached only by a process of inference which necessarily excludes a rigorous demonstration and which is inevitably exposed to sophistical negation, just as there must also be doubtful and uncertain cases with respect to which different persons incline to one side or another, depending upon their particular habits and the point of view in which they are placed.

On the occasion of the perception of an object which pleases us and which awakens in us the idea of the beautiful, critical philosophy may be led to a different solution of the same fundamental problem, depending upon whether it approaches it from the point of view of esthetics, or from the point of view of knowledge unadorned, and independent of the feeling of pleasure which accompanies it. For example, an architect who understands the effects of perspective will purposely alter the proportions of a building so that from the position from which the spectator views it the perspective rectifies these alterations, and the object appears as it must if it is to please us and is to present to us the characteristics of beauty. In the same way the tragedian or pantomimist may exaggerate certain effects in his performance in order to make allowance for the distance of the stage from the audience. Now, in such cases, if we consider the idea that the sensible impression gives us of the external object, in so far as it is representative of the object itself, that idea is certainly falsified by subjective conditions. Consequently the characteristic of beauty that we impute to the object in reality belongs only to the image as the senses present it to us. But it by no means follows from this that the conditions of the beauty of this image are purely relative to our sensibility and that the image, as we conceive it, or the object which would bring it into being externally, has no intrinsic beauty which subsists by itself, whether or not we may have been so formed organically that we may be affected by it, as light would subsist even though we had no eyes with which to learn that it exists.

175. Before entering into more detailed explanations, let us make some general remarks. Not only do a multitude of natural objects please us and seem beautiful to us, but the world itself, considered as a whole, presents us with the characteristics of beauty in an eminent degree, and the very name the ancients gave it, if we may believe their own testimony,[3] expresses that eminent beauty. External reality is not only an inexhaustible source of methodical observations for scientists, of calculations for mathematicians, and of meditations for philosophers. It is also a marvelously fruitful source of poetic beauties and of ravishing ecstasies. Now, what would be likely if man were to get his ideas of beauty only from the comeliness of his own nature and from the particular circumstances of his organic structure; if, for example— as many men have claimed—we were to judge the beauty of forms and proportions only in so far as they are related to the proportions and the forms of the human body? Would it not be an entirely unprecedented and improbable chance that starting from this arbitrary model, we should continuously find in external nature, to the degree that we probe its depths and scrutinize its details, not only some objects which fortuitously bring together the conditions of this relative and entirely human beauty, but also innumerable beauties of detail and comprehensive beauties which infinitely surpass the most beautiful productions of human art in the way each part falls into place? Do we not see that with respect to other ideas, for example, with those of the good and the useful which are actually related to our nature and needs, we do not observe a similar agreement? As a result, in speaking of the

---

[3] " As far as I am concerned, I am in agreement with the peoples of the world. The Greeks have designated the world by the word ' cosmos ' which means ' ornament,' and we have given it the name ' mundus ' because of its perfect finish and grace." Pliny *Natural History*, ii. 3.

many marvelous works that nature seems to produce simply for the joy of producing them—according to our human ideas—it is almost impossible for us to say what they are of service to, in what way they are good and in what way useful. Therefore, that human idea of the good and the useful must not be carried over, or, at least, we have no authority to carry it over into the realm of natural facts, and we run a great risk of going astray if we seek in them the reason for the order and harmony of phenomena. But inversely, since the beauty of the works of man appears to us to be only a weak reflection and an image of cosmic beauties, there is reason to infer from this that the idea of beauty does not originate in purely human convenience. And just as when we see that the universe is subject to mathematical laws, we infer that mathematical ideas and relations subsist independently of the mind that conceives them and must not be put in the same category with artificial abstractions but with the rational principles of things, so the beauties which are so profusely scattered throughout the world as a whole and in its details must lead us to believe that the principles and the reason of beauty do not arise in the particularities of man's organic structure, but are of a sort which is clearly superior to that of purely human facts.

At least at first sight, it might even be supposed that the influence exercised on man by the spectacle of nature is what has fashioned his tastes for the purpose of producing an object agreeable to him and of making sure that he finds beauty in it, when this object recalls to him the proportions, forms, assortment of colors, and so forth to which the daily pageant of the world has early accustomed him. In general, all the hypotheses which were discussed in Chapter V, and to which one may have recourse for the explanation of the diverse harmonies of nature, may be called upon to explain the harmony between the order of the world and our taste with respect to beauty, except for the more thorough examination of the question as to which has the higher degree of probability, according to the force of inferences and the spread of analogies which work in its favor. But the supposition that our ideas and our tastes in matters of beauty depend upon the peculiarities of our individual or specific organizations and that nevertheless they are found to agree fortuitously with the general arrangement of the world seems improbable and inadmissible from the very first.

176. This is the place to make another general remark completely analogous to that which has been suggested to us concerning moral ideas (173). In the products of human art, the discovery of the rules and conditions of beauty is the fruit of patient research or of the inspiration of genius. If we reach them methodically and progressively as a result of the development of civilization and of the culture of individuals and peoples, and in such a way that ideas and tastes which differ radically among themselves in primitive periods and during the infancy of a people tend to approach the same types as a result of the intercommunication and progress which civilization introduces, then there is reason to believe that man has not forged these types but has discovered them and that he perceives them more clearly in proportion as his eyes are better prepared to be opened to the impressions of a light from without. But what, on the contrary—and this seems to be more in conformity with historical evidence—is the situation where the influence of the individual genius plays the most important part in the discovery of the beauty in matters of art; where the masterpieces of genius, which are objects of continual imitation and study, exercise an ineffaceable influence upon the ideas that men form of beauty? Can this fact be reconciled with the general law which supposes that every accidental and isolated occurrence leaves only passing traces, at least to the point of admitting that the individual genius has revealed to humanity permanent forms, the knowledge of and feeling for which cannot be lost once it is acquired, without a return to a condition of barbarism which would abolish every impression of them?

177. " There is in art," says La Bruyère, " a point of perfection, as there is one of goodness or of maturity in nature; one who perceives and loves it has perfect taste; one who does not perceive it, or who likes only this part or that of it, has defective taste. There is, therefore, good and bad taste, and tastes can reasonably be disputed." [4] But what is this point of goodness or maturity in nature which must be regarded as the ground and the reason, or at least as the model of perfection in art? Let us take an example, and discuss from this point of view the idea that we have given of the specific prototypes and of the conditions of ideal perfection in organic beings, fashioned according to these types.

[4] *Des ouvrages de l'esprit.* Chap. I.

For greater simplicity, let us consider at first only that which pertains to dimensions, contours, and sensible forms. Is it necessary to believe that we measure all the dimensions, all the lines, and all the angles in a great number of individual objects which may serve to determine their individual forms, that we take the mean for all these particular quantities, and that the system of these mean values determines the form, the εἶδος of the specific type? Modern statisticians seem to have understood the matter in this way, but in doing so they have failed to take account of a grave theoretical difficulty. In fact, it could well happen, indeed in most cases it must happen, that these mean values do not agree among themselves and may as a whole be incompatible with the essential conditions of the existence of individuals and of species. To take a different comparison, but one whose geometrical simplicity enables our mind to grasp it clearly, let us suppose that we are interested in the problem of a right-angled triangle whose sides vary accidentally, within certain limits, from one to another, without maintaining exactly either the same absolute dimensions or the same proportions. Let us measure a great number of these triangles, let us take the mean values for each side, and let us construct another triangle having these mean values that, in one sense, we may call an *average triangle*. Yet this mean triangle will not be the specific prototype of any of the individual triangles, for—as geometry proves—it will not be right-angled and thus it will not possess the essential characteristics of the class of right-angled triangles. Let us admit that this essential condition is taken account of by postulating that the prototype of the triangles be right-angled, and that we succeed in fixing the limits of it by using as the length of the two sides which form the right angle the average length of these two sides, as obtained from the series of individual triangles. The two acute angles of the triangle so constructed would not be the averages of the corresponding angles as these would be given by the series, and the base would not conform to the average base. In a word, no matter from what point we approach the problem, it will be mathematically impossible to construct or to define a triangle in which we find realized at one and the same time, and interconnected among themselves, the average values of all the dimensions which, for each individual triangle, have yielded perfectly determined and perfectly compatible values. If this is the case for the most simple geometrical figure, the triangle, there are good reasons why, without restriction or arbitrary conventions, we cannot define the form or the specific type in a group as complex as that of the aggregate of the organs of a plant or an animal by means of a table of averages. What would the situation be if we wished to take account of a great many other physical or physiological characteristics, such as weight, muscular force, power of the senses, and so on? Evidently, the mean values of such different elements would agree among themselves only by a very great chance; and the synoptic table of all those values, not being intended to be regarded as the definition of a possible individual, is still less the definition of the specific type. We still seek the idea and the approximate description of this sort of definition no matter how much difficulty we experience, and even though it may be impossible to give it a perceptible image and an adequate expression through the use of methodical and rigorous procedures.

178. But let us move along, and come back to the principles of esthetics from which these preliminary remarks have diverted us. Even though the collection of individuals were to furnish a system of perfectly reconcilable mean values, it would by no means be necessary to conclude that this system exhibits the image of the specific type, or that it is suited to give us an idea of what this type is in itself, independently of the influence of external and accidental circumstances which alter and deform it. No doubt if these accidental features affect one of the elements of the species, for example, the length of a line, sometimes in one way and sometimes in another, by exaggeration or by diminution, and with the same facility and the same intensity, then the average furnished by a great number of individual cases would be precisely the value which belongs to the prototype, and all the alterations due to accidental and external causes would exactly compensate for one another. But should this exact compensation not occur, or even though the causes of deformation always acted in the same manner,[5] it would not follow that they lose their character of

---

[5] Thus the average height of men in France, and probably everywhere, is far from being what might be called the ideal height, for the very simple reason that the accidental causes of stunted growth, which result from defects in diet and the unwholesomeness of common occupations, greatly exceed both in in-

accidental and unusual causes, nor does it cease to be necessary to think of the effects that they produce as alterations of the original type, alterations that must be put aside if we wish to conceive this prototype in its ideal perfection and in its essential beauty. Such is the object or one of the objects of art. In default of the methodological procedures of science, it is to this object that the indefinable feeling which we call taste applies; and it is that which, when exercised by a particularly sensitive organism, turns the results of attentive study and observation as much to account as does science, although in a different way.

179. No doubt the conditions of the perfection of specific types of ideal beauty do not attract the attention of laymen and artists in the same degree and with respect to all species. There are two reasons for this: one is relative to man and causes him to take a preferential interest in species which are most closely allied with him, which best serve his needs or his pleasures, and which are his natural friends or enemies; the other is based upon the very nature of the various objects displayed for human observation and is

---

tensity and in frequency those which tend to result in excessive height. With even more reason, the average span of life, or that which statisticians call "life expectancy," is clearly well below the ideal we set up for the natural span of life, abstracted from accidental causes of destruction, or from the ideal that we may refer to as the " longevity " of the species. The average life is so essentially different from the specific longevity that there are species in which, most of the individuals having perished before becoming adults, the average life does not reach the age at which the individual beings are capable of reproduction and of perpetuating the species, since the greater part of them perish before they become adults. This is an instance of the discrepancy indicated in the preceding section. When statisticians tell us that in a certain country or for a certain period of time the average length of human life is roughly 25, 30, or 40 years, no one will understand this to mean the natural and normal longevity or life span of man in the country or during the period. On the contrary, we understand that the average span of human life may vary in two thoroughly different ways. It may vary either because external conditions of hygiene, policy, customs, and social economy have undergone changes influencing the chances of mortality, while the physical make-up of the species remains the same in other respects; or it may vary because the very formation of the species has gone through a long period of hereditarily transmissible modifications which, from the naturalist's point of view, are the only ones of which it is necessary to take account to fix the longevity of the species or of the race. But to go into greater detail on this matter would carry us far from the considerations with which we are dealing in this book.

rooted in the fact that certain specific types, in comparison with others, thoroughly combine, to a very high degree, the conditions of perfection and of ideal beauty. In fact, why may we not say of species of the same genus what is said with reason of individuals of the same species? It is true that we know even less about the causes which modify the fundamental characteristics of genera and of the way in which species become differentiated than we do of those causes which continually act to modify the fundamental characteristics of species or of races in such a way as to produce individual varieties. But this ignorance does not keep us from understanding very well, in both cases, the subordination of modifying and accessory causes to those which give rise to fundamental characteristics. Therefore, in the case of each of those genera that we call natural because the kinship of the species which compose them is so strongly marked, as in the genus *Felis,* for example, no naturalist fails to pick out one species, such as the lion, which is, as we say, the prototype of the genus. In other words, it is the species in which we find combined in a more excellent manner than in any other the distinctive characteristics of the whole genus. For this reason, whether it be clearly understood or confusedly perceived, this is the species that we find beautiful or excellent in comparison with the other species, and yet no arbitrary element enters into such a judgment.

It is possible to go higher in this hierarchical progression. Then the prototype of the genus *Felis* will also be the prototype of the order of carnivorous mammals, provided the genus in question is the one in which the power, harmony, and perfection of the characteristics of carnivorous mammals are found in their maximum development. For the harmony without which no work of nature could continue to exist does not reveal itself to us in such well-defined characteristics, and does not actually exist in the same degree in all works of nature. Some imperfections may be and are incompatible with the conditions of the existence of individuals and of the perpetuity of the species. Among strongly indicated types of intermediary and indistinct forms, imperfect sketches or less perfect models may be and are recognized. In their way these give evidence of the unfailing fecundity and infinite resources of nature, but they can neither excite our admiration in the same degree nor stimulate the imagination of the artist, because, actually they do not, like the others, have an ideal type and form of beauty which is characteristic of them.

180. Suppose man sets about to modify nature; that he creates new races and to some extent new species which are suited to his needs and enjoyment. No one would deny that these more recent and less stable species have certain conditions of perfection and harmonious adaptation, an ideal type and a kind of beauty which is different from those which belong to natural and uncultivated species, even though they are derived from a common source. If in addition we assume that these types which cultivated or uncultivated nature offers him are seized upon imaginatively by the artist to express symbolically a moral or an abstract idea; if the lion becomes for him the symbol of force, the horse the symbol of tractable impetuosity, we shall allow him a certain exaggeration of fundamental characteristics. Judged from the point of view of art, his work will be beautiful or not in spite of the fact that there are in nature no individuals of the sort he has portrayed, but also in spite of the fact that the existence of such individuals would be incompatible with the organic traits of their species. This is the reason why the beauty of works of art may be distinguished from the beauty of works of nature, and why the conditions of ideal perfection are not necessarily the same for both, in spite of community of origin.

181. Next to the works of art made in imitation of the works of nature are placed the works so peculiarly adapted to the needs of civilized man that nature offers no model of them. Nevertheless, here, too, we still encounter the beautiful and the ugly, that good and bad taste of which La Bruyère speaks, and the reason for which must always be given by a theory of esthetics. For example, take what is perhaps the simplest of the products of plastic art, a vase that is required by the conveniences of manufacture as well as those of use to have a rounded form. In such a case the question is no longer that of tracing its contour or, as geometricians say, its generating curve. The sole question is to know why one contour pleases and seems more beautiful to us than another. Now, if we consult those who have dealt with the problem thoroughly,[6] we shall find that after all ordinary forms have been

---

[6] See particularly Ziegler, *Études céramiques*, Paris, 1850. [Claude Jules Ziegler (1804-1856), a French painter and ceramic artist known, among other

put aside, in the use of which the only purpose has been the construction of a utensil without any pretense of satisfying the conditions of plastic beauty, the remaining forms arrange themselves naturally under a rather small number of specific types, each determined by the combination of a very small number of elements and by simple relations among their principal dimensions. The scheme of those conditions can be represented as determining, for each type of species, a system of points through which the curve of the form is required to pass, and which can be united in a continuous line which serves to determine the contour of the vase according to the taste of the designer, and which stamps it, so to speak, with the seal of his individuality. Now we understand that in order to answer to the idea that one must know the perfect form of the object under consideration, four things are necessary: first, its form must indicate clearly the use to which it may be appropriate, even though in reality it serves as an ornament or as an image of something rather than the thing itself; second, the physical conditions resulting from this usage as, for example, the conditions of stability, should obviously be satisfied; third, the subordination of incidental parts to principal parts must clearly result from the way in which they are associated and from their relative dimensions; and fourth, among the various relations capable of satisfying the above conditions, we choose by preference the simplest,[7] relations which are more pleasing, and not just because our mind is better able to grasp them, but because reason is offended by unnecessary complications by virtue of the same principle which makes it take offense at an absence of symmetry when there is no intrinsic reason why the symmetry of something should be disturbed, and because this manner which the human mind has of looking at things constantly finds its confirmation in the study of the phenomena and of the laws of nature. So much

---

things, for his work in decorating the Madeleine in Paris. In his later years he was the director of the Academy at Dijon and the curator of the museum in that city.]

[7] Men have been struck by the role played by the number *three* in esthetics as elsewhere, and mysticism has always been engrossed with this observation. Really nothing is less mysterious, as is demonstrated simply by the fact that the number *three* is the simplest number after *unity* and the number *two,* and that we seldom compute it although it is easy to compute.

for the explanation of the fundamental conditions of what is done and for the reasons which, in a species, fix the points of reference of the profile. It would not be so easy to say what guides the taste of an artist in the apparently arbitrary sketching which must connect these points; nor would it be easy to say what leads us to prefer one sketch to another as more correct, more elegant, or more pure. But observation teaches us that in this part of his task the artist has to avoid two extremes: on the one hand, a stiff or formal style, and on the other, an affected or distorted style. Moreover, we understand that one of these extremes is offensive to us because it seems to reveal a servile constraint; the other because it indicates a capricious complication. That is sufficient to convince us that, independently of any arbitrary system, there must be a mean and normal form between these extremes. In short, the history of art informs us through a mass of examples of all sorts that the natural progression of the human mind in the arts is to begin with inflexibility and to end with affectation in performance. This is a subject for subtle analysis and one which involves some of the most interesting problems which remain to be resolved, problems of such a sort that it would seem very strange if an algebraist were to attempt to find their solution.

182. The poverty of language need not lead us to confuse affections which are of different sorts and are thoroughly distinct, even though they are closely associated in those complex phenomena which we call sensations and feelings. The sentiment we have for the beautiful is one thing; the pleasure of agreeable emotion which the spectacle of beauty gives rise to in us is something else. Just because a tragedy or an opera, although often mediocre, may affect us more than the sight of a picture, a statue, or a great piece of architecture, we must be on guard against concluding that there are in the play or in the opera beauties of an order clearly superior to anything that the art of Phidias or Raphael is able to produce. This is principally a matter of the ability to react to agreeable or voluptuous impressions which depend upon very changeable particularities of structure to the point that often what pleases one person displeases another, and what has given pleasure at one time does so no longer. Intellectual taste, as it has been called, which is only one way of spontaneously judging the conditions of beauty, and catching sight of that in which it exists, has much more con-

stancy and fixity. But so that the distinction may be clearer it will be convenient to make a new approach to this problem and to start with the more homely effects of physical sensibility.

An odorous or savory object acts upon the olfactory nerves and on those of taste in such a way as to give rise to a characteristic impression that we recognize as being the same, even though we may have acquired an aversion to the taste or odor which originally was agreeable to us, or vice versa. This follows from the fact that the sensory nerves may be affected in the same way and still provoke in the rest of the organism sympathetic reactions which are entirely different, according to the general disposition of the system as a whole or to that of some of the great sympathetic centers. One man will face with courage or even with serenity a physical pain which would cause another to faint. This is not because both do not experience the same sensation of pain in the nerves affected, but because the general make-up of one is formed in such a way as to resist the disturbance caused by local pain, or else the excitation communicated by moral causes produces the same results as an increase of physical force. In all these cases we see clearly that it is necessary to distinguish local and special sensations from the attractive and repulsive feeling which is connected with them, and which, being a very much more complex phenomenon, must have much less constancy and fixity.

A similar statement might be made concerning colors, about which the adage assures us it is no more necessary to dispute than it is about tastes. A color which has pleased us may come to displease us, although the special function of the optical sensation certainly remains the same so long as the eye remains normal. Likewise, after having preferred the sound of a string instrument to that of a wind instrument, we may come to have an opposite preference, although the impression *sui generis* that the timbre of each instrument produces upon the auditory nerve is always the same. There is no distinction to be made in this respect between the two superior senses of sight and hearing and the inferior senses of taste and smell. But if the ear perceives a succession of different tones, or if the eye is struck by an assortment of colors, then harmonies and contrasts appear which are based—as physics brings us to understand—not on the particularities of organic form which vary from one person to another or even on anatomical or physiological

traits native to the species, but on the very nature of the phenomena, the perception of which comes to us through the sense of
hearing or of sight.[8]   This explains well enough why the notion of
beauty is connected with the sensations that we obtain from these
two superior senses, while it is never associated with sensations of
taste and smell.   Consequently, the preference for the brilliant colors
of a Flemish painting or the more somber tints of a Spanish canvas
will depend on individual taste, according as one or the other is
more in harmony with the condition of the nerves and the state of
the mind.   There will be, so to speak, a chromatic scale which will
change from one master to another, or from one school to another,
but whatever the influence of the master or the school on the
general tone of the coloring, it will be necessary to apply the same
rules to the relations of the colors among themselves, to their harmony and to their contrasts, and the observation of these rules will
establish the beauty or perfection of coloring in all systems.   Likewise there is a beauty and perfection in an air in music, which belongs essentially to the melody, that is to say, to the succession of
sounds and to their relative intervals of tone and duration, no
matter what the absolute value of the fundamental note may be,
and even though there are preferences of individual taste in regard
to the timbre and quality of sounds, depending upon the instruments and the voices used in playing the piece.

183.   These individual preferences, upon which depend what in
art is called style or manner, and what in matters that do not aim
at esthetic beauty is called fashion, must therefore not be confused
with taste, which follows from the essential conditions of beauty,
in conformity to a certain ideal.   And it is no more necessary to

---

[8] " Music charms us even though its beauty consists only of the agreement
of numbers and of the counting of the beats or vibrations of sounding bodies
which are recognized by certain intervals, a calculation of which we do not
catch sight and that the mind does not cease to make.   The pleasures that
sight finds in proportions are of the same nature, and those that cause the other
senses to reflect something similar, although we may not be able to explain it
in a direct manner."   Leibniz, Dutens edition, II, p. 38.   In the last part of
the sentence, Leibniz manifestly confuses sensuous or purely neural sensation
with the perception of appropriateness and of beauty, which, whether it remains
obscure, or whether it becomes distinct, certainly assumes the intervention of the
mind and the concurrence of cerebral functions.   It does not seem possible to
admit with Leibniz that a perception of this sort may be induced by the
sensuous functions which the inferior senses of odor and taste produce for us.

confuse the perception of beauty according to a type which is constant and independent of the organism with the sensuous excitation which is connected with it. The vivacity of the latter is deadened by habit—although it may serve to give more persistence to our judgments of intrinsic beauty—and is so variable in other respects, depending as it does upon temperaments, upon the nature of the agents or of the materials used in the arts, and upon the way in which they act upon the organism. Now, when thought has abstracted in this way all the accessory and variable feelings which are connected with intellectual taste or with the perception of beauty, what remains, if not a faculty of pure reason, a way of judging and discerning in things the relations of order, fitness, harmony and unity? *"Omnis porro pulchritudinis forma unitas est"* [Moreover, every form of beauty is unity], Augustine said in a phrase that everyone cites, and which in fact would be the best definition of beauty if it were possible to define it and to include in one general formula that which presents itself under such varied aspects, in our eyes at least, and at the distance we are removed from supreme principles. We still prefer it to those definitions which are more modern and more mystical than they are philosophical and which make beauty consist in a pretended relation between the finite and the infinite, about which it is very doubtful that the great artists had ever thought, and the study of which would, in any case, characterize the quest of a particular school rather than answer to the idea that men have always had about beauty.

When this degree of abstraction is reached, morality itself may be considered, and often has been considered a branch of esthetics. In fact, some actions are morally beautiful as well as morally good, since they conform to those ideas of fitness, order, and harmony of which human reason is capable of conceiving the model and of proceeding with the application. As Plato insisted, it is at this point in particular that ideas of the beautiful and of the good are confused or tend to become confused. For if we prefer to reserve the epithet of the good for actions which presume a rare virtue, a generous devotion, and which excite in us a sentiment of admiration that we do not experience as a result of ordinary acts of honesty and beneficence, it is plain that it would be difficult to draw a clear line of demarcation between those falling on one side and those on

the other. It is also plain that the feeling of duty and the satisfaction that is experienced in carrying it out, or the remorse at
having failed to do so, are affections of the mind which can neither
be identified nor compared with the attraction that is felt for the
beauties of nature and of art or with the disgust that ugliness in-
spires. But as soon as these diverse affections of the feeling subject
are removed, in order to consider only those aspects of the acts by
means of which we apprehend the qualities of things which have a
representative value for the understanding, we see that each depends upon the same superior faculty which seeks and finds order,
harmony, and unity everywhere, and which, having found what it
seeks, is convinced in that way of the legitimacy of its pretensions,
and of the conformity of the general laws with the laws of its own
nature.

CHAPTER XIII

ON CONTINUITY AND DISCONTINUITY

184. As soon as our intellect begins to distinguish among perceptions, it acquires the notion of distinct and similar objects, such as the stars in the sky, the pebbles on the seashore, and the trees or animals across an expanse of open countryside. In this way we derive the idea of " number," the simplest and most common of all our abstract conceptions, and the one which contains in rudimentary form the most useful as well as the most perfect of the sciences.[1] Even though man, deprived of all or of some of his senses, were to have no knowledge of external objects, it is conceivable that the idea of number might be suggested to him through his consciousness of what occurs within him, as a result of his attention to the intermittent reproduction of identical or analogous internal phenomena, provided his faculties were not condemned to inactivity in other respects.

Number is conceived as a collection of distinct *units*. That is to say, the idea of number implies three things at one and the same time: the notion of the individuality of the object; the notion of the connection or *continuity* of its parts—if it has parts; and the notion of the separation or *discontinuity* of individual objects. Even though the numbered objects are physically contiguous, reason must distinguish them, and we must be able to consider them separately, in spite of the fact that this contiguity or continuity is accidental and in no way inherent in the nature of the numbered objects. Stones do not cease to be naturally distinct objects simply because they touch one another, and the cement which sometimes holds them together does not prevent the recognition of fragments of rocks of different sorts and origins which existed prior to their being mixed. When specific numbered objects and, consequently, collections of those objects can be compared with respect to quantity, the quantities established by such collections are called

---

[1] " To this was added number, a thing not only necessary to life, but also unchangeable and eternal." Cicero *De republica* iii, 2.

" discrete " or " discontinuous." When one of the objects which make up such a collection is added or removed, the collection passes abruptly from one state to another without passing through intermediate nuances and without imperceptible gradations.

While we understand that this property of individuality or discontinuity is characteristic of a great many objects which we perceive, others have a contrary characteristic. For example, just as in the case of a pile of pebbles, we think of the water which fills a vase as a quantity which may be increased or decreased. But whereas the pile of pebbles necessarily undergoes abrupt changes in its volume, weight, and form when pebbles are added to or removed from it, the flow of water into or out of the vase results in a continuous variation of the weight, volume, and height of the liquid in the container. This variation is such that these different quantities pass from one condition to another only after having passed through an *infinite number* of intermediate stages, however nearly adjacent to one another one may imagine the different quantities to be.

185. In the example we have selected, the continuity may be only apparent and relative to the imperfection of our senses, for perhaps the liquid is only a collection of particles which differ from larger pebbles and whose individuality eludes our senses only because their dimensions are so extremely small. But in other cases, the notion of continuity comes from a rational examination independent of sensible experience. It is indeed only through such an examination that the idea of continuity and, consequently, the idea of continuous mass can be understood in their absolute exactness. Thus we necessarily conceive that the distance from a moving body to a body at rest, or that between two moving bodies, can change only by passing through all the unlimited or infinite number of intermediary distances which separate them. The same thing is true of the time which elapses during the passage of bodies from one place to another. All geometrical quantities—lengths, areas, volumes, and angles—are called continuous quantities because they evidently have the property of increasing and decreasing continuously. The same thing is true of the quantities dealt with in mechanics, such as velocity, force, and resistance.

In general, when a physical quantity varies because of the passage of time or only because of changes in the distances between molecules or material systems or as a result of both these factors, it is contradictory to suppose that it passes from one determinate value to another without having gone through all the intermediary values in the interval. But, in the imperfect condition of our knowledge about the structure of material media, we are justified in admitting, for certain physical quantities, as we can conceive and define them, disruptions of continuity resulting from the sudden passage from one finite value to another. Thus when two dissimilar liquids, such as water and mercury, are superimposed, we regard density as a quantity which varies sharply at the point of contact between the surfaces of the two liquids. This is the case even though all inferences lead us to believe, and although it may be philosophical to admit, that the disruption of continuity would disappear if we could take complete account of the structure of the liquids and of all the modifications which occur near the surfaces where they are in contact.

Physicists and mathematicians no longer admit the existence of what were called discontinuous forces, that is, of forces to which were attributed the ability to change suddenly the direction of the movement of a body and to make it acquire or lose a finite velocity instantaneously. It is generally recognized that the forces in question, which are developed on the occasion of the impact of two bodies, for example, are not unlike the other forces of nature, such as gravity, which require a finite time to produce a finite effect. Today forces which were formerly called discontinuous are no longer distinguished from others except by the property of using up their energy in a very short time, which is ordinarily imperceptible to us because of the inadequacy of our senses and of our means of observation.

For example, when an elastic band strikes an obstacle, the sudden change which seems to us to occur in the direction of its movement and in its speed only appears to be sudden. In reality, the body is insensibly distorted and gradually loses the speed with which it was propelled. After that, molecular reactions restore it to its original form and impress on it another speed in a different direction. All this occurs in such a brief interval of time that it is beyond our ability to estimate it and hence we cannot grasp it, although we cannot doubt the succession of the different phases of the phenomena.

In the same way, when a ray of light seems to us to be bent suddenly in passing from one medium to a second contiguous medium which has a different density, it actually changes direction without discontinuity. The new direction is connected with the old through a portion of a curve whose dimensions are too small for us to detect.

186. In the idea we form of lines, angles, forces, durations, and so forth, the attribute of continuity is found to be associated with that of magnitude; and we think of the latter as a homogeneous whole which may be divided, at least by thought, into any desired number of perfectly similar or identical portions. This number may be increased more and more, for there is nothing to limit its being increased indefinitely. The notion of magnitude is immediately connected with that of *measurement*. A magnitude is considered to be known and determined when we have indicated the number of times it contains a specified magnitude of the same sort which has been posited as a standard of comparison or as a *unit*. All magnitudes of the same sort, of which the latter is an aliquot part, are then represented by numbers. Since the unit may be divided and subdivided according to some law into as many aliquot parts as may be wished—parts which in their turn may be regarded as derivative or secondary units—it is clear that, after the standard unit has been chosen arbitrarily and the law according to which its successive divisions and subdivisions are to be made has been arbitrarily fixed, any continuous magnitude whatsoever allows a numerical expression of any desired degree of approximation. This follows from the fact that it necessarily falls between two magnitudes which are capable of an exact numerical expression, and the deviation from which may be reduced as much as we may wish. Continuous magnitudes, expressed numerically in this way by means of an abstract or conventional unit, become quantities, or are what we call "quantities." Thus although the idea of quantity is altogether simple, and although it has generally been considered as a fundamental category or an original idea, it is not really one. The idea of quantity is constructed by the human mind by means of two truly irreducible and fundamental ideas—those of number and size. Not only is the idea of quantity not primary; it also implies something artificial. Numbers exist in nature, that is to say, they subsist independently of the mind which observes and conceives

them; a flower has four, five, or six stamens, with no possible intermediary, whether it does or does not occur to us to count them. Continuous magnitudes are likewise found in nature. But quantities appear only because of the arbitrary choice of a unit, and because of our need—a need which springs from the nature of our minds—to have recourse to numbers for the expression of magnitudes (153).

When numbers apply to the measurement of continuous magnitudes, the term " unit " evidently takes on another meaning than the one it has when it is applied to the numbering of objects which are individual and truly *single* in nature. Philosophically these two meanings are quite properly opposed to one another. This is an inconvenience which results from the nature of ordinary language, but it is a lesser inconvenience than having recourse to a term that is not sanctioned by usage.

On the other hand, both the philosophic meaning and the analogies of language are simultaneously offended when the name quantity is applied to pure numbers and to numbers which designate collections of truly individual objects, by calling them " discrete " or " discontinuous " quantities. The merchant who delivers a hundred feet of wood or twenty horses does not deliver quantities but numbers or quotas. If we are interested in twenty hectoliters or a thousand kilograms of wheat, the delivery will actually concern quantities and not quotas. This is so because we compare the grains of wheat considered collectively to a continuous mass with respect to volume or weight and have no interest at all in separating or numbering individual grains. A sum of money may also be considered a quantity because it represents a *value* which is a continuous magnitude; and then counting the pieces of money, a process that gives a number that may change even though the total amount remains the same, is only an auxiliary operation devised for the purpose of measuring the value of the money more quickly.

187. In conformity with the common definition, we give the name " quantity " to anything which is capable of being increased or decreased. But a great many things which are not magnitudes, and which consequently cannot be quantities, are capable of being increased and decreased, and even of being increased and decreased in a continuous manner. A painful or a pleasurable sensation in-

creases or decreases and goes through various degrees of intensity with no sudden transition from one to another. Yet it is impossible for us to fix the precise instant when it first begins to appear and when it becomes completely extinct. At least, this is undeniably the way things happen in a great many cases. If, at other times, a pain seems to begin or end suddenly, to increase or diminish by definite increments, there is every reason to believe in this case too—just as there was in regard to the impact which suddenly changes the direction in which a body is moving—that the discontinuity is only apparent, and that in reality the phenomenon is always continuous. This is the case even though we may confuse at any moment the duration of phases whose sequence escapes us because of the imperfection of that internal sense that we call psychological awareness. Sensations of pleasure and pain, however, have nothing in common with the mathematical notion of magnitude. It cannot be said that a more intense pain is the sum of weaker pains. Yet in going through its continuous modification, sensation often passes from pleasure to pain or inversely from pain to pleasure, after passing through a neutral condition—which, in many respects, recalls the decrease of certain magnitudes in the passage from positive to negative quantities. But this neutral condition cannot be regarded as resulting from an algebraic sum or from a balance of pleasures and pains.

188. It is true that through the study of anatomy and physiology we come to see how the continuous variation of intensity in a sensation of pleasure or pain may be connected with the variation of certain measurable magnitudes and depends on the continuity inherent in extension and duration. For we recognize that the greater the size of a nerve [2] among members of the same species the greater the intensity of the painful experience which results from the irritation of the nerve. A certain intensity of pain corresponds to each unit of area of the cross-section of the nerve, provided all other conditions remain the same. But this correspondence or relation is in no sense mathematical, since the attribute of measurable magnitude which belongs to the area of the transverse section does not belong to the sensation.

[2] For the measurement of cross-sections, we take account only of the sum of the cross-sections of the elementary nerve fibres and not of the tissues which protect and cover them.

If a person plunges his hand into a bath having a temperature of forty degrees and leaves it there for a sufficient length of time, he experiences at first a sensation of apparently sudden cold. Following this, assuming that the bath does not become cooler, the sensation weakens gradually and without shock, in such a way that it is impossible to fix the precise moment at which the sensation ceases. All other things being equal, the intensity of the sensation depends upon the time elapsed since the moment of immersion; and the continuity in the passage of time gives sufficient reason for the continuity of the variations of the intensity of the sensations produced. But this sensation must not be thought of as a measurable magnitude that can be related to a unit and expressed numerically.

Since the velocity of the vibrations of a sounding body or of the ether are measurable and continuous quantities, a sufficient reason is seen why the passage from the sensation of one tone to that of another, from the sensation of one color to that of another takes place with continuity. But that does not mean that there are numerically assignable relations among the different sensations of tone and color, as there are among the velocities of the vibrations which correspond to them. The sensation of the tone G is not equivalent to one and a half times the sensation of the tone C simply because the velocity of the vibrations which produces the sound G is one and a half times that which produces the sound C. The sensation of *orange* is not five-sevenths or any other fraction of the sensation of *violet,* simply because the velocity of the vibration of the ether in the case of the orange ray is nearly five-sevenths of what it is for the violet ray.

Continuity in the variation of the intensity of the power of attention or of sensuous appetite will be fully explained by the continuity in the variation of certain physical and measurable magnitudes, such as the velocity and abundance of the blood, or the electric charge on or the temperature of certain organs, which have or may have an immediate influence on other vital forces. But this affords no basis for concluding that the attribute of being a measurable quantity belongs either to these vital forces themselves, or to the phenomena which they determine.

189. Just as the continuity of certain purely physical quantities is sufficient to enable us to submit the forces, affections, and phenomena of organic and animal life—which are not measurable

quantities—to the law of continuity, so we understand that these forces or phenomena, which are continuous but not capable of being measured, may introduce continuity into variations that permit forces or phenomena of a higher order which even more obviously lack the character of measurable quantities. If, among men in particular, the phenomena of the intellectual and moral life depend upon or presuppose those of animal life, as the phenomena of animal life in their turn depend upon or presuppose the general phenomena of the physical order, then the continuity of the fundamental forms of space and time would be sufficient to make us presume the continuity which is habitually observed in matters relating to the web of organic forms, of life, and of thought, with respect to matters pertaining to the intellectual and moral orders, which are liberated most mediately from the conditions of animal sensibility and those of materiality. In a word, the continuity of space and time would suffice to make reasonable the old scholastic adage, as invoked by Leibniz: *Natura non facit saltus* [Nature makes no leaps]. This does not hinder us from supposing, if we wish, that the continuity in the cases of the intellectual and of the moral order has still other bases or grounds than the continuity of space and time, nor does it keep us from admitting, with Leibniz, that continuity in all things comes directly from a higher law of nature, of which the continuity in the phenomena of extension and duration is only a particular manifestation.

190. In the development of intellectual faculties, sensations which are accompanied by perceptions, that is, representative sensations which are capable of giving rise to images which persist or which the mind can reproduce after external objects have ceased to act on the senses, come after purely affective sensation. Now for the very reason that sensation is representative or that it gives rise to images, it is perfectly clear that a continuity or discontinuity in the intellectual phenomenon of the image corresponds to the continuity or discontinuity of the object. If I think of the constellation called the Big Dipper (Ursa Major), the image present to my mind is that of seven twinkling points, clearly distinct from one another and arranged in a certain pattern; but if I recall the view displayed before my eyes when I had reached the summit of a mountain, my imagination is no longer presented with a collection

of a determined number of points. What I remember is a continuous and harmonious whole, into the details of which I cannot go without finding other details in them, and so on to infinity.

The same thing is true of sensations that come to us through other senses than those of sight, and to which, by extension, we also give the name "images" (109 and 110). Thus after having heard a musical air, I may be able to recall perfectly the series of notes which compose it. In this case my perception will be made up of a system of distinct and discontinuous perceptions. But, if my recollection involves all the impressions that I experienced on hearing this piece performed by a talented singer, including the timbre, accentuation, and modulations of voice that cannot be written into a score, I should once more catch a glimpse of the infinite nuances in the harmonious and continuous whole. All this has been verified and expressed a thousand times in all the forms of language.

Discontinuity or continuity is found in the data of memory not only because of the nature of the objects upon which recollection rests, but also because of the nature of the forces and conditions upon which the acts of memory depend, whatever these are and whether they are organic or hyperorganic. It is often noticed that after making a long effort to recall a name, a date, or a historical fact, the recollection of the forgotten item occurs suddenly and by surprise. At other times we have a vague and confused remembrance, the features of which clear up little by little until they attain a perfectly definite form.

191. We say of an image that it is faithful, of an idea that it is true. By this we mean to express the conformity between the object or the type perceived and the image or the idea presented to the mind. If the conformity is rigorous, the idea is said to be exact or adequate; but modifications of the idea which alter this rigorous conformity may, depending upon the case, admit of continuity or discontinuity. In this way there may be a sudden transition from truth to error or, on the contrary, a continuous degradation of truth.

Everyone knows that the portrait of a person, or the picture of a bit of landscape may be more or less faithful; and that there are infinite nuances in this resemblance. This does not make it possible, however, either to achieve a perfect or exact resemblance or to draw a line of demarcation between what resembles, although imperfectly, and that in which the resemblance disappears completely.

We say that a portrait is or is not a true likeness, and we point out certain portions that are better than others; but we would not think of making a tally of the truths or errors found in the portrait.

A geographic map may be regarded as a sort of portrait. However, it is a part of the daily task of geographers to reveal and compute the errors in a map. This is why their attention rests so exclusively upon a certain number of prominent points which may be fixed exactly, at least within the limits of precision consistent with our instruments and our observations. These points are noted or forgotten; they are or are not in the exact locations that accurate observations assign to them; where they are concerned, there is a place for an enumeration of truths or errors. However, when we turn our attention to the continuous lines which connect the landmarks and which serve to show the course of rivers, the indentations along a coast, or the contours of mountains, resemblance is only more or less approximated. It is no more possible to count and balance truths and errors mathematically in this sort of portrayal than it is in any other.

In the memory I have retained of a musical air, I may mistake one note for another, an *F-natural* for an *F-sharp*; and should I play the air on an instrument like the piano having definitely determined intervals between its notes, I shall make a mistake or an error, because there are no intermediate degrees between consecutive keys on the keyboard. But when an artist wishes to imitate the playing of a rival on the violin or on the French horn, the imitation will be found to be only more or less faithful. It will be said that an auditory image of this sort is or is not true. But no one would dream of trying to calculate truth or error in such a situation.

192. The fidelity of a portrait, the resemblance of an image to its prototype admit of progressive variations, and their progression falls under the law of continuity, but that does not mean it is something measurable. There is no standard of measurement for that species of truth to which we properly give the name resemblance. Let us reduce this analysis to the simplest and the most mathematical terms. If, in order to make a copy of an ellipse,

I draw another ellipse having the same relation between the long axis and the short axis that is found in the first, the resemblance or —to employ a technical term used by mathematicians—the similarity will be perfect. Now, if we conceive of a series of ellipses in which this relation, which is a measurable quantity, varies continuously, the individual figures will resemble the first one less and less in proportion as they become more extended and more flattened. This resemblance will depend upon the smallness of the deviation between the fixed value of the relation in the ellipse taken as the prototype and the variable value of this relation in the series of copies. We shall nevertheless be unable to determine this error quantitatively on the basis of a mathematical law, which we like to think of as the measurement of similarity or dissimilarity, except by means of a purely conventional and arbitrary rule. With still greater reason, if, in order to duplicate some sort of an oval curve other than an ellipse, and one which is not amenable to mathematical definition, a curve is traced which more or less resembles the one in question, the resemblance being such as to permit innumerable intermediate degrees, then it would be impossible to measure or to evaluate the resemblance numerically. It is the very nature of things and not simply the imperfection of our theories and our methods which puts an obstacle in the way of such an evaluation. Similarly, if we compare one triangle, which remains constant, to a series of triangles in which the angles and the relations of the sides undergo progressive and continuous alterations, it would be impossible to assign a function of the angles and the relations of the sides which would be the natural measurement of the resemblance to the invariable type without using an arbitrary stipulation.

193. We are faced with an entirely different problem when a question comes up concerning the picture of a living creature and of the expression of that indefinable characteristic that we call physiognomy. We are always struck by the singular fact that a silhouette, a photograph or a bust cast from life may show less resemblance than an engraved or pencil portrait by an artist. This reflection takes full account of the superiority of the version obtained by art over the version to which it seems that nature has made the major contribution. For example, a picture sketched on a plane surface is a projection of the object in relief, and it is possible that in the most wisely chosen projection almost insensible

nuances of form which characterize physical individuality and more especially moral individuality are effaced and obliterated in such a way that the artist who is endeavoring to express just these nuances can do no better than feign a projection which is geometrically impossible. He will be able to strengthen or exaggerate traits in such a way as to have, however, only the exact expression of what he must portray. He will not be reproached for making these exaggerations, in the sense in which this term is used by artists, if he actually overdoes not the lines which identify the form of the drawing, but the physical, intellectual, and moral characteristics that the strokes must depict. This is a resemblance of another type than geometrical similarity or resemblance, but one of such a sort that in portraits which resemble one another in a similar way, the *technique* or style of painting will be very clearly recognized. Each painter attains the same degree of resemblance in his own way and by substantially different techniques. In short, this type of resemblance is even much less susceptible to measurement and evaluation than is a purely mathematical construct, although its progressive alterations may always be subject to the law of continuity.

If the painter is commissioned to make not a family portrait, but one of a historical personage whose physical features reveal little of value except that they have the merit of indicating forcibly the most notable characteristics of an intellectual and moral personality, he will have to be satisfied with other circumstances of resemblance. He must make the picture less a geometrical or physical and more an ideal reproduction (180). This progress toward the ideal will become even more marked when, in representing an allegorical type or a sacred figure, the common forms of humanity appear only to the extent necessary to give substance to the idea that the artist must present and wishes to present.

194. The tendency of art toward the expression of an ideal conceived by the mind, without either a logical formula to define it or a mathematical method for approaching it, is so obvious a thing that no one has ever denied it and that it may perhaps have been exaggerated in the subtle refinements of modern criticism. We have reached the point of making the artist too philosophical, and, on the contrary, we have not sufficiently noticed that in the expression of a pure idea, in so far as it is only an object of knowledge independent of any intention of pleasing or impressing, the

philosopher is also and cannot help being an artist after his own fashion. Deceived by the nature of the institutionalized symbols to which they are forced to have recourse, men have pictured their ideas to themselves as so many units, ciphers, or monads, and they have supposed that the whole business of thought consists in systematically combining or grouping these individual objects. It always seems possible to count the truths and the errors scattered through a book, in the same way that an astronomer makes a catalogue of the stars, or that a public official takes the census of a city, or, even more, in the same way that the propositions contained in a treatise on geometry or the errors of calculation missed by an editor of tabular data are counted. However, although falling outside the realm of sensible phenomena, if the object of the idea is one of those which permit continuous modifications, the character of truth which consists in the conformity of the idea to its protoype and of the expression of the idea with the idea itself will likewise admit of continuous gradations. In that case, it will be clearly possible to say that a given person had come closer to the truth than another; but it will not be possible to enumerate the new truths of which he is the inventor. Everyone will evaluate the worth of this approximation in his own way and will form an opinion about this sort of resemblance, without being able to refute with precision those who do not adopt his evaluation and who contradict his opinion.

The inaccuracy of the picture of an animal is obvious to a naturalist if the number of toes, teeth, feathers, fins, and so on which characterize the species are not found in it. In such cases we find errors which may be taken account of and established without debate because there is no intermediary, no gradual shading among three, four, and five toes. On the contrary, a painter, whose attention is never fixed on the characteristics which serve in the methodological classification of species, finds the physiognomy or *external aspects* of the animal portrayed with more or less fidelity. If his judgment is contested, all he can do is to appeal to those who, like himself, have a feeling for the form of the animal and for the skill shown in the portrayal. He is no more in a position to appeal to a formal proof than I am able to prove to a man that a painting has failed to catch his likeness, if illusion or caprice leads him to find a resemblance in it.

195. A botanist has committed an error in his description of a

plant; two imperfectly developed stamens have escaped his notice because of their smallness. As a result he has put a species which should have been grouped with those which are heptandrous into the pentandrous class of Linnaeus. A more attentive observer or one who is not misled by an accidental monstrosity will correct this error with the use of nothing more than his eyes and a magnifying glass. Once this has been done, the error should not appear again, for the botanical description of the species will have been definitely cleared up and, in return, it will have been enriched by a precise, positive, and incontestable fact. But let us imagine that the flower is subject to an imperfect development which is constant, normal, and specific; that the two stamens are so modified in the course of their development that their functions are further and further removed from those of the ordinary stamen. Let us further imagine that because of this fact one botanist may put the plant in one of the families in which the presence of five stamens is a distinctive characteristic, while another botanist evaluates the relative importance of the characteristics differently, and, distinguishing what is essential and persistent from what is accessory and variable in the structure of the organs, again places the plant in one of the families having seven stamens. How will the difference between the two men be settled? No doubt on the basis of the judgment of the botanists whose opinions are considered most authoritative. How is that judgment formed? It cannot result from an experimental demonstration that falls within the ken of the senses and still less from formal arguments like those which are used by logicians and mathematicians. For if, on the one hand, there are cases in which this transformation of organs shows itself only as a secondary phenomenon which should not escape the eye of a naturalist more experienced in closer resemblances, on the other hand, as we go from metamorphosis to metamorphosis, we should not know when to stop, and in the end we may confuse things which are most unlike. Here truth and falsity tend, so to speak, to be based upon one another; truth does not show itself as a light uniformly illuminating a clearly circumscribed area, but rather as a flash of light which becomes weaker after leaving its source, and the track of which the eye is able to follow for a longer or shorter distance depending upon the tone of its sensibility.

And no one can say, for the purpose of invalidating the example, that the problem of knowing whether to classify a plant or an animal in such and such a family is purely a matter of nomenclature and of method. A truly natural classification, and indeed any classification in so far as it is natural, can be nothing but an expression of affinities which connect organic beings closely among themselves and of laws to which nature is bound in varying or modifying organic types. These laws subsist independently of our methods and of our artificial procedures just as do the laws governing the movements of inert matter, although they may neither be expressed in the same way in terms having a rigorous exactitude, nor established by measurements which are precise, or the precision of which has no limits other than those which follow from the imperfection of our instruments. In general, as we have tried to demonstrate in the next to the last chapter, alongside of artificial abstraction, which is only a fiction of the mind and one which is suited to its instruments and needs, is placed rational abstraction, which is simply the conception or the ideal representation of the bonds which nature has placed among things and of the subordination of phenomena. But, because of the continual efforts of the mind to achieve an understanding of phenomena, there is almost always a mixture of the two sorts of abstraction and a continual transition from one to the other. The bonds of solidarity, of kinship, of harmony, and of unity which we try to grasp through rational abstraction may be more or less stretched or eased, yet our mind experiences the same need for classification, regularity, and method in dealing with all objects of nature. As far as possible, a critical philosophy must begin with artificial and rational abstraction, by basing itself upon inductions and probabilities. Now, as we have already explained, it is the essence of philosophic probability to yield to continuous alterations or progressions, without implying that this probability can be evaluated numerically, and without its becoming a numerical quantity after the manner of mathematical probability. Thus, under whatever aspect the subject may be viewed, we find that the law of continuity prevails in that intelligible world in which the thought of the philosopher searches for the principles and the reasons for sensible phenomena, no less than it does in the material world which falls under the senses.

196. In the sphere of moral ideas, nothing is more evident than the continuous transition from one idea to another and from one quality to the contrary quality. Murder inspired by a passion of malevolence or cupidity is one of those great crimes which arouses a general public censure, and to the repression of which every member of society, according to the nature of his functions, readily lends his co-operation, except for perversions of custom of which we cannot take account here. On the other hand, if only the natural feelings of a person are consulted, moral sympathy and approbation will continue to be won by any person who avenges by murder, and at the risk of peril to himself, the offended honor of persons of whom he is the natural protector. Purely human laws will be unable to triumph over that natural feeling. Between these extreme cases there are murders that some will censure and others will excuse. A human authority will find it impossible to fix the exact point at which criminality ceases and devotedness, not to say virtue, begins.

Even when the status of the act is not doubtful because of the circumstances under which it is perpetrated, we are aware that the moral responsibility of the agent and the perversity the act presupposes may permit an infinite number of nuances, depending upon the age, the sex, the temperament, and the education of the culprit. The interest which attaches itself to the defense of the accused among civilized and humane people must not be permitted to disregard that truth when dealing with cases of great outrages which call for severe repression by penal laws. But the case is the same with respect to the notions of equality, of honesty, and of decorum, as it is with that of criminality.

It is legitimate to draw a profit from one's work and from one's capital, to prefer to do business to this end with those persons who offer the most advantageous conditions, and even to make one's profit larger in proportion to the risk of loss which one runs. The most honest man can do all this without in any way prejudicing his reputation. Yet we justly stigmatize with an odious name a person who makes a business of speculating in goods necessary to life during calamitous times, or who lends his money at excessive rates on coming into contact with those whom wrong conduct, lack of foresight, or misfortune force to submit to his power. Now, is it possible to say precisely at what point usurious profits begin, whether we are dealing with wheat, money, or any other sort of merchandise? Is there a line of demarcation such that it suffices to stay on one side of the line in order to lay claim to scrupulous integrity, and which one needs but to cross to come to be likened to the most dishonest persons? This conclusion is obviously distasteful. It must be admitted, on the contrary, that with a more delicate feeling for the morality of his acts, such a trader will restrain the temptation of personal interest more rigidly and will have a right to a higher place in our esteem, without making it necessary to condemn absolutely anyone who goes beyond the former's self-imposed limits.

When the rate of interest on money is fixed by a positive law, we fully understand that a formal censure should, or may, overtake anyone who exceeds, by however small an amount, the legal rate. In such a case moral censure follows from the infraction of a higher law, namely, that by which citizens are morally bound to submit to the positive laws of their country in matters which follow from the discretionary power of the legislator. The intervention of this power must be considered as having precisely the purpose of introducing an artificial discontinuity where none has been placed in the nature of things, as will be shown later.

In the histories of any people we read of governments which have been established through the misuse of force and through the violent upsetting of certain institutions which had long prevailed. Power gained in this way is said to be usurped, in contrast with legitimate powers that create and maintain the regular operation of the institutions of a country. But, on the other hand, institutions are continuously being modified. Changes that the course of events produces in these institutions, even when quite sudden, create new rights and exclude outdated claims, and yet it is impossible to indicate in any other way than by legal fictions or the needs of the parties involved, where illegitimacy ceases and legitimacy begins. Even here the nature of human affairs, in contrast with certain theories for the use of speculative minds, maintains continuous transitions among terms which remain perfectly distinct so long as attention is fixed only on the extreme cases. The abuse of logic and of casuistry in politics as in morality consists in not taking account of the continuity of transitions and in wishing to apply the exactness of definitions, formulas, and logical deductions to things which are incompatible with it because of this very continuity. On the contrary, the practical good sense of people and statesmen consists in accurately apprehending the relations between things at the point at which they have insensibly introduced forces the nature of which is to act progressively, slowly, and without intermission or discontinuity, and in protesting against the absolute systems of some arrogant minds whose error is not in setting up a theory, but a false theory, and who believe they are making use of logic when actually they are abusing it by applying it to things to which it is impossible for it to be adapted.

197. We hope to demonstrate that the distinction which is best suited for clarifying the theory of human understanding is that of the continuity and discontinuity in the objects of thought. This is the case whether we are dealing with sensible phenomena or with qualities and relations which are purely intelligible but which subsist among things or in things independently of the mind which conceives them. We claim that this distinction gives the key to the most common acts of the mind, as well as to the methods the use of which is reserved to philosophers and scientists, at the same time that it takes account of a great number of the details of social organization. In short, we maintain that, as a general law of nature, continuity is the rule and discontinuity the exception in the intelligible and moral order as much as it is in the physical, and for ideas as much as for images, and that, if this important fact has been misunderstood or if one does not apply oneself sufficiently to developing its consequences, lack of application must be attributed to the nature of the symbols which are for us indispensable instruments in the work of thought. The remaining parts of our study will be especially devoted to the development of these consequences, with which logicians generally have been so little concerned.

We will say that continuity is *quantitative* or *qualitative* according as it concurs or does not concur with measurability. But in thus opposing *quantity* and *quality,* it is not necessary to consider them as general attributes, as Aristotle and his successors did, that is, as predicates or categories of the same order. On the contrary, in justice to the idea, it is necessary to understand that the relation between these predicates or categories is that of species to genus, or of the particular, or, rather, of the *singular* case to the general. It follows from this that, if the singular species were to be abstracted in order to put it in opposition to all other species, by conserving the generic name in this collection, the reason is that, because of its importance, the singular species acquires for us a value comparable to that which the generic idea, in contrast, has preserved by its extension, or by the innumerable variety of specific forms that it is able to cover.

198. Thus, to use a comparison, the circle may be considered as a kind of ellipse. It is a species of ellipse in which both axes are equal and in which, consequently, the two foci coincide in the center. But it is not simply a particular species lost, so to speak, among the innumerable multitude of all those which can be obtained by varying the relations of the axes in any manner whatever. It is a singular species, and one which it is convenient to consider by itself, for two reasons: first, because the properties common to the whole genus of ellipses undergo very remarkable modifications and simplifications when we come to the case of the circle; second, because all ellipses may be considered as projections of a circle seen in perspective, and because by thus connecting the generation of ellipses with that of the circle—as the ancients did—we find in the properties of the circle the reason for all the properties of the curves included in the genus of ellipses. In the same way, the continuous variations of that singular species of quality which we call " quantity " are amenable to the regular methods of determination which are possible in the case of no other quality. Furthermore, in the present state of our knowledge it is permissible to think that the continuity of every qualitative variation is a necessary consequence of the continuity inherent in the quantitative variations upon which the others depend. No doubt, variations characterized by qualitative continuity depend in addition on other principles whose action, on being applied to the forms of space and time, leaves its specific mark on each of these variations. And it is possible (189) that these elements may themselves be capable of continuous variation, but of a sort which is not quantitative or measurable and which is entirely independent of the quantitative variation inherent in the forms of space and time. So the qualitative continuity in subordinate variations does not proceed solely from a quantitative continuity

in certain primary data. This is indeed, if you will, probable, but not demonstrable; and, for our purposes, we are not required to detain ourselves for a discussion of this hypothesis.

199. Depending upon the circumstances, a variation of quantity may be thought of either as the cause or as the effect of a variation of quality. But, in either case, in so far as the question depends upon it, the human mind tends to reduce every variation in the quality of things to a variation in quantity—for which we have regular methods of determination and expression. For example, in almost every case it would be impossible to measure the pleasures and enjoyments or the inconveniences and disadvantages which are connected with the consumption of any kind of goods, or with the possession of property of any sort, in comparison with advantages or inconveniences connected with the consumption of another good, or with the possession of property of another sort. At first, all such factors have a very irregular influence on the discussions between buyer and seller; later on, when the transactions are more numerous and more frequently repeated, they quickly have a mutual influence on each other; a current price is established, and a definitely measurable quantity, namely, the purchase value of real estate, of a commodity, or of a service is found to depend upon non-measurable qualities. This dependence follows from the development of social organization and from the need that man feels, as a result of the peculiar structure of his faculties, to reduce to numbers and to processes of indirect measurement things which by their nature are least susceptible to direct measurement. Are we not led to make use of numbers even in cases in which the problem is to classify numerous candidates on the basis of their knowledge and intelligence by means of examinations and competitions? As if it were possible to evaluate the erudition, sagacity, and keenness of the mind! Indeed, we are well aware that the small number of judges is reason enough why the figures which they choose are very unreliable. On the other hand, we realize that if we were able to bring together a sufficiently large number of competent judges so that anomalies of individual judgment would offset each other, we should obtain an average result which would give, if not the exact measurement, at least the exact gradation of the excellence of the various candidates, as this is revealed through their examinations.

Nothing varies more with circumstances and is less directly measurable than the criminal character of an act or the moral responsibility which attaches to the commission of a crime. But although legislators have wished to leave to the judges the power to take account of all the possible gradations of the offense and of deciding the severity of the punishment, they must indicate the kind of penalties to be imposed, such as a fine or a temporary imprisonment, and these are perfectly measurable quantities. The gradation of penalties would then accurately reflect the gradation of the offenses—at least as these appear to us and to other men like us—if the number of judges were sufficient to compensate for any fortuitous errors among individual decisions.

Even though it is sometimes clumsy and premature, the prodigious development of what is called statistics in all branches of the natural sciences and in social economy stems from the need of measuring in a direct or indirect way everything that is measurable and of establishing by means of numbers everything that permits of such determination. This is true even though statistical numbers most often measure only very complex effects, and effects which are very far removed from those which would have to be known in order to have a rational theory of phenomena.

It is because they did not recognize this law of the mind that philosophers from Pythagoras to Kepler (153) vainly sought for the explanation of the great cosmic phenomena in ideas of harmony which they mysteriously connected with certain properties of number considered in themselves and independently of the application that could be made of them to the measurement of continuous quantities. True physics was founded on that day when Galileo rejected these long-sterile speculations and thought not only of the idea of examining nature by means of experiments—which Bacon also proposed in his own way—but of stating precisely the general form to be given to experiments, by setting up as their immediate object the measurement of everything in natural phenomena which is capable of being measured. A similar revolution was made in chemistry a century and a half later when Lavoisier ventured to submit to balances, that is to say, to measurement or quantitative analysis, properties to which chemists before him had applied only the type of analysis which they call qualitative.

200. What is the source of this singular prerogative which the ideas of number and quantity have? On the one hand, it comes from the fact that the symbolic expression of numbers may be systematized in such a way that with a limited number of conventional symbols—for example, our written numeration, which consists of only ten characters—we are able to express all possible numbers, and, consequently, all quantities commensurable with those that have been taken as units. On the other hand, it arises from the fact that, although incommensurable quantities cannot be exactly expressed in numbers, we have a simple and regular technique for giving them a numerical expression approximating whatever limits our needs require. From this, it follows that the continuity of qualities presents no obstacle because we express them all by means of combinations of a limited number of distinct symbols, and because, in this way, we submit them all to the operations of calculation. The error resulting from this can always be indefinitely reduced, or have as limits only those which are produced by the imperfection of our senses in the exact determination to primary data. "Metrology" is the most simple and the most complete solution, but only in a particular case, of a problem on which the human mind never ceases to work. This problem is to express qualities or relations in terms of continuous variables, with the aid of syntactical rules applicable to a system of individual or discontinuous symbols and necessarily limited in number because of the convention which institutes them. The three great innovations which, in modern times, have successively extended the domain of calculation, namely, the decimal system of numbers, Descartes' theory of curves, and Leibniz's infinitesimal calculus, are fundamentally nothing but three great steps taken in the art of applying conventional symbols to the expression of mathematical relations governed by the law of continuity.

201. The matter needs no further explanation where the invention of our decimal arithmetic is concerned. Descartes' idea was to cease making the distinction in algebraic formulas between known and unknown quantities—as had been done prior to his time—but to make it rather between quantities which are constant, because of the nature of the problem, and quantities which are variable without being disconnected. In this way the essential purpose of an equation or algebraic relation became that of estab-

lishing a dependence between the variations in the one and the variations in the other. This was promoted by means of abstraction, for, whereas in the older algebra, with nothing to specify with respect to the numerical values of certain quantities, mathematicians had always had in mind quantities which had become fixed and in some degree stationary, now what the mind grasped, embracing as it did a continuous series of values which includes an infinite number of values, made evident the law of the series rather than of the values themselves. Simultaneously with his discovery that algebraic symbols, which were originally designed to represent numbers or discrete quantities, were thus suited to exhibit the law of a continuous series, Descartes invented another device which made this law clear and gave it a form and a description. By means of the path of a curve, he depicted the ideal law already defined in algebraic terms. As a celebrated modern writer has said poetically, he was not content to apply " algebra to geometry as the word to the thought," he reciprocally and figuratively applied these two great mathematical *ideas* or theories to one another; and he obtained from both symbolic expressions which are singularly appropriate, each in its own way, to sustain the human mind in its investigation of more hidden truths and of still more general and abstract relations.

Above all, we are indebted to this invention by Descartes for clearing the ground for the third important discovery we have mentioned, that of the infinitesimal calculus. This was destined to replace the complicated and indirect methods based on the *reductio ad absurdum,* or on the consideration of *limits.* What is called the method of limits consists in initially positing a fictitious discontinuity among things actually subject to the law of continuity; for example, by substituting a polygon for a curve, a series of impacts for a force acting without intermission. Then the limits are sought which the results obtained continuously approach when we restrict the abrupt changes so that they follow one another by smaller and smaller intervals, and, consequently, since the total variation must remain constant, each individual change becomes smaller and smaller. The limits found are exactly the values which hold in the case of continuous variations; and these values are found to be determined in this way in conformity with an exact, though indirect, procedure, since this passage from the discontinuous to the continuous is not based on the nature of things, but is simply a logical device suited to our methods of demonstration and calculation.

The complication of this artificial scaffolding was hindering the progress of the sciences until Newton and Leibniz conceived the possibility of directly fixing the view of the mind with the aid of suitable notations; the former fixes this view directly on the unequal rapidity with which continuous quantities tend to vary, while other qualities on which they depend undergo uniform variations; the latter fixes it directly on the relations among elementary and infinitely small variations of diverse quantities which depend upon one another, relations whose law contains the true reason for the progression that the variations of these same quantities conform to, so that we can observe them at the end of a finite interval. From this was developed the infinitesimal calculus, the peculiar virtue of which is that it enables us to understand directly the fact of the continuity in the variation of quantities, and which consequently is adapted to the nature of things, yet not to the way the human mind proceeds, since nothing but finite variations are perceptible and capable of being grasped directly by it.

Thus, when in cooling a body continuously emits heat, the loss of temperature that occurs in any interval of time whatever, no matter how small we may assume it to be, is really a complex fact, resulting, as if from its cause, from the law according to which the body continuously emits an infinitely small quantity of heat during each infinitely small interval of time. The relation between the elementary variations of heat and of time is the reason for the relation which is found between the variations of these same quantities when they have acquired finite values.

Similarly, the space through which a freely falling body passes in yielding to the action of gravitation varies proportionately with the square of the time elapsed since the start of the fall because the infinitely small space traversed during each moment is proportional to the acquired velocity, and, through an evident result of the continuous and constant action of gravitation, this velocity is proportional to the time elapsed from the moment the body began to move. From this very simple relation between the elements of the elapsed time and the space traversed is derived, as though from its cause, the less simple law which connects both the finite variations of these two quantities. In this sense it may reasonably be said that the infinitely small exists in nature. This statement is legitimate not because infinitely small quantities can fall within the domain of imagination or of sensory perception in any way, but because the abstract and purely intelligible notion of infinitesimal elements, far from being an abstraction whose origin is artificial (156) and adapted to the nature of the mind and to the way in which we conceive and imagine things, is rather quite the opposite of this, since it is directly applicable to the way in which natural phenomena come into being and to the expression of the law of continuity which governs them. This is the reason why Leibniz's calculus, which gives to the infinitesimal method the assistance of a regular system of symbols, has become so powerful an instrument, has changed the appearance of pure and applied mathematics, and constitutes by itself alone an invention of capital importance, the credit for which belongs entirely to this great philosopher.

202. The methodical and indefinite approximation of the continuous through the use of the discontinuous is possible not only in questions dealing with relations among quantities in the strict sense. It is equally adapted to relations of position and configuration in space, which have the additional property of being capable of being implicitly defined by means of the relations among the quantities. Thus whether or not we have considered the length of a curve and the area of the surface it circumscribes, we shall be able to determine with an unlimited approximation their bearing, their deviations, and their sinuosity, or, in short, all the accidents which relate directly to form and not to size, provided we have exact methods for determining as many points on the curve as we wish to choose and provided these points may be brought nearer to one another indefinitely. Indeed, when we wish to unite these independently determined points by a continuous line, the hand of the person drawing the line will be guided by a feeling for the continuity of forms which cannot be translated into fixed rules and which does not permit of a rigorous analysis. This will be a matter of art and not of method. But the closer the reference points are to one another, the more restricted will be the limit of deviation among different figures drawn by differ-

ent persons, depending upon whether their hands are more steady and more skillful, or whether they follow the direction of a mind endowed with surer and clearer perception of the continuity of forms (46 and 181).

Everyone knows the process used to copy a design or an image of the two dimensions so as to preserve or change its scale. We break up the surface of the model and the surface on which the copy is to be made into corresponding squares. Then we copy square by square in such a way as to restrict the possible deviations of the copy between limits that approach each other more and more as the number of squares is increased. In this way less and less is left to the skill and the taste of the artist, to the clarity of his perceptions and to the steadiness of his hand. Workers who reduce the block for sculptors have an analogous technique for reproducing somewhat methodically and mechanically in marble the figures for which they have the clay model made by the artist, by *sharpening up* the figure, as we say. This does not exempt the artist from later giving his work those final skillful and physically scarcely noticeable touches, in connection with which method is of no help, and of which the artist alone has the secret.

Basically, and however bizarre this comparison may seem at first glance, the administration of justice and of public affairs constantly depends upon a similar device. Rules are established —as we shall explain later on—limits are set up so that the conscientious evaluation of an expert, an arbiter, a jury, a judge, or an administrator is kept within more or less narrow bounds. Evaluation rebels at analysis and consequently breaks loose from rigorous control. But since it is no longer a question either of magnitude or of extension, or, in short, of any quantitative continuity, the nature of things is averse to whatever would systematically organize a method of progressive and indefinite restriction; and the nature of things is also averse to whatever, at each step taken by a method of systematic restriction, would make a precise accounting of the approximation obtained by such a method.

203. It is evident that every logical rule which promises or seems, in theory, to promise an unlimited exactness, leads in practice only to a limited exactness since it requires, if it is to be applied, the intervention of faculties or the use of instruments

whose perfection is limited. It is possible to carry to any desired length the whim of indefinitely extending such a series of decimals as that which results from the calculation of the relation of the diagonal of a square to its side, or that of the relation of the circumference of a circle to its diameter. Once the rule for such a calculation is discovered, we say its application is mechanical. This does not mean precisely that it could be carried out by an automaton. It indicates, rather, that since the rule prescribes a series of perfectly distinct and determined acts, both may be controlled by persons who apply it in such a way as to give quasi-certainty to the correctness of the resultant (78). Now, if on the basis of this rule we are dealing with the problem of expressing numerically the length of the diagonal of a square the side of which has been measured, the precision of the measurement is necessarily limited, since the perfection of the senses and of the instruments used are necessarily limited. Hence, in the application of calculation or of logical rules, it would be fanciful to exceed the limit of precision imposed by the operation of measuring. If it is not possible to agree closer than a decimeter in the measurement of the length of a side, it would be unreasonable to carry the calculation of the diagonal into millimeters or fractions of a millimeter. When we reach fractions of this order, the lack of precision in the data would take away any significance so far as the precision of the calculation is concerned. This may seem like a very simple thing to point out. Yet it has very frequently been lost sight of in the application of mathematical processes to the physical sciences. Without considering in detail all the circumstances which must have an influence on the limit of precision of often very complicated observations and measurements, an illusory precision has been effected in calculations or in certain details of experiments. This apparent precision is inconvenient not so much because it involves useless exertion and work, as because it gives the mind a wrong idea of the result obtained.

An illusion of this type, but one much more difficult to discover and destroy, may deceive us as to the range and results of those administrative and judiciary rules by means of which it has been proposed, and not without good reasons, to limit the discretionary use of certain powers and the arbitrary latitude of certain appraisals. Because reason might be entirely satisfied with a system of such rules, it would be necessary that the appraiser checked at one door—if we may be allowed this trivial image— might not go out through another; that in imposing one part of the rules of procedure or of minute consistency, a latitude may not be left to the judge in the appraisal of certain facts, or to the accountant in the administration of certain affairs, which destroys the guarantees purchased as a consequence of the fulfillment of annoying or expensive formalities. In a word, it is necessary to be forewarned against the abuse of formality in public affairs, as well as and for the same reason that it is necessary to be cautioned against the abuse of calculation in physics. The reason is that there are limits to possible precision; that in matters related to practical or experimental determinations, rules would only be empty forms, dead letters, without the intervention of those forces which spring from the principle of life, the continuous develop- ment of which eludes measurement, rule, and control. No doubt, there is much that is vague in these generalities, as in a great many other logical rules. In what follows we shall try to indicate some applications that can be made of them in a more special and better established order of facts.

204. If one of the habitual devices of mathematicians is to assume initially a fictitious discontinuity where there is actually continuity, once this artifice has put them in possession of rules for measuring the continuous they quite frequently have had re- course to a contrary device, which is to posit a fictitious con- tinuity where there is actual discontinuity, for purposes of re- ducing calculation and making it more convenient. By such means, they obtain only approximations of true results, but these are so arranged that the approximation is sufficient, while exact calculation, although theoretically possible, would in fact be im- practicable because of the excessively long operation it would require. This device of the mathematicians, which is especially useful in the calculation of chances and of mathematical proba- bilities, resembles fundamentally that which is practiced every day in the most ordinary circumstances. Thus, instead of count- ing grains we measure them as though the grains formed a con- tinuous mass. This is done because it is recognized that if the grains are of the same kind, the relation of volumes does not differ appreciably from the relation between the two large numbers

which would express the number of grains in the measured volumes, if one had sufficient patience to count them. This is also true when, in banks, sacks of money are weighed instead of counting the coins, although the value of the coins, when they circulate as money, is legally counted by the piece and not measured by weight, or may be independent of variations in weight from one coin to another provided these variations, which are continuous in their nature, do not exceed certain legally established limits.

In general, if man's mind is bound, by the nature of its organization and by the form of the tools that it uses, to substitute habitually an artificial discontinuity for the continuity in things and, as a result, to note degrees, break up lines, and set up divisions on the basis of artificial and, to a certain extent, arbitrary rules, then there is also reason to put into practice the opposite device and to treat the discontinuous as though it were continuous, by breaking away from systematic and rigorous procedures whose application is impossible because of the time and energy they require. Thus, although we had exact methods for putting in perspective an object like a machine or an architectural decoration, all the parts of which are capable of being defined mathematically, the designer, the painter, or the theater decorator will apply these long and difficult procedures only to the principal points which will serve him as landmarks, and he will leave the rest to his dexterity as artist. Similarly, in social affairs, at each moment we come to conclusions about chances the exact evaluation of which, while not theoretically impossible, would in fact be impracticable because of the immense amount of calculation it would entail, or, rather, about chances the evaluation of which, while not requiring much time, would demand much more than the habits of society and the rules of the game would be disposed to allow it. In that case it is necessary that such an estimation of chances be made instinctively and spontaneously by a sort of sense whose acuteness springs from a natural aptitude or from exercise, and that constitutes what is called the spirit of the game, or skill, or knack. This applies not only to games but also to trading, to military tactics, and to a great many other things in which man needs to be guided by the light of a sudden inspiration, even in things which by their nature would not absolutely rebel against an exact analysis and rigorous thinking.

CHAPTER XIV

ON LANGUAGE

205. A language is a system of symbols, necessarily limited in number, which must be associated and combined according to certain rules, and which are intended to furnish man with means for making known his sensations, ideas, feelings, and passions. On the basis of this simple statement, plus what has been said in the preceding chapter, we should understand that in most cases the purposes of discourse can be attained only imperfectly. The work of the orator, and consequently that of the writer, is analogous to that of the mosaic artist who has been given nothing with which to copy a natural object or an ordinary picture but an assortment of stones whose colors and dimensions are fixed in advance. It is clear that such an artist will be able to reproduce only approximately the colors and the contours of the objects which he uses his talent to imitate.

Vocal articulations and the representation of these articulations in ordinary handwriting are not the only symbols nature has put at man's disposal for the communication of his thoughts. The advantages of spoken language over written discourse come primarily from accessory signs used in speech, such as accent, intonation, gesture, movements of the eyes and face, and the speeding up and slowing down of delivery. These lend themselves to the needs of infinite shadings, similar to those of the thoughts being expressed, and to some extent they fill in the intervals and breaks in language and, to use a common expression, " paint word pictures." That is to say, these accessory signs re-establish continuity of the sort that would be found in that kind of image which is the most perceptible, and to which, consequently, we like to compare all others. For these reasons we should not be surprised at the pre-eminence of spoken language, not only when dealing with description, narration, and exhortation, but even when, in the mouth of a skillful teacher, it is used with the intention of bringing to light abstract truths and of making known relations which permit of an infinite number of nuances and continuous gradations, and doing so as effectively as the lines of a pencil sketch or the tones of a picture. When we read a discourse, a lawyer's plea, or a written lesson we are not surprised that we find in it only part of the emotions, the images, or even the purely abstract conceptions that its oral presentation would excite.

On the other hand, it is clear that all these signs which make up the accessories of spoken language, and whose ordinary, every-day use is the object of the art we call "oratory," remain limited for most men to the translation of the simplest affections of the sensibility. They are left with what must have been the first rudiments of language, and which make up what grammarians call "onomatopoeia." It is true, the art of gestures has been perfected and systematized for the use of deaf-mutes. This systematization, however, is the work of persons whose whole education has been gained under the influence of ordinary language. That alone is sufficient to indicate that the figurative and conventional language of which they are the authors is simply a translation of oral speech, and that while it has kept the advantages of conventional language to a large extent it has also retained its imperfections.

206. Language is so intimately connected with the products of our intelligence that the Greeks used the same word to designate both language and reason. At first, it must seem to be impossible to distinguish what depends upon the nature of our intellectual faculties from what depends upon the form of the instruments they use. How can we judge what form of development our intellectual faculties would have taken had they used instruments and symbols of another sort, of which we can form no precise idea? Would the loss of language result in the perfecting of other means of communication, other systems of representative symbols, as happens when the loss of eyesight ordinarily leads to the perfecting of the sense of hearing or of touch? The example of what would happen to deaf-mutes if they were left to themselves is not conclusive; for they live in the midst of men accustomed to speech, whose efforts cannot correspond to theirs; and above all, among these persons placed in an anomalous position, we do not find that transmission of efforts from one generation to another which is the essential condition of all human progress. But in-

stead of setting up systems upon empty fictions, we may at this point make some general remarks which are basic to the subject.

207. A language would be quite poor if it consisted only of onomatopoeia or of vocal signs having natural relations to the things signified. Any other type of perceptible signs would also offer little assistance if we used only those which naturally have the property of calling up the idea of the thing signified, and if we had no recourse to symbols which have been established or given a conventional value. But the number of such symbols cannot be unlimited, so that they correspond to every object of thought. There must be laws of combination or syntax for such symbols having prescribed forms which can be kept in the mind until they become habitual, so that attention may proceed to the matter of meanings without being distracted by syntactical form. Now, how can these syntactical laws be adapted to anything other than individually determined elements, and how could the products of such a combinational synthesis vary without discontinuity? From this it must be concluded that the radical imperfection of language comes from the discontinuity of its elements, which in turn, results primarily from the abstract nature of conventional symbols and not from the physical characteristics which they specify. In other words, the imperfection of language is connected with a property of the form and not with what may be called the matter and perceptible content of the symbol (107).

On the one hand, nature has sought to subordinate the operation of human thought and the development of human intelligence to the use of perceptible signs (112). On the other hand, the only development possible for a system of discontinuous signs is one which parallels the development of thought. It follows that one of the greatest handicaps to intelligence must result from this clash between the essential nature of signs and that of most ideas. This is a handicap against which the mind has struggled ever since it began to develop, and a handicap from which it has sometimes been fortunate enough to free itself, but which in other respects holds it in eternal infancy. A meditative mind will recognize in this discordance between ideas and signs one of those details in which nature seems accidentally to deviate from its general plan of continuity and harmony. For philosophy and the human sciences, those eminent products of thought upon which we rightly pride ourselves, are, after all, only incidents in the history of nature and even in that of humanity, the result of a somewhat exaggerated development of faculties which seem to have been given to man for a much less ambitious purpose.

208. This is not to say that conventional symbols other than words may not have been superior to language in other respects. In fact, man invented writing only to remedy one of the greatest inconveniences of speech, that of being a short-lived symbol. The period in which the invention of writing occurred may be regarded as the critical period in the history of the human mind. The direction impressed on the later development of thought had to depend on the form taken by this great invention. We are beginning to lift the veil which has hidden those distant times, and to recover vestiges of the elaboration by which the system of graphic symbols was definitely fixed, at least among the great families of peoples in the midst of which philosophy and the sciences were destined to emerge from their infancy. Thanks particularly to the ingenious work that has been done in Egypt since the opening of this century, we are beginning to understand how writing—which consisted at first only of general signs to which analogical symbols were soon added, then of symbols which were purely conventional but were still independent of language, and finally admitted conventional symbols—has tended more and more to become an indirect symbol, a simple conventional representation of spoken language, until this revolution was systematized by the invention of letters and of the alphabet. From then on writing has been nothing more than language made permanent and stripped of some of its sensible accessories.

We are tempted to wonder if this complete subjection of the graphic symbol to the word, which was completed with the invention of a written alphabet, has been more favorable to the development of the human mind than the coexistence of two systems of independent symbols. Our numerical figures and algebraic symbols are inventions which give evidence of the utility of ideographic writing independent of language. Descartes' conception, to which we have already called attention (201), furnishes a no less remarkable example of the importance of a graphic and conventional symbol especially adapted to the nature of the thing signified. It has been said, and it is natural enough to believe, that written Chinese makes possible certain niceties of expression, certain beauties of style to which nothing in the spoken language corresponds. We must, however, consider a number of factors in this connection; the ideas expressed by means of arithmetical or algebraic characters are only a small number of those which permit an exact determination; the continuity of forms of the understanding could never be adjusted according to a regular and systematic method in the conventional representation of qualitative variations; for this reason all ideographical writing would remain an art rather than a method, or would become a method only by losing its special advantages, and by permitting the existence of the inconvenience of two independent and heterogeneous languages, the habit of using which would have to be acquired, and which would continuously have to be translated into one another. If we keep these things in mind, we shall understand how the invention of a purely phonetic language, by simplifying teaching, must at least have facilitated the raising of the average level of minds, and contributed powerfully to the progress of what is properly called civilization.

209. Without pushing this discussion further, let us make some examination of how language which is, so to speak, our sole mode of expression with respect to abstract matters, and which results essentially from the association of discontinuous elements according to certain syntactical laws, may lend itself more or less to the production of types which are modified with continuity. Let us also examine how in general the expression of the continuous is accomplished by the discontinuous, a thing which becomes so simple in the particular case of quantitative continuity (200).

The reason of philosophers has not proposed this question; men have solved it without being aware of doing so during the slow course of the formation of languages. Most of the constitutive elements of language have not been given a fixed, determined value, like that of each number or each musical note, considered in its tonic relation to a fundamental note. Words which are essentially distinct may assume identical forms as a result of a fortuitous coincidence, especially in very mixed languages. Furthermore, the same words may be used with a determined number of clearly distinct meanings because of the original poverty of language, or because of our need not to weigh memory down with too great a number of different forms.[1] Finally, if the same word is considered in each of its several meanings, we shall most often find that these meanings vary between limits that it is sometimes possible and at other times impossible to assign, or in other cases we shall pass through imperceptible nuances in going from one meaning to another.[2] Now, the trick in the use of

[1] " The nuances of even the most perfect language can never be equal to those of human thought. Modifications of speech are necessarily confined within certain limits; otherwise they would exceed the capacity of human memory. Consequently, it is necessary that, in all languages, a sort of economy makes a single idiom serve many different ends, just as the dagger of Hudibras, which was made to stab and to break heads, was used for many other things besides." Reid, Vol. V of the French translation of his *Works*, p. 331.

[2] In spoken French, the words *fin* and *faim* are phonetically confused, but their spelling clearly distinguishes them in the written language.   Chance has confused, in both sound and writing, two words *fin* concerning whose distinction etymology permits of no misunderstanding: one is derived from the Teutonic word *fein,* meaning *délié* in French [and " fine " in English]; the other comes from the Latin *finis* and has retained all its different meanings.   The word *fin* (*fein*) and its derivatives have many sets of meanings, some referring to physical and others to moral matters, some of which are clearly distinct, while others assume indefinite shades of meaning.   When we refer to an embroidery of gold, the word *finesse* ["fineness"] expresses clearly separate ideas, depending upon whether we refer to the work of the embroiderer, or to the quality of the metal; but if we speak of the *finesse* of a design the discourse must be reasonably well developed to enable us to discern the meaning of this expression without ambiguity; and if we go on to the moral meanings of the term, it will sometimes be necessary even to watch the facial expression of the person using it in order to catch the shade of meaning he attaches to the idea.   In English, in which the same Germanic root is found again in the form " fine," it usually designates beauty or elegance; thus its principal meaning is one which is only accessory and strained in French and in modern German.   The other French word *fin* (*finis*) also has two principal groups of meanings; the one is used to refer to the limit or extremity of something; the other is related to the end in view of which a thing is done; and in these two sets of meanings it is possible to point out nuances which cannot be determined exactly or can be approached only in the context of the discourse.

The lexicographer needs only scrupulous attention in order to enumerate all the distinct and determined meanings that a word acquires in his language; his toil becomes a work of art when it becomes a question of indicating, through the happy choice of examples, the most important shadings in a series of meanings in which the transitions are insensible.   The same ingenuity is indispensable in indicating the nuances of what we refer to as synonymous terms; they are not exactly equivalent, yet their meanings are not so distinct but that the

language consists principally in determining the precise value that each element should receive, or at least in reducing to some extent the field in which indetermination is found, so far as the nature of things makes this possible. This is done by referring to the context of the discourse, and by means of mutual reactions among the elements which compose it. It is, therefore, necessary to amend the comparison made at the beginning of this chapter (205), and to assume that the mosaic artist, wishing to represent a flower, or some other object, has at his disposal fragments whose colors are changeable, and which are able to affect us differently depending upon the reflections and contrasts of surrounding colors, instead of fragments whose colors are fixed and constant. In such a case, the skill of the artist would consist in placing the colored bits in such a way that their mutual reflections and the resulting contrasts would reproduce as faithfully as possible the shadings characteristic of the object imitated.

210. In order to correct the essential imperfection of language which results from the impossibility of expressing exactly ideas which are subject to continual modification and expressing them by means of combinations of distinct artificial symbols, the most common expedient is to multiply symbols or create new words. As a matter of fact, it is easier to increase the number of keys of an instrument the tones of which are fixed than it is to imitate the skill of an artist who knows how to draw out of a few strings all the tones possible within the musical range of the instrument. But while a rough approximation may sometimes be obtained in this way, this advantage hardly ever compensates for the demands on memory and for the amount of work necessary to make new words familiar. Finally, reason, in harmony with taste, recognizes that the true resources of language consist in that elasticity of its elements which gives them more or less extension and in the reaction these elements exercise upon one another in the exact determination of their individual values.

person using them may often be free to substitute one for the other for the sole purpose of giving greater smoothness or harmony to a phrase, as in this series cited by Voltaire (letter dated January 24, 1761): pride, imperiousness, haughtiness, ostentation, stand-offishness, loftiness, disdain, arrogance, insolence, renown, vainglory, presumptuousness, bumptiousness, to which may be added vanity, conceit, self-importance, boastfulness, bragging, and so on.

When the process of thinking bears on precise objects or relations whose variations are not continuous, when it deals with fixed ideas and definite combinations of fixed ideas, it would be unreasonable to have recourse to devices of approximation instead of the rigorous methods that may be used. Consequently, in that case the creation of new symbols and new terms is as legitimate and profitable as it is of little use when it tends merely to establish one arbitrary interpolation in a series in which there are an infinite number of possible intermediary shadings between one term and another. All sciences which give precision to ideas that remain vague among laymen or which give rise to combinations of ideas unnecessary in the ordinary conduct of life must, on that account, use special or technical terms. But it must be pointed out again that the sciences draw a good deal less help from the creation of technical terms than they draw from a technical method for the derivation and association of terms. This amounts to saying that the establishment of syntactical rules for combining symbols is generally a more fruitful institution than the creation of new symbols. Thus, the invention of a syntactical form as ingenious as it is simple produces all the advantages attached to the use of our arithmetic. The clarity of modern chemical nomenclatures is derived in this way from the system of associating root words, which make evident, in the expression of each compound substance, the presence of the chemical radicals which enter into it and the way they are associated in the compound. At least this is the case to the degree that the nomenclature is understood.

211. Furthermore, it must be kept in mind that language is not only used as an immediate symbol of thought—for if this were the case, its utility would be quite limited. Language is also used as a mediate symbol, in so far as it evokes other symbols which are more suited to the immediate expression of thought. In fact, what do we call figurative language? It is not solely, as rhetoricians have always believed, a means of making an impression on the sensibility, and of stirring the passions by means of images. For, were this the case, when we have recourse to cold reason, when we speak to the understanding about purely intellectual things, all figures would have to be given up. And yet it is easy to see that philosophical language is no less figurative than that of orators and poets. Philosophers constantly proceed by means of

comparisons with sensible objects, and those who have wished to make this a matter of reproach in their predecessors have in their turn fallen into the same fault, if indeed it be one. But, far from being a fault, it is a fruitful device by means of which we remedy defects which are native to language and enable it to contribute indirectly to the representation of abstract ideas to which it is not directly adapted. Since the fundamental law of the human mind is that it can raise itself to the conception of the intelligible only by relying on sensible symbols, as soon as language, in itself, ceases to be appropriate for the representation of the intelligible, it clearly becomes necessary for us to call other symbols to our assistance. We select these symbols from among the phenomena of the external world and those which occur within us. We choose them above all, on the one hand, from among phenomena having extension and movement because these are the most simple, the most fundamental, and those whose image has the greatest representative clarity among all the phenomena subject to the law of continuity, and because the obstacles to the direct expression of our ideas by means of language arise primarily from the discontinuity of vocal symbols. On the other hand, we select them from among the internal phenomena of desire, will, and passion, that we do not imagine at all in the way we do external objects perceived by the senses, but of which we have an internal awareness. When looked at from this point of view, discourse is no longer simply a system of special symbols, a distinctive set of symbols more extensive than algebraic language, but devised for analogous functions; it is rather a framework set up to bring together the most diverse symbols, not directly and, as it were, personally, but through representation by means of vocal symbols which recall them.

This property of language seems to be taken account of when the name poetry is reserved pre-eminently for the art of depicting nature and of arousing the passions by the aid of a language whose exalted form distinguishes it from ordinary speech. Yet the poetic conception may be the essence and, as it were, the soul of all art, and in that way there may be a close kinship among all the arts in spite of the diversity of physical techniques by means of which they are carried out, and in spite of the variety, or rather, the heterogeneity, of sensible materials that clothe poetical thought.

But if it is true, as people come to agree, that the artist may propose some other end for his endeavor than pleasing or exciting, that he may be motivated by a philosophical idea and seek to reproduce it in his works under the forms and by means of the expressions which are best suited to him, then there are the very strongest reasons for recognizing that the poetical and figurative forms of language are often a means and sometimes the unique means for the expression of philosophic ideas.   In this way philosophy is related to poetry and art, although, on the other hand, it is closely related to the body of scientific knowledge, as will be explained later on.

What is more, the strictest science also has its poetic and figurative language, images which we could not prohibit it to use without doing essential harm to its conciseness and neatness and to the clarity of its statements.

212. If the same thing is true of poetry, eloquence, music, and the plastic arts that is true of those sisters of whom the poet speaks,

> not one face in all,
> Nor yet diverse . . . ,

everything that is remarkable in the distinctive characteristics of the art of speaking must also be noted attentively.   If chemistry is able to furnish painters with new colors which are more vital and more durable, if the invention of new instruments enables the musician to conceive new orchestral effects, what is this increase in material riches to the genius of the artist, when compared to the resources that the poet, orator, or writer may find in a more harmonious, richer, or more flexible language?   The artist, like the writer, is shaped in the school of his master.   He carries on its traditions and is inspired by the works of his predecessors; but he makes use of material which is, as it were, raw and inorganic in comparison with that wonderful organ which is called language into all parts of which life and thought penetrate.   The genius of the poet or the writer must control this organ, this living machine.   He must take it as it is when destiny offers it to him, in its infancy or in its senility, or better, he must know how to make use of its youth or of its vigorous maturity.   Not only does language extend or restrict its vocabulary, gain or lose idioms, purify or corrupt its syntactical rules. Its words are also like coins the imprint on which is effaced, worn away, or depreciated in circulation; the proper meaning of words falls into oblivion; analogies that have successively introduced various figurative meanings are lost from sight; the agreement that reason demands is no longer found between ideas and images, between thoughts and their sensible expression, and between the material construction of the elements of the language and their representative value. Neologism and archaism, bizarre combinations of words, forced and affected figures of speech, are born in the search for an expressive power which language, in its pure condition, seems to have lost by too long usage. These remarks, which are of so much interest to the philosopher, have often been made; but nowhere will they be found more cleverly expressed than in the elegant Preface put, in the name of the French Academy, at the beginning of the sixth edition of the *Dictionary*. Perhaps it would indicate bad form for us to insist at further length on matters which seem to require a delicacy of feeling, a cultivation of literary taste having little compatibility with the dryness of our ordinary studies and with the didactic exactness which we hoped to approach in this work, so far as the subject makes it possible.

213. At least, we are able to see, from the preceding discussion, what must be thought of the project of a philosophic and universal language about which the greatest geniuses of the seventeenth century, Bacon, Descartes, and Pascal, dreamed but which Leibniz in particular projected, according to his own testimony, to the point of giving serious consideration to the means of carrying it out, as is indicated by passages which have been cited many times.[3] This philosophical language or *caractéristique universelle*,

---

[3] See Bacon, *The Advancement of Learning*, Book VI, Chap. 1; Descartes, *Letter to Mersenne*, dated November 20, 1629, Vol. VI, p. 29 of the edition by Cousin and Vol. IV, p. 128 of that by Garnier; Liebniz, *Historia et commendatio linguae charactericae universalis, quae simul sit ars inveniendi et judicandi* [*The History and Recommendation of a Language of Universal Signs which would serve as an Art of Discovery and of Forming Opinions*], in Raspe's collection. For other citations consult two inserted articles in the *Moniteur* of August 23, 1837, and February 12, 1838, and Reid's comment on the attempt by Wilkins, Vol. V, p. 199, of the French translation of Reid's *Works*.

Long after this chapter and the one preceding it were written, and long after the wording had been submitted to well informed persons, we have found in a piece of writing by M. Bordas-Demoulin, entitled *Théorie de la substance*, which is appended to his monograph, *Du Cartésianisme*, ideas which have, in many

as he called it, which was to be based upon a catalogue of all simple ideas, each one being represented by a symbol or by a catalogue number, would have had this advantage over all ordinary languages: it would have used elements having only fixed, determined, and invariable values, with the result that as a consequence of its very perfection, it would have been entitled to the claim of universality. Algebra would have been simply one branch of this symbolic system. All the work of thought would have been manifested by combinations of symbols. The art of reasoning, which would have been to arithmetical or algebraic calculation as the genus is to the species, would, in its turn, have had to be regarded as nothing but a special application of the combinative synthesis, or of the art of forming, classifying, and enumerating combinations.

This comparison itself should put us on the trail of an important error with which the idea of a universal symbolism is tainted. How limited the application of arithmetical or algebraic calculation would have been, if they had been concerned only with quantities capable of being expressed exactly by numbers, and freed from the law of continuity! The nature of the idea of magnitude allows us to apply to continuous magnitudes, within whatever limits of approximation we may desire, the processes of calculation directly applicable to discrete quantities or quotas. But disregarding this singular case, how can qualities and relations which vary in a continuous manner be expressed in general with suitable approximation, by means of discontinuous or distinct symbols, which are limited in number, but which have fixed and determined values? In any case, how would the approximation obtained be defined?

Condillac and the logicians of his school—whose ideas agree on this point with those of Descartes and Leibniz, a fact which should be noted because it is so rare—by perhaps exaggerating the power

_____

respects, a great resemblance to our own. He proves by the same reasons (Vol. II, p. 446) that the construction of a universal symbolism is chimerical; for what he calls " ideas of perfection " as over against " ideas of magnitude," are evidently ideas capable of the same sort of continuity that we believe should be called " qualitative continuity." While we congratulate ourselves on agreeing in some important points with so distinguished a thinker, we may nevertheless point out that, in other respects, our theory differs completely from that of M. Bordas-Demoulin, both in its principles and in its development.

of the institution of language in general, exaggerate above all the imperfection of individual languages, as usage has fashioned them, by contrasting them endlessly to that ideal type which they call " a well-made language." Now on the contrary, it is in its abstract nature or in its general form that language must be considered essentially defective, while spoken languages, slowly formed under the lasting influence of infinitely varied needs, have warded off that radical disadvantage, each in its own way and depending on its degree of suppleness. In accordance with the genius and the destiny of races, under the diverse influences of zones and climates, languages are more especially suited to the expression of images, passions, and ideas of a certain kind. This is the source of the difficulty and often the impossibility of making translations, a difficulty which is just as great for metaphysical passages as it is for pieces of poetry. What would augment and perfect our intellectual faculties by multiplying and varying the means of expressing and transmitting thought, if it were possible, would be to arrange all spoken languages to suit our liking, and in accordance with the needs of the moment and not to find already constructed this systematic language which would be, in most cases, the most imperfect of instruments.

Of all the works of man, languages most closely approach those of nature in the way in which they are formed, in their slow growth and their ties of kinship, and in the periods of maturity and decadence through which they pass. They participate in some way in the life of a race or of a nation. Between languages formed in this way and the symbolic language the project for which has intrigued philosophers, there is, so to speak, the same difference as there is between the eyes and an optical instrument, between the organ of speech and the harpsichord, between an animal and a machine. Certainly, when, as in manufacturing, we are concerned with the problem of producing an effect which is determined, precise, and measurable, and of such a nature that it can be divided or broken up into a system of distinct operations, work done by machines will advantageously replace the work not only of animals, but also of man himself.[4] On the contrary, the most ingenious machinist will never be able to replace the hunting dog with an

---

[4] Adam Smith, *The Wealth of Nations*, Book V. Chap. 1.

automaton, that is, with a system of engines and wheels. In general, as soon as we have to deal with nuances, with continuous modifications, what creations of human genius could hope to equal the creations of nature?

214. In addition to algebra, which is, as everyone should recognize, the most extensive application of the principles upon which a universal symbolism would rest, the chemical nomenclature whose foundations were laid by Guyton [5] and Lavoisier, and the more ideographic than phonetic notation which Berzelius adapted to more modern chemical theories offer other remarkable applications of these principles. Thus, in its present form, chemistry is the most simple and the best defined of the natural sciences. It deals with ponderable bodies only in so far as they are reducible to a small number of fixed, determined, indestructible, and unalterable radicals. It combines these radicals in ponderable proportions which are fixed and determined in the same way. In a word, all the relations, the study and systematic co-ordination of which are the object of the speculations of the chemist, consist in combinations of elements which are discontinuous or are treated as such.

But when we pass on to the study of the infinitely varied phenomena that life produces among organic beings, there is no more discontinuity, no more possible reduction to systematic combinations; or at least such combinations present themselves only exceptionally and by some sort of accident. Therefore, there are no absolute theorems, no more precise and exact methods, no more invariability in the value of elements of discourse when they must be adapted to the expression of facts of this order. The same

[5] [Guyton de Morveau, 1737–1816. An eminent French chemist and lawyer. In 1755 he became advocate general in Parliament. In 1773 he discovered the use of fumigants and succeeded in checking an epidemic in Dijon with the use of chlorine gas. In 1782 he proposed a methodological nomenclature for chemistry and later worked with Lavoisier in setting up the system of nomenclature which has been in general use since that time. His " Chemical Dictionary " written for the *Encyclopédie méthodique* (1786) is notable for its erudition and good judgment. He played an important part in the establishment of the École Polytechnique, where he taught for eleven years. He contributed many articles to the *Institut*, of which he was a member, and to the *Annales de chimie*, of which he was editor. He was a Fellow of the Royal Society of London.]

thing is true, with even greater reason, when we pass from the description of the phenomena of organic and animal life to the phenomena of the moral and the intellectual life, or to the study of relations which arise from social life.

215. Moreover, it must be recognized that even in the context of discourse words often keep all or part of the indeterminateness that they would have in isolation, and yet it cannot be said that this leads to an imperfection in language. Are we not aware that the power of algebraic language is due in part to the graduated indetermination of the symbols it uses, and that, on account of this indetermination, the order of difficulties is found to be inverted in an advantageous manner? Geometricians habitually represent by the Greek letter π the relation of the circumference of a circle to its diameter, a relation that cannot be exactly expressed in figures, although the true value may be approximated as closely as anyone desires. This abbreviatory symbol is convenient because it makes unnecessary the writing of a rather long series of numbers in all the algebraic calculations in which the relation in question enters, and because it allows us to hold off until the end of the calculations the mathematical operations which may rest on the numerical value of this relation. This makes it much simpler to estimate any error that may have been committed, according to the extent to which the numerical value is approximated. But the principal advantage of the use of such a symbol comes about because it frequently happens that although it enters into the expression of quantities that we are comparing, it does not enter into the expression of their relations, and thus disappears from the final result that we had in view. The surface of a sphere having a radius of one meter and the surface of a great circle of this sphere are two quantities which cannot be exactly expressed numerically, because the number π, which is, as mathematicians say, a *transcendental* number, and which cannot therefore be exactly defined or expressed by means of numbers, enters into the expression of them both. But the first quantity is always just four times the second; and the *transcendence* which is found in each is not found in the relation of one to the other.

In the same way, terms whose meaning cannot be, or at least up to the present time have not been clearly circumscribed, do not cease to be in current discourse, with the indetermination that is inherent in them and with advantage to the movements and the manifestations of thought. Take the word " nature," for example, understood in the active sense of *natura naturans*—as the School-men understood it. No attempt can be made to fix its meaning exactly, by religious faith or otherwise, without resolving the most difficult problem of transcendental philosophy. Nevertheless, it is evident that its use can no more be dispensed with in science than it can be in ordinary conversation. On all sides, this warp of mysterious finality, of which neither the origin nor the goal can be scientifically demonstrated, appears to us like a guiding thread, with the help of which order is introduced into observed facts and which puts us on the track of things to inquire into (71). Even when it fails to apply, for the purpose of explaining the general harmonic system of the world or any particular har-mony, to the intervention of final causes, to the principle of mutual reactions, or to that of the weakening of fortuitous combinations, we do not for that reason have any less need for expressing the idea of that harmony and its necessary consequences, and for ex-pressing it in a language common to all, independently of every philosophical hypothesis and of every religious belief. This is why the most religious person, like the most zealous partisan of materialistic fatalism, uses this idea, subject to supporting on the basis of philosophic debates and religious disputes, the definition of the idea *transcendent,* wrapped and, as if deliberately, veiled under the expression destined for common or scientific usage, for which the definition is not required. And since we do not believe that the words to which these remarks apply are limited in num-ber, we could just as well have taken the terms " matter," " force," " substance," " right," or any one of a great many others whose definitions, being essentially problematic, are the concern of trans-cendental philosophy and yet the uses of which cannot be dis-pensed with in scientific circles, before political tribunals and assemblies, in the practice of the arts, and in the most ordinary affairs of life. Yet the transcendental indetermination with which these ideas are affected never causes the least ambiguity with re-spect to questions in which we seek to reach an idea of relation, as in the comparison of one right to another right, one force to another force, or one elementary substance to another elementary sub-stance. In fact, in such cases it is understood that an elimination must be made of the philosophic or transcendent ideas implied under these terms, the scientific definition of which is impossible. And it will be a great day when we have come closer to the notions which, as we see it, are indispensable for the full understanding of what is the essential character of philosophic speculations, in opposition to the positive sciences which are capable of indefinite progress, perceptual verification, and practical applications.

216. The observations we have made concerning terms with which essentially philosophical or transcendental ideas are connected may be applied equally well to terms whose value may some day be scientifically and incontestably defined but the exact definition of which has not been found up to the present time. It may be that a day will come when we shall define the idea of organic species positively. It is also possible that this precise determination implies the solution of the problem of origins, which decidedly lies outside the field of scientific observation and investigation. In all these cases this precise determination has as yet had no place, since the possibility of the mutation of species or of certain species is still being disputed, and because one naturalist sees a difference in species where another finds only racial variations, with neither the one nor the other being able peremptorily to distinguish the variation from the species.[6] But,

___

[6] No doubt, up to a certain point, the idea of vegetable or animal species is made clear by the form of propagation which causes all the individuals of the same species to seem to belong to the same family and makes it possible to regard them as all having come from the same ancestor or from the same couple, while individuals of different species either never mate, or, if they do, form only sterile unions, or produce only sterile offspring which disappear without leaving any trace in the permanent order of things. But it is not necessary to take, as the elemental and essential basis of the idea of species and of diversity of specific types, the basis of the distinction of the species in the order we observe at present. Let us suppose in the beginning a creation of specific types, all clearly distinct from one another; but let us also assume that some of these types, although perhaps not those in which the organic structures taken as a whole have the greatest resemblance because of certain secondary similarities of organization, form fruitful unions and produce hybrids which are fecund. The consequence of this tendency will certainly be that, at the end of a period of time which is long enough to introduce the evolution of all the fortuitous combinations, the species having this capacity will be intimately mixed, and that new types will arise from the fusion of these older types, in which the traits of the original types will be found to be diversely combined

on the other hand, this ambiguity is found in only a small number of cases in comparison with those in which all naturalists are in agreement on the distinction of specific types, although they do not have the same opinion as to what constitutes the origin and the essence of this specific distinction. Whence comes the basis for the obligation to speak a common language and to apply without disagreement in science terms whose definitions remain problematical among the schools? The solution of the problem may be postponed without postponing on that account the progress of botany and zoology. We discover, catalogue, and classify new species every day, and leave until a later date the matter of exact definition of zoological or botanical species, if that should become possible. In most cases, and in the present order of things, it is sufficient that the distinctions between the species be clear, whatever may be true in other respects of their origin and their fundamental cause.

217. When, in what has preceded, we speak of terms which it is impossible to define, we have not been thinking of definitions as understood by lexicographers and logicians. This latter will be the object of a special discussion in the following chapter. The words " number " and " angle " are not definable according to the notion we commonly have of definition; and yet the corresponding ideas are exactly defined or determined. Everyone conceives of them in the same way, even though it is impossible to set up with respect to this matter philosophic discussions worthy of serious attention. On the contrary, philosophy discusses and deliberates ceaselessly not only about the terms " nature," " force," " right," and so on, but also about the ideas connected with them, the determination or definition of which would imply the solution of problems which it raises, but for which it cannot find truly scientific solutions, because it is possible to apply neither experience nor logical demonstration to them.

---

and modified. Therefore, reciprocally, it is clearly necessary that the species between which we now observe a distinction be those having a peculiar structure which, from the beginning, has not ceased to be incompatible with fruitful unions or, at least, with unions whose offspring are able to reproduce themselves. But the consequence must not be taken as the principle, nor must the resultants of the particularities of specific structure be taken as the basis for the definition of the idea of species.

CHAPTER XV

ON LOGICAL ROOTS AND DEFINITIONS

218. By "roots," or radicals, grammarians and philologists mean a small number of words—in comparison with all those that enter into the making up of the vocabulary of a language—which ordinarily have a simpler phonetic form, being even most commonly monosyllabic, at least in languages which are not formed from the remnants of many others. These roots are words each of which, through its ability to receive inflections, word endings, and different modifications, becomes the stock of a family of words whose various accepted meanings may tend to approximate or differ from the full meaning of the radical. It is generally known in our day that the study of linguistic roots is one of the most interesting that can be proposed, and one which throws the most light, not only on the origins of peoples, but also on the development and the processes of the human mind.

Nevertheless, such a word is a root in one language, while its equivalent is derived in another. A radical of this kind manifestly can and must be defined by a system of derived words. Therefore, a kind of inquiry is conceived the purpose of which would be to indicate, on the one hand, the words which must be regarded as primitive, and to include, on the other hand, the words which must be thought of as playing the part of secondary symbols on the ground that they are equivalent to a combination of primitive symbols. Such an inquiry would be independent of the material form of the elements of discourse in the different languages. Words of the first category are what we call "logical roots."

219. For example, the word *plan* ["plane"],[1] *eben* in German, is a root in the grammatical and linguistic sense; and in the language of mathematicians, it designates an elementary and essential notion, but not a logical root or a logically irreducible idea, for

[1] [In this section, words which are italicized and used for purposes of illustration in the original are left untranslated. The English equivalent of each such word follows it in brackets.]

" plane " is very clearly defined by saying that it is a surface upon which a straight line can be drawn in any direction. In Greek, a compound word, ἐπίπεδον, is the scientific word used for expressing the same idea. To indicate the sort of solid that we call a " cube," the account of which serves as a point of departure in the theory of the measurement of the volume of bodies, a frivolous game has provided from another very natural source the comparison of dice, and the word κύβος (dé), [" die "], which was regarded in Greek as a root, has become the technical name of this solid. In German *Würfel* indicates both the piece used in games and the geometrical figure, but this word is connected with the root *werfen* (*jeter*), [" to throw "] and is not itself a root. From πρίω (*scier*), [" to cut off "], the Greeks have formed the word πρίσμα (*prisme*), [" prism "], a word which, on the basis of its etymology, should be given to all bodies bounded by a plane surface or to all polyhedrons. Actually this name is applied to only one group of polyhedrons, in which the cube is found to be included. Contrariwise, the Greeks resorted to the excessively compounded word παραλληλεπίπεδον (*parallélépipède*) [" parallelopipedon "], to designate a class of solids included within the class of prisms, and including the subclass of the cube. Yet there was no reason to raise it to this degree of complexity in one case rather than in the other.

220. Common languages, with which technical or scientific languages are necessarily connected, at least with regard to the material of which they are composed, must bear the imprint of man's relations with things, rather than the mark of the relations of things among themselves. Only after long effort, if at all, does man reach an understanding of the true principles underlying things. What is simple in things must often be represented as compound in the material of expression, and reciprocally. For the same reason, if we are to avoid a vicious circle, the role of logical roots, or of words that we give up the idea of defining, should, in general, belong to different words, depending upon whether we refer to the intrinsic structure of things, or take as our point of departure the first impression they make on man, without seeking to free ourselves from the conditions in which nature has placed us in the study of them.

Thus, chemistry teaches us that oxygen and hydrogen are two non-decomposable gases which, when they unite in the proportion

of two volumes of hydrogen to one of oxygen, give rise to a compound substance to which the complex name " oxide of hydrogen '
should be given, according to the rules of systematic nomenclature. But this substance is not one of those which is known
only in the chemical laboratory. It is indispensable to the existence of man. Indeed, it plays the most important role in nature
with the result that philosophers long considered it as one of the
four elements, and even as the one from which all the others were
derived. In a word, this substance is *water.* So chemists have
not fallen into the ridiculous pedantry of designating it by other
than its common name. But this name, as used by them, is a
simple symbol given to a complex idea which can always be broken
down, and which actually is expressed in the complex term " oxide
of hydrogen," whenever the need for this arises. In the language
of the chemist, the words " hydrogen " and " oxygen " are simple
symbols, root words by means of which the chemical meaning of
the word " water " may be completely defined.

On the contrary, in common languages which must be adapted
to the totality of natural relations that man has with external
objects, and not to a certain determined order of special phenomena, the terms " water," " *eau,*" " *Wasser,*" ὕδωρ are the true
radicals from which the derivatives " hydrogen," " *Wasserstoff* "
have been derived. We will be consistent both with etymology
and with the natural development of the human mind, if we define
" hydrogen " as the chemical element, the combination of which
with oxygen produces *water,* and if, after having shown that air is
made up of a mixture of two gases, we define " oxygen " as that
one of the two constituent parts of *air* which is required for
respiration and combustion, and which acidifies objects with which
it combines in certain proportions. So it is that, depending upon
whether we look at the matter from one point of view or the other,
the words " hydrogen " and " oxygen," on the one hand, and those
of " water " and " air," on the other hand, exchange their roles of
primitive and indefinable symbols, or of derived and definable
symbols.

Instead of a compound like water, with which everyone is so
well acquainted, and of substances like hydrogen and oxygen,
whose discovery in very modern times could have resulted only
from experiments which were conducted with great learning, we

could just as well have taken cinnabar, mercury, and sulphur as our examples. Everyone knows sulphur and mercury at least as well as the compound to which we give the name cinnabar, which is formed by the combination of these two simple substances. Consequently, it would be unreasonable to define either sulphur or mercury by means of cinnabar; and, contrariwise, we could hardly give a definition of cinnabar that would not at least allude to its chemical composition.

221. Note carefully that the chemical definition of water, "oxide of hydrogen" or, better still, "the combination of oxygen and hydrogen, in the proportion of two parts by volume of hydrogen to one of oxygen," is complete in the sense that it excludes every substance other than water. That is, it fixes and determines unambiguously the thing to which the word applies, for there is no other body which, when analyzed chemically, could be broken up into hydrogen and oxygen in the proportions indicated. But the definition by no means makes known the whole group of properties and physical characteristics the idea of which the word "water" calls to mind. It would not excuse anyone who knows hydrogen and oxygen from, so to speak, making the acquaintance of water, if he wished to form an idea of the role this substance plays in any other matter besides chemical combinations. It is only in relation to this special kind of phenomena that the word "water" is identically equivalent to the definition of the chemists.

What has been said of compound substances may equally well be said of chemically simple substances which we set over against them. Oxygen, hydrogen, sulphur, and mercury have a number of properties which can be made known only through experience. The number of these properties increases for us in proportion as these substances are observed more attentively. In the present state of knowledge, it is impossible to summarize all these properties by means of a definition which potentially contains them all. However, if by definition we understand a description with the help of which it is possible to distinguish a given object from all others or to recognize without ambiguity the object to which the word applies, then it will be possible to give a definition of simple substances as well as of compound substances. And that summary definition or description will allow more or less precision and brevity, depending upon the circumstances. Thus, by regard-

ing the general notion of " metal " as being fully established, we
shall be able to define mercury as " a metal which is liquid at ordinary temperature," [2] while it would be difficult to find so brief and
trenchant a definition as this for other metals such as iron, copper,
and silver.   Mercury is clearly distinguished from other metals
and is adapted to a number of uses for which it can be replaced by
no other substance because of its property of being liquid at
normal temperatures [and pressures].[3]   Yet, on the other hand,
our knowledge of physics permits us to regard this property as
only secondary and accidental, in the sense that no one can discover any connection between this property and other characteristics which everything leads us to consider essential and dominant,
when we study the composition of substances considered by themselves, and apart from the fact that they are appropriate to our
uses.

Descriptions of the sort employed by Linnaeus, of which
naturalists now make general use, are definitions of the same sort.
They are designed to make known the species to which a certain
name applies, or conversely, the name which is applied to a certain
species, by the enumeration of characteristics which pertain exclusively to a natural species.   But even the most extensive of
these Linnaean descriptions would still be far from being sufficient
to enable anyone to restore by a pencil sketch or to form a mental
image of the animal or plant with which they are concerned.
This is so because it is simply impossible to portray the infinitely
varied modifications of form, color, and structure by means of
words.   Except for the triviality of the comparison, these summary definitions or descriptions might be compared to the description which appears on a passport.   The description is sufficient
to establish the identity of the individual, especially when his
appearance presents a combination of very unusual traits or some
accidental deformity.   But, on this basis, it would be impossible
to make a portrait which would give an idea of the likeness and
of the expression of the person's face.

222.   We have been drawing our examples from concrete and
tangible objects.   But if philosophers are greatly concerned with

2 [" and pressures " should be added.]
3 [See note 2, above.]

the theory of definition, it is principally because they have in view the ideas in the conception of which reason makes use of its power of generalizing, abstracting, associating, dissociating, and elaborating in various ways the materials furnished it by the senses. The hierarchy of genera and species, upon which Aristotle based his theory of the syllogism should, by a necessary correlation, have served as the basis for a theory of definition, the systematic apparatus of which had the greatest importance in the eyes of the peripatetics of antiquity and of the Middle Ages. Thus Porphyry, followed by Boethius and all the scholastics, has given as the introduction to logic his treatise on the *Five Modes,* or the five rubrics on which the theory of definition and consequently the whole of logic depends. These five modes are *genus, difference, species, property, accident.*[4] *Difference* is what is added to the idea of *genus* to form the idea of the subordinated *species.* *Property* designates a quality which belongs exclusively to the species, and which is found in all the individual members of the species, but does not constitute it, or is not its essential characteristic. *Accident* is that which fortuitously distinguishes one individual from another. The *species* is defined by genus and difference (*per genus et differentiam*). In this definition—" Man is a rational animal"—animality is the genus, humanity the species, reason the difference. "Laughter" is peculiar to man, but the essential and specific characteristic of humanity must not be made to consist in laughter. When it is said, "Achilles is blond"; "Socrates is pug-nosed"; "Caesar is bald"; "blond," "pug-nosed," and "bald" designate individual accidents. The categories which form the apex of the hierarchy of genera and species, and the idea of which consequently may not be included in a more general idea, cannot be defined and characterized by genus and difference. The same thing must be said with respect to individuals, below which there is nothing, and which are distinguished from one another in the same species only by simple accidents. The enumeration of properties and accidents does not constitute a definition, in the true sense of that term, but rather a description; therefore, individuals may be described but not defined. True definition (*la définition proprement dite*) and

---

[4] Γένος, Διαφορά, Εἶδος, Ἴδιον, Συμβεβγκός.

description apply to things: in this sense they can be opposed to nominal definition (*la définition de nom*) or to interpretation. The object of the latter is to use etymology, synonyms, or translation, to make known the meaning of a word to anyone who may be ignorant of it, simply by making known the thing to which the word applies. The word which has been interpreted or which does not need interpretation points out the object, but in a veiled (*enveloppée*) way and by a sort of synthetic intuition; definition unveils (*développe*) and breaks up this notion, by distinguishing the matter and the form, the genus and the difference.[5]

223. Very briefly, this is the basis of the peripatetic teaching about definition. In it we see appearing the germ of the ideas that modern science has given us on the different degrees of importance and on the subordination of characteristics in genera, species, and individuals. But instead of being clarified by the notion of the independence and of the interdependence of causes, these ideas are found to be inopportunely complicated by chimerical hypotheses concerning a sort of progressive refinement or purification of substances and essences (166 and 167). This idea may also be taken in a sense which is much too systematic and absolute and which is contrary to the habitual continuity of nature's plans. It is clear from the first that the hierarchy of genera and species can be applied naturally neither to all the external objects of knowledge, nor to all the conceptions of the understanding. The ideas of " number " and " angle " are indefinable (217), and they are neither categories nor ideas of individual objects. In the methodical arrangement of chemical combinations, water and cinnabar (220) may be considered as individual objects, and these objects are perfectly definable; chemical definition lays hold of their essential and constitutive character; and description is applicable to secondary properties whose existence does not follow for us from the chemical definition, although they are doubtless a necessary consequence of the chemical composition of these substances and of the qualities which belong naturally to the substances which go to make them up. On the other hand, description should apply not only to individuals but also to species and genera. It often happens that we do not perceive among the

---

[5] See especially the extracts from the dialectic of Abélard, given by M. de Rémusat, pp. 338, 438, and 474 of the work mentioned above (167).

many characteristics which invariably belong to all the individuals of the same species or genus any rational subordination which is so well marked that we are in a position to say of one, rather than of another, that it is not only a property but also an essential and constitutive trait of the species or the genus. Depending upon the case at hand, such a trait may be regarded as an individual accident or as the attribute of a type. It is an accident that Achilles is blond and Socrates pug-nosed, while the Negro type characteristically has woolly hair and flattened noses.

In those cases in which the setting up of genera and species rests only upon the artificial choice of characteristics which are otherwise clearly marked, definition *per genus et differentiam* will be appropriate to make the object to which a certain name is given distinctly known, just as we find books in a library by means of a catalogue in which the symbols indicating the sections and the shelves are written beside the title of each book. But such a definition will yield no knowledge of the intrinsic nature of the object, or at least it will not indicate a rational subordination among characteristics of unequal value. From this point of view, it will be a verbal definition (*une définition de mots*) rather than a real definition (*une définition de choses*). In sum, where it is impossible to establish a clean-cut demarcation between natural and artificial groups (160 ff.), there will be an imperceptible passage from real definition to verbal definition. As an example of an artificial arrangement of characteristics, for the use we are indicating, the *method* of dichotomy used by botanists may be cited. Through its use, the name of a plant which we wish to identify will be found promptly in a manual of botany by successively dividing each group into two subordinate groups on the basis of the presence or absence of a conveniently chosen trait. Yet this will tell us nothing about the true nature of the plant in the scheme of natural affinities, nothing about the relative importance of the characteristics which have successively directed us to the name for which we were searching. Yet a person has learned something who has found the name of the plant in a systematic catalogue which, while not designed to meet the needs of research, has been constructed in such a way as to take account of the intrinsic value of natural characteristics and affinities as far as possible.

Some scholastics have gone so far as to pretend that if it is to be regular, that is to say, if it is to be based on the differences and species which are closest in the descending order in the series, every division into genera and species must be two-limbed. This would make the hierarchy of genera and species, and the definitions based upon this hierarchical distribution, entirely comparable to the method of dichotomy of the botanists. But then, without suspecting it, they would be shown to be exaggerating the purely artificial side of their theory; in most cases they would reduce their definitions to nothing but verbal definitions.

224. The notion of the defined object that is given us by definition *per genus et differentiam* varies according to the nature of the genus. The statement that an animal belongs to the genus or the class of birds already tells us a great many things about its structure; it acquaints us with its fundamental traits, in comparison with which those traits which differentiate species have a highly secondary importance. In fact, the type "bird" is one of the most natural generic types (163). If, without being natural to this degree, the genus is nevertheless clearly distinguished by properties common to the congeners, when the genus is indicated to us by recalling the common properties observed in the congeners already known, we are informed that these characteristics are found again in different degrees in the object defined. Thus to say that an object belongs to the genus of metals indicates to us that it acts on light, heat, and electricity in the way we know metallic bodies generally act. But such characteristics may actually come and go by degrees. And if only an artificial bond or certain negative characteristics establish the genus as a pure logical entity, definition *per genus* will give us no positive notion of the object defined. To tell us that a body belongs to the genus of non-metallic things, is to tell us what it is not, but not at all what it is. The scholastics called such genera "infinite," that is to say, indeterminate.

Briefly, the same thing is true of definition *per genus* as of syllogistic deduction, of which we will speak later on. Its utility consists in extending our knowledge or simply in putting in order the knowledge we have acquired. This is possible to the degree that the generic idea or the general truth contained in the major premise is grasped by the mind as implying the reason for the joint possession of characteristics or the identity of the conclusion in each specific variety. Contrariwise, it is possible to the extent that these generic ideas and general propositions are only the logical summary of what is given by observation for each particular case.

225. The precept of the scholastics that definition must be made *per genus proximum* and without skipping degrees in the hierarchy of genera and species is of value in an artificial and purely logical system of classification. But it may be found to be defective if we propose to express the rational subordination of characteristics by means of the definition. For example, take the definition already cited: man is a rational animal; in order to apply the scholastic rule, it would be necessary to say: man is a rational, vertebrate animal, or better yet, man is a rational mammal. But, apart from the pedantic pretentiousness and poor taste implicit in this manner of speaking, it offends us for two reasons. It offends us in the first place because, in the present state of our knowledge, we have no sufficient reason for admitting that the quality of being rational necessarily depends upon organic traits on the basis of which man is classified in the order of mammals. It also offends us because what strikes us with respect to the nature of man is the union of the faculties of reason with the fundamental characteristics of the types of animality, and not at all their union with the organic characteristics typical of the order of mammals.

226. It is by a vague recollection of the scholastic theory of definition that we speak of things *sui generis,* or things which belong to no genus whether natural or artificial; those things which we are unable to make known, even imperfectly, either by means of a definition *per genus et differentiam* or through the use of a description of the sort that the scholastics pretended belonged to individual objects (222). A sensation of color, taste, or odor unlike those with which we are familiar, or the articulations of a strange language which have no equivalent in our own, would be an example of what is meant by this. However, it is clear that these things which may reasonably be regarded as indefinable and indescribable may, nevertheless, be classified by genera and species quite as well as a great number of others that we are in the habit of defining. An odor *sui generis* is a sensation included in the very

natural genus of odors. But it is little better known to us on that account, if we have not experienced it, than is any other sensation which does not fall in the category of the sensations received through the special senses, and which, for that reason, we do not know how to distribute according to genera. Such, for example, is the nervous irritation that the rustling of certain substances produces in some persons, but not in others.

When things spoken of as *sui generis* have a measurable character, they admit for that very reason of a rigorous definition which permits us to recognize them without ambiguity. This is true of musical tones, that we can define by the number of vibrations that a sounding body emits in a given interval of time. Thus, there is no other way to define the meaning attached in France to this expression, " The *A* of the Opera," than to indicate the number of vibrations which correspond to this musical tone. Then, by means of a similar definition—which, as we see it, is quite foreign to the scholastic method of definition—we shall be able to reproduce the same tone at any time and to be certain whether the tone designated in this way always remains the same, or whether, in accordance with a very probable opinion, it becomes progressively higher with the passing of time. Moreover, it is evident that the measurable character connected with the sensation or with the phenomena which produce the sensation does not make the sensation known but simply furnishes the means for recognizing or reproducing the sensation which we have intended to designate by the imposition of a name. If we are concerned not with a musical tone, but with the timbre of a sonorous sensation or with that other little known process by means of which the vowel-sounds of one language differ among themselves and differ from the vowel-sounds of another language, then no definition is possible, since we know of no measurable characteristics which distinguish them.

When the celebrated optician Fraunhofer discovered in the solar spectrum (128) obscure rays which are extremely fine, and unequally spaced, and which follow one another in a constant order no matter what substance is used in the refracting prism by means of which the white light of the sun has been broken up, he furnished points of reference which are extremely useful for defining with precision the colored ray on which one works in an optical experiment. This is so in spite of the continuity with which the colors shade off, or seem to shade off, into one another in the spread of the spectrum. With respect to the mixed and confused colors which constitute what are called the natural colors of bodies, we set up similar points of reference and characteristics which make it convenient to define them numerically, after the manner of musical tones, or through the use of some other exact method. This is the reason for the vagueness that came to be attached not only to the designation of the colors of particular objects, but also to the names of the colors taken abstractly, or to the expressions which are used in different languages to indicate a certain number of principal shades. And a person well versed in the ancient languages has difficulty in recognizing the shades to which the names adopted by the peoples of antiquity are applied, shades which may very well not be those that modern peoples, who live in a different world and who are fashioned by other habits, have found more striking and have wanted to name.

227. There are a great many generic definitions which may be called " correlatives," which imply, or seem to imply, each other mutually. Chemists are familiar with *acids* and *bases* which, in general, are endowed with contrasting properties, and the properties of which neutralize each other respectively when substances are combined in such a way as to form the compounds that we call " salts." At first, the name " acid " was based on the property that certain of the most conspicuous substances in the first category have of producing in us tastes analogous to those of wine which has been soured by fermentation. But this is only a secondary quality of variable strength which is not a property of some substances that are evidently chemically analogous to the principal acids. It is necessary to say as much of many other characteristics, such as that of changing blue vegetable dye to red, which are often used, because of their convenience, to enable us to recognize at first glance the acidity of substances. Finally, chemists have been led to recognize as the fundamental property of acids that of combining with bases to form salts in which the properties of acids and bases are found to be completely or partially neutralized, depending upon the strength of the contrasting forces. Similarly, they have been led to accept as the fundamental

property of bases that of completely or partially neutralizing acids by reacting with them to form a compound salt. This results in a circle which cannot be called vicious, because it belongs to the nature of things. It does not allow the mind to pass, by a series of exact definitions, from the idea of acid to that of base or inversely. Rather, it forces us to admit both these notions at the same time as correlatives and as sustaining one another, after the elimination of variable and secondary properties which had originally given rise to the distinction of the two kinds of substances. The categorical division into two kinds even ceases to be possible after we have noticed that a given substance, which reacts like an acid in relation to powerful bases, reacts like a base with substances which have acid properties in a still higher degree. The idea of a distribution into two groups then gives way to that of the division into a single and linear series, in which the contrast between the end terms, which are like the opposite poles of the series, is weakened in proportion as the mean or neutral portion is approached. Each term of the series plays opposite roles in relation to what precedes and to what follows it. But this is a conception that we do not know how to translate by means of logical definitions *per genus et differentiam*. Moreover, the same thing must be said, and with even more reason, of the more complicated conceptions to which we are brought by the study of the natural relations of living things.

Thus, in general, it will be possible to define each piece in the skeleton of an animal by its form and its functions, and to give them names suitable to designate the traits, whether functional or formal, which they imply. This can be seen by looking at books which deal with human osteology. But later on progress in comparative anatomy may bring us to see that forms may change and functions may differ completely in the passage from one species to another, while the relations among the pieces of the skeleton nevertheless indicate traces of a fundamental plan that persists throughout the metamorphoses and the modifications which the organism undergoes in adapting itself to the surrounding environment and to the conditions of animal life. In this case it must be fully recognized that the essential definition of each of the skeletal pieces—a definition which is valid not only for one particular species but also for all the species included in the same genus or in the same class—must be based upon fundamental and persistent connections.[6] Out of this arises a new sort of generic ideas, which

[6] " In the anatomical sciences, to determine is to fix the principles according to which an organ or a system of organs must be distinguished. Determination is the basis for the philosophy of these sciences, as facts are the basis of their material part. Up to the present time, naturalists have clung to the idea of determining the parts, at one time solely on the basis of functional consider-ations, at another time on those of form, and at still other times on the basis of form and function taken together. The position and connections of the parts were almost entirely neglected." Serres, *Principes d'organogénie*, Part I.

There are sometimes (154) peculiar relations between the most disparate things. Thus, remarks could be made about the letters of the alphabet which would be perfectly analogous to those which modern anatomists, and especially Geoffroy Saint-Hilaire, have made concerning skeletal pieces, and which have become the basis of a new anatomical theory. At first glance, it seems that the identity of a letter cannot be better recognized than by its name, its form, and its phonetic value. But from the outset the names of letters, like all other words, change in passing from one language to another, or may even disappear completely. This is illustrated by the fact that although the Romans drew their alphabet from a Semitic source, as did the Greeks, they did not, like the Greeks, retain the names of the adopted letters, but simply gave the name *A* to the letter that the Greeks knew by the name of *alpha*, and the Orientals by *aleph* or *alif*. As for the form of the letters, the merest smattering of paleography is sufficient to reveal that nothing is more likely to vary according to the prevailing fashion and the fancy of the copyist than the configuration of letters as they pass from one alphabet to another which is derived from it, even among the same people, speaking the same language. The phonetic value is also subject to the gravest sort of alterations, as can be verified by comparing the Greek β with the Latin *B*, the Greek γ with the *G*, the η with the *H*, although these are incontestably the same letters and even occupy the same relative positions in an alphabetical series derived from the same source. It is similarly certain that the Latin *F* or the Eolian *digamma* is the same letter as the Hebraic *vau*, that there is an essential identity between the Greek *o* [omicron], and the *aïn* [*ayin*] of the Hebrews, between the ξ and the *samech*, in spite of the great changes these very letters have undergone, so far as their phonetic value is concerned, in passing from one people to another. Among peoples who, like the Greeks and the Hebrews, gave their letters a numeral value, this value is a better established trait, precisely because it is more immediately connected with the relative position of the letters in alphabetic series that has been handed on from one people to another. So it is that the Greeks, by discarding from the Semitic alphabet the *vau* and the *qaf,* which the Romans retained, invented special symbols to take the place of them in the table of numeral letters, so that the accident of a vacant place in the series would not upset the order of the whole series of places any more than the accident of a piece of bone which has atrophied and disappeared would derange the general system of anatomical

cannot be defined in isolation, but the definitions of which, on the contrary, imply and support one another in a system of which the mind may be able to form a clear picture, in spite of the fact that it is impossible to apply to them the ordinary rules of logic, and notably the definition *per genus et differentiam.*

228. People were wearied by the subtleties of the scholastics. As a reaction against the doctrine of peripateticism, Pascal gives some reflections, marked with the stamp of his vigorous genius, in one of the most interesting of his posthumous fragments, called *De l'esprit géométrique,*[7] on the usage and utility of definitions, on the complete impossibility of defining, and on the distinction between *verbal* definitions (*des définitions de mots*) and *real* definitions (*des définitions de choses*). These reflections have been put to use by his friends of Port-Royal in their *Logique,*[8] and have since been repeated without anything very important having been added to them. However, we believe the subject not only may but must be more thoroughly investigated, and we shall attempt to present for consideration some observations that it suggests.

Among lexicographers, the purpose of verbal definitions is to make known the meaning of a word to those who do not already have a more or less clear or obscure, superficial, or profound notion of the thing this word signifies. When we are dealing with a new word, the purpose of the verbal definition is to designate the word by which an author has chosen to indicate a physical thing or a simple or complex idea that has not yet been named. Attention is then entirely directed toward the introduction of the new symbol which is an arbitrary symbol, except for the proprieties of language and etymology. On the contrary, the purpose of conceptual definitions (*des définitions d'idées*) is to make the nature of the thing indicated better known to those who are already supposed to know to what thing the word applies (222). As the

---

connections. Thus the identity of each letter, like the identity of each piece of bone, may be deduced only from the whole group of relations. The definition of each part, so far as what is essential to and persistent in it is concerned, implies the definition of all the other parts of the system which mutually sustain and determine one another.

[7] Volume I, page 123, of the edition of the *Pensées* following the autograph manuscripts, published by Prosper Faugère.

[8] Part I, Chaps. 12, 13, and 14; Part II, Chap. 16.

Port-Royalists have said, they must *be confirmed by reason*. This is not intended to mean that their correctness must be proven by means of logical demonstration, for, on the contrary, in most cases the feeling for the fitness of a definition or for the true perception of the essential and fundamental characteristics of the idea will result from a whole group of relations and inferences which convince reason, but which could not be attributed to the setting up of a rigorous demonstration.

229. Some words have been imagined simply for the purpose of taking the place of a periphrasis, and for the convenience of discourse. Nothing is more simple than the role of definition in relation to terms of this category, from among which Pascal has drawn his examples of *verbal* definitions. Thus, since the study of the properties of numbers continuously furnishes the occasion for distinguishing numbers divisible by 2 from those which are not, the former have come to be called " even " and the latter " odd," for the sake of brevity. It is always permissible to replace a word by the periphrasis which explains the meaning of it, or to substitute mentally the definition for the word defined. In the same way, it has been found convenient to call by a single word, "hypotenuse," the side of a right-angled triangle which is opposite the right angle, and it has always been possible to substitute the periphrasis for the term which is the conventional abbreviation of it, without altering in any way the basis of the ideas, although the expression of the thought is more difficult. Algebraists use a perfectly analogous device when, in order to facilitate certain calculations, they replace a complex expression, except for reinstating, at the end of the calculation, in all the places where the auxiliary letter is found, the complex expression for which it was substituted.

But, without going outside the field of geometry, which always furnishes the clearest and simplest examples in matters of this sort, it is possible to cite technical terms whose role is entirely different and whose definition must not be looked at in the same way. Thus, when we study the properties of the oval curve which results from the intersection of a cone by a plane, and which we call an ellipse, we find that inside this curve there are two points so situated that the sum of the distances from each point of the curve to these fixed points equals a constant length, and that this length is that of the long axis of the curve. These two points are called the " foci " of the ellipse. But the utility of this technical term is not limited to dispensing with the use of a periphrasis, as was the case in the preceding examples. For the foci of the ellipse are characterized not only by the property which serves to define them, but also by a great many others for which they might furnish as many other different definitions as the case requires. In giving these remarkable points a name, we have indicated their existence and called attention to them, at the same time making anything arbitrary or fortuitous in the choice of the characteristic property by means of which they were originally defined disappear entirely. If I say that the earth moves in an ellipse of which the sun occupies one of the foci, I put into convenient terms a truth the statement of which retains no trace of the particular ideas which may have led the author of a treatise on geometry to take a given trait as a definition or as a point of departure in the exposition of the theory of foci.

What has just been said about the foci of an ellipse may be said about the ellipse itself, since this curve may be generated in an indefinite number of ways and possesses an indefinite number of different traits any one of which could serve to define it and to characterize it without doing any damage to logic.[9] To give a name to an ideal object, such as the curve of which we are speaking, at the same time that one of its characteristic properties is stated, indicates that the object must be conceived in itself, independently of the circumstances which are responsible for the choice of this property for its characterization; or at least, it has put the mind on the track of a similar conception. Such definitions, which call attention to an object and make it known to us, clearly deserve to be called conceptual definitions or real definitions in contradistinction to the definitions which we considered above, which have only a verbal or logical value. But

[9] " The mind can determine in many ways the ideas of things which the intellect forms from other ideas. For example, to determine the plane of an ellipse it imagines a pencil which is fastened to a string to be moved around two centers or it conceives an infinite number of points all having the same relation to a given straight line, or a cone cut by an oblique plane so that the angle of inclination is greater than the angle at the vortex of the cone, or in infinite other ways." Spinoza, *Treatise on the Correction of the Understanding*, XV, sect. 7.

Pascal and the Port-Royal school seem to have confused both in the class of verbal definitions, in so far as their object was to impose an arbitrary name on the thing defined.

230. Inversely, there are very common and very widely accepted definitions which, when first considered, might be taken as conceptual definitions, but which are really only verbal definitions. When we read in a dictionary or in the beginning of a treatise on mathematics that arithmetic is the science of the properties of numbers and that geometry is the science of the properties of extension, we have not gained a clearer or more extensive idea of the object of arithmetic or of geometry than we had before knowing the definition. But whatever notion we may have had already concerning the properties of numbers or of figures, after the definition, the idea causes the terms " arithmetic " and " geometry " to cease being words devoid of any meaning and like the articulations of an unknown language, as they would be for an infant, and as infrequently used technical terms are for educated persons. The acquisition of new terms facilitates the interchange of ideas, even when the new terms are not immediately accompanied by the acquisition of new knowledge or by the clarification of knowledge previously acquired. Consequently, whenever they are possible, summary definitions of this sort, although having no great utility, must not be regarded as being entirely useless, and it is well to distinguish terms which permit of such a summary and, fundamentally, purely verbal definition from those which do not.

For example, while, as we have seen, it is very easy to give a definition of arithmetic and of geometry worthy of being included in a dictionary or being placed in the beginning of a textbook, it is, on the other hand, very difficult to define algebra. The very different definitions of it that have been given are obscure and unintelligible to those to whom algebra is not already familiar, and they present great disparities according to the systematic views of those who have adopted them.[10] It is when definition plays essentially the role of conceptual definition, and when it is very difficult to grasp by thought and impossible to express in a concise statement, that the eminent and distinctive characteristic

[10] Cournot, *De la correspondance entre l'algèbre et la géométrie*, Chap. IV.

of algebra, in its successive developments and in such diverse applications, is revealed.

231. Some definitions, being originally only definitions of words, have the virtue of leading to conceptual definitions. In that way they render a service which has an entirely different importance than that which would consist in the abolition of a periphrasis. Thus, after having given the technical name " multiplication " to the arithmetical operations which consist in repeating one number as many times as there are units in another number called the " multiplier," men were not slow to perceive that, in the use made of numbers in the measurement of continuous quantities, the multiplier might very well not be a whole number, that it is a whole number only by chance, because of the arbitrary choice that has been made of a certain determined quantity to serve as the unit in the measurement of quantities of the same kind. Consequently, it is necessary to look for another definition of multiplication which fits the case of fractional as well as of whole multipliers. This generalization is not conventional or arbitrary. On the contrary, it is necessitated by the obligation to do away with what was originally too particular and too limited in the manner in which we subordinated to the arbitrary choice of the unit, the notion of a relation between quantities which are known not to depend upon this arbitrary and conventional choice. But in the particular circumstances in which we are placed at first because of the natural order of the operations of the mind, the imposition of the name is what gave rise to the perception of the idea with the generality which intrinsically belongs to it and, consequently, to the generalization of the original definition.

If a choice is to be made between different definitions of the same idea, for the purpose of giving the idea the proper extension it should have, and for the purpose of grasping what is truly essential and dominant in it, we believe that the disputes over this matter are not disputes about words, as has often disdainfully been said. Consequently, we believe that one is justified in calling such definitions conceptual or real definitions, as opposed to verbal definitions.

232. No one could fail to understand the immediate utility of the definition of an idea so far as the clarification of acquired

notions and the profitable direction of the subsequent efforts of the mind are concerned, when the purpose of the definition is to get away from the vague and indecisive meanings a word has in common language. This is particularly true when the definition does this in such a way as to fix precisely the idea to which this word should be applied in the language of philosophy and of the sciences by assigning to this idea its fundamental and essential traits. Everyone continually uses the word " chance " in connection with events which involve something irregular, fatal, extraordinary, or unforeseen. Ordinary conversation, dramatic dialogue, and even the accounts of history, as it is ordinarily conceived, being decked out in animated and picturesque forms, may adapt themselves to everything obscure and vague in all those accessory ideas which are arranged in such a way as to give value to poetic imagination rather than to austere reason. But how can a scientific theory be built on the notion of chance unless that which is fundamental to it has been established? And how could the applications of the theory not fail to differ depending upon whether we started from Hume's definition, " Chance is our ignorance of true causes," [11] or from the position we stated earlier (30ff.), " The idea of the concurrence of many series of independent causes in the production of an event," as being that which is essential and characteristic in the notion of chance?

The idea of chance, as we understand it and as we undertake to define it, is not an artificial product of the ability to make abstractions. It is the expression of a fact whose consequences, being theoretically foreseen, are continually established by means of the observation of phenomena. But conceptual definitions may also have the purpose of stabilizing artificial abstractions. The merit of these definitions will consist in leading abstractions of the sort that we reach by the most direct or the most convenient route to an exact or approximate understanding of the things which we may wish to study and cannot study directly in their natural state of

[11] [This appears to be a paraphrased quotation of Hume's statement: " Though there be no such thing as *Chance* in the world, our ignorance of the real cause of any event has the same influence on the understanding . . ." David Hume, *An Inquiry Concerning Human Understanding*. Sec. VI, " Of Probability." The Liberal Arts Press, New York, 1955, p. 69.]

complexity. From this point of view, the theory of conceptual definitions becomes confused with the theory of abstract ideas, the outline of which we have given in Chapter XI.

233. It is not necessary to conclude, on the basis of what has been said in section 229 about the variety of definitions which may be given of the same ideal object, such as a geometrical curve, that all the characteristics of an object are equally suitable for defining it. In the rational concatenation of different truths, a concatenation which does not conform to a linear series which is equally tight throughout, like, for example, that of cause and effects (25), some belong to the series without dominating it, so that there is no determinate reason for making one take the role of antecedent, the other the role of consequent. But there are also some among which a subordination is evident and in regard to which an inversion of roles, while compatible with the exactness of logical demonstration, would not agree with the idea we have formed of things and of their natural relations.

The curve which geometricians call a " parabola " may be very clearly defined by saying that it is given by the intersection of a cone and a plane which is produced parallel to one of the generating lines of the cone. But among the multitude of definitions that can be given of a parabola, some will appear to be as convenient, or even more convenient than the one just stated, depending upon the order of abstractions in which it will be placed. Others will be rejected without hesitation, as expressing derived or secondary properties, because the statement of them is complicated, or still more because they are subject to limitations and restrictions which are incompatible with the idea we have formed of a fundamental and primitive property.

234. It is regarded as a perfection of logical order to bring into the definition only the properties absolutely necessary to characterize the object. But this perfection, which has to do only with form and is connected with a systematic view of the mind, may have real inconveniences, if it arbitrarily subordinates to one another characteristics which are presented together and are of the same degree of importance in the notion that we have naturally of the object. For example, without having studied geometry, all men have naturally the idea of the similarity of two figures, whether they are plane or in relief, and recognize at once that one

is a larger or smaller copy of the other. This notion includes two others, namely: first, that all the lines of the figure are reduced in the same proportion when we go from the large to the small; and, second, that all the lines are equally placed and inclined in relation to one another whether in the large or the small copy. Thus it can be said that two polygons are similar when all their angles are equal and all their corresponding sides are proportional; it can be said that two polyhedrons are similar when corresponding faces in the two polyhedrons are similar polygons formed in the same way, and when the planes of each are inclined to one another at the same angle. Now, although all these concepts enter simultaneously into the idea we naturally have of the similarity or the resemblance of figures before any scientific study has been made, geometricians have noticed that a certain number of these conditions are sufficient in order to infer the others. Then they have taken it upon themselves to find definitions which bring together only that number of conditions which is strictly necessary, and this can be done in various ways. But, in this artificial arrangement of premises and consequences, we separate that which is united naturally in the idea,[12] and we reject, among the secondary and derived notions, a truly primitive idea, not without prejudice

[12] Only a mediocre knowledge of elementary geometry and a little reflection are necessary to convince us that the *imperfection* of the theory of parallels—to use the accepted expression—arises from the refusal to admit as natural and primitive the notion of similarity, or the idea that a figure being given, it is always possible to imagine another which differs from the original figure only because the scale has been changed, or because all the lines of the figure have been proportionally increased or decreased. From this primitive notion—for which it would be fanciful to seek a pretended analytical demonstration such as Legendre sought to give in a note attached to his *Eléments de géométrie*—it follows immediately that equiangular triangles have proportional corresponding sides, or conversely. Once this proposition is admitted, there is no need in the theory of parallels for Euclid's famous postulate, nor for any other which is substituted for it. Inversely, the necessity of this postulate or of an equivalent, which has been proved by the uselessness of efforts which have been made to rid geometry of it since it became a science, well indicates that our artificial systems are unable to prevail over the nature of things, and that, if a natural and primitive idea is mutilated in the interest of pretended logical perfection, it will never be possible to connect by simple logical links the part cut off with the part retained. It will always be necessary to fall back upon immediate intuition and to admit as a premise in one place or another, under one form or another, a logical equivalent to the portion inopportunely rejected.

even with respect to the logical exactness in view of which the artificial argument was set up.

235. Coming back to the remark made at the beginning of the present chapter, we see how important it would be to have a methodical catalogue of logical roots, that is to say, of indefinable words which serve to define other terms, and for the understanding of which, as Pascal has said, "Nature sustains where discourse is wanting," whether it is a question of notions derived from sensations or from ideas which are grasped only by the mind. We should need this catalogue in order to take account of the functions of language, and before any effort is made toward a universal symbolism. To carry out such a project, a number of things would be necessary. We should need to take account of the diverse roles of definition; to distinguish verbal definitions from real definitions; to distinguish those which indicate the essence and which consequently implicitly determine all the properties of the thing named; to determine those which designate it only by means of some distinctive trait without being exempt from recourse to the intuition of the thing for knowledge of its other fundamental or derived properties; and, finally, to indicate those which tend only to substitute a simple symbol for a complex expression. It would be necessary to pay attention to the conditions which surrounded the circumstances of the formation of each particular language, to those which result from the general laws of the formulation of language and from the natural development of man's faculties, and, finally, to those which are given in the intrinsic nature of things. It would be possible to set up as a goal a purely artificial and logical perfection which would consist in reducing to a *minimum* the number of roots or the degree of complexity of the definitions for the words defined by means of radicals; just as we should be able to lay aside that goal of perfection in form for the purpose of being able to represent things and their relations as faithfully as possible, in proportion to the degree of simplicity or complexity that nature has given them. But that itself is sufficient evidence that the achievement of such an enterprise would be something of a summary of all science and of all philosophy, and we must not be surprised if the preliminary outline of it has not even been attempted.

CHAPTER XVI

ON THE LINEAR ORDER OF DISCOURSE. ON LOGICAL CONSTRUCTION AND THE SYLLOGISM.

236. We move on now to some considerations of a different sort, but equally important, if we wish to take account of the influence exercised on the development and on the conduct of thought by the form of the instrument which molds it and by means of which it exercises its influence. These new considerations bear on the necessity that the use of discourse imposes on us to present our ideas *in a linear order;* and so that the reader may better understand the meaning of this statement, we must return to some fundamental notions which we have already mentioned, but only incidentally.

One of the most general ideas, perhaps the most general of all those to which the mind can attain, is the abstract idea of " order," which Bossuet said (17, note) is the *natural object* of reason. This idea becomes specific and receives particular forms which are more determined when it is applied to time and space, or when it is associated with other ideas which are more abstract and further removed from sense impressions, like those with which algebra deals.

When we consider a series of successive events, we are able to take account of the intervals of time which separate them, but we are able to abstract in this way only for the purpose of considering the order according to which they succeed one another; the order which accounts for the fact that event *A* has priority over event *B,* and that the latter has priority over the event *C,* with the result that *A a fortiori* has priority over *C*, and so on.

Similarly, if we consider a series of points arranged in a straight line, or even laid out along an open curve, it is possible to take account of their distances. Once more, the order according to

which these points succeed each other, and which determines that
if the point *a* is behind *b*, and *b* is behind *c*, then *a fortiori* the
point *a* is behind *c*, and so on, can be abstracted and considered
by itself.

For this comparison it is sufficient to present and set up a basis
for abstracting what is common to the series of events *A, B, C, . . .
N* and the series of points *a, b, c, . . . n*. This common arrangement is called arrangement in linear series, or more simply, " linear
order," because in stating an idea, we prefer to particularize it in a
form which yields an image. In this case the image is a series of
points arranged along a straight line.

If the points *a, b, c, . . . n* do not lie along a straight line or
along any other line capable of being infinitely extended, but lie
along a circle or some other closed curve, we shall have the model
of a particular order that may be called " circular " or " closed."
It is sufficient to recall such proverbial expressions as these, " the
extremes meet," " we move in a circle," and similar ones, to make
us aware that the geometrical image, in this case as in the preceding, does nothing more than give a tangible form to an abstract and general idea capable of being realized in other forms
which in other respects have nothing in common with the representation of things in space.

Carried over into time, circular order becomes what is called
" periodic order," and becomes one of the most remarkable and
most frequent forms of the succession of natural phenomena.

237. The nature of the idea of time is so simple and so little
suited to generating diversified combinations that the idea of
order cannot be realized in time under very diverse forms. But,
for the contrary reason, a multitude of different orders are represented in the countless variety of geometrical conceptions (141).
This is the basis for genealogical and encyclopedic trees, of
chronological diagrams, of historical atlases, and of synoptic tables
of all sorts.

In questions having to do with an artificial classification, or
with an order artificially set up among given objects for the purpose of facilitating research or study concerning them, we are often
content to arrange the objects in a linear series on the basis of
certain characters or symbols which recall them. This is shown
by the very frequent use of the alphabet for this purpose. Each

term of the series may be represented by a number which gives its position in the series so that by assigning a number the corresponding object would be implicitly assigned. Such a series is what mathematicians call, in their technical language, a "single entry series." If the objects dealt with are too numerous, or if, for any other reason, the linear order cannot be conveniently used, it is possible to arrange the objects by squares marked out on the surface of a plane, as on a piece of graph paper. To fix the ideas, if we assume that the squares are arranged on a vertical plane, each division will be determined by the serial number of the horizontal and of the vertical sections in which it falls. This obviously allows us to take account of two distinct characteristics in one process of classification; for example, to classify books according to the subject matter with which they deal and the language in which they are written. A series of objects classified in this way or arranged according to two series of indices or recall numbers, is called a "double entry" series. A "triple entry" series is one in which each object is determined by a system of three indices or serial numbers. It can be represented geometrically by imagining tiers of compartments placed horizontally and superimposed upon one another in such a way that, in order to indicate a specific compartment or object, it would be necessary to give the serial number of the tier in the vertical series, plus the two positional numbers which fix the position of the compartment with respect to the horizontal plane of the rack. Finally, there is no reason why we should not conceive of quadruple or quintuple entry series and so on, except that the relations of order or the ideal situation that such a conception assumes could not be represented by a geometrical construction analogous to the preceding cases, since we are unable to find more than three dimensions in extension.

238. In order to understand how relations of order, arrangement, and ideal construction, as complex as, or even more complex than those of which we have been speaking, may be proposed for our study and be understood by reason, we need only to refer to what already has been said about the relations between specific types and the objective existence of such relations, independently of every artificial method (160ff.). But it will not be out of place to make some more detailed explanations with respect to this matter.

We are familiar with what Bonnet [1] and other naturalists of the last century understood by the " chain of being " (*la chaîne des êtres*),[2] an idea the germ of which is already present in Leibniz and Aristotle and which is even the basis for the doctrine of emanations with which so many theological and theurgical systems born in the orient and in Greece are connected, in every period of man's history.[3]   In order to restrict us in the application of this idea to that which concerns specific types of organic beings, Bonnet and the other naturalists of whom we have spoken admit that each specific type is conceived between two others, one of which is more simple and the other more complex: the plan of nature having consisted in developing gradually from the simplest being, such as infusoria, to the most perfect being from the point of view of organic complexity, namely man, just as in following a chain we move gradually from one link to another.   If sometimes the chain seems to be suddenly interrupted, it is so either because the intermediary links correspond to types that have been destroyed by particular or general causes, such as the ancient geological upheavals of the earth, or because discoveries are yet to be made by us which will some day fill in the gaps.

It has not been difficult to point out all the aspects of this

---

[1] [Charles Bonnet, 1720–1793. An eminent French naturalist and philosopher whose highly-praised *Treatises on Insectology* (1745), and *On the Use of Leaves in Plants* (1754) were followed by four philosophical works: *Considerations Concerning Organic Bodies* (1762), *Contemplation of Nature* (1764), *Philosophical Palingenesis* (1769), and *Philosophical Researches Concerning the Evidences of Christianity* (1770).]

[2] [No American philosopher will read this phrase without having called to mind Arthur O. Lovejoy's important and illuminating study entitled *The Great Chain of Being*.]

[3] " Accordingly, since Mind emanates from the Supreme God, and Soul from Mind, and Mind, indeed, forms and suffuses all below with life, and since this is the one splendor lighting up everything and visible in all, like a countenance reflected in many mirrors arranged in a row, and since all follow on in continuous succession, degenerating step by step in their downward course, the close observer will find that from the Supreme God even to the bottommost dregs of the universe there is one tie, binding at every link and never broken. This is the golden chain of Homer which, he tells us, God ordered to hang down from the sky to the earth." Macrobius, *Commentary on the Dream of Scipio*, I, 14, xv. [From the translation by William Harris Stahl in *Records of Civilization*, 48, (New York: Columbia University Press, 1952), p. 145.]

theory of a continuous chain which are arbitrary and inexact. It is even evident prior to any discussion that this is not a question of continuity in the proper sense of the term, since this would require that there be an infinite number of species which could not be distinguished from one another by any clearly marked trait, which is in contradiction with the notion of organic species itself. Moreover, we clearly understand that if the dimensions and forms of a multitude of structures which have to enter into the organization of a living being were made to vary simultaneously, there would have to be a whole series of combinations incompatible with the preservation of life in the individual, or in which the individual is subject to such great chances of destruction that the species would have no chance of survival and would necessarily or very probably disappear as a final or stable form (Chap. V). Thus we can account perfectly, *a priori*, for the gaps in a series of specific types. We understand that there may have been sudden jumps from one type to another, not only because intermediate forms have disappeared, but because they are not possible. This is very much like the case in which there is a sudden change from one position to another when a body capable of assuming different positions of stable equilibrium is turned in various ways on a plane surface. Yet, if in the series of specific types we were to recognize only breaks depending upon causes of this sort, we should still be able to say that the series is continuous in the sense that nature had brought into existence all the possible intermediaries or had submitted an identical basic form to all the possible modifications. This is probably the way the defenders of the philosophical idea commended by Bonnet would argue the matter. But even in this sense, it is contradicted by observations which give evidence of a very great inequality between intervals, and sometimes of such large gaps that the hypothesis of regular gradation, which recalls or simulates continuity, must be completely given up. On the contrary, breaks in the chain seem to depend upon capricious and irregular causes of the sort that have brought about the continents and the oceans, and which reveal, even in their regular and permanent order, the fortuitous circumstances of their original formation.

239. Although the idea of continuity in transitions is disposed of in this way, the idea of arrangement in linear series always re-

mains. This is the idea to which modern classifiers have been obliged to cling, not only because it brings the most general traits of the plan into clear relief, but still more because it is better suited than any other to the needs of didactic exposition.

When we look particularly at what has been done in connection with the vegetable kingdom, which is less complicated than the other, we see that the work of the Jussieus [4] and of botanists of their school was to form groups called genera, families, and classes, on the basis of a certain evaluation of organic characteristics as a whole, but an evaluation which cannot be submitted to fixed rules. These groups were then arranged into a general catalogue or into a linear series, the links of which may be thought of as being differently spaced without disturbing what is properly called the order of the series. However, to fulfill the absolute conditions of order, this series would have to be such that each term resembles in more numerous and more important ways the two terms between which it falls than it would the terms which precede and follow these two. As a matter of fact, each term must have a greater resemblance to some particular term than it has to any other term which precedes or succeeds it in the series. In this situation, it makes no difference whether the degree of resemblance has been made to depend upon the number or upon the importance of the characteristics.

In order to satisfy the conditions of this abstract *schema* in the best possible way—whether these conditions are formulated or not —modern botanists have taken full advantage of the tentative efforts, and have successfully rectified the mistakes in classification made by their predecessors. But they all recognize that it is impossible to satisfy these conditions completely. Linnaeus was aware of this when he proposed to give up comparing the vegetable kingdom to a chain or linear series, in favor of the analogy of a geographical map on which botanical families would be represented as so many large states, the genera as provinces, and the species as centers of population. It is not necessary to give up

---

[4] [A family of French botanists consisting of three brothers, Antoine de Jussieu (1686-1758), Bernard (1699-1777), and Joseph (1704-1779), a nephew of the third brother, Antoine-Laurent (1748-1836), and his son Adrien (1797-1853).]

hope that some day this idea of a great master may be realized in a useful way, as Candolle [5] believed.

240. Reduced to its clearest terms, and devoid of every metaphorical image, the idea suggested by Linnaeus amounts to replacing a linear or single entry series by a double entry series. If we preserve the metaphor and continue to represent each term as a link, it is as though the centers of the links, instead of being limited to maintaining a certain alignment, may be distributed over a surface like the meshes of a net. It is clear that with this new arrangement the points of contact and the ways in which they are related would be multiplied in such a manner as to facilitate, by a sensible image, a truer conception of the methodical arrangement of the resemblances among the different terms.

It must be clearly recognized, however, that Linnaeus' geographic map, or any other conventional device of this sort, would still only imperfectly represent relations as varied as those found among plants when we take account of the structure and the functions of their numerous organs, their habitation, their geographical distribution, the properties of the substances that can be extracted from them, and all the aspects of their history. In other words, the natural order of resemblances among plants has no geometrical analogy among all the sorts of methodological arrangements which may be worked out on an extended surface having two dimensions, although these may be infinitely more varied than those which can be represented by a linear series.

241. If we wished to submit this conception of Linnaeus to an analogical extension, after having substituted the comparison of a plane or a map for the common one of a chain, it would be necessary, in turn, to substitute for this latter the image of a relief map made of a transparent substance so that its internal sections could be studied. This would make the three dimensions of space col-

---

[5] [Augustine Pyramus Candolle, 1778-1841. A distinguished botanist and naturalist whose work was associated with that of Saussure, Cuvier, A. von Humboldt, and Lamarck. His first important work was his edition of Lamarck's *Flora of France*, a work which he revised at the request of the author. The importance of Candolle's own work was first made clear in his *Theory of Botany* (1813), in which he developed his new classification of plants according to natural systems.]

laborate in the representation of relations of order for which we seek a sensible image. Putting the matter in purely abstract terms, a system of triple entry would replace that of double entry which has already taken the place of the linear series.

But in order that the method of co-ordination and the *system* of different types of the organic structure of plants may be expressed perceptually, or may be portrayed or modeled truthfully, it would be necessary that extension permit not of two or three, but of an infinite number of dimensions. On the other hand, if we are dealing simply with objects among which the possible relations are restricted by a limited number of aspects it would sometimes be possible to express the scheme of their relations completely by means of double or triple entry series, although they could not conveniently be expressed by means of a linear series.[6]

[6] When, instead of embracing a kingdom of nature as a totality, fragments of this vast whole are separated in such a way as to distribute systematically only beings clearly united in all the fundamental traits of their organic structure, the idea of a series of double entry or multiple entry is confused with that of *parallel* or *collateral series*, of which my old colleague and learned friend, Isidore Geoffroy Saint-Hilaire, long ago gave some ingenious applications, by relying upon some considerations the basis and, in part, the form of which are the same as those we have been summarizing in the last three sections. For example, all naturalists since Buffon have divided the family of apes into two great divisions: one branch includes the apes of the old world whose dental structure is formed according to the same pattern as man's and among which are found species having no tails, and which most clearly resemble man in all their physical characteristics as well as in the development of their intelligence; the other branch is made up of the American apes which have 36 teeth rather than 32, and a tail which is generally developed to the point of being a prehensile organ. Each of these divisions constitutes a partial series in which the species can be easily ranked in a linear order and without too sudden transitions from those species which, because of the nature of their brain, their facial angle, and their whole organic structure, are most closely related to man, to those species which, on the contrary, as a result of this same group of traits, most resemble the flesh-eating types of animals which are lower in the scale in the great class of mammals. Between the species of each branch there is a kinship which follows from geographical association and from primordial facts the reason for which is hidden from us. This is a relationship of such a kind that it is impossible to separate these species by inserting between them species from the other divisions without bringing confusion into everything. At the same time there are laws of a higher order, by virtue of which the most general traits of animal structure shade off into one another in spite of the diversity of

242. Now that the meaning of our expressions has been made known, we will return to our point of departure by saying that one of the radical imperfections of written or spoken discourse is that it constitutes an essentially linear series, that its mode of construction obliges us to express successively, by a linear series of symbols, relations that the mind does or should perceive simul-

---

original accidents. These laws determine the passage from the human type to the animal type in both series of apes, by intervals or degrees which very nearly correspond, although nature does not limit itself to an exact correspondence, and although the terms in each series may be equally numbered and spaced. In order to express these different circumstances, M. Geoffroy says the old-world apes and American apes compose two parallel or collateral series and that nature *repeats itself* in the two. Another very striking example of parallelism or of repetition in the series of mammals is found in the series of *rodents* and of *insectivora*. While the dental system, the type of nutrition, and certain structural traits allow us to mistake neither the essential differences between the two types nor the kinship of the species falling within each series, nor, consequently, to confuse the two series or to mix them to form a single series, we observe, on the other hand, a manifest correspondence between the groups which follow one another in the two series; and the analogy of the types is most often marked by an analogy in the terms used to designate them, both in common and in scientific nomenclatures. Thus, among insectivora, the shrew-mouse (*sorex*) is analogous to the mouse among rodents; the mole, to the mole-rat; the hedgehog, to the porcupine; the climber called taupaïa, to the squirrel; the leaper called rhinomys, to the jumping-hare; the aquatic animal called desman, to the ondatra and the beaver. The analogy rests not only upon traits which appear to be necessarily related to the mode of life and to the habits of the various animals, but also to accessory or accidental characteristics of the sort that must be traced back to an original parent and to a primitive pattern rather than to the tendency to preserve acquired habits and the influence of the surrounding environment, such as the trait of having a *distichous* tail of the sort found in both the *tupaïa* and the squirrel, and which, so far as we can see, has no relation to the habits of climbing animals. The reproduction of such characteristics seems clearly to indicate a tendency on the part of nature to repeat concordant modifications in distinct types. Arrangement in parallel series is an image or an expression which is very well suited to such relations. In spite of the conjecture presented by Candolle and cited in the text, perhaps that is the only advantage gained by representing natural relations between organic beings through images of two-dimensional extension, or tables of double entry. But such representation is obviously applicable only to fragments detached from the general system and not to the system itself taken as a whole. See a note written by Isidore Geoffrey Saint-Hilaire in Serres, *Principes d'organogénie*, I, 205.

taneously and in another order. For this reason we dislocate in speech what is connected in thought or in the object of thought. This matter will be apparent to everyone when an attempt is made to use words to describe not a picture or a landscape—for we have already found in the continuity of forms, nuances, and magnitudes, another cause which makes exact translation by means of discontinuous symbols impossible—but a system which is made up of discontinuous parts, like a timepiece. From whatever point of view we undertake to describe the parts of the timepiece and their reciprocal play, and whatever order we follow, we shall find the greatest difficulty in making the mechanism understood as a unity by the use of discourse alone, and we shall be able to give only a very imperfect idea of it. The reason for this, manifestly, lies in the necessity of decribing the pieces one by one and in the impossibility of passing from one of them to another which is immediately connected with it, without at once losing hold of all the other pieces which are also immediately connected with the first.

Now this simultaneity of connections, these mutually dependent relations, are found not only in things which are extended, material, and perceptible, but also in everything that becomes the object of speculations of the understanding. How often we have experienced difficulty in putting *in order,* as we say, ideas which present themselves to us simultaneously. And after many attempts, we often find this order which has cost us so much effort is not a faithful reproduction of the order whose prototype we think we possess subjectively and which we seek vainly to indicate to others or to stabilize for ourselves with the aid of symbols, handicapped as we are by the nature of symbols, by the law of language, and by the sensible form of this instrument of our thoughts.

No matter on what scale we work or what form of abstraction we use, the same influence makes itself felt in the same way. Our treatises, our scientific methods, our histories, our systems of law are so many attempts whose purpose is to co-ordinate in linear series, to link—this is the proper term—facts, ideas, phenomena, and relations which most often do not submit to such linking without being forced. The result of this is that some things are found to be separated which really are intimately connected; that the description of some relations cannot be sufficiently complete without causing confusion or disturbing the general plan of the work. Everyone wants to substitute a better plan for the one in which he recognizes defects; everyone seeks artifices of diction best suited to disguise incoherencies, just as a composer of music seeks to mitigate a necessary dissonance. In seeking the solution of an insoluble problem we use up energies which often could be more profitably employed.

243. In certain cases, however, we have fully recognized imperfections inherent in the form of discourse, in this alignment to which it submits ideas. We try to correct these imperfections by constructing synoptic tables, trees, and historical atlases: types of tables of double entry, in the outlining of which we are more or less successful in representing two dimensions of an extended surface, so as to indicate systematically relations which are difficult to disentangle within the concatenation of discourse. The difficulty of material execution is opposed to our being able to turn the three dimensions of space to account for the same purpose, although, in the order of abstract conceptions, analogy may lead in that direction.

But it must be remarked that these synoptic tables have had a thoroughly real utility only when they have been adapted to relations which lend themselves to enumeration, and which consequently are connected with an order of ideas or facts which is not subject to the law of continuity. When the opposite is true, these tables only serve to bewilder thought by introducing a determination and a fictitious discontinuity into what is fundamentally undetermined and continuous. Thus, a genealogical tree is the simplest and best device in the world for expressing clearly the system of relations which connect among themselves all the members of a family, while an encyclopedic tree or table cannot avoid giving only an imperfect picture of connections of the sciences and their mutual dependence.

244. In discussing the consequences which result from the linear construction of discourse, we have had in view the massive constructs which constitute oratorical or didactic order, rather than the detailed construction of each separate sentence which makes up grammatical order. If the formalistic dryness and precision of scholastic argumentation were used in ordinary language, f a discourse of any length and of any importance could be formed

by a superimposition of syllogisms, enthymemes, and dilemmas, as a column can be by the superimposition of courses of stone, then there would have to be a correlation between grammatical phrasing and the natural division of the operations of thought. But usually the grammatical phrasing of forms, which is subordinated to the exigencies of the ear, to the attentive capacity of the mind, and to other accessory circumstances, has no more fixed relations with the general ordering of discourse than the dimensions and form of an undressed stone, in a piece of masonry, have to the architectural lines and proportions of a building. When we examine discourse in detail and sentence by sentence, we not only find that the words of each sentence are bound to succeed one another in a linear order, as do the sentences themselves. We also find that their numerical sequence is assigned with more or less fixity according to the genius of the language and by virtue of certain syntactical rules. The languages that permit inversions, those which are the object of our admiration and of our classical study, are those in which freer syntax allows the writer to choose more readily among all the linear forms of construction the one which is most graphic, that is to say, the one which most faithfully portrays the order according to which the mind of the speaker or writer has perceived the ideas that each word, in its turn, should awaken in the mind of the hearer or reader; if, keeping entirely within the linear order which prescribes the essence of language, we find so many differences as to the force of expression as we go from one language to another, and if these differences are due above all else to certain syntactical liberties, what would the situation be if the writer were free to go outside the circle of combinations limited by the linear form of grammatical construction?

245. We must not regard these two characteristics of discourse —being composed of discontinous elements and developing in a linear series—as being of the same order of importance. The first characteristic is found in every other system of artificial symbol (206); the second is more particularly determined by the organic conditions of the activity of speaking. Even though the graphic symbol of thought had not been formed on the basis of the spoken word, it would still have developed the imperfections pertaining to the first characteristic that appears in discourse, and to the same

degree. But it would differ essentially from discourse in the way in which elementary symbols would be arranged and consequently would open up other avenues to the development of thought. Thus, an algebraic formula is better suited than ordinary writing for presenting, in the form of a synoptic table, the idea of symmetry with which the elements of a formula are grouped and combined. Moreover, any system of graphic symbols, together with any special resources it may be able to offer, in the end would always be found to be inadequate to represent all the kinds of orderliness and connections that nature reveals to us and that reason conceives. It is for this reason that mathematical images soon cease to hold the attention of the analyst through their correspondence with the conceptions of pure analysis. Therefore, mathematical analysis rests only on ideas of a very particular and even of a very special nature, among all those to which thought applies (200).

246. In a didactic treatise, it is possible to give an exposition of a science only by following a linear order which presents obstacles to the exact representation of the relations among the various parts of the science. But it is not necessary to conclude from this fact that the imperfections of the didactic order may not be corrected so far as the ideas that have already made this science familiar are concerned, and still more in the thought of a man whose genius enables him to master it. On the contrary, it must be admitted that such a man, throwing away the scaffolding by means of which his genius has reached its present attainments, or preserving only the useful portions of it, correctly perceives the system of relations of all the parts among themselves, and is guided by this mental picture in the investigation of facts and relations as yet unknown. But since this entirely subjective perception cannot be translated into discourse, or can be so only imperfectly, it constitutes an individual, changeable and perishable trait. Only ideas capable of being fixed by means of symbols may be transmitted identically and enter into the system of science which instructs and grows continuously.

247. We have already learned to distinguish between rational order and logical order (24). This is a very important distinction, and one that, contrary to what is usually the case, etymology alone would fail to recognize. We know that rational order is

adapted to the order of things and that logical order is related to the order of our faculties. We also know that logical order is essentially linear, whereas we perceive no necessary limits to the variety of forms rational order may assume. Moreover, since these forms are most usually capable of sensible representation, our idea of them presents this peculiar characteristic: we do not know how to give adequate expression to the idea, yet it is useful as a term of comparison, for the purpose of choosing those from among the forms capable of sensible expression which lend themselves least imperfectly to the translation of the natural relation of things. It is like the ideal type that the artist possesses, and to which he tries to give adequate expression through his pencil or chisel, but without success, since, although the ideal form does not exist among tangible forms, he may run across things that resemble it and that he can use, up to a certain point, for communicating his idea to minds which are capable of understanding it and to spirits which are in harmony with his own.

It is always possible to prove that we have been mistaken in the rules of logical deduction, and that a given demonstration cannot be accepted because of a vicious circle or because of incomplete enumeration. Yet there is no way in which we can rigorously demonstrate that a given arrangement among the theoretical axioms conforms to the rational order or deviates from it (156). This is the point at which philosophical insight comes into play. The judgments which rest on this insight can neither be precisely confirmed or refuted, nor can they be imposed on the reason of other persons.

We may guess that a glimpse of the most important truths has first been caught sight of by means of this philosophical insight which precedes rigorous proof. As a result, we should not be surprised if, in such sciences as mathematics, where logical exactness is prized above everything else, it often happens that even though we accept the discoveries of inventive minds, we are not always satisfied with the demonstrations they have given for their discoveries, *as though they had poorly contrived what they had so well discovered,* to use the piquant expression of an ingenious mathematician.[7] This tendency toward exactness has certainly

---

[7] M. Poinsot, *Théorie nouvelle de la rotation des corps,* sect. 2, *in fine.*

been abused. And in any case, the worth of the discovery, even when accompanied by an imperfect demonstration, always carries with it much more worth than a subsequent improvement which gives, or seems to give, greater logical exactness to the proof. But nevertheless, the process by means of which the mind grasps new truths is often very distinct from the procedure by which it logically and demonstratively connects truths with one another. That which results from the order imposed by the forms of logic is not always the order that best expresses the reason of things and their mutual dependence.

248. We shall have reason to make some observations about the logical order which are perfectly analogous to those which we made in the preceding chapter concerning the theory of definition. Just as it is necessary to admit logical roots, or indefinable words which serve to define others, so some principles or axioms which serve as the basis for all later demonstrations must themselves be admitted without demonstration. The perfection of the logical order consists in reducing as far as possible, on the one hand, the number of principles and axioms admitted without demonstration, and, on the other hand, the number of links or intermediate propositions by means of which one proposition is logically connected with another.[8] It is toward this perfection of logical order

[8] In Volume VIII of the *Anciens Commentaires* of St. Petersburg is to be found a memoire by Euler concerning a problem of no importance in its statement, but curious because of its relation to the theory of combinations, of position, and of order, and which consists of indicating the course that must be followed in order to traverse one after another—when this is possible—all the bridges which connect a system of islets and lagoons, with the essential condition that we cross each bridge only once, although we may set foot on each islet many times. A correlative problem would consist of using a sequence and selection among the bridges in such a way as to cross all the islets one after another, but to be on each islet only once. By retaining in the statement of these problems only that which makes up their abstract form, we see the analogy with the problem of the concatenation of a linear series—a *moniliforme*, as the naturalists say—by means of transitions that suggest natural relations among those objects which have been linked together in this way. Depending upon the number of objects we wish to arrange in series, and depending on the number and the distribution of relations which may serve as connections or passages, the problem may give rise to many rigorous solutions, or may lead only to approximate solutions in which a certain number of objects whose presence makes the solution impossible will not be taken account of, except to

that mathematicians, and particularly ancient mathematicians, have especially striven; and the taste for this kind of perfection is regarded as one of the important traits of the mathematical mind.[9] But so long as a person applies himself to perfecting the logical order in this way, he must expect to disturb frequently the essential relations, the analogy, the symmetry, and, in a word, the rational order among the different parts of a scientific construct.[10]

The degree of evidence which determines that a proposition may conveniently be taken as an axiom does not always belong to the original truth, to what reason conceives as being the principle and the source of the others. On the contrary, in many cases it pertains to something far removed from this primordial truth, which, by the aid of accessory circumstances, is presented to the mind in a more favorable light. It also happens that the simplicity or the complexity of demonstrative arguments is not always proportionate to the simplicity or complexity of the ideas on which hang the propositions which are the objects of demonstration. In other

---

connect them irregularly, independently, and in the form of appendices with the general series.

These considerations are equally applicable to the theory of logical roots and definitions (Chap. XV), to the logical sequence of a series of propositions or truths which depend on one another, to classifications of the naturalists (239), and, finally, to hypotheses concerning the genealogy of ideas and to didactic expositions in conformity with what is called the method of invention, which, as we are well aware, most often differs greatly from the course followed by the inventors, a matter on which the history of the sciences clearly informs us.

[9] See the fragment from Pascal already cited in the note to sec. 228.

[10] For example, when Euclid, and after him Legendre, demonstrated this fundamental theorem of geometry—that corresponding sides of equiangular triangles are proportional—by basing it on the proposition which gives the measurement of the area of a triangle, they had in mind the harmony of the logical order; for the passage from the commensurable to the incommensurable having occurred with respect to the measurement of the area, we find we can dispense with the rebuilding of that inconvenient scaffolding with respect to the theorem concerning the similarity of equiangular triangles; but, on the other hand, the theory of areas is so placed that it follows the other, and if it is a proposition and almost an axiom (234, note) which should serve as a natural introduction to geometry, which was actually discerned first, according to the evidence of history, and which expresses the most important property of the straight line or of a system of many straight lines, the entire harmony of the rational order is disturbed; and most authors have rightly thought that the advantages of pure logical form do not justify such a reversal of natural relations.

words, it sometimes happens that demonstration takes an easier or more expeditious turn when we go from the compound to the simple than when we go from the simple to the compound. Thus, although the complexity of constructions, calculations, and demonstrative arguments generally increases as we pass from the propositions of plane geometry to their analogues in solid geometry, it has long since been noted that, by exception, construction in space and the arguments about three-dimensional figures are sometimes more easily introduced to demonstrate certain properties of plane figures. The object of statistics is less simple than that of geometry, since in it the consideration of forces is added to the consideration of forms. Nevertheless, it happens that we have been able to put statistical notions to work ingeniously in order to establish certain properties of figures more simply and more briefly than could be done by means of geometry alone. Similarly, we apply geometry to algebra and the different parts of the calculus to one another in such a way that an increase in construction or complication in the rational order results in a logical simplification. The examples of such inversions are innumerable. But, without having to go into technical details which lie outside the province of this book, we may note the variety of methods followed by mathematicians in establishing the same axiomatic principles, and the numerous verifications to which they never fail to submit each important theorem. These instances show clearly enough that some of these principles, which proceed from one another and recur in accordance with divers series, and which are all logically connected, are almost like those forms which mineralogists relate to the same crystalline system, each of which, when taken as a point of departure, reproduces all the other forms when it is submitted to a series of regularly defined modifications. These observations hold even though some members of the group may be much more simple, and some may reasonably be taken as points of departure in the successive transformations.

249. The importance of the syllogism [11] in peripatetic logic results from the role that the doctrine of universals and of definition *per genus et differentiam* plays in it (222). The genius of Aristotle

[11] Compare, on the theory of the syllogism, the Port-Royal *Logique*, Part III; the *Lettres d'Euler à une princesse d'Allemagne*, Part II, Letters 35ff., in our edition, Paris 1842; and, among recent works, de Morgan's curious book entitled *Formal Logic, or the Calculus of Inference*, London, 1847.

admirably grasped and co-ordinated all the parts of the system, and we must follow it even when combating his theories in so far as they are too absolute or too extreme. There is reason to make the same distinction in regard to the *major premises* or general truths, from which we wish to deduce a particular truth by means of the syllogism, that we have been making concerning generic ideas and abstractions of every sort. Some are artificial or purely logical; others are natural and are based upon the rational subordination of things. If the general proposition contained in the major premise is nothing but the summarized expression of particular judgments based on each species of the genus, the syllogism is an artificial construction which may be useful for putting acquired knowledge in order, but which is ineffective for the purpose of extending or developing our knowledge. If, on the contrary, the general truth is conceived as entirely independent of the particular and the concrete forms that a general and abstract idea can assume, and if it is also conceived as being not the logical resumé of, but the reason and the basis for the particular truths, then the syllogism becomes a method of extending our knowledge and of increasing our understanding of the properties of particular things, by suitably applying general ideas to them. This is so because the syllogism expresses the subordination of our judgments in a manner conformable to the intrinsic relations of things—or in a manner conformable to the order according to which facts give support to one another and truths emanate from one another.

"All metal is opaque." This is a proposition which cannot be used as a major premise in a syllogism which can increase our knowledge in certain respects. We acknowledge the truth of this general proposition only because we have verified the fact that every kind of metal that is known has the property of being opaque. It must be directly verified in the case of any metals that may subsequently be discovered. Before the discovery of sodium and potassium, it might just as well have been said, "all metals are heavier than water," or, better still, before the discovery of mercury, that "all metals are solid at ordinary temperatures." On the other hand, the proposition, "all mammals breathe by means of lungs," can very well be used as a major premise. As a major premise it will serve to increase our knowledge or to prove that any extinct animal breathed by means of lungs even though we know nothing else about it than that it was a mammal (49). This follows not simply because there are no exceptions to this observed fact, that all animals provided with mammary glands have lungs, but even more, and principally, because the knowledge acquired as to the mutual adaptation of parts in the total organic structure of the animal permits no doubt as to the fact that being viviparous and suckling the young, are conditions of existence which are subordinated to a condition of greater importance, namely, that of breathing by means of lungs. It is in this way that logical proof, which results from the connection of propositions in the syllogism, may have a philosophical inference as its preliminary condition and a probability for its foundation, but a probability of the sort to which reason is forced to assent.

250. Nevertheless, this is not true in every case. The principle of morality and of law, that " no one has the right to enrich himself at the expense of others," is a maxim that reason finds in itself. It does not need to be established on the basis of repeated observations or by invoking analogies or inferences of the sort we have called philosophic. Therefore, when a lawyer argues from this general principle or from this major premise in order to prove that the owner who recovers his goods ought to reimburse the possessor fairly for any expense he has had, in proportion to any improvements which have been made on them, or to decide any other more subtle question whose solution is less evident, his deduction will be free from any previous induction. He will pass directly from the general truth to the particular truths which must really be considered as so many emanations from this general truth, as distinct from cases in which the general proposition exists only as a logical expression and as a collective symbol of particular truths.

The distinction of major and minor premises and the syllogistic construction which follows from this is shown particularly in arguments before the bar, in which the major is ordinarily called " a question of law," and the minor, " a question of fact," or sometimes " the case in question," as a remnant of the influence of Scholastic traditions. It is perfectly clear that in a lawsuit between an owner who reclaims his goods and the person who has possession of them in good faith, the solution of the question of law of which we have been speaking, or of any other analogous one, is independent of the knowledge of the particular case and of the proof of the fact that the possessor acted in good faith, that he has incurred some expenses, and that his expenses increased by that amount the total value of the things reclaimed. The syllogistic distribution, prescribed in this case by the nature of things, is entirely to the advantage of the orderliness of ideas and the clarity of proofs.

On the contrary, sophistry and bad faith will be favored, if the argument hangs on ideas whose value it has been impossible to determine invariably because of the continuous modifications which these ideas permit. The sophism could be resolved by distinctions in the case of simple equivocation, that is to say, if the terms have only a certain number of distinct, determined meanings. But this is most frequently not the case, and it is then that the apparatus of logical forms disturbs and falsifies judgment instead of clarifying and strengthening it (196). We shall delay no longer the development of these considerations, although we shall apply our principles especially to the theory of judicial questions.

251. We are able to abstract, as we say, the " matter " from the nature and the origin of the ideas and judgments which make up and constitute the syllogism, so as to consider only the " form " or the kind of propositions; that is to say, their characteristics as general or particular, affirmative or negative. For, as these characteristics are variously combined in the three propositions which make up the syllogism, it will be possible to distinguish many moods and figures of the syllogism, and to set up rules in order that a syllogism may be valid and conclusive, or, contrariwise, may be deficient in form regardless of the intrinsic truth or falsity of the conclusion or of the premises. Hence, arises a curious theory, every aspect of which is perfectly rigorous, and the invention of which long precedes that of algebra and the general theory of combinations, although it calls attention to this latter theory, and although it bears a very close analogy to the elementary rules of algebra. In fact, although the species may not be contained in the genus in the same way as one quantity is contained within another, there are, nevertheless, principles of such generality (154) that they may be applicable to either mode of comprehension or of extension. We are able to say that from the two propositions, A includes B, B includes C, the proposition A includes C results. This is true, whether A, B, and C designate homogeneous quantities, or whether these same letters are used to indicate generic terms which are subordinate

to one another in the hierarchy of universals. The rules of synthetic combination, adapted to the syllogistic series, must therefore have the greatest resemblance to the rules of the calculus that is called in mathematics the "calculus of inequalities," and consequently they also resemble very much the rules for the calculus of equalities or equations. But this great resemblance is found only at the starting point. The fruitfulness of mathematical principles and the loftiness of the structure for which they furnish the materials result from the great simplicity of their forms and from the great regularity of their syntax. The calculus of inequalities, compared with that of equations, rests upon principles which are less simple and which are subject to more restrictions. Thus, this branch of mathematics has remained in what may be regarded as a rudimentary state as compared with the vast developments that have been made in the theory of algebraic equations. For a similar reason, the theory of the syllogism would in no way include scientific developments comparable with those of algebra, even though it were not to have a practical use which is as limited as the applications of algebra are numerous and important.

252. Can all reasoning be reduced to syllogistic forms? Aristotle himself did not think so, and he early noted that mathematicians do not use the syllogism because, as he put it, they make no use of the notion of genus and species. But there is a better reason than that. For, how can the theory of combinations be established syllogistically when the rules of the syllogism themselves are dependent upon the very theory of combinations? And how can the rules of algebraic calculation be made to depend upon a syllogism, since the relation of the conclusion and the premises in the syllogism is supported neither by more nor by less evidence than the rules of algebraic calculation? The algebraic argument, A is equal to B, and B is equal to C, therefore A is equal to C— which is not a syllogism since we can distinguish in it neither a major or a minor premise, nor a major, middle, or minor term—is as self-evident as the relation of the conclusion to the premises in this common example:

> All men are mortal;
> Kings are men:
> Therefore, kings are mortal.

If the probative force of algebraic argument came from the fact that it could be converted into syllogisms, it would be necessary, on the same basis, that we should be able to prove by syllogisms the very rules of the syllogism, and we could not escape from a vicious circle.

The device which consists in making evident the relation of two terms or ideas through the mediation of a middle term or of an intermediary idea is not something unique in syllogistic argumentation. Nor can it be found only in the rules of algebraic calculation. It is one of the most general procedures to which the human mind has recourse to assist its weakness, and the mind uses it continuously under the most varied forms. In order to compare the surfaces of two rectangles having neither the same base nor the same height, we conceive of a third rectangle having the same base as the first and the same height as the second, the surface of which serves as a term of comparison between the surfaces of the two proposed rectangles. Such a figure has a function analogous to, although not precisely identical with, that of the middle term of the syllogism. In general, in order to compare quantities we measure them. That is to say, the unit of measure is the middle term by means of which we obtain the relations between quantities of the same sort, which, in most cases, can be neither physically related nor completely compared. In commerce, the precious metals are the middle term by means of which we compare the exchange value of objects which are unlike in every other respect. In a geodetic survey, objects which are too distant from one another to be seen from the same station are connected by means of signals and intermediate stations, and so on.

253. In treatises on logic we find rules concerning the conversion of propositions by means of which it is permissible, under certain conditions, to draw rigorous conclusions by going from the particular to the general. This manner of drawing a conclusion is often called "induction," as opposed to "deduction," which consists in going from the general to the particular, and which is the procedure that we practice in the syllogism properly so-called, *per genus et speciem*. Thus, it often happens that the observation of a single fact is sufficient to upset a general theory. But the same induction, submitted to formal conditions which are as clear and precise as those of syllogistic deduction, is, like the latter, a

logical demonstration, a rigorous calculation, which must not be confused with philosophic induction, the nature and role of which we have stated in Chapter IV. The same thing must be said of a feat of understanding which is very often used in mathematics, and which is also called induction. This consists in proving that if, in a series which may include an infinite number of terms, one or more consecutive terms are brought under a certain law, the next term can be similarly brought under the same law, and so on from term to term. As a result, it is sufficient to establish the law for one term or for a finite number of terms. Then a rigorous induction extends it to the whole series of consecutive terms, which are infinite in number. As a matter of fact this is clearly an instance of going from the particular to the general. But this feat is not the only one mathematicians delight in performing. It continuously happens to them that when they wish to establish the properties of a class of figures, they first consider the figure in a particular or rather a singular state (198), in that it introduces simplifications which are exclusively characteristic of it, and they then show that the general case may be brought back to the particular—from the genus to the species, as Aristotle would put it—when they deal with the general case first of all, in order to deduce the particular case from it as a corollary. Passing in this way from the general to the particular or the singular, there is ordinarily nothing which recalls syllogistic construction. Finally, and still more frequently, the terms and propositions maintain the same degree of generality throughout the series, as happens in the type of arguments of which the description of an algebraic theory represents the typical notation. On this basis it cannot be said in such a case that the mathematician proceeds by means of deduction rather than induction. Analogous observations can be applied to the course of reasoning, whatever objects we reason about, although mathematics always enjoys the privilege of furnishing the clearest examples of this sort, and of being, in all respects, the best school of formalist logic. The truly essential distinction in all things is that which consists in contrasting logical or demonstrative proof and philosophical induction, the origin and nature of which we have previously tried to make known. As for logical or demonstrative proof, it sometimes proceeds by deduction and sometimes by induction, depending upon the case involved, from known to unknown, depending upon whether there is more or less generality in the known truth than in the unknown truth. Sometimes, by means of a contrary course, it proves hypothetical premises by falling back upon known consequences. Finally, in a great many cases, the premise and the conclusion have the same degree of abstraction or generality. And in order to include at one time all the different cases, the most suitable term is " construct," the meaning of which will be still better fixed by remarks contained in the chapter which follows.

CHAPTER XVII

ON ANALYSIS AND SYNTHESIS. ON ANALYTIC AND SYNTHETIC JUDGMENTS

254. In speaking of the formation of abstract ideas and of the progressive elaboration of the primary materials furnished to the mind by the sensibility (147), we have already had to say something about *analysis* and *synthesis*. We have even had to establish an important distinction between the abstract ideas resulting from the analysis or the breaking up of sensory images and those which come from the need that reason feels for bringing together and co-ordinating scattered impressions. From the point of view of logic, these are the processes of elementary analysis and synthesis, since they result in the formation of ideas or of terms which serve as elements in the series of logical propositions, arguments, and constructions, no matter what position the terms may have in the scale of generalization or abstraction. Logicians have dealt with analysis and synthesis principally with respect to method. That is to say, they have considered these two acts of thought more especially in so far as they include a complete series of judgments and arguments which are capable of formal statement and of logical co-ordination. We shall also treat analysis and synthesis from this point of view by attempting to describe and explain, on the basis of our principles, the different meanings these terms acquire in ordinary language and in various scientific dialects, but principally in the language of mathematics, in which they are so frequently used, and in which they acquire a technical value which it is interesting to compare with the value which logicians attach to them.

255. If we go back to etymology, the word " synthesis " simply means *combination* or *construction,* while the word " analysis " means *decomposition* or, better still, *resolution.* But it is well to observe that the two correlative terms, in passing from scholarly language into ordinary language, have not had the same fate. Ordinary language has completely taken over the one, the use of which has become most common; the other has never emerged

from the didactic style, or, if it is tending to lose this style at the present time, it is the result of a pretentious neologism. Since there must be a reason for everything in the organization of language, we have already been led to infer from this comparison that one of the two terms recalls an operation which is more natural to man, the other a more reflective and scholarly operation. Nevertheless, in many cases, the term " composition " may be considered as the true correlative of that of *analysis;* we speak of the composition of a discourse, of a poem, or of a picture, just as we speak of making an analysis of a tragedy or of a discourse.

In general, the term analysis is commonly taken to mean a breaking up of intellectual objects into their elements by thought, as contrasted with material or mechanical separation into parts. Here its use depends simply upon a perpetual metaphor. We see clearly that the idea connected with the word analysis cannot lend itself to a rigorous definition when we are dealing with a subject which is as complex, as abstract, and sometimes as vague as any system of ideas, of arguments, or of images which a book is intended to produce. Among the ideas to which this word gives rise are many which no more call to mind decomposition in the strict sense of the term than they do composition. Finally, unless we propose to stop speaking, it is necessary to use this term as well as others with the indetermination with which it is affected, by endeavoring not to carry this indetermination over into the essential forms and characteristics of thought.

To the degree that the subject matter becomes more simple and better defined, the metaphorical allusion to the common notion of composition and decomposition acquires a greater precision. Thus, when a grammatical phrase is analyzed for a child, a real decomposition or resolution is made of the statement into the grammatical elements which compose it. These isolated elements are dictated to him, and he must join them together again according to the laws of grammar. This latter process is a real instance of composition, a synthesis or a syntax, for these terms may be considered as synonyms, except that in the use of the latter term there seems to be a more direct allusion to the relations of order and position among the associated elements. Finally, the child becomes more skillful, and as we discover by acquainting him with a dead language, the study of which lends itself to the methodical development of his mind, he is able to do what we call translations and exercises with the assistance of a two-language dictionary. But this has already become too complex an operation for a metaphorical expression to be perfectly adapted to it. Yet, when we reflect on it, we shall see that the work of translation is more like what we call analysis, and that synthesis or syntax is dominant in doing composition exercises.

There is another science in addition to grammar in the vocabulary of which the terms analysis and synthesis always have a precise meaning which never gives rise to any equivocation. This science is chemistry. The reason for this is evident, since this science deals with physical objects to which the ideas of composition and decomposition may be directly applied without the use of figurative style.

256. In his writings, Condillac incessantly reconsiders analysis and synthesis, to prove that the analytical method is the only proper one and the only one founded on nature. Indeed this seems to be the sole object of all this philosopher's writing. However, it has long been noted that Condillac has made an entirely different use of synthesis and that his treatise *On Sensation* in particular, in which he attempts to reconstruct a man out of whole cloth by successively giving his statue each of the five senses, is pre-eminently a work of synthesis; and we may observe that the faults of the work follow precisely from the use of synthesis in a subject with which it is incompatible.

If a watchmaker wanted to explain the mechanism of a watch, it would make absolutely no difference whether he began by taking it to pieces, i.e., by making an analysis of it, or, conversely, by putting together disassembled pieces, i.e., by first making a synthesis of it. Probably the student would have a clear and lasting idea of the mechanism only after both these operations had been repeated many times. The same thing might be said of all theories relating to complex objects whose parts may be put together and taken apart in this fashion. It is of little consequence whether they are stated by means of analysis or synthesis; or rather, when they are presented, it is almost impossible to follow exclusively either analytic or synthetic processes. Even when this difficulty is overcome in the editing of a didactic treatise which has been thought about for a long time, the theory would really include only that with which the mind has become familiar by means of these alternative compositions and decompositions, and with respect to which it would have become equally easy to proceed analytically or synthetically.

Instead of the watchmaker who explains the mechanical structure of a watch, we might just as well have considered a teacher who explains the skeleton or the bony structure of the human body, a structure composed of a known number of distinct pieces whose natural relations to one another are maintained reasonably well by means of artificial ligaments. In this common case as in the other, we say that it will make little difference whether the teacher separates connected parts or puts together isolated parts, that is, whether he makes an analysis or a synthesis of the skeleton.

But in living animals, the skeletal structure is related to other systems of organs, in the study of which we can move only from the whole to parts, and not, conversely, from parts to the whole. This is so not because these systems of organs lack stability—for with patience and skill the circulatory system and the nervous system can easily enough be isolated, and the art of injections is specifically intended to make the vascular system visible—but results rather from the fact that these organs are endlessly ramified, from the fact that innumerable flexions and inosculations exist, in a word, from the fact that they form a continuous whole and not an assemblage of pieces. Thus, as etymology indicates, anatomy is simply the analysis of organs; the corresponding synthesis is not known, for obviously physiology is something altogether different from this synthesis. Human osteology is complete and the synthesis of it can be made; the opportunity for anatomical discoveries will never be exhausted so far as the vascular and nervous systems are concerned. New optical instruments may push back the boundaries of these fields indefinitely, but that shows the impossibility of making a synthesis of these systems of organs.

257. Note well that, when synthesis is impossible because the real elements are never reached, strictly speaking analysis is also impossible. Yet there is this essential difference: the analytical or anatomical procedure tends toward the indefinite attenuation of errors, to a more and more distinct and true knowledge of the object; while the synthetic method, or the reconstruction of the

whole out of parts which are not its true elements, tends toward the accumulation of errors; and should it happen that errors do not accumulate, but, on the contrary, compensate for one another, this compensation is purely fortuitous. When we are working with objects which are capable of continuous modifications, whether these objects belong to the physical world or to the world of ideas, the construction of a theory implies the substitution of an artificial system for a real one, that is, a system of the sort that is presented to us by nature. But the artificial system, the product of an analysis which is performed on the whole, may constantly be brought to resemble the real system, yet the latter may deviate indefinitely from the one which we might construct by bringing together artificial members or elements.[1]

[1] " On the analogy of the method used by physicists to assure themselves of the accuracy of an experiment, Reinhard * set himself the task of rebuilding the whole fabric of moral existence by bringing together again the elements of our nature which were the outcome of the resolution effected by Kant's philosophy. This attempted rapprochement did not see a rebirth of that admirable and harmonious whole in which all our powers mutually assist one another, each contributing in its own way, without conflict or superfluous activity, to the end indicated by our own physical and moral needs. Rather, there came out of this effort, which was amended at various times, a whole so incoherent, so lacking in accord among its constituent parts and in traces of that wise order, of that far-sighted concern, which is so conspicuous in all the works of nature, that he felt the strongest aversion to adopting principles which led, by the testing of the synthesis, to results so little in conformity with the needs of man and the paternal purposes of his creator. He believed he was justified in suspecting in the analytical work of Kant some secret defect, some important gap, that the skill of the master and the prestige of his talent had hidden from his attention, almost like a chemist who is unable to succeed, by combining anew elements which he had obtained in the decomposition of a substance, in order to reproduce it as it is found in nature, would remain convinced of the imperfection of his experiments." *Lettres de Reinhard*, translated from the German, with a prefatory note by Stapfer.

The " secret defect " suspected by Reinhard consists in replacing a harmonic whole in which the law of continuity presides over the arrangement and the modification of the constituent parts by an artificial system, with discontinuous pieces; and the *important gap* consists quite obviously in the suppression of the vital principle [*principe de vie*] which produces continuity, unity, and harmony, and which disappears as much under the scalpel of the logician, as in the crucible of the chemist.

* [Franz Volkmar Reinhard, 1753-1812. A Protestant theologian whose most important single work was his *System of Christian Morality*.]

The different parts of a machine may very well be made by as many different workmen as there are different parts; and the parts will be made in this way with greater rapidity and with greater precision, according to the well-known principle of the division of labor. On the contrary, if we wished to have a copy of a statue, it would be a very bad policy to have one artist make an arm, another a leg, a third the trunk, and so on. Each of these parts might be copied with sufficient exactness and yet the statue formed by putting the various pieces together would be an unsatisfactory piece of work. The same perception of the unity among relations which guides the artist is what guides, or should guide, the analyst; the same accumulation of errors or of deviations from the real relations which is manifest in the piecing together of the separate members is a consequence of synthetic procedure based upon theory.

258. It has been said that analysis is the method of discovery and synthesis the method of exposition. The fact is there is no method of discovery; and we should not consider, and actually we do not consider, anyone a discoverer who does nothing more than apply a method. We tried above to make the reason for this understood; only rarely may the method of exposition be exclusively analytical or synthetical. This has been observed by everyone who is not dominated by the systematic temper. When rigorous synthesis is possible, it ordinarily surpasses analysis in conciseness and clarity. If philosophers and those who have wished to become the interpreters of nature have more frequently been misled by the synthetic method, the reason is that, so far as we can judge, continuity is the rule and discontinuity the exception.

Nevertheless, since progress in the knowledge of things results principally from the operation of the mind by means of which we distinguish, in the complex data of perception, that which originates in the nature of things themselves from the circumstances which result from our point of view or from the conditions in which we are placed when we observe them, it is clear that in this sense the method of discovery is necessarily analytical. On the other hand, we shall conform to the order of phenomena, and consequently shall simplify the exposition of them, by first making things known with respect to the circumstances of their intrinsic organization, then by examining the circumstances in which we observe them, and by finally explaining through a combination of the two processes, the complex phenomena which are the objects of our immediate observation. Because of the perfection which it has acquired and because of the simplicity of the magnificent object with which it deals, astronomy offers us a very clear example of the use of both methods. In the less perfect sciences, in which we are not sure that we have grasped the fundamental laws and the true theoretical elements, synthetic construction may lead to the accumulation of errors, and through that itself become the touchstone of the theory when the results of direct observation are compared with those which had been deduced from the synthetic construct. In this way synthetic hypotheses and constructions contribute to the perfecting of our knowledge, even though they are premature and faulty.

From these remarks it is seen that when the term analysis is applied to the sciences, it refers less frequently to the breaking up of a whole into its integral parts than to the distinction of the principles whose combination furnishes the reasonable explanation of a complex phenomenon and not simply the description of the phenomenon. When we make an analysis in this sense it is not, as is customarily said, because of the weakness of our intelligence, which does not permit us to embrace the different parts of a whole all at once in a clear view. On the contrary, it is for the sake of using the most noble attribute of our minds: our ability to know the reason and to discern the fundamental truths of things. These remarks hold whether the analysis is made by the aid of reason alone, or with the assistance of intelligent experimentation, which brings into prominence the essential circumstances of the production of a phenomenon by separating from it the accessory circumstances or disturbing causes.

Consequently, logicians say that we proceed analytically when we go from the particular to the general, from the concrete to the abstract, by seeking to recover the general principles or the abstract truths which should result in its solution through the examination of the particular and complex question that we wish to resolve. On the contrary, they say that we proceed synthetically or in a doctrinal manner when we go from the general to the particular, from the abstract to the concrete, or at least from a more simple abstraction to a more complex one, by combining general principles and thus by pushing the theoretical construction until we reach the combination which, because of complexity, is immediately suited to the particular and concrete question. The former course is the procedure followed by a lawyer in his brief or in his pleading, or by a judge in his official account on the matter of litigation which has been submitted to the decision of a court; the other method is that which is followed by a professor of law in his course, or by an author in the redaction of a treatise on jurisprudence.

259. When we consider a series of connected propositions which taken together constitute a proof, it may very well be that the last term has exactly the same degree of generality as the first (253). It is possible to reverse the series by adapting another didactic plan for the whole theory with which this series of propositions is connected, and to deduce the first proposition from the last without changing the general character of the method. And this may be done without having any reason for distinguishing the two coherent orders by speaking of one as analytical and the other as synthetical. To find examples of such inversions of partial series, and of transposition of this sort among propositions having the same degree of generality, it is sufficient to compare elements of geometry organized according to different plans. But, on the other hand, from the point of view of logic, there will be an essential difference between the methods. In the one method we always go from one demonstrated proposition to another which is a consequence of propositions previously demonstrated, and which in its turn is found to be demonstrated that way. In the other method we go directly from the proposition to be demonstrated, as an hypothesis, by pursuing the consequences which result as much from this hypothetical proposition as from other accepted truths, until we reach a proposition which has already been recognized as true, and whose truth involves that of the hypothetical proposition which serves as the first term in the particular series of deductions. This is the inverse method of which the Greeks thought Plato was the inventor, and to which they gave the name analysis,[2] that is to say, the method of resolution to first principles,

---

[2] "There is a way of seeking truth in mathematics, which Plato is said to have first discovered, called by Theon 'analysis,' and defined by him as the assumption of that which is sought as if it were permitted by its consequences, in order that truth may be yielded. 'Synthesis,' on the contrary, is the as-

or of inverse solution, applying the term solution to the method of synthesis or construction by the direct method. In this latter method we move from one truth to another that the preceding one affirms, just as in constructing a building we go from the lower courses of stone to those which are superimposed upon them, without appealing to hypothetical concessions which, in this kind of construction, have the function of temporary shoring, similar to that used to support a building when we wish to repair its underpinning.

260. In this last meaning of the terms synthesis and analysis,

sumption of that permitted by its consequences for the definition and understanding of that which is sought." Viète, *Isagoge in artem analyticam*, in the beginning.

"Analysis is the method which, starting from the matter in question, that we agree to at the moment, leads through a series of consequences, to something known previously or included in the group of propositions recognized as being true. This method makes it possible for us to work back from a true statement or from a proposition to its antecedents, and we call it 'analysis' or 'resolution,' as if one were to speak of a solution in reverse order. In synthesis, on the other hand, we start with the proposition which is the last term discovered by analysis, arranging later on, according to their nature, the antecedents which previously appeared as consequences, and we reach the sought for end from which we started out in the first place, by combining these antecedents among themselves.

"We distinguish two types of analysis: in one, that may be called 'contemplative,' we set out to discover the truth or falsity of a proposition that has been suggested; the other is related to the solution of problems or to the investigation of unknown truths. In the former, we posit the subject matter of the assumed proposition as true or as already existing, and then we proceed, by means of the consequences of the hypothesis, to something known; and if the conclusion is true, the suggested proposition is also true. After that, direct demonstration is made by going back over the different parts of the analysis in an inverse order. If the conclusion we come to last of all turns out to be false, we conclude that the analyzed proposition is also false. When we deal with a problem, we assume first that it is solved, and we follow out the propositions which derive from it until they lead us to something known. If the ultimate conclusion is obtained, if it is included in what mathematicians call a 'known quantity,' the question proposed can be resolved, the demonstration, or rather the construction in this case, is set up again by taking the elements of the analysis in an inverse order. The impossibility of the ultimate conclusion of the analysis will obviously prove in this case, as in the preceding one, the impossibility of the thing asked for." Pappus, Prolegomena to Book II of *Collections mathématiques*.

which is very different from those which have previously been set forth, there is an allusion to two modes of construction, direct and inverse. But reference is not made either to the assembling of parts and to the breaking up of a whole into its integral parts, or even to the passage from the general to the particular and from the particular to the general. For, as we have already observed, it ordinarily happens that in the series of geometrical propositions to which the definitions of Theon and Pappus—which are cited above—apply, the different propositions of the series have the same degree of generality. Therefore, it should be possible to change the general method of doctrinal exposition in such a way that the order of the terms of a partial series may be reversed in it, and that the same order, which was direct or synthetic on the basis of the first method, would become indirect or analytic in the second. The proposition which originally served as a hypothesis—analytic method—or as consequent—synthetic method—now serves as the known quantity or the antecedent, with respect to propositions which now precede the partial series, and which, in the original method, were consequences of it.

The analytical procedure of the Greek mathematicians was that of *reductio ad absurdum,* in which, in order to demonstrate the truth of a proposition, one started by setting up the contradictory proposition as a hypothesis, in order to reach, consequent by consequent, a proposition of recognized falsity, or one which contradicted a proposition recognized as true. This entails the absurdity of the hypothesis, and consequently, the truth of the contradictory proposition. Even now, in doctrinal exposition, the analytic method of the ancients is still used in this form. It is used in order to avoid the logical difficulties which the passage from discontinuity to continuity in the synthetic or direct method involves (201).

But this same analytical procedure, which is indirect in doctrinal exposition and in questions which have to do with the logical sequence of theorems, becomes the direct method in dealing with problems to be solved. For then, nothing is more natural than to consider the problem as having been solved. For example we assume mathematical objects upon which the problem rest with respect to the relations of magnitude and situation which we wish them to have. Then we follow out the consequences of this hypothetically until we have reached constructions we know how to make, calculations we know how to perform, or, quite the contrary, constructions or calculations which are impossible and absurd, and which will prove that the problem cannot be solved.

261. The language of algebra, as it has been established by modern algebraists, is marvelously suited for dealing with mathematical problems in this way, since it furnishes means for submitting quantities to the same operations of calculation without distinction between knowns and unknowns. Consequently, we have related to the contrast of the two logical methods set forth by the Greek mathematicians the contrast between analysis and synthesis that is laid down in the mathematical language of modern writers. The term analysis now means the science of calculation, taken in its complete extension, whether considered in itself, or as applied to geometry, mechanics, physics, and so on, while the term synthesis is applied to geometry as this was dealt with by the ancients, without calculation but with the aid of figures and graphic constructions. Nevertheless, those who concern themselves with the philosophy of the sciences have not failed to observe that, when understood in this way, synthesis and analysis no longer form two logically distinct procedures. They know that it is possible to make an analysis with figures and graphic constructions, and to carry through a synthesis with algebraic symbols and calculation. This is true whether we attribute to the terms analysis and synthesis the logical meaning the ancients attached to them, in the statements of Theon and Pappus, or whether we take them in another of the logical meanings already indicated.

These comments are fair, and yet we believe the meaning which the terms " analysis " and " synthesis " have assumed in the idiom of modern mathematics, simply as a result of usage and without anyone ever having attempted to take full account of it, may very well be explained and justified theoretically. This meaning is connected with a logical distinction which is more profound and more important than those with which we have already been concerned in the course of this chapter; this is the distinction that Kant made between analytic and synthetic judgments, which will appear lucid and simple if it is separated from the Scholastic forms in which the great logician too often becomes involved.

262. In fact, when we study an object we may start out from certain properties of the object as expressed by definitions. Then, without needing to fix our attention further on the object so long as we are careful not to violate the rules of logic, we reach the conclusions or judgments which Kant called analytic, and which, properly speaking, clarify and develop our knowledge of the object rather than extend it. These are analytical judgments because we were thought to have been given implicitly, in the notions expressed by the definitions with which we began, all the consequences that logic is able to deduce from them. Or else, on the contrary, we may need to leave our attention fixed upon the object itself, in order to find, either through experience or through some consideration or construction that the nature of the object suggests to us, a property of this object which was not implicitly contained within the terms of the definition, and which could not have been deduced from it by the force of logic alone. The judgments by means of which we affirm the existence of such properties in an object are those that Kant called synthetic and that truly express our knowledge of the object. The synthesis is empirical, if it is necessary for us to have recourse to experience in order to obtain this increase of knowledge; in the contrary case, the synthesis is *a priori*, and this latter synthesis is that of which pure mathematics makes use.

For example, I propose to prove that two triangles are equal when they have a side and two adjacent angles equal respectively. To accomplish this, I imaginatively place one of the triangles on the other in such a way that the equal sides and the equal angles coincide. When this has been done, it is enough to go back to the notion or to the definition of a triangle in order to recognize that this entails the relation of all the similar parts. The synthesis or ideal construction is necessary in order to establish the proposition set forth on the basis of the definition of the triangle; it forms the essence and the probative force of demonstration.

263. We should call synthetic method that which consists in successively drawing, from the special nature of the object, the constructions which are suited to make known the truths we seek to establish. Similarly, by contrast, it is convenient to call analytic method that which consists in defining an object once and for all and then drawing from the progressive development of this defi

nition all the properties of the object. In this, it is irrelevant whether this progressive development requires only that we commit ourselves to the rules of universal logic, or whether it requires that we borrow the rules of development of a science whose object is more special than universal logic, but less special than the science we wish to construct with its help. We have such a case when we use rules, in dealing with geometrical figures, which apply not only to geometrical quantities, but to any quantities whatsoever.

This is precisely the sense in which modern mathematicians have been led to make use of the terms analysis and synthesis. For them analysis includes, in addition to algebra properly so called, all branches of the calculation of quantities in which we work by means of general symbols in which every trace of the concrete, special, or particular in the nature of these quantities has been made to disappear. Once the rules of calculation have been established for a small number of fundamental properties of abstract quantities, calculation becomes a language, a logical instrument which functions by itself, so to speak, and in such a way that attention need not be fixed on anything but the observance of the rules of calculation.

The advantage of the analytic method as thus defined lies principally in the generality and the regularity of its procedures. On the other hand, the synthetic process, which never makes us lose sight of the special object of our research, enables us to grasp the characteristic most immediately applicable to the manifestation of the property we have had in view, and often has the advantage of simplicity and brevity as compared with analytical procedures. Moreover, it ordinarily happens that this characteristic, which simplifies and abridges the demonstration, is what contains the immediate reason for the fact to be demonstrated. Thus, the superiority of the synthetic method consists rather less in the convenience of the course the demonstration takes, than it does in the manifestation of the true relations according to which the abstract truths which are the object of our speculations are rationally connected. Finally, the analytic method is suited to the nature of our mind in such a way that the regularity of the procedure and the symmetry of the operations gives rise to a continual application. But the synthetic process is amenable to the variety of relations that nature has established among things, without being limited to systematizing them according to laws of combinations which better suit our minds and which are the basis of our logic.

264. After having attempted to emphasize the importance of this distinction perceived by Kant, we must be permitted to criticize the use he made of it in order to set mathematics, which, according to him, is always based upon an *a priori* synthesis, over against metaphysical speculations, which he says should consist only of analytic judgments. From the beginning, mathematics has had no less need for analysis than it has had for synthesis, even in the meaning he gives these terms. The distinctive character of the *corollary* is that it is implicitly given in the proposition or propositions from which it follows and can be deduced from it analytically, without a new synthesis. Yet the task of making evident certain corollaries of a proposition is no less important because of that. The results of a calculation are implicitly contained in the information given in the calculation. Yet a rare sagacity is often required to deduce them from it, and a talent of high order is needed, as is the case with every other language, to write the language of calculation with the elegance and the simplicity which it allows. The purpose of the organization of methods in mathematics, as well as in other sciences, is to reduce the work of synthetic judgments and to dispense with talent for discovery (258) which tries to find new constructions and to work out unforeseen relations, in order to bring to light an unknown truth or to demonstrate a known truth more easily. We find in mathematics the most beautiful examples of such methods, which are powerful auxiliaries to our limited intelligence.

On the other hand, if it is true that philosophy is unlike the sciences primarily because the work of the scientist tends toward the growth of our knowledge and toward new discoveries in the inexhaustible domain of nature, while the work of the philosopher tends to clarify the principles of our knowledge, and to carry out the analysis of the laws of the human mind—to which the ignorant submit quite as much as the philosopher—then it is also true that induction, analogy, the feeling for order, and all the elements of what we have called philosophical probability, supply judgment, so far as philosophical speculation is concerned, with the bases that give it the ideal construction or the synthesis *a priori* in pure mathematics and experience or empirical synthesis with respect to speculations about the nature of the laws of the sensible world. Kant did not take account of the latter because, in his skeptical system, he regards as null and void all truths which do not permit formal demonstration, because of which we are able to reduce sophistical contradiction to absurdity. All the reasons we have put forward against this system militate against the consequences that the author has drawn from it, so far as concerns its impact upon analytic and synthetic judgments and the applications which they make possible.

265. We have already had occasion to take exception to Leibniz's idea that the contrast between mathematics and metaphysics follows essentially from the fact that the one is founded upon what he called the principle of identity and the other on the principle of sufficient reason (28). We have shown that the notion of the reason of things, and that regulative principle of the human mind, namely, that nothing exists or appears unless there is a reason for its being or appearing, find their application in mathematics quite as much as in other sciences. This is the place to point out that it is by no means true that mathematical reasoning consists essentially of a series of identities, although Leibniz has said that this is the case—and Condillac has charged himself with the development of the doctrine in regard to this point. This proposition, which is the counterpart of Kant's, itself fails for being too absolute. No doubt mathematics includes series of propositions with respect to which the whole art of reasoning consists in showing the identity of different expressions of the same quantity; in showing that each proposition, in spite of differences of statement, is nothing but a transformation of that which precedes it. But every time the operation of synthetic judgment intervenes, the mind grasps new properties which belong to the object by the direct contemplation of it. A science founded upon a series of operations of this kind can no more be said to be based on the principle of identity than this can be said of physics, or of any other science constructed through the aid of observation and experiments, and successively enlarged, either by the progress that has been made in the art of conducting experiments, or by the discovery of new instruments which extend the power of the senses.

266. To clarify this matter still further through the use of other examples, suppose the point in question is to establish the proposition known as "Archimedes' principle," namely, that a solid body immersed in a liquid loses a part of its weight which is just equal to the weight of the volume of liquid which it displaces. We can gain knowledge of this truth or demonstrate it to others by means of a direct and easily imagined experiment which is described in all treatises on physics. This is a very simple example of empirical synthesis. But we can also give a purely rational demonstration of the same truth. For that purpose we imagine that the fluid mass whose place the body occupies after immersion becomes solidified or congealed with no change in density. That is to say, the molecules of which the mass is composed, while maintaining their positions, are thought of as being connected among themselves in an invariable way, like the molecules of a solid body. The fluid mass, which we have assumed to be solidified in this way, would be in equilibrium in the midst of the surrounding fluid, for there is no reason why the connection among the molecules, which would undergo no change except that of losing some of their mobility or having their movements more restrained, should upset an already established equilibrium. Therefore the pressure which is exerted by the surrounding fluid against the solidified portion, and which tends to raise it, is exactly equal to the weight of the solidified portion which tends to make it sink. But this pressure depends only on the dimensions and the form of the solid body against which it is exerted and has nothing to do with the internal structure, the density, or the other properties of the matter of which it is composed. Therefore, if we come back to the real solid immersed in the fluid, the pressure which the surrounding fluid exerts on this body is exactly equal in intensity to the weight of a mass of liquid having the same form and the same volume, or simply having the same volume, since form has nothing to do with weight. Therefore, the body loses part of its weight as a result of being immersed in the fluid and as a result of the pressure which the surrounding medium exerts upon its surface, and this part of its weight is just equal to the weight of an identical volume of displaced liquid. This is the principle of Archimedes which we set out to demonstrate. One need not be versed either in mechanics or in physics to follow this

series of propositions, and this is what allows us to enter into the details of this example, which is as clear as anyone could wish.

[267] In fact, in this case, the ideal construction or synthesis obviously consists in imagining that a portion of the fluid, having the same volume and even the same form as the immersed body, is solidified, although its molecular structure undergoes no other change than that which a change of density may induce. Moreover, we see an application of the principle of sufficient reason intervening in the judgment by means of which we state, on the one hand, that there is no reason why the solidification should disturb the already existing equilibrium, and on the other hand, that the effect of the pressure of the surrounding fluid depends only upon the dimensions and the form of the immersed solid, and has nothing at all to do with the internal constitution and the molecular structure of this solid body. But the intervention of the principle of sufficient reason is connected in a necessary way only to this method of demonstration and synthetic construction, and not as the very basis of the proposition which we are trying to establish, and which can be reached by entirely different arguments.

268. Mathematics is peculiar in that every rational or, to use Kant's term, *a priori* synthesis may be checked by means of an empirical synthesis (28). In this way it is distinguished, on the one hand, from the physical and natural sciences which are based primarily upon experience and upon induction which generalizes the results of experience; on the other hand, from the sciences which rest upon the ideas and relations that reason conceives, but which do not come within the ken of the senses. After a lawyer has analyzed a contested question with the greatest care, after he has stated the principles of its solution in a form whose clarity is most satisfying to reason, he is not able, if it is necessary to do so, to furnish experiential proof of the validity of his reasonings and the exactness of his deductions, as a mathematician can do.

There is another and also a very important difference between the doctrine of the lawyer and that of the mathematician. This arises from the fact that the latter deals with very simple ideas and relations, while the former is concerned with very complex ones. Therefore, while a mathematician may connect to a primary synthesis, or to a fundamental property which is given by the immediate contemplation of the object, a great number of consequences, which can be deduced from it by logical combinations or by using the devices of calculation, the lawyer is frequently obliged to interrupt the series of logical deductions in order to derive new theorems or rules, which come to be combined with principles already admitted, from the direct contemplation of the object.[3] Mathematics, because of the simplicity of its subject matter, and jurisprudence, because of the complexity of the matter with which it deals, are so many types by means of which we are able to take account of the organization of the sciences and of the nature of the processes the mind follows in raising them to the position of a body of doctrine. Moreover, it may be true that in the study of the intellectual faculties of man, as in that of all the functions of life, it is not necessary to start from the rudimentary condition in which everything is confused but, quite the contrary, from the condition of perfection in which the development of all the parts distinguishes them from one another (97). This being the case, might it not be better to throw light on the theory of judgment than to study the organic development that the faculty of judging must have undergone in becoming appropriate, not to the government of the individual, but to that of the collective body of the citizens, in becoming one of the molding or regulative forces of the body politic, and in creating institutions in which the individual activity of man is exercised, but always in such a way that the stamp of individual personality is wiped out so far as possible in the system of doctrinal decisions or in the series of particular judgments? Etymology should be sufficient to indicate that the judgment of the public official, charged with making decisions about laws and the fate of citizens, is what more than anything else deserves to be called judgment. Etymology should also indicate that, in some degree, it is the formal standard, borrowed by philosophers from the life of the community, in order to

---

[3] It is with this important restriction that the following quotation from Leibniz must be understood. " Jurisprudence is like geometry both in other respects and in that both have elements, both have occasions. The elements are simple—in geometry, figures, triangles, circles, and so forth; in jurisprudence, performance, promise, transfer, and so forth. The occasions are developments of these, which in both cases are infinitely variable." Dutens edition, Vol. III, 362.

fix more precisely the idea of what goes on within man, and in the order of wholly personal phenomena.[4] Therefore, following certain reflections on the scientific organization of law, we will introduce others on the nature of decisions and on the organization of judicial bodies. If one wishes to regard it as such, this will be a digression; but a digression for which we shall ask to be pardoned, since it will give us a perfectly natural opportunity to clarify, by applying them, most of the notions of general logic we have sought to bring into prominence. Should any lawyers glance at these pages, they will also pardon us for any ideas that may appear strange to them, either because of matter or of form, by remembering that we have been unable to do more than outline the subject, and to treat it in a way that is not too much different from our study as a whole.

[4] This is just what Plato declares when, in many passages in the *Republic*, he says that his purpose, in studying the organization of the city, is to arrive more easily and more surely at the knowledge of the organization of the human mind.

CHAPTER XVIII

APPLICATION TO THE MODE OF THE ORGANIZATION OF LAW AND JURISPRUDENCE

269. Nothing varies more among different peoples than the sources of their national laws, or than the historical origin of the fundamental maxims on which, from a certain period in the civilization of these peoples, the reasoning of the lawyers is carried on in such a way that it gives rise to a body of doctrine which assumes the characteristics and proportions of a science, and to which we give the name jurisprudence. The elements of law do not come only from natural equity and from the physical and moral needs that all men experience; they are also derived from the habits and the beliefs of every people, from the instincts and the fortunes of every race; they may be connected with an interest in public order or the financial system, with the need to strengthen a political institution or to wipe out the vestiges of a proscribed institution. Most of the principles or the materials of law may and actually do change with the society. In spite of the influence of the spirit of imitation and the authority of the traditions of learned men, our legislation has hardly anything in common with that of the Rome of the patricians and of the emperors. Yet the scientific form of law has varied no more than have the laws of the human mind, and the jurisprudence of Rome has remained the model for ours because of logical form.

To know the laws of a country completely, including all its legislation, written or customary, is to possess a body of knowledge that is very useful in the affairs of life, and that can never be acquired in a country such as ours, in which each decade gives birth to an enormous mass of legislative and regulatory provisions. Nevertheless, no one would dream of putting a study whose objects are so variable and whose applications are so particular among the sciences properly so called. There are positions in the academies for doctors, scientific agronomists, and economists; there are none for the man of public affairs and the barrister.

Anyone who does not restrict himself to learning the laws of his country and of his time for some practical purpose, but who studies ancient and contemporary legislation in order to compare them with respect to their common subject matter, and in order to recognize the general forms and to lay hold of the analogies between the different forms, sets up a veritable science in doing so. Finally, the study of law is conceived in its highest terms and in the manner which is of the greatest interest to men by anyone who seeks the reason for bodies of laws, for the causes of their decadence and their fall, and reciprocally, for the manner in which legislatures influence the genius and the history of people, in the history of man, in his moral and physical conditions, and in his relations to the external world. Such a person has raised himself to the philosophy of law (22). But from whatever point of view we may wish to consider the science of law, we must not confuse it with the techniques or logical procedures which serve to develop systematically the application of legal provisions, and which apply in the same way to external laws born of revolutionary turmoil or of the whims of a despot, as well as to the laws which give rules for the greatest interests of society and to the regulations invented in a scrupulously careful spirit of financial red-tape or monopoly. Therefore, in jurisprudence as elsewhere, the problem is to distinguish matter and form, the material which is organized, and the organizing force.

270. In order to make this distinction easier to understand, we shall turn to some examples and shall mention the rules which, in our French law, govern the status of property in the association of man and wife. From what source are these regulations derived? They spring from a clearly simple usage like the mores of the peoples who have adopted them.[1] All profits are held in common during marriage. At the death of one of the spouses, his heirs

---

[1] " The men having reckoned the amount of goods received as dowry from their wives add to it an equal sum from their own wealth. Account is kept of the total sum and of the profits which accrue. Whichever of the two survives receives the portion of both with the profits accumulated during the time." Caesar, *Gallic War*, Book VI, Chap. 19.

We must not commit the gross error of confusing the type of joint possession indicated in this famous book with that which is established by our customs. Yet it makes little difference, for the object we have in view, whether a higher or lower origin of these customs is shown. See Pardessus, *Loi salique*, p. 675.

take the real property which belongs to him in his own right. The surviving spouse, in turn, does the same with his personal goods. Then the goods acquired as a result of the labor and the savings of the couple during the marriage are divided into halves, each party taking half. Finally, the personal goods which furnish the home or the common habitation and which have been worn out and replaced from the joint revenue, become the common property of the couple. Whatever the origin of these goods, they too are divided in half.

Among other peoples, whose mores impose on the woman a greater degree of dependence and restrain her outside life more, things are not done in this way. Sometimes the husband buys his wife. More often the wife gives a dowry which must be returned in case of divorce or of widowhood. Sometimes the wife is received as a slave, or sometimes as a guest, under the marital roof. In such a situation she has no interest in the labor and profit of her husband. Here we see the beginnings of dotal law.

When, as a result of the progress of civilization, of industry, and of luxury, the mores have lost their original simplicity, matters affecting interests become complicated. Custom can no longer foresee every circumstance; yet it will still provide principles of solution, but it will be necessary to combine them systematically in order to adapt them to the indefinite number of combinations or *cases* which will require a solution.

271. Thus it is clear that nothing is more natural in the mores of a poor people than the difference established by the French common law between real property, such as fields and houses, that belongs to each spouse, and the furnishings that outfit the home. The former remains separate; the latter intermingle in the very nature of the case. But suppose a people becomes industrial and commercial. Then they will create great assets, such as working capital, credits, public funds, and shares of vast enterprises. Will the mores or customs, which express them, teach us what law must be enacted in regard to these newly created goods? Undoubtedly not; and if we wish to go back to the natural origin of the custom and enact laws in keeping with its spirit, circumstances alone would not determine arbitrators endowed with a sense of the fitness of things to decide sometimes in favor of separating,

and sometimes in favor of lumping together goods of this sort, to include them in the joint possessions or to exclude them from them. However, in most cases, those making the decisions would be unable to give an exact account of the motives behind them.

Moreover, and without having to suppose a very advanced development of civilization, many of the different combinations will become more complicated and make it more difficult to apply the customary principle. The couple will inherit fortunes encumbered with debts or easements; they will have real estate to reclaim or to seize, debtors to prosecute in person. This is when the law is systematically organized and becomes the object of an abstract science called jurisprudence so that it will be possible to fix exactly the respective rights of each spouse. Reasoning by analogy, legal experts conceive of fictitious real and personal property. Then, once in possession of certain abstract ideas or forms, they follow out the logical consequences of these ideas indefinitely and in this way derive from the simple and the abstract the solution of the most complex and the most particular cases.

272. Throughout Roman law we find examples of the simple, crude or even barbaric customs which have given way to the development of a scientific and systematic jurisprudence, so that already nothing remains in the mores of that which the original custom determined. We shall limit ourselves to indicating the distinction between things *mancipi* and *nec mancipi*.[2] It was natural that among a poor, agricultural, and illiterate people, custom would require certain ceremonies or symbolic demonstra-

---

[2] [" The Res Mancipi of old Roman law were, land, . . . slaves and beasts of burden . . . . It is impossible to doubt that the objects which make up the class are the instruments of agricultural labour, the commodities of first consequence to a primitive people. Such commodities were at first, I imagine, called emphatically Things or Property, and the mode of conveyance by which they were transferred was called a Mancipium or Mancipation; but it was not probably till much later that they received the distinctive appellation of *res mancipi*, ' Things which require a Mancipation.' By their side there may have existed or grown up a class of objects, for which it was not worth while to insist upon the full ceremony of Mancipation . . . . Such commodities were the *res nec mancipi* . . . , ' things which did not require a Mancipation,' little prized at first, and not often passed from one group of proprietors to another." Henry Sumner Maine, *Ancient Law* (New York: 1888), p. 269.]

tions such as the assistance of the *libripens* [3] and witnesses when the ownership of the most precious goods, such as plots of ground, slaves, horses, and cattle, was transferred.   Later, when the Romans had developed the use of gold and silver, when bronze and marble were used in the decorations of their dwellings, when writing, having become common, gave them the most convenient and the surest means of recording their transactions, they were no longer in the condition of society which would give rise to this distinction.   But a religious respect for the institutions of their ancestors caused this practice to be preserved as a principle of law, and the abstract consequences which were deduced from an outmoded practice by rigorous logicians gave birth to one of the most ingenious theories of the Roman Law.   Even today this theory taxes the sagacity of interpreters.

273. Finally, to cite a last example not taken from particular customs or special needs, we shall notice that the laws have sanctioned a principle of public and universal morality by excluding from the inheritance of a deceased person any heir who has turned against him by violent acts or by other capital offenses.   On the other hand, it conforms to humanitarian feelings and to good will—fruits of the softening of the mores—by not punishing innocent children for the crimes of their fathers.   How does it happen then that, according to legal experts, the children of the disinherited will receive the inheritance of their grandfather if their father is an only son, or if their uncles have been renounced, and that in the contrary case they will be excluded from this heritage?   According to public morality, humanitarian feelings, and natural equity, is not the position of these children the same whether there are or are not other heirs having the same status as their culpable father?   Yes, certainly; but jurisprudence intervenes with its abstractions and legal fictions.   In the case in which the persons disqualified do not have joint heirs of the same rank, the children become the principal heirs, as being the closest relatives after the exclusion of those disinherited.   In the contrary case, according to the abstract rules of law, they should be allowed to share jointly with the closer relatives only in so far as they may represent the

---

[3] [The *libripens* was a neutral person or balance holder who was present at a transfer or conveyance of real property.]

disinherited person and may succeed to his rights; but, on the one hand, no one succeeds a person who is still living, and, on the other hand, the disqualified person is unable to transmit to them rights which he has lost through his own fault.

274. In order to understand better what jurisprudence is theoretically, it is necessary first of all to have its practical purpose in mind, the purpose without which it would be merely an ingenious game, or rather, would not have undergone the scientific development that we know.

Jurisprudence is organized for the practical purpose of determining the solution of a judicial question, and of imposing on the judge a rule of decision. The correctness of this maxim stated by Bacon that, notwithstanding the imperfection of the rule and its frequent inconveniences, it is still better than the arbitrariness of the judge, was perceived at an early period.[4] Now, if there are certain decisions which no one would hesitate to declare iniquitous and certain statutes which everyone would regard as pernicious or immoral, it happens still more frequently that we find no clear line of demarcation drawn between honesty and dishonesty, between equity and inequity, between utility and injury (196ff.). Innumerable nuances fill the interval between the extreme terms. These are the nuances the conscientious evaluation of which should be submitted to the enlightenment and integrity of an arbitrator, whether his function is a public one or not. But such an evaluation would neither be capable of a rigorous analysis, nor, consequently, of control. It would have all the inconvenience of an arbitrary decision. It has, therefore, been necessary (202) to outline artificial divisions, to establish an ideal discontinuity in this system in which nature proceeds through continuous transitions. For this purpose, certain fixed principles have been set up

---

[4] Plato, *The Laws*, Book IX; Aristotle, *Rhetoric*, I. Nevertheless, the common opinion is that civil law was born among the Romans much less from the desire to strengthen public liberties by protecting the citizen from the arbitrary power of the magistrates, than from the policy of the patricians who wished to maintain the superiority of their caste by giving to civil law—as to pontifical or religious law—a learned or systematic development to which they alone held the key. But whatever the political end served by the scientific development of jurisprudence by means of juridical abstractions and logical constructions, the consequences to which we wish to call attention remain the same.

and certain legal fictions have been invented, by means of which
it is possible to reach a determinate decision in each case through
the force of reason and analogy alone. Jurisprudence is organized
in this way. Nevertheless, whenever the logical consequences
plainly do violence to equity, to the mores, or to public opinion, as
may be the case in the last instance cited above, it has been felt to
be necessary to reduce what is called the severity of the law, by
returning once more to its natural sources, that is to say, to feelings of equity, honesty, and the public good. This is done for the
sake of drawing from them new juridical principles which must be
taken into account in subsequent analyses and constructions, as
well as of the anterior principles from which they detract. This
accumulating of principles, some of which are more general and
others more circumscribed in their application, creates all the per-
plexities of jurisprudence and places it in a theoretically inferior
condition *vis-à-vis* the other abstract sciences, which derive only a
very small number of primordial data from the common notions,
and which are regarded as being more perfect in proportion as this
number is still further reduced (268).

275. In practice the terms " law " and " jurisprudence " are included among those whose meaning may be extended or restricted
as the case requires. But from the point of view of good logic and
for the correct understanding of the scientific organization of law,
all the juridical principles, no matter what their source, that science admits as so many primitive and incontestable data must be
regarded as law, or at least assimilated to law properly so called,
that is to say, to the will of the sovereign expressed in a written
formula. This clearly includes the maxims of equity, morality,
and public benefit, which are sanctioned through a unanimous
consent, as well as the traditional customs which the Roman jurisconsults said resemble the law [*legem imitantur*]. Actually, however, positive laws may have been created on the whole in imitation of traditional customs, which have preceded them everywhere.

The term " jurisprudence " is used especially to designate the
body of doctrine established by the decisions of courts and by the
opinions of renowned lawyers. However, if the courts and the
lawyers have proceeded not by the method of logical analysis and
deduction from principles already laid down in the law, but by the
method of estimation or evaluation in a case with respect to which

the law declares nothing more implicitly or explicitly; if the authority which attaches to their decisions and to their opinions is based not on the presumption that they have been well thought out, but on the presumption that they have properly weighed these decisions and opinions after they have acquired the authority necessary to set up legal principles, as we say, then they really share in the nature of the written or customary bodies of law that they supplement or which they complete.[5]

On the contrary, if the legislator does not limit himself to the proclamation of principles, but charges himself with combining them and deducing their logical consequences, he really plays the role of the legal expert. In this part of his work a real science of law clothed in legislative forms must be recognized.

276. Jurisprudence attained the high degree of scientific perfection that it did among the Romans only because of the small number of their laws and because of the religious respect they so long maintained for the code of the Twelve Tables, the simplicity of which was well suited to the rude and strongly pronounced traits of the first Roman civilization. Until the fall of the Republic, one finds only a small number of laws or plebiscites for the modification of the civil legislation supervening, and the very form of these acts, which necessarily produced their extremely concise wording, obliged one to limit oneself to writing some principles implicit in them. Thus the jurisconsult could concern himself entirely with the rigor of logic, use reasoning and analogy to follow out the ultimate consequences of an abstract principle, and finally erect that systematic and learned edifice that Leibniz admired so greatly that he placed it just below geometry.[6]

The legislation of the Twelve Tables fell into disagreement with the mores only too soon. When this happened, it was, if not abrogated by the Pretorian Law, at least nearly annihilated, and jurisprudence should have become more complex as a result of the loss of its regularity. However, since in the long run the Pretorian Law was formed through successive developments and

[5] "A law is a general rule decreed by wise men." Papinian, *Digest of Laws*, Book I, Chap. II, Sect. 1.

[6] " I have often said that, after the work of the geometricians, there is nothing comparable to the writings of the Roman jurisconsults for force and simplicity—so much strength and so much profundity is in them! "

amendments of a text which was sacramental, we understand why it placed fewer obstacles in the way of the scientific organization of jurisprudence than would have been set up by a voluminous collection of written laws, each of which was drawn up out of whole cloth, and the text of which would have had a sacramental value.[7] When, in turn, the Pretorian Law finally ceased to be a vital thing, and when the Emperors, with that facility which is the attribute of absolute power, had thrown away legislative provisions, Roman legislation was still quite capable of receiving important improvements, several of which have been worth carrying over into the law of modern peoples. But jurisprudence was tending toward its decline, and finally, the science of the jurisconsult gave way to the erudition of the commentator and compiler.

277. The moment the legislator attempts to stabilize jurisprudence by means of a systematic code, either he intentionally derogates the logical consequences of the principles laid down by it, and as a consequence of doing this lays down new principles which give rise to all the more numerous and more complex combinations which must lead to a judicial decision; or his intention is to perpetuate the logical consequences of the fundamental principles. But he is liable to make errors; and even though he makes no mistake, he may be misunderstood because of the way in which he expresses his thought. In either case he hinders scientific development; he alters the regularity of relations; and, what is more serious, he increases the number of contested points by the very provisions which he thinks are appropriate to reduce the number of them.

In fact, when the number of written laws is small, and they perpetuate only principles or affect only the simplest cases, it is necessary, if they are to be extended and made the basis of a system, to take their spirit and not their letter, to convert into a

---

[7] " It has rightly been said that no law seems to have been promulgated that has been from the beginning, sufficient for all things, but that it stands in need of much correction in order that it may suffice for the variety of nature and her contrivances." *Novellae*, Chap. III.

In the dialogue called *The Politics*, Plato says, " It is not the law which is more exactly known and prescribed that is more just, better, and more useful for everyone. Men and their actions resemble one another so little, and there is so much instability in human affairs, that it is impossible for any art to be provided with simple, universal, and perpetual rules."

general and abstract idea the particular and concrete idea which is directly connected with the terms used by the legislator. Once this preliminary work is carried out—and it is ordinarily favored by concise or proverbial forms or by archaism of style—the task of the legal expert is no longer simply that of combining ideas. On the contrary, the more a code is extended, the more it encroaches upon doctrinal discussion, and the more jurisprudence tends to descend from the combination of ideas to the comparison of words. As for the task of the judge, if codification makes his work easier as far as the anticipated and apparently the most frequent cases are concerned, it complicates it in relation to unforeseen cases, increasing the number of texts to be related and the number of elements to be combined.

278. The predilection of moderns for the written word, a predilection which makes them regard as null and void all that is not in written form, has already been pointed out many times. In the present state of civilization, we attribute to the written symbol that sacramental value which the ancients attached to ceremonial words, to symbolic gestures, to ceremonies, and to religious rites. Thus, if we were to believe some modern philosophers and legal experts, the perfection of jurisprudence would require that we recognize no other rules of law than the written texts and official formulas. This would be the means of avoiding everything arbitrary in the administration of justice and of warding off the abuse that has long been made of such vague expressions as " equity " and " natural right." Jurisprudence would finally take on the characteristics of positive science, which is so esteemed in our times. But, for the same reason, would it not be necessary that the law should officially establish the rules of good sense or of logic according to which the judges would have to interpret and apply written laws? Can we believe that article 1315 of the Civil Code [8] has added to the force of this natural law: *Onus probandi incumbit actori; reus excipiendo fit actor?* On the contrary, it has hidden it by applying it only in a special case. The legal expert and the judge must free the abstract principle from the wrapping the legal text has given it. If it never occurs to anyone to

---

[8] " Whoever demands the execution of an obligation must prove his case.
Reciprocally, whoever demands to be freed from one, must prove the payment
or the fact which has brought about the extinction of his obligation."

contest the generality of this maxim, if it is certain that among us a judgment would be quashed for having refused to apply the principle in any other matter than that of obligation and payment, has not the article become superfluous, and does it not offer the disadvantage of promulgating as a rule of positive law something which is only a particular application of a general maxim, the force of which exists independently of the written law, and that we have reasonably judged unnecessary to insert into it?

It is not so much legal texts which limit the arbitrary power of the judge, as it is the proper organization of judicial powers, their hierarchical relations, and the manner in which they are controlled. Later on we shall have occasion to examine the way in which our judicial organization tends to fulfill this end, and how among us the nature of things has pretty much made the Court of Cassation [9] the judge of the logical application of the principles of justice, under whatever form they are expressed in law, and even when they are not expressed in it. Yet, in the minds of the founders of this institution, the Court of Cassation was to have been only the guardian of a sacramental text.

279. There is no more proverbial adage, not only among courts but also in the world at large, than this: " The letter kills and the spirit makes alive." For all that, I do not know how the true principle of decision may be accounted for with sufficient clarity in arguing, according to the need of the case, sometimes from the text and sometimes from the spirit of the law, trying to make the scales of Justice tilt alternately to one side or the other. Are we to understand that under the legal phrase, under the combination of words by means of which the legislator has stated his thought more or less happily—for he does not possess the qualities of the philosopher and the writer to the highest degree—it would be necessary to know how to recognize in all its generality and abstraction the principle which he has sought to establish, and the ideas he has attempted to relate and the nature of this relation? If that were the case nothing could be more true and more important than the adage we have just cited. But do we want to say that it is necessary to go back to the spirit of the law in this sense: that if the legislator has been guided by motives of equity,

---

[9] [The Supreme Court of Appeal in France.]

public benefit, or anything else, and that in the particular case these motives are found not to apply, or to apply in a contrary sense, he would fail to take account only of the motives and not at all of the terms of the precept? Such a doctrine would be subversive of the law and would tend precisely to replace a judgment capable of surviving in the crucible of dialectic by an uncontrolled evaluation. As a result, in the interest of a good decision in a particular case, we should not attempt to take account of motives of public order which have made the judgment of strict law prevail over the judgment of equity.[10]

[10] Thus, to take one of the simplest examples, although the French legislature has said, in Article 1326 of the Civil Code, " That the bill or promise under privy seal by which *a single person* obligates himself to *another* for the payment of a sum of money or something else of value must be written entirely in the hand *of the person* who subscribes to it, or that he must at least, besides his signature, write with his own hand *an acknowledgment of indebtedness* or *an approval,* entering in every letter the sum or the quantity of the thing contracted for," the courts have held that the terms of the disposition should allow the rule to be extended to the case in which *many persons* contract a similar engagement with one or with *many others.* For evidently, the rule which the legislators wished to establish was this, " Every obligation to pay a sum of money or to transfer something valuable will be proved juridically by *unilateral* acts under privy seal, only insofar as the writing be entirely in the hand of the person under obligation, or, and so on." Because this principle was not drawn up with sufficient exactness by the legislators, who had suited their language to the most common particular case, it is no less necessary to *understand* the article *sanely* in the manner handed down by the Court of Cassation (decision of May 5, 1816) to restore to the principle its generality and to set forth the idea contained in it without being bound to the letter. This is a case in which the adage exists in full force.

But if we go back to the intention that dictated this article, it is clear that the legislature proposed to guard against deceits and the abuse of signatures, to oppose a barrier to fraud and bad faith. Therefore, when, as is often the case, circumstances remove all suspicion of fraud and demonstrate that the signator knew fully the extent of the obligation, would not annulling the judicial proof have a tendency to favor fraud, and to violate the spirit of the law by a blind respect to the text? Nevertheless, this is the case in which the adage cannot be applied without express violation, not so much of the terms of the law, as of that which gives the essence and the force to the juridical principle laid down in it. Otherwise, it would become lawful for the judge to evaluate the circumstances which should result in the proof that the signator fully understood the exact extent of the obligation. The conclusions of the evaluations would depend upon variable impressions and would thus fall outside all anticipation and con-

280. By means of these principles it is possible to throw light on the theory of interpretative laws and on the practical application of the famous maxim: *Ejus est interpretari legem, cujus est condere* [It is for him to interpret the law whose function it is to make it].

Interpretative law is either only the logical conclusion or the analogical extension of principles laid down in earlier laws, or it lays down new principles which may become the root for subsequent deductions, or, finally, it attributes a meaning to words that present none or present an ambiguous one.

In the first case, interpretative law is useless, and it must be blamed for the defects of judicial organization, if the rigorousness of deductions and the fitness of analogies do not prevail in the end, after some transitory aberrations of slight importance from the point of view of general interest. In all countries having good judicial institutions, after the formality of the proofs which ordinarily precede interpretative law, it would be unreasonable to expect to find in the man or in the groups of men in whom the legislative power is vested, no matter what it is called, greater guarantees against logical error than is offered by judicial proofs.

But there is more to this matter. The legislative power makes the law; it lays down principles as a result of an evaluation which is supreme and beyond control; it habitually rules on the basis of what seems good to it, and not on the good or bad, but logically necessary, consequences of what has appeared good to other groups at other times. Therefore, for these very reasons, the legislative power is less fit to interpret a pre-existing law logically than would be a judicial body in the functions and practices of which, on the contrary, such an interpretation reappears essentially.[11] This ineptitude will be all the more apparent when the legislative power

---

trol, contrary to the wish of the legislator who thought it was his duty to limit the discretionary power of the judge in this respect through the institution of a juridical rule. See what will be said in the chapter which follows on the distinction between questions of fact and questions of law.

[11] After many variations, these principles were virtually sanctioned in France, by the law of April 1, 1837, which by abrogating that of July 30, 1828, gave back to the Court of Cassation (the sections meeting conjointly) the supreme decision on points of law, in case of conflict with the appellate courts.

is exercised by a larger body, under the forms which lend themselves better to instinctive estimation and less well to decisions based on abstract ideas and logical construction. On the contrary, and whether the point in question concerns general laws or special laws, a particular kind of functionary or the citizens as a whole, where the legislative power is concentrated in the hands of a less numerous council—in which case its form is more like that of judicial bodies—we are able to obtain a truly interpretative decision. And we shall have to consider such an action as having some of the characteristics of opinions and of law: of opinions in so far as it is only the logical consequence of earlier laws and as it governs past events; of law, in so far as it makes rules of a general sort, and for the future.

281. There remains to be examined the case in which the interpretative decision, which is appealed because of the inadequacy or the obscurity of existing laws, should contain new principles of juridical solution or give a determinate meaning to a text whose meaning is confused or ambiguous. This is just the case in which the maxim we mentioned should be applied, for to interpret the laws in this way is really to make a new law which can no more be enacted except for the future, without being sullied by the defect of retroactivity, than can any other law.

However, this is not the opinion of most lawyers. But an estimable writer has refuted it so well that we will limit ourselves to quoting his words.

It is difficult enough (he says) to give a reason why we have seen in an interpretative law something other than a new provision, subject to all the forms and conditions of a previously unknown law. . . . Each person is the best interpreter of his own expressions, in the sense that he knows what the ideas are which he wishes to express. But this is the case only so long as his ideas appertain to him. No one who has entered into a contract would be allowed to pretend that he has attached to his statement a meaning contrary to what it had given to the party who serves as his security. Still less could legislators claim the right to declare the meaning of the law, should this meaning be contrary to the terms in which it is stated. After its promulgation, the law becomes public property. . . . The rights that individuals have acquired as a result of this law, the obligations it has imposed on them, are existing facts and cannot be made to depend upon a declaration which, in explaining the meaning of the terms employed, distorts, alters, or modifies the state-

ment. . . . It should not affect those who have acted, agreed, or contracted as a result of faith in the law, in its original state. To put the
matter in a word, it can only affect the future, just like a completely new
law.[12]

" If judgments," says Montesquieu, " became a matter of the particular opinion of the judge, we would live in society without
knowing exactly what obligations we had contracted in it." [13]   But
what would be worse than ignorance of this sort would be to be
submitted, too late, to an imperative provision concerning things
in regard to which we had been led, at the moment of performing
an act or entering into a contract, to count on the benefits of a
conscientious evaluation and on all the latitude of an application
suited to the circumstances of the particular case.

282. In the interpretation of laws, it is not always easy to distinguish the part reserved for the legislative power and that which
reverts back to the authority of the judiciary. It is still less easy,
for reasons embedded in the nature of things, to define clearly the
prerogatives of the power which makes the laws, and those of the
power which has as its essential task to provide for the carrying
out of the laws and is, for that reason, invested with the right to
prescribe regulative stipulations which are equally obligatory for
all citizens and which have the greatest sort of affinity to laws
properly so called. Indeed these stipulations often have even more
affinity than there is among the resolutions to which official form
allocates the common denomination of law, because these resolutions emanate entirely from a like power which is officially called
legislative power. To vote a loan, a levy of men, a declaration of
war, a treaty of peace, a budget, an approval of accounts, a tax, or
a political, administrative, penal, or civil law, is to fulfill functions
among which there is a greater difference than there is between the
functions of legislative power taken in their entirety and those of
other public powers. When, in regard to all these matters, the
legislature must consult the nation as a whole or those assemblies
which are constitutionally provided to represent it, we willingly
acquiesce. But must the manner of consulting them and of getting their response be the same? Is it convenient to apply the
same forms of procedure to acts so essentially different?

[12] Meyer, *Esprit des institutions judicaires*, Bk. VIII, Chap. 4.
[13] *Esprit des lois*, Bk. IX, Chap. 6.

Today everyone clearly understands that the judicial decisions gathered together under the common name of judgments and decrees do not have the same nature. In civil matters, the institution of recourse to appeal, in criminal matters, the introduction of judgments by juries make us aware of a distinction which is none other than the distinction between conscientious appraisal and dialectical judgment, as we shall try to show in the following chapter. It is understood that judicial functions of different sorts are no less distinct for being entrusted sometimes or even most ordinarily to the same person, as is the case with us in our inferior and appellate courts in both civil matters and misdemeanors. Similarly, our legislative assemblies reconcile functions analogous to those of judge and jury, with this difference, which comes from the nature of the object and not from the nature of the intellectual act, namely, that more often than not they make rules that apply generally and not to particular cases. To organize the legislative procedure in such a way as to separate these functions, to surmount so far as possible the obstacles which in practice tend to maintain confusion, is, we believe, one of those essential improvements with which the public mind will have to concern itself if such questions are ever to be treated without partisanship, except in those periods of political crisis which seem to be diminishing gradually in intensity if not in frequency. At least this will be necessary provided we admit that, in the economy of human societies as well as in the physical world, everything must, in the long run, tend toward a stable condition and a permanent order.

283. Let us see how authority, guided by public good sense, proceeds in these matters outside of politics. Let us suppose the question at issue is to judge the worth of a work of art, a statue, picture, or musical composition. Shall we get the opinion of a celebrated artist or of an academy of artists? No doubt we could do this, and this is a course administrators sometimes take; but it is often found to be bad. The judgment of experts is upset by the public, as we say, and these are decisions of the public which posterity—a judge from whose decisions there can be no appeal—usually confirms. On the other hand, if the question is one of judging the worth of a discovery in mechanics or chemistry, we address ourselves to the Academy of Science, which is a more

competent body than the public: the Academy will refer the memorandum to a committee or to a section which is better able to appraise it than the Academy as a whole would be; the section, in turn, will select to make a report those of its members who have specialized training in that branch of science with which the memorandum seems to be concerned. If we admit, in the case of a scientific problem as in that of a work of art, that personal passions or interests do not obscure the reason or pervert the taste of judges, the guarantee of the excellence of the judgment seems to be increased, in one case by augmenting, and in the other case by reducing the number of judges.

It might be possible to say that special instruction is necessary to understand anything related to questions of mechanics and chemistry, while only eyes or ears are necessary to find a picture beautiful or ugly, or a musical composition melodious or strident. The first part of the assertion is incontestable, and this is an excellent reason why we never dream of making the reputation of a mechanic or a chemist depend upon the masses. But it is false to assert that the degree of training of the poetic or musical sense has no influence on the worth of the appraisal made of a tragedy or an opera. If I must take a chance on the opinion of a man of the street, of a man of the world, or of an academician, I should go to the last rather than to the two others, provided only that I have reason to believe he is neither interested nor partial. But although the opinion of an artist, or of an expert, as we say, may prevail over each of the individual opinions which concur to form the general or common opinion, this latter opinion must prevail over the opinion of the artist. This is so because in these matters, in which the question is one of evaluation and of inexplicable judgments—as Leibniz put it [14]—rather than of dialectical or *explicable* judgments, it may be said with good reason that *no one person has*

---

[14] " Reasons which persuade us are of two sorts: some are *explicable*, the others are *inexplicable*. Those which I call explicable may be proposed to others by means of distinct reasoning; but the *inexplicable* reasons subsist uniquely in our consciousness or perception, and in an experience of an internal feeling which we do not know how to share with others unless we happen to find a way of making them feel the same thing in the same way." Dutens Edition, Book I, p. 679.

as much wit, or even better, *as much discernment as all persons.*[15]

284. A conclusion of this sort may be accounted for in two different ways.

Where things that do not permit of logical precision are concerned, the opinions of men influence each other reciprocally; errors on the part of one counteract the errors of another. We look for the sound evaluation of things within the limits between which we find most of the persons making a judgment, and thus we reach a common evaluation which is the result of all the individual gropings. It is in this way that, through the mutual reactions of a mass of molecules disturbed by confused movements, a common movement is definitely established which tends to perpetuate itself with regularity (54).

Disregarding the reciprocal influence that the appraisers exert on one another, even when we consider their appraisals as so many isolated and independent facts, there would be good reason why the average evaluation among a large number of individual evaluations would, in general, give more protection than any single evaluation, taken at random from the group which most frequently makes them. This reason, which is connected with the mathematical theory of chances, is the same as that which determines that the value of an angle will ordinarily be obtained with greater precision, by taking the arithmetic average of a great number of observations made with a somewhat imperfect instrument, than if we were to limit ourselves to one or two observations made with even a very perfect instrument. Obviously, the situation would

---

[15] " For the many, of whom each individual is but an ordinary person, when they meet together may very likely be better than the few good, if regarded not individually but collectively . . . . Each individual among the many has a share of virtue and prudence . . . . Hence the many are better judges than a single man of music and poetry; for some understand one part, and some another, and among them they understand the whole . . . . Individually they may be worse judges than those who have special knowledge—as a body they are as good or better . . . . Any member of the assembly, taken separately, is certainly inferior to the wise man. But the state is made up of many individuals. And as a feast to which all the guests contribute is better than a banquet furnished by a single man, so a multitude is a better judge of many things than any individual . . . . ' Hence, the many are better judges than a single man ' (Διὰ τοῦτο καὶ κρίνει ἄμεινον ὄχλος πολλὰ ἢ εἰς ὁστισοῦν)." Aristotle, *Politics,* Book III, Chaps. 11 and 15.

be entirely different if we had reason to believe that the construction of the instrument caused it to tend to make more errors in one direction than in another, for example, errors of excess rather than errors of deficiency (82 and 88). If this were the case, by multiplying observations with a defective instrument, far from decreasing the probable error of the average value, we should increase it. Similarly, if we have reason to believe that the judges who make the evaluations are under the influence of a strong feeling or of a prejudice which acts in the same way upon most of them, then the more numerous the judges, the greater the chance that the conclusion of the average evaluation will be notably erroneous.

285. If we return to the subject that gave rise to these explanations, our attention will be focused upon one important point among many others which might give rise to innumerable comments.

In provisions pertaining to legislation, the legislator evaluates or he arranges.

For example, he evaluates when he lays down the principles of civil law, when he elevates into imperative maxims for the magistrate the rules suggested to him by equity, public utility, the mores, and the customs of the country; he arranges when he applies himself to the work of codification.

He evaluates when he fixes the condition of the age or property qualifications that must be met to be a voter or to be qualified to be elected to office; he arranges when he organizes the forms of the election.

He evaluates when, from a feeling for the needs of his country, he decrees a loan, a tax, or a bill of indemnity; he arranges when, in a system of political economy, he sets up a schedule of duties or when he organizes a tax schedule on the transfer of property which gives rise to complicated jurisprudence.

What has reason to say about this, when freed from prejudices or interests which may be set up to sustain what is called the prerogative of such and such a power? In the first place, that the legislative mandate is a chimerical fiction in the present condition of peoples; that a legislative assembly does not represent the country from which it has been elected in the way the representative represents his constituency; but that in accordance with the conditions of election, there may be more or less probability that the opinion of the majority of the assembly squares with that of the majority of the inhabitants of the country, or at least with a majority of those eligible to vote, where all the citizens without distinction are not called upon to take part in the election of members of the assembly.

That, all things being equal in every other respect, this probability is increased as the assembly becomes more numerous.

That, in the hypothesis—which, perhaps, is destined to apply only to the utopian state for a long time—according to which the decision of the assembly will be protected from the influence of political passions and prejudices, the probability of the excellence of the evaluation which motivates it increases, at least within certain limits, as the number of voters increases.

That if, coming back to the actual situation, we take account of this influence, increasing the number of voters will have the effect of making more probable, not exactly the excellence of the evaluation, but its conformity with the evaluation which would be made by inhabitants of the country under the same circumstances, if they could be directly consulted.

Therefore, we completely understand the intervention of large assemblies in legislative declarations, every time these declarations are connected with a judgment of evaluation which is not broken down into a series of abstractions and logical constructions. But if the question concerns the matter of grouping, of logically organizing the consequences of a principle, what need is there to insist on what has become the object of daily observations by everyone, namely, that large assemblies have to surmount inextricable difficulties in order to bring such a task to a successful conclusion?

We have seen that for magnanimous resolutions, for acts of lofty politics, large assemblies find the vigor, the decision, and even the surety of judgment and the clarity of views which may be missed in a gathering of a few capable persons seated around a council table. But why speak of assemblies when the public square offers us an illustration of the same facts on a much larger scale? Has not the common sense of the people become proverbial? And what do we mean by common sense if not the habitual correctness of its instinctive judgment, when it is not given up to those violent emotions which can make it change momentarily into a blind instrument of destruction?

286. The ancients understood this very well, and, among the Romans especially, the populace would respond yea or nay, like our juries, to the proposition [*rogatio*] of a magistrate from whom the law takes its name, because he was regarded as the author of it. No one would think of discussing amendments in the forum; the people accepted or rejected the whole instrument, as is reputed to have been done sometimes in our own day, in the case of those laws which are regarded as fundamental and which we call constitutions. Would not this role of national jury be the natural role of modern political assemblies, which, as we must fully recognize, have at one and the same time the qualities and the faults of the great popular assemblies among the peoples of antiquity? Would not the proper organization of legislative power require that we should concentrate the logical work of editing and co-ordinating the laws more on it and that we should separate this more completely from the function which consists in an instinctive and conscientious evaluation? No one should expect to see such grave and such complex questions, which we deal with here only incidentally, treated with the serious intention of demanding innovations or of indicating reforms. No one respects more than we do the practical spirit and no one is more struck by the immense gap found between the conception of an idea or an abstract form and its application to the realities of life. But, on the other hand, every abstract or philosophic conception is capable of practical consequences which may serve as tests and proofs in the judging of the very value of the idea, and it is not unprofitable for us to pay attention to that claim at any rate. It is for the same purpose and with the same caution that we move on to deal rapidly with some no less important questions that underlie the organization of judicial power, by showing in what way these questions are connected with the points of logic which interest us.

CHAPTER XIX

APPLICATION TO JUDICIAL ORGANIZATION, AND PARTICULARLY TO THE DISTINCTION BETWEEN QUESTIONS OF FACT AND OF LAW

287. We shall speak first of criminal justice, the administration of which more nearly resembles the form of political organization than does that of civil justice. Not only are the life, the honor, and the liberty of citizens more precious goods than those which are the object of civil disputes, but also the confusion which has prevailed for so long between public crimes and private crimes [1] would have surrendered to the judicial power all the institutions of the country, if in free states we had not assumed the task of limiting the jurisdiction of the criminal judge. On the contrary, wherever despotism is identified with the national mores, the administration of criminal justice has been a pure emanation of despotic power. The deputies of the ruler inflict corporal or pecuniary punishments, yet we cannot believe that their ignorance or their partiality in carrying out this function, which is prejudicial to certain individuals, alters the political constitution.

Everyone knows that the English have always regarded the intervention of juries in criminal matters as the measure of their political freedom, and that one of the first acts of the Constituent Assembly was to establish (or, if one prefers, to re-establish) in France the institution of the jury in cases involving major crimes. This institution is among those which, having persisted in spite of all political reactions, may now be considered as being a part of the constitutional or fundamental law of the country.

288. However, for our present purpose, which is to examine from the point of view of certain abstract and theoretical relations an institution on which so much has already been written, let us begin by setting aside the idea of judgment of peers, which is an artificial resurrection of the traditions of another age, which per-

---

[1] [See H. S. Maine, *Ancient Law*, Chapter 10.]

haps never conformed exactly to a historical reality. But, as a matter of fact, at the same time that the institution of the jury was being perpetuated in England, the other nations of western Europe, which are also descended from Germanic tribes or at least were moulded by their domination or impregnated with their spirit for a long time, quietly allowed the bench to assume a sovereign power in regard to criminal matters. If this was so, it was the natural result of the modification of the mores and of the progress of society. It was no longer a question of murder, rape, pillaging by armed forces, peculation, or patent corruption on the part of the most important personages as it had been in barbaric times, or even in the republics of antiquity. For some time past, the class of malefactors has not been recruited from the upper classes of society, except in rare instances. Where the political passions had been allayed, where the inquisition did not succeed in setting up its formidable tribunals, the chance of falling victim to a slanderous and capital accusation was so slight that most citizens cared little about the need for obtaining guarantees against the magistracy armed with the sword of justice.

Now the same causes are continuing to act. And from that day when the distinction between offenses against the political order and those against the moral order shall have been introduced into the laws as completely as it exists in opinion, we shall be interested in the system of criminal procedure and jurisdiction because of love of humanity, order, and justice, rather than from motives of safety or of personal guarantees. We shall seek possible improvements in this part of legal organization, just as we seek those that permit the regulation of prisons and penitentiaries without being seriously afraid that we may ourselves be placed in the condition of condemned men or convicts.

Therefore, let us put aside every political consideration and suppose that we are dealing simply with offenses against the general order of society. Let us see what the guarantees of the accused and of society, of humanity, and of justice may and should be in terms of this hypothesis.

289. Three elements concur in the formation of a judgment in a criminal matter; there is a fact to be established, an action to be evaluated, and a penalty to be determined. Ordinarily the accused denies certain deeds which are imputed to him. Collaterally he

contests the character of criminality that is attached to it; and, finally, he raises questions as to the application of the penalty. When efforts were made to allot the judicial functions in criminal matters on the basis of theoretical considerations, one of the benches of the court was given the first two points to decide, under the heading of questions of fact, at one time distinguishing them clearly, and at another time confusing them by the way in which the questions are stated.[2] The third point was given to the other bench to settle, under the heading of questions of law.

As we see it, all three questions are by the nature of things questions of fact and not of law, in the sense that in most cases they cannot be settled in accordance with a system of fixed and general rules by means of abstraction and logical construction. But for each particular case submitted to the judges, the solution of the three questions depends upon a conscientious evaluation in which the most learned may err, for want of scientific rules.[3]

The physical fact which is the ground for the accusation either has or has not taken place. Nothing can be more categorical than this disjunction. What is susceptible of continuous variation is the probability of the fact. This is almost never certain with a certainty comparable to that given by the testimony of the senses. And, since in most cases of this sort we must be content with high probabilities or run the risk of paralyzing the action of justice or of seriously compromising the security of the citizens and of the social order, it is impossible to assign the point at which probability ought to determine the affirmative response of the judge about the truth of the fact. A twofold appraisal necessarily occurs at this point: first, an appraisal of the probability of the fact, affirmed by the indictment and denied by the defense. This probability, which is capable of all degrees of variation, is certainly not one of a type which results in an evaluation which can be submitted to the general processes of calculation. In the second place, there is the judgment that this probability either is sufficient or is not sufficient,

[2] Law of Sept. 16, 1791, chap. 7, art. 21ff. Code of the 3rd of Brumaire, fourth year, art. 374ff. Code of the Criminal Institute of 1808, art. 345. [Brumaire, the second month of the calendar of the first French Republic.]

[3] " Right can be and should be limited; the interpretation of a fact also frequently deceives the most circumspect." Neratius, Book V, *Membrane* (D. lib. XXII, Chap. 6, fr. 2).

depending upon the circumstances of the case, to lead the judge to declare the fact to be established. It is incontestable that the judge will and should show himself more or less exacting with respect to the evidence of the fact, depending upon whether the verdict will have more or less serious consequences for the accused, and whether there will be more or less danger to society in not being content with evidences of the sort which are found in the particular case. For example, the case of domestic theft, the unrest that it introduces in families, and the difficulty of proving independently the declaration of the master will be sufficient to lead the judge to declare a fact of domestic theft indubitable on the basis of testimony which would appear insufficient for facts of another sort.

It is apparent that the evaluation of the moral character of the action, based upon inferences drawn from different circumstances, may no more be submitted to invariable rules than may judgments of evaluation relative to the establishing of physical facts (196).

Finally, the scale of penalties can only be graded in such a way that everyone senses its insufficiency for the exact correlation of the penalty with the offense.

290. Nevertheless legislators have sought to establish scales of penalties, and in general they have thought it their duty to seek to proscribe arbitrary penalties, a laudable aim if it can be attained. In the same way, in regard to matters of civil interest, it frequently happens that the legislator would rather offend natural equity than leave the solution of all the disputes which may be raised among the citizens to the arbitrary action of the civil judge. It is better to run the risk of there not being a perfect correlation between the harshness of the punishment and the seriousness of the offense, than it is to run the risk of error or prevarication by the judge in criminal matters, in the dispensation of his discretionary powers. Moreover, since penalties are instituted only in the general interest of society, and rather less to punish the culpable than to assure the maintenace of laws through salutary fear, the matter of estimating the penalty which should make this fear efficacious and of deciding to what extent the interests of individuals need to be sacrificed to a general plan of social organization falls within the functions of the legislator rather than those of the judge.

But, in that case, the system must be complete. Rules, that is to say, legal presumptions, must also be established both to determine the physical fact which gives rise to criminal proceedings, and to determine its criminal character.

It is necessary to come back to the system which has so often been attacked and so reasonably so: because the life, the honor, and the liberty of the citizens is not to be put at the mercy of legal presumptions; because even considerations which touch upon the public interest should yield before considerations which touch upon individual interests of this kind; because a criminal judgment which has moral effects should always be invested with a character of morality which may not square with the logical and artificial combinations of a system of legal presumptions.

To wish to split up such a system, to accept one part and reject another, to proscribe here fixed rules and legal presumptions for the purpose of giving the judgment a moral character, to establish there fixed rules for the purpose of excluding the arbitrary, is a chimerical pretension (203) whose illusory character is perfectly demonstrated by the practice of the courts and by the history of our modern legislation concerning the organization of criminal justice. When we entrust the function of declaring the culpability and that of applying the penalty to different courts or to distinct members or distinct benches within the same court, it always happens, or at least it should generally happen, that the verdict of guilty adjusts itself to the legal formula of the penalty in such a way as to satisfy the conscience of the deciding judge as much as possible or to deviate as little as possible from the end his conscience would have wished to attain. As gaps in the system of penal legislation become more perceptible, decisions of the discretionary power will present more incoherence. The judge will say yes or no in nearly identical cases; he will acquit purely and simply or he will condemn to capital punishment, because a slight weight will incline the balance to one side rather than to the other and because the law has left to him no mean term between yes and no, between acquittal and a death sentence.

291. Because of the common notions that are connected with the terms " fact " and " law," it is supposed that the simplest and in some way the plainest of the elements which concur in forming the judgment of guilt has to do with establishing the material fact.

This is so because it is fancied that neither experience nor perspicacity beyond the ordinary are necessary for this, and that ordinary good sense is sufficient. It is also thought that this is the reason why it is convenient to give over the task of establishing the fact to citizens chosen by lot. On the other hand, the question of law, that is to say, the application of the penal law, requires special knowledge and professional study and must, for these reasons, fall within the prerogative of permanent magistrates.

What is surprising is that such a doctrine not only has been set forth by speculative minds, but also has been admitted and stated with approbation or at least without debate by men versed in the practice of criminal matters, whereas the experience of legal disputes daily contradicts it.

In most cases, would not the task of the defense be completed after the jury had given its verdict, had the law not left to the judge a certain latitude in the fixing of the penalty? Do we not often see that after this verdict contests begin between the lawyer for the defense and the public prosecutor—contests like those in which they engage prior to the verdicts of the jury, discussions like those which could not fail to give birth to a civil suit, however unimportant they might be?

The fact is that, in most cases, barring a very imperfect editing of the penal law, nothing is more simple and gives less occasion for controversy than the application of this law, while, except for cases of flagrant offense, a combination of rather unusual circumstances is necessary before it is impossible for the accused, overwhelmed by the weight of the evidence, to try to arrange the facts and the witnesses in such a way as to leave doubt and hesitation in the mind of the jurors, at least momentarily.

292. If the interests of the accused and those of the accuser are presented on the same basis; if the only thing we have in view is to reach the judgment most protected against the chances of error, whether these turn out to be to the detriment of the accuser or of the accused; if, nonetheless, in order to protect the citizens against the excess of power or of the overeagerness of a permanent magistrate, we proceed to separate the judicial functions in criminal matters, then certain conclusions are clear. The best thing would be to entrust chosen persons, real experts, with that function which is ordinarily most difficult, namely, that of evaluating the probability of the physical fact and of deciding if it ought to lead to an

affirmative verdict, except that afterwards adventitiously chosen judges would make a decision about the moral character of the actions, of which every man of upright mind makes a just evaluation, or about the determination of the penalty, which no longer requires special knowledge or aptitudes.

But in making the accidental and inexperienced judge responsible for the task which is ordinarily most difficult, the legislator has had another end in mind. He has wished to turn the very inexperience of the judge and the diffidence which results from it to the profit of the accused. He has wished that there may be many undeserved and poorly-grounded acquittals; and he has prejudged the human heart by thinking that nearly always when the juror senses that the verdict exceeds his courage, when his judgment has been confused by the eloquence of the lawyer, when the length and complexity of the pleas has exceeded the limits of his attention, he would nearly always render a verdict of acquittal. However paradoxical this assertion may seem to be, we have no hesitation in saying that what moves juries or inexperienced judges to intervene in a criminal action in order to establish the facts of the case is not the easiness but, on the contrary, the difficulty of the question submitted to them. At least this is the case when this intervention is independent of all political considerations.

Moreover, in judging the fact, it is natural that jurors indirectly judge what is called the law. That is to say, they virtually decide, although in a roundabout way, on the application of the penalty, on the basis of the correlation that they do not fail to establish between their verdict and the prescribed form of the penalty. The legislators of the Constituent Assembly and those in France who have gone on with the acclimatization of the jury have vainly wished that the verdict should be given without regard for the penalty by making it the object of ceremonial instruction to the jury. Their efforts have foundered before the superior laws of the human heart; and the omnipotence of the jury, proscribed by the text of the law, has become an actual truth which is made more venerable with each passing day. Whatever a person may have done, the jury has the penalty in mind in the writing of its verdict. If, on occasion, something is lacking or it is mistaken in its views, it hastens to make a protest against the mistake in the only way the laws leave open to it, by an appeal for mercy.

Instead of contending against an irresistible predisposition, leg-

islators should have seen that the morality of a judgment of guilt
consists in evaluating the culpability and fixing the penalty, rather
than in the preliminary declarations of the truth of the physical
fact. And they should also have seen that the character of morality, sought for in such a judgment, can result only from a conscientious evaluation, protected from invariable rules.

293. But in admitting the incontestable fact that jurors virtually decide on the penalty, there are still good reasons why their
power in that respect can only be indirect. On the one hand, the
ideas concerning the organization of public power, ideas which we
became familiar with long ago, would be harmed if the sentence to
punishment were to come directly from temporary judges, ordinary
citizens, and not from permanent judges who are invested with a
public character. On the other hand—and this consideration is still
more decisive—the ordinary citizen who wishes unequivocally to
lend assistance to the law and who knows the consequences of his
vote with respect to the application of the penalty could not easily
persuade himself to be the direct instrument of the severity of the
law or to appear to fix directly the penalty of which he is, nevertheless, the true arbiter through the correlation he establishes between his verdict and the legal scale of punishment. There is
something painful and often harrowing in fulfilling such a function, for which the honors of the magistracy are a poor compensation, and from which it is well to spare ordinary citizens, especially
for fear that their all too natural repugnance may paralyze public
prosecution.

Therefore, there can be no question either of abolishing the
scale of penalties or of transferring to the jury either the fixing or
the direct application of the penalty. It is important only to convince oneself that, once the jury has been admitted into criminal
matters, it becomes the natural and necessary judge of everything
that falls within the scope of conscientious evaluation, and that it
falls to its lot to determine the degree of culpability as well as the
decision as to the penalty. Then, far from contradicting this natural tendency by written laws, we shall adapt the laws to this
tendency in such a way as most nearly to approach the desirable
end and especially to escape the inconsistency of judicial decisions.
We shall see in the graduated series of penalties only a framework
offered to the jury to excuse it from a dreaded initiative, and in

the sentence of the magistrate only a solemn sanction given to the opinion of the jury. This solemn sanction should be of the same sort as that which the president of the Court of Assize gives to the verdict of not guilty, as a result of his order of acquittal. This sanction should require no further review or logical deduction on the part of permanent judges. Indeed this assumes, on the one hand, that the law does not charge the permanent judges with the task of fixing the penalty within certain limits; on the other hand, that the forms of questions and answers for the jurors are limited to being found in such simple correlation with the penal formulas that there can be no doubt as to the consequences of the verdict. Now, it seems to us that the legislator should work out this double hypothesis. In the first place, when in a great many cases we have trusted the judgment of the jurors for such delicate and difficult evaluations as that of establishing a physical fact or even that of fixing the degree of moral responsibility, it is strange that later on we should fear to trust them to estimate, between the legal limits, the severity of the penalty, a kind of evaluation for which experience and special knowledge are least necessary.[4] In the second place, the jury, uncertain of the consequences of its decision because of the latitude of evaluation left to the permanent judges, rarely fails to adapt its verdict to the case in which the latter would assess the *maximum* penalty. This will render the moderating power of the permanent judges illusory or will upset the organization of penal law. For the same reason, if the jury feels uncertain about the correlation of its verdict with the rule of punishment, it will formulate its verdict in view of the hypothesis which is most unfavorable to the accused, which will often decrease the penalty even below the standard to which the humanity of the jury wishes to reduce it.

294. But, if these principles are admitted, what part will be left to the permanent judges in the disposition of criminal justice? It will remain for them to decide questions concerning jurisdiction and procedure, pleas and prescriptions which, in criminal cases, are

---

[4] It must be well understood that this gradation of the penalty within legal limits must always be done by the jurors in an indirect manner only; for example, as was happily attempted in the law of April 28, 1832, for the reform of the Code of Criminal Instruction and of the Penal Code, by means of the declaration of extenuating circumstances left to the native feeling of the jury.

the only questions which may properly be called questions of law. Great as is the discretionary power that the legislator should leave to the judges concerning the matter, it is equally important to maintain in judgments concerning the issues involved that character of morality which can result only from a conscientious evaluation, protected from the authority of fixed principles; and it is just as indispensable to define with certainty and to maintain invariably the conditions which give rise to the occasion for this discretionary power and which indicate the bounds beyond which it must not go. Everyone senses that the social order would be upset if it were left to the discretion of the judges to extend or curtail their jurisdiction, to invert the methods of accusation or defense. Everyone senses that there is no middle ground between being under and not being under the jurisdiction of a court, between observing and transgressing judicial forms; that the solution of all questions that may be raised on this subject should come from a system of rigorous deductions; that these are questions of public order to which that character of morality fundamentally inherent in matters of criminal judgment is not connected.

From the moment that accidents of procedure may affect the issues involved, when, for example, they tend to extend or to restrain the means of defense, they clothe the issue with an inherently moral character; they fall indirectly within the domain of the evaluator. Thus it would be fruitless for the law to forbid to the accused certain discussions of principles and the production of certain witnesses. It would be useless for the law to say fictitiously that certain facts and documents that have a real and physical existence are null and void. It would be vain for the magistrate, in obedience to the law or in the use of a discretionary power with which it had invested him, to draw up similar limits for the defense. The jury would judge as if the alleged facts had been proven, and as if the witnesses to whom audience had been refused had testified in support of the statement of the accused.

295. The question of the number of judges advisable for the composition of the permanent bench of the criminal court has often been discussed. The solution evidently depends upon the nature of the functions that are attributed to this bench. If these functions are limited to clothing the evaluations of the jury with a ceremonial ratification, to pronouncing the verdict of acquittal or

the penalty which has a necessary correlation with the verdict, one magistrate is sufficient, since there should be nothing for him to deliberate upon. If the permanent bench is vested with a power of evaluation in regard to such matters as fixing the penalty within limits which are often very wide, there is reason, as there is in all judgments of evaluation, to demand the intervention of a greater number of judges. This is so because, in general, the guarantees of justice in judgments of this sort are increased with the number of those who have a share in them and whose opinions reciprocally influence each other. When the permanent section is formed of three judges, as it is according to the French law in force at the present time, we should be appalled at the exorbitant power with which two men are invested in certain cases of extending or shortening the imprisonment of a citizen many years, of deciding the amount of a fine between limits one of which is minimal while the other would completely exhaust the fortune of the condemned person, if it were not true that for other reasons we are nearly always sure that the verdict of the jury is formulated in anticipation of the *maximum* penalty, and that the person has been virtually condemned to this maximum by the agreement of at least seven judges, who are naturally more inclined toward indulgence than toward severity.

296. The distinction between fact and law, in the judgments rendered in civil suits, is seen more or less clearly depending upon the nature of the judicial institutions and the forms of the procedures in use in various countries. In Rome, where the magistrate charged with affirming the law and dispensing justice would have been inadequate for the examination of the particular circumstances of each suit, he would make a distinction between law and fact, between the abstract and the concrete, between the rule and the arbitration at the beginning of the proceedings. The pretor, in drawing up a prescribed form that is comparable to the English " writ," would deduce the consequences applicable to the case being tried on the basis of the law or of the maxims of jurisprudence which he had set up as a rule. He would subordinate the application of the prescribed form to the verification of certain facts, to the appraisal of certain circumstances, for which he would send the parties before the judges, arbitrators, or experts, to whom the prescribed form is addressed, and who should by limiting them-

selves to the wording of the prescribed form, condemn or free
the party against whom the action was directed. *Judex esto,*
or *recuperatores sunto. Si paret . . . , condemnato; Si non paret,
absolvito.*[5] This was a syllogism (250), of which the magistrate,
or the chief justice, laid down the major premise and deduced the
consequences by presenting the minor under the form of a hypothesis, the truth or falsity of which would have to be declared
by the judge delegated by him, who was a veritable jury with respect to civil matters.   Quite unlike our jury in criminal matters,
the *judex* would derive from the delegation of the supreme magistrate the power of himself pronouncing condemnation or acquittal;
but this difference holds less in regard to the substance of the issue
than to the form, and to the ideas of each people as to the organization of public authority (293).   Moreover, by force of circumstances, the Roman *judex,* like our jury in criminal matters, was
called upon not only to establish facts but also to recognize the
laws resulting from the facts.   The prescribed forms given by the
pretor were sometimes relative to the law (*in jus conceptae*), sometimes relative to the fact (*in factum conceptae*); just as in English
procedure the parties " join an issue " either of law or of fact.   The
forms may even be *in jus* and *in factum* at the same time.[6]   Some-

---

[5] " Let there be a judge, or let there be a board of arbitrators.   If
proved . . . , convicted; if not proved, acquitted."   As a technical term in Roman
law, *recuperatores* refers to a board of three or five members, originally only
named in cases between Romans and foreigners, but later called for in all cases
requiring a speedy decision, especially in suits concerning property.

[6] " §47.   Some actions may be instituted by formulas either of law or of
fact, as for instance the actions of Deposit and Loan for use.   The following
formula: ' Let C D be judex.   Whereas Aulus Agerius deposited a silver table
in the hands of Numerius Negidius, whatsoever it be proved that Numerius
Negidius is on that account bound by good faith to convey or render to Aulus
Agerius, do thou, judex, condemn Numerius Negidius to convey or render, unless he makes restitution; if it be not proved, pronounce his acquittal: ' is a
formula of law.   A formula thus framed: ' Let C D be judex.   If it be proved
that Aulus Agerius deposited a silver table in the hands of Numerius Negidius,
and that by the fraud of Numerius Negidius it has not been restored to Aulus
Agerius, do thou, judex, condemn Numerius Negidius to pay Aulus Agerius
whatever shall be the value of the table; if it be not proved, pronounce his
acquittal: ' is a formula of fact.   And there is a similar alternative in the case
of loan for use.

" §48.   The condemnatio is always to pay a pecuniary value.   Even when

times the prescribed forms suppose the fact to be established; sometimes they leave it to be established by the judge; sometimes they determine exactly the amount of the pecuniary punishment that the judge should pronounce, after he has recognized the truth of the fact or the correct application of the law, without being able, at least on the basis of personal responsibility, to sentence to the payment of a greater or lesser sum; sometimes they only assign a limit that the sentence by the judge should not exceed; sometimes they leave it entirely to his evaluation. Therefore, the functions of the *judex* are not limited to the establishment of a material fact. He is also charged with giving an opinion concerning what is good and equitable (*ex bono et aequo judicare*). He even passes on the strict right, in this sense, that he discovers, according to the elements which gave him an understanding of the action, and in conformity with the indications of the prescribed form, the right to reclaim the ownership of goods, the right to demand a sum of money or the rendering of a service. Yet he has not, according

---

we claim a corporeal thing, an estate in land, a slave, a garment, an article of gold or silver, the judex condemns the defendant to deliver not the thing itself, as in the elder system, but its value in money.

"§49. The formula either names a certain sum in the Condemnatio or an uncertain sum.

"§50. In a condictio, when it names a certain sum, it concludes as follows: 'Do thou, judex, condemn Numerius Negidius to pay Aulus Agerius ten thousand sesterces; if it be not proved, absolve him.'

"§51. An uncertain sum is either named with a limitation or maximum, for instance, thus: 'Do thou, judex, for that (or, of that sum) condemn Numerius Negidius to pay Aulus Agerius not more than ten thousand sesterces; if it be not proved, absolve him:' or it is named without a limitation, as when we demand our property from the possessor in a real action, or demand the production of a person or thing in a personal action, where the conclusion runs as follows: 'Do thou, judex, condemn Numerius Negidius to pay Aulus Agerius whatever shall be the value; if it be not proved, absolve him.'

"§52. But whatever the claim, the judex must condemn the defendant to pay a definite sum, even though no definite sum is named in the condemnatio. When a certain sum is laid in the condemnatio, he must be careful not to condemn the defendant in a greater or lesser sum, else he makes himself liable to damages: and if there is a limitation he must be careful not to exceed the maximum, else he is similarly liable; but he may condemn him in less than the maximum." Gaii *Institutionum Iuris Civilis Commentarii Quattuor*. Edited with a translation and commentary by Edward Poste. Oxford: Clarendon Press, Second Edition, 1884. Book IV, §§47-52, pp. 528-530. Cf. p. 422.

to the ideas the Romans were used to, the power of restoring the ownership of goods, or of compelling the performance of a service, but only that of fixing a pecuniary compensation or, as we say, damages. As a matter of fact, he has a right to pass judgment on everything not decided in advance, in a general way according to fixed rules, independently of all knowledge gathered from the circumstances peculiar to the suit. His decisions as to the law, motivated by the evaluation of circumstances which are particular and essentially variable from one case to another, are those which cannot be made by jurisprudence or which do not concur to give its systematic development and its rational construction to the interpretation of legal statutes and judicial aphorisms. This latter belongs to the jurisdiction of the pretor. Consequently, since the greatest number of lawsuits, as much among the Romans as among us, should not have presented those difficulties which follow from the scientific development of law and which are concerned with doctrine, it would most frequently happen that the intervention of the pretor, for the purpose of stating the prescribed rule and the designation of the judge, was an act of authority rather than a judicial decision. It was an act comparable to that by which, among us, a court names experts or submits the case to arbitrators for whom the function is reserved (unlike the Roman pretor) of confirming the decision, in order to give them executory force by bestowing on them the seal and the bond of public authority. Now, it is perfectly clear that when among us, for example, in accordance with the direction of the law, a court sends the judgment in a suit that arises between commercial partners back to the arbitrators, the latter not only have the duty of establishing the facts but also of evaluating such reciprocal rights and obligations between the parties to an action as result from the facts which have led to the suit. It is no less evident that, if the legislator has prescribed that suits of this nature be sent back to the arbitrators, the reason is that it is because he wished the decision in each case to be the result of an evaluation which is both conscientious and made in good faith and is made by persons acquainted with business affairs. He did not want the decision to result from a rigorous application of certain logical rules, as is the habit of ordinary judges, for whom making decisions has become a science and an art, at the same time that it is a profession.

297. M. Ortolan says,[7]

Fundamentally, the system of prescribed forms is nothing but a clever means of setting up a jury in civil matters. It is necessary to proceed from the principle that the judge is not a magistrate, but simply a citizen; that he has, consequently, no other prerogative than those which are conferred upon him by the magistrate. Outside of the terms of the prescribed forms, he is without power. The wording of the forms is, therefore, the principal point in the procedure. Juridical science directs all its attention to this and works on it unceasingly. The analysis and concatenation of its conclusions, the conciseness and correctness of its terms are admirable. Every law, if it needs ever so slight a shading, is foreseen; for every law requires a special prescribed form for its operation. The prescribed forms are drawn up in advance, incorporated into jurisprudence, inscribed on the *album* and laid open to the public for examination. The plaintiff, coming before the court of the magistrate (*in jure*), indicates what he asks for. The elements of the case are debated between the parties, the form adapted to the particular case, and finally delivered to the pretor (*postulatio, impetratio formulae, vel actionis, vel judicii*).[8] Then the judge, *the appraiser of fact or of law*, depending on the case, listens to the parties, makes proper verifications, resolves the problem which has been presented to him, and renders his decision (*sententia*) within the limit of the powers conferred upon him by the prescribed form.

Thus, so far as we are able to judge, following the discovery of valuable fragments, and especially that of the manuscript of Gaius, which have thrown their light on this previously very obscure point of Roman institutions, the organization of the *judicium* and of the formulary procedure was naturally adapted, not to the distinction between fact and law, in the narrow sense which has for so long a time falsified our ideas as to the role and functions of the jury in criminal matters, but rather to another distinction which has its fundamental cause in the nature of the objects of thought and in the laws of thought itself, and which does not permit judgment by means of logical construction and judgment by means of conscientious evaluation to be confused, any more than it is possible

---

[7] *Histoire de la législation romaine*, second epoch, sect. 3, no. 48.

[8] [Complaint, the requesting of the prescribed form, either for a suit, or for a judgment.]

to confuse the power of calculation and the reliability of taste. Is this to say that, in the time of the Scipios and the Gracchi, the legal experts of ancient Rome had taken account, through profound logicians and philosophical generalizers, of the different paths the human mind follows, and the different aptitudes that it develops depending upon whether it reasons or evaluates? Far from it; nothing is more characteristic and, from our point of view, less liberal and more strict than the prejudices and the traditions of city or of caste which have progressively introduced this philosophical and learned organization. This follows from the very simple reason that the laws of logic and the permanent conditions of the human mind in general come, in the long run, to prevail over accidental causes and traditional influences in institutions, by imprinting a definitive form on the structure of which the materials alone bear the trace (which is often indistinct) of the earliest beginnings.

The formulary procedure was substituted in the sixth century of Rome for the ancient procedure of " actions at law " (*actions de la loi*), true ritual by means of which the patrician class was able to get back into its hands the tie which connected the political and civil institutions with the religious institutions. The author we have just cited says,

The prescribed form was a simplified derivation of what was important and fundamental in the actions at law. The *demonstratio*, which indicated the object of litigation, replaced in a purely intellectual way those pantomimes, gestures, deposits of objects or of symbolic remains, the purpose of which is to make this demonstration tangible in the action at law. And it may be noted that the *intentio*, which indicated the claim of the plaintiff, was evidently enough copied from the very words pronounced by the plaintiff in the action at law. For example, the plaintiff said, " *Hunc ego hominem ex jure quiritium meum esse aio*," [9] in the *sacramentum* in actions concerning things rather than persons, by touching with the lance, the *vindicta*, the man against whom he complained; " *Si paret hominem ex jure quiritium Auli Agerii esse*," [10] said the pretor in his prescribed form for real action. These are the same ideas materialized in the action at law, spiritualized in the prescribed forms of the pretor.

[9] [" I declare that this man belongs to me according to the law of the Quirites."]

[10] [" If it is proved that this man belongs to Aulus Agerius by the law of the Quirites."]

But there are some ceremonial acts, symbols, or signs by which people express and fix their ideas, when they give to these ideas the value of political and civil institutions, like the ordinary symbols of language with respect to ideas which are in no way *official*, or which do not touch upon the government and administration of the city. Words are impaired by circulation, meanings are misused, traces of etymology are lost, and language becomes perverted (212). In the same way, legal and juridical ceremonies, whether they consist in symbolic demonstrations, as is the case in the early periods of the life of peoples, or in ceremonial words, or in written formulas, lose their primitive force and their original meaning little by little. They are weakened by use, fall into disuse, or are abolished when they become nothing more than awkward formalities. Therefore, once the pretorian law was fully established, it was in the nature of things that in the course of time the delivery of the prescribed forms by the pretor became something analogous to what we call, in our modern fashion, letters of chancery, that is to say, a pure formality. Furthermore, the change in the political constitution, by leading to the abolition of the pretorian jurisdiction, was bound to involve that of the system of procedure which is no less worthy of the most careful attention. Its full vigor corresponds to the time during which law among the Romans, already freed from its religious trappings, and no longer shackled by the prolix or incoherent laws of an absolute power, had attained its highest degree of perfection as scientific doctrine (276).

298. Using the accepted terminology, what the institutions of the Romans seem to have done, at a certain period in their history, was to distinguish questions of fact and questions of law, that is to say, actually, to distinguish judgment by means of conscientious evaluation, valid only in the particular case, and dialectic judgment, valid as a doctrinal decision. What their institutions did, we suggest, to introduce this distinction from the first stages of the proceedings, institutions of an entirely different sort have done among us, but in such a way that the determination of these two elements might be, on the contrary, the crowning achievement of the proceedings and might appear with clarity only in the highest judicial trials to which a civil suit can be submitted. When our kings, in the interest of their power, organized under the name of *parlements* permanent courts of final appeal which called them-

selves supreme courts, they came to fear the independent spirit of these bodies of legists, which had served them by throwing down feudalism. For the purpose of maintaining the supremacy of the royal power, the kings authorized litigants to enter before their privy council their appeal against judgments which might include infractions of the royal ordinances. When the ancient regime fell, new legislative authorities showed themselves to be no less jealous in maintaining the subordination of the judicial bodies. A supreme court was created whose principal function was to annul, on the request of the parties involved, all proceedings in which the forms had been violated, and any decision which involved *an express contravention of the text of the law,* but which was without power to deal with the issues involved in lawsuits, and on condition that decisions as to these issues would be returned to the ordinary courts, after the appeal of the proceedings or of the previously entered judgment.[11]

In establishing, or rather in strengthening this great institution, the beginnings of which go back to the royalty of the Middle Ages, the founders of the new order of things did not have as their sole purpose the defense of the legislative power and of the text of the law against the slight inclinations to independence and opposition on the part of judicial bodies. In keeping with the spirit of the times, they were bound to propose to strengthen national unity more and more, and, in order to do this, to prevent through the regulatory action of a supreme tribunal, the diversities of jurisprudence which could not fail, in the long run, to alter seriously the uniformity of the law from one jurisdiction to another, in spite of the uniformity of the statutes and codes. But it is clear that this latter purpose would have been attained only indirectly and imperfectly by the institution of a court of appeal, if it were held to the letter of the law which, in order to attain an altogether different political object—that of preventing the courts from resisting the legislative power—confers on the court the right of review only in case of an express contravention of the text of the law. Either contraventions of this sort rest on a manifest and accidental error, which is not of such a nature as to perpetuate itself or to become a precedent, or they arise from the

---

[11] Law of November 27, 1790, art. 3.

fact that the law does violence either to public opinion or at least to the opinion which prevails in the magistracy. Such errors are little to be feared in ordinary times and in the midst of a well-ordered society. In general, therefore, the diversity of jurisprudence, from one jurisdiction to another, can be established only in doubtful cases in which the text of the law may be differently interpreted or completed by persons who lack neither insight nor good faith, without giving rise either to an express contravention of the law or to an opening for an appeal, in the strict sense of the legal definition. In such a case, it should happen that the supreme court would reject the petition, no matter what decision has been reached by the courts to which the decision had been deferred. The highest court should always have the power to correct judicial errors, but it should not have a regulative power in the true sense of the term. Its intervention should place no obstacle in the way of differences of jurisprudence which gain strength and tend to disturb the uniformity of law, from one jurisdiction to another.

In fact, the Court of Cassation has understood and accomplished its mission differently. This mission has been magnified in public opinion, to the degree that the highest court, declaring the law on doubtful points and clarifying obscure aspects of legislation, has made decisive, although reserved, use of this powerful jurisdiction, which the literal provisions of the constitutive law do not seem to have conferred upon it. Although there have been many more denials than appeals—because a favorable prejudice is attached to the case that has been decided, and because we should more frequently encounter obstinacy among the litigants than errors among the judges—the number of cases in which the Court of Cassation has refused to settle doubtful questions, by rejecting the appeals for one reason or another, is surely very small as compared to those in which it has resolutely taken part. This is contrary to what should have happened if it had held rigorously to the legal definition of its functions, and if it were not in the nature of things that the logical principle or general idea, the rudiments of which are contained in a judicial institution, finally freed itself from the arbitrary restrictions of the original definition.

299. It has been admitted not only that the highest court checks express contraventions of the text of the law, but that it also regularizes the interpretation of it, by establishing principles

and subordinating them to one another on the basis of their degree of abstraction and generality, by deducing their logical consequences, proceeding in these matters by way of analogy and similarity. In a word, the court participates in the systematization of the law just as the Roman pretor did in other times, although it does so by different means. Therefore, it was necessary to distinguish juridical error, subject to the censorship of the highest tribunal, from faulty decision (*mal jugé*), whether as to law or as to fact, which escapes this censure and which rests on points that the judges of the lowest court have power to decide without appeal. Now, here again we find that the radical distinction is not between fact and law any more than it is between matter and form. A question of form or of procedure may depend at one time upon a juridical interpretation, and at another time on the evaluation of proofs or upon the significance of a fact, just like a question concerning the matter of litigation. The law might have left to the conscience of the judge the evaluation of the right of the parties, as this is established on the basis of the averred facts or on the basis of precise clauses or authentic actions, as well as the interpretation of the intention of the acts, the will of the parties, and the evaluation of the means by which the proof of a fact or of an intention can be established. Or, on the contrary, the law might have defined and set forth the proof of a fact, the worth of a clause, the intention of an act, as well as the nature and the consequences of the law to which the fact, the agreement, or the contract furnishes an opening. Whether it is a question of law or of fact, of form or of matter, there would be no violation of the law in the case in which the judge, given the authority to evaluate on the basis of his conscience, has made use of a necessarily discretionary and consequently sovereign power, after one has gone through all the degrees of ordinary jurisprudence. On the contrary, the law will be violated if the judge has evaluated the issue contrary to the legal definition of evidence, contracts, and facts which are defined and characterized by law, as much as if he had infringed upon a rule of procedure, or misunderstood the stipulations of the law as to rights properly so called.[12] But, on the other hand, the basis

[12] We shall cite briefly in this note a few authorities without entering into detailed and technical explanations which are beyond the scope of a book in which this subject is touched upon only because of its connection with questions

for appeal does not require an explicit contravention of the text of the law, the establishment of which would really fall within the powers of a political bureau rather than within those of a court on which we wish to bring together so many luminaries. It follows that the appeal would have occurred not only if the judge

---

of philosophy in general. "To be acceptable, a plea for appeal must rest upon the violation of a law." Cassation, July 17, 1827. — "The false application of a law gives an opening to appeal only when it results in the formal violation of some law." Cassation, Nov. 14, 1826. — "In general, a bad decision in the lower court does not give an opening to appeal, even when the law seems to be applied contrary to its spirit." Opinion of the Council of State, Jan. 31, 1806. — "Above all when the bad decision consists only in a false evaluation of the facts or the acts or the suit." Cassation, 19th day of Nivôse, year XII. [Nivôse, the fourth month of the calendar of the First French Republic]—"Still less when it does not appear that, in order to evade the law, the judges have reached an erroneous decision on the facts submitted for their evaluation." Cassation, Jan. 5, 1809. — "In case of the silence of existing laws, there is ground for appeal in the violation of the Roman laws governing a point of law or equity." Cassation, April 10, 1821. — "Because of the violation of principles established by the law of nations." Cassation, March 29, 1808.

"There is no ground for appeal for the violation of a contract." Cassation, Feb. 13, 1827. — "Nor for error in the nature of a contract that the law has not defined. That is nothing but a faulty decision." Cassation, Feb. 2, 1808. — "But there is more than a faulty decision, there is ground for appeal in falsifying the title of a contract defined by the law which distinguished its elements." Cassation, July 26, 1823. — "In general, the mistaken interpretation of a contract gives no ground for appeal." Cassation, Mar. 18, 1807. — "But there is ground for appeal when the judges affirm that there has been a renunciation of the law by making it follow only from presumptions, in the case in which proof by witnesses is not admissible." Cassation, May 1, 1815.

"It is not the function of the Court of Cassation to evaluate the evidence and the witnesses which have produced conviction in the mind of the jurors, when the law does not attach to certain acts or to facts a special and necessary characteristic of evidence." Cassation, June 11, 1825. — "As a general rule, the Court of Cassation, charged solely with checking the contravention of the law and of maintaining the observation of the essential formalities that it prescribes, does not enter into the examination of points of fact; it takes the facts as they are established by the judgment or the sentence that is attacked and concerns itself with the point of right on which judgment has been made only in regard to its conformity or nonconformity with the law." Cassation, Oct. 13, 1812. — "Nevertheless when a decision of law rests upon an error of fact, denied by the very instrument which is basic to the action, the Court of Cassation may be able to verify the error and therefore, consequently, to grant the appeal." Cassation, Feb. 16, 1813. — "The decision of ordinary judges as to a

had gone against the express terms of the legal definition, but even if he had gone against the logical consequences, whether these consequences are proximate or far removed, that can be derived from this definition or from a system of similar definitions by means of deduction and doctrinal discussion, independently of a conscientious evaluation of the particular circumstances of the suit. From this it follows, finally, that the principle which will guide the highest court in the determination of its competence, when this is controversial (that is to say, whenever the issue in question is not an express contravention of a formal text) presupposes the distinction between judgments delivered after conscientious and sovereign evaluation and those arrived at by the logical application of principles that the law proclaims. The judgments may be amended, if a better logical discussion demonstrates the falsity of the application. Whether we can or cannot clearly take account of this fundamental distinction, from the point of view of theory, whether it is or is not expressed in a general way, we believe that if the particular motives of each decision are zealously analyzed, it will be discovered that the feeling for the logical distinction on which we insist has constantly guided the highest court, always careful to leave to the judges the free exercise of their discretionary power wherever this is authorized by the law, and to circumscribe the exercise of it within the limits fixed by the law.

If it happens that the Court of Cassation substitutes its evalu-

---

fact, on which contrary evidence results in a legitimate action, may be annulled by the Court of Cassation, which, even in civil matters, evaluates certain legal facts, or facts defined by law, in order to deduce their legal consequences; thus it has recognized its own competence to decide, contrary to the verdict of the judges of the lowest court, the constitutive characteristics of a right of way." Cassation, June 13, 1814. — "Of a dismissal of a writ." Cassation, August 3, 1819. — "Of an adjustment concerning feudal tenure." Cassation, Feb. 15, 1815, and so on.

The Court of Cassation seems to have determined its competence more particularly, especially with respect to financial matters, for the purpose of evaluating the acts and the circumstances of suits. Thus, in the matter of registration, it decides that proof of a change of property, giving an occasion to resort to law, follows from certain acts or facts. Cassation, Aug. 2, 1814. — In matters of indirect taxation, when the judges of the lowest court have committed an error of fact in the interpretation of an official report. Cassation, March 25, 1825, and so on.

ation for that of the judge, wherever the law has not stepped in
with precise definitions, and has left the field open to conscientious
evaluation, it is in exceptional matters in which there is reason to
fear not the error of good faith, but the revolt of the judge against
the authority of a law considered harsh, or which is contrary either
to common law or to natural equity. This is also the case in the
matter of the imposts, in which the interest of those who are
liable to taxation usually is given more consideration than that
given the treasury. Then the supreme court, in which doctrinal
distinctions and general principles may not act to obliterate the
political commission with which it is invested, enters into the
evaluation of the matter, because this is the only way to maintain
the authority of the law and to forestall its systematic evasion.
But the exception itself proves the rule. In everything which
touches upon the government of societies, there are some necessary
impairments in the most firmly established maxims, the rigorous
application of which on every occasion would be convenient only
in those ideal republics created by philosophical imagination. But
it is no less important to grasp in their intelligible purity the rules
toward which the conduct of human affairs tends, and which it ap-
proximates more closely in proportion as societies have become
more perfect as a result of the progress of time and of the general
reason.

300. The question of introducing a jury in civil matters has
often been raised in this country. If this innovation, that we are
far from wishing to advocate and that seems repugnant to the
secular traditions, were ever attempted, it seems to us to follow
from all the preceding explanations that, in the organization of the
procedure adapted for such an institution, the legislator should
have in view primarily the distinction between dialectical judgment and the judgment through conscientious evaluation, whether
of fact or of law. All questions which depend upon a conscientious
evaluation should be given over to the decision of a jury and the
decision in all others should be confined to the sagacity of magistrates or permanent judges. This fixing of the prerogatives would
certainly give greater clarity to judicial discussions. Perhaps this
would expose it to more chances of error on the part of inexperienced judges, and in any case it would require an unusual care on
the part of the legislator to regulate in advance, according to

special functions, not the cases but the substance of disputes and, by way of general decision, the distinction of powers. At least, we should not wish to confer upon a magistrate of inferior rank a power comparable to that of the supreme magistrate, such as the Roman pretor who not only could resolve points of law in his own way but could even circumscribe the competence of the juror or of the judge delegated by him.

We cannot pursue these remarks further, and we shall terminate at this point a digression which perhaps is already too long, in order to return to questions of pure philosophical speculation. We have concluded what we wished to say about the functions of the understanding and their logical instruments. Now it is necessary to take up again the series of general considerations with which we began and to submit to a new and rapid examination the system of our knowledge concerning the world in the midst of which man is placed and concerning man himself. This should be done in such a way as to set forth throughout this system the distinction and the contrast of three elements: the historical element, the scientific element, and the philosophical element. This will be the object of the chapters which follow.

CHAPTER XX

ON THE CONTRAST BETWEEN HISTORY AND
SCIENCE, AND ON THE PHILOSOPHY
OF HISTORY

301. When the genius of Bacon undertook to summarize in an encyclopedic table the classification of human knowledge and to point out its principal connections, he arranged them first under three great categories or rubrics: History, Poesy, and Science, corresponding to the three principal faculties of the human mind: Memory, Imagination, and Reason. Later on we shall have to come back to this celebrated classification, which has been praised and criticized so much. We shall also have to consider the modifications that d'Alembert made in it in the essay put at the beginning of the eighteenth-century French *Encyclopédie*. We cite it here simply as evidence of the contrast of two elements, the one historic, and the other scientific or theoretical, which enter into the arrangement of the general system of our knowledge, and we attempt to grasp its nature and its distinctive traits with precision. Moreover, these two elements are not combined solely in the system of our knowledge. We shall come across them again, sometimes combined with each other and sometimes opposed to each other, when we leave literary or philosophical speculation to enter the field of practical application and the world of actual experience. When these two elements are looked at from the vantage point of our contemporary understanding of social institutions and of the conditions required for the life of peoples, we recognize that one part of these elements goes back to traditional influences and to the circumstances of their origin, that is, to facts to which history alone holds the key; another part goes back to the conditions bound up with the permanent nature of things, which constitute an object of study for reason independent of all historic precedent. This contrast is so striking with respect to everything which refers to the law and juridical institutions that it has led

naturally to the formation and the antagonism of two schools of legal experts: the historical or traditional school and the school that, in opposition, is called rational or theoretical. A tendency of society to organize itself according to a systematic and regular plan, in conformity with theoretical conceptions, and a struggle against the obstacles that historical precedents put in the way of the realization of the systems and theories is most clearly evident in the revolutionary crises of modern times. Sometimes society is seen to be entirely freed from pigheadedness; sometimes inevitable reactions have given the ascendancy to the guardians of historic precedents and of the traditions of the past; and, finally, sometimes these have sought to come to terms with the new spirit, by upholding on the grounds of theory what could only have real force through the influence of historical precedents. The excesses of thought (to say nothing of excesses of another and more regrettable sort) have consisted chiefly in a cult of intolerance, in an exclusive partiality for one or the other of the two elements which must be taken into account and to which a legitimate part of authority in the organization of societies will always come back. Besides, it is evident that, in proportion as less attention is paid to individual lives, absolutely or comparatively, as inequalities of all sorts are leveled, and as ideas and passions become generalized, the influence of historical precedents must lose its strength. At the same time, the course of events should tend to conform increasingly to a certain theoretical order that is no longer disturbed to the same extent by accidents arising from superiorities of classes, of talents, and of temper.

But we must not understand by history only the account of political events, the description of the destinies of nations and of the revolutions of empires. The intervention of moral causes, the play of liberty and of human passions in the struggle with external necessity, which enflame the imagination of the historian, color his pictures, and give dramatic interest to his narrations, do not belong to the essence of history. The sciences, the arts, and literature also have their history. Great natural objects and phenomena of the physical order likewise admit in many cases of a chronology, of annals, and of historical narration. For example, one can write the history of a volcano just as one can write the history of a city. Therefore, let us try to find out what the essential characteristics of the historical element are, without fearing the dryness of abstract conceptions, by trying to disentangle the fundamental idea from the accessories which complicate or embellish it.

302. Let us take the simplest illustration among purely mechanical phenomena, that of the movement of a billiard ball rolling on a table as a result of an impulse it has received. If we consider this ball at any instant while it is in motion, we need know no more than its position, its present velocity, the nature of the frictional and the other resistances which act on it, to be able to assign its position and its velocity at either an earlier or a later time than the one we are considering, and, finally, to be able to predict the spot and the moment when these various resistant forces will have brought it again to a state of rest from which it cannot move, at least not until new forces act upon it. But if we start with one of the instants after the ball has come to rest, it is clear that neither its present state nor even the state of surrounding objects which gave it its original propulsion any longer offers traces of the phases through which it has passed while in a state of motion. However, it would still be possible to indicate the phases of rest and of motion through which it will pass in the future, from the knowledge of its present state, as much in the case of the billiard ball as in that of objects which, as a result of their present motion, may later strike against it and communicate a new impulse to it.

In general, and without spending any more time on the conditions of this example, which may perhaps be too homely, we understand that the conditions of theoretical knowledge are not the same for past and for future events. This is due essentially to the fact that, among series of phenomena which are linked together because they are successively effects and causes of one another, some series come to an end and others are prolonged indefinitely; just as, in the sequence of human generations, some families die out and others perpetuate themselves (29).

In order to avoid becoming involved immediately in scholastic disputes concerning what is called man's free will, let us limit ourselves to begin with to the consideration of natural phenomena in which, as everyone acknowledges, causes and effects are linked together according to a rigorous necessity. In this case it will certainly be true to say that " the present is large with the future," and with the whole future, in the sense that all subsequent periods are implicitly determined by the present one in accordance with the action of permanent laws or external decrees to which nature is obedient. Yet, we shall not be able to say without qualification that " the present is similarly large with the past," for there have been periods in the past of which no traces remain at the present time and which not even the most penetrating intelligence could recover on the basis of theoretical knowledge of permanent laws and the observation of the present state of affairs. On the other hand, that would be sufficient to enable an intelligence provided with faculties analogous to, although more powerful, than those of man, to discover in the present state of things the sequential order of all future phenomena, or at least to take in a larger part of this order in proportion to the greater perfection of its faculties. Thus, however bizarre the assertion may appear at first glance, reason is better adapted to know the future scientifically than the past. The obstacles to theoretical prediction stem from the present imperfection of our knowledge and of our scientific instruments, and may be surmounted as a consequence of progress in observations and theory. Lost in the past are a multitude of facts whose nature is essentially hidden from any theoretical investigation based upon the verification of present facts and on the understanding of permanent laws, and which therefore can only be known historically, or which, for want of historical tradition, are and will always be for us as if they had never happened.[1] Now, if theoretical knowledge is capable of indefinite progress, the references of historic tradition, relative to the past, necessarily have a limit that all the researches of antiquarians will not be able to push back. Here we have a contrast of prime importance between theoretical knowledge and historical knowledge, or, if this way of putting the matter is to be preferred, between the theoretical and the historical element in our knowledge.

303. Astronomical tables enable us to predict eclipses, the conjunction and opposition of the planets, and all the circumstances of the movements of the stars which compose our planetary system for periods very far removed from our own times. As these tables

---

[1] " Wherefore future zeal cannot restore to the Silent and Dead that of which fleeting oblivion has robbed most of us already." Varro, *De ling. lat.*, at the beginning.

become more perfect, we shall be able to penetrate further into the future and to make the calculation of future phenomena more precise. In this matter we may do the same thing with respect to the past that we can with the future. Once these tables have been perfected as much as possible, we shall no longer have to borrow anything from historical knowledge in dealing with phenomena of this sort. On the contrary, we shall be able to apply and, in fact, we already apply science to history, by using it, for example, in calculating the time of an eclipse which occurred in the distant past in order to fix the precise date of an event that historians tell us was contemporaneous with that eclipse.

But if we wish to deal with an astronomical phenomenon like the appearance of the star of 1572, which suddenly disappeared without leaving any trace, it is quite necessary that history come to our aid. The most highly perfected theory, even when it can explain to us the causes of such a phenomenon as this, and even when it enables us to say which among the stars now shining are doomed to such a fate and when these events will occur, could not make known to us, without historical evidence, the existence of stars which attained great brilliance in the past but are now extinct and forever excluded from our view.

Perhaps some day we shall have a sufficiently clear understanding of the constitution of the earth and of the theory of the forces which move it to assign in advance the date and the phases of such geological phenomena as a volcanic eruption, an earthquake, or a great melting of polar ice, just as we can predict now the date and the phases of an eclipse. Yet this theoretical knowledge, no matter how perfect we may assume it to be, unless it is accompanied by historical evidences, will always leave us insuperably ignorant as to a great many geological phenomena which have left no traces, or which have only left traces which are insufficient to reveal all the essential particulars about the upheavals which the earth has experienced.

304. It is not within the power of an intelligence such as ours, or of any finite intelligence, to embrace in a single system the phenomena and the laws of the whole of nature. But even though we were able to do this, we should still distinguish in this whole the parts which are detached from one another and which are the object of theories that are independent of one another, although

they may be connected with a common origin (29ff). Here we find another cause of the insufficiency of theoretical knowledge, and another part necessarily reserved for the historical element of knowledge. For example, the planetary system, including the comets whose periodic return has been proved, is the subject matter of such a perfected theory that we are able, as we said above, to calculate all the actions that the bodies composing it exert on one another, and to do this in such a way as to predict the future and to go back into the past without the assistance of historical documents. But suppose that in the distant past a comet traversed this system and appreciably disturbed the motions of the bodies which compose the system; and let us further suppose that it then vanished into celestial space, or that, if it did not vanish, it passed forever from our view and beyond any influence on the planetary system in describing a hyperbolic curve. Neither theory nor observation of the present state of the planetary system could tell when and how such a disturbance took place; nor could they lead us to suspect the interference of this disturbing cause. Finally, all calculations that we may make with respect to periods prior to that in which the disturbing cause occurred will, without our being aware of it, inevitably be tainted with error, in the absence of historical information about the disturbing event. Similarly, the accuracy of the applications that we may make of the theory to future phenomena will rest upon the hypothesis that no unforeseen event of the sort we have indicated will occur to disturb the condition of the system. Indeed, if we were to know perfectly the present state of the whole universe, and not only that of the bodies which make up our planetary system, we should be in a position to predict theoretically a similar occasion or to affirm that one has not occurred. But over and above the fact that it would be chimerical to pretend to a universal knowledge, grounds would still remain for considering the planetary system as forming a separate system in the universe, a system having its own theory; and events whose sequence is determined by the laws and by the particular organization of the system should not be confused with accidental and adventitious disturbances whose causes lie outside of this system. These are those irregular and fortuitous external influences which must be regarded as becoming a part of knowledge under the heading of historical data, in contrast with that which is for us the regular result of the permanent laws and of the nature of the system.

It must be well understood that the contrast we are pointing out here clearly depends less upon the nature of the faculties by means of which we acquire historical and scientific knowledge, than it does on the very nature of the objects of knowledge. The external, irregular, and fortuitous influences of which we have been speaking would preserve no less of this triple character, even though we had some way to foresee them and to calculate their effects *a priori*, without the help of observation and of historical testimony (36). And while maintaining this triple character, these influences could not, at the same time, be considered as forming the subject matter of a science properly so called; that is to say, a body of systematic and exact doctrine.

305. In fact, we have a mass of knowledge, coming from different sources, to which we do not and should not give the name of science. The processes involved in metallurgy and in many other industries to which the discoveries of modern chemistry have given the secret were already well advanced long before chemistry deserved to be called a science. Meteorology is barely beginning to take on a scientific stability; but for many centuries country folk have had their meteorological predictions and proverbs that scholars can no longer despise even when they can explain them. Science is only one form of learning or of knowledge, and it appears only as the belated fruit of an advancing civilization, after poetry, after the arts, after historical, moral, and philosophical compositions. It is contemporaneous with erudition; but erudition must not be confused with science, although the world often honors scholars and scientists with the same name and treats them more or less as though they were on the same footing. When a person masters the exact topography of Athens at the time of Pericles or has untangled the genealogy of a Median dynasty, we should have more reason to esteem him than we should if his knowledge were about the plan of a modern capital or the alliances of a ruling house, for the one is more difficult than the other and presupposes a rarer conjunction of previous knowledge. Nevertheless, it is not because items of knowledge are more difficult to acquire or because they relate to distant objects or to ancient events that they must be regarded as being essentially different from analogous bits of

knowledge about contemporary events or objects which are close to us. One may have been able to collect in one's memory a great number of historical facts, to gather a mass of notions about the mores and the customs of the peoples visited in one's travels, to become familiar with the vocabularies and the idioms of a great number of languages. In this way one will acquire great learning; one will know much, without taking a place among those who cultivate the sciences, although a great many of the facts gathered may be capable of serving as materials in the construction of a scientific system. In the study of a language or of an art, such as music, we distinguish very clearly between that which serves as the object of a scientific theory and that which does not admit of scientific exposition, although it may still be an object of study, of learning, or of knowledge.

306. It was an accepted maxim among ancient philosophers that there is no science of the individual, the particular, the contingent, or the variable; that the idea of science refers to the idea of knowledge in so far as it applies to general notions, to necessary conceptions, and to permanent results. But in the present state of the sciences, we can no longer be content with these commonplace views. It is necessary to examine thoroughly and to make sure by means of examples how, under what circumstances, and by means of what conditions those bodies of doctrine which truly deserve the name of science take shape and become organized.

And, to begin with, is it true that science has only immutable truths and permanent results in view? By no means. There are sciences, like geology and embryology, which, on the contrary, rest essentially upon a succession of variable states and transitory phases. And even when we consider natural objects in what we call a stable and permanent condition, everything leads us to believe that this is still only a matter of relative stability, and that we regard as permanent what changes only very slowly and in such a way as to offer appreciable variations only over periods of time which exceed those that we are able to comprehend. Thus the stars which we call fixed really have proper motions which alter their reciprocal distances and the configuration of the groups that they appear to us to form on the celestial sphere. Yet these motions, which are inquired into only on the basis of scrupulously questioned observations which are performed with instruments of extreme delicacy, have not altered the appearance of the sky noticeably since the most remote historical times. The same thing is probably true of phenomena of every sort. The specific types of wild and untamed nature, whose variability from the beginning of historical times up to the present era has not been established, may well be subject to slow modifications. These modifications should be no more prejudicial to the dignity of the natural sciences than the slow perturbations of the planetary system or the still slower displacements of the stellar systems are to the scientific perfection of astronomy. The rapid changes that time brings about among the facts which fall within the scope of social order, by making more difficult the study of the economic sciences, as far as those aspects of them which are positive and determinable are concerned, do not strip from them the character of sciences. In a word, nothing requires that the objects of a scientific theory should be fixed and invariable and suited to all times and places.

307. Although we may not conceive of scientific organization without rules, principles, and classifications, and, consequently, without a certain generalization of facts and ideas, it is no longer necessary to accept to the letter the aphorism of the ancients to the effect that the individual and the particular have no place in science. Nothing is more unequal than the degree of generality of the facts upon which rest the sciences which in other respects are equally susceptible of the order and classification which constitute scientific perfection. In zoology and botany, specific types are dealt with which include myriads of individuals, all different from one another, and in which science is not interested. From the point of view of chemistry, each element or definite compound is a particular and individual object, absolutely identical in all particles of the same stuff, whether element or compound. Nature might have fashioned only a single instance of a crystal of a certain sort, and yet it would have its place among mineralogical types as genuinely as the species having the greatest number of individuals. In astronomy, the celestial bodies are considered as so many individual objects; some, like the rings of Saturn, up to the present time appear to be the unique example of their type; it was possible to regard our moon as such an object until the discovery of the moons of Jupiter; and the profoundest inquiry into celestial

mechanics rests upon nothing but the motions of a system limited to only a small number of bodies. Finally, geology is simply the careful study of the form and structure of one of those bodies of which astronomy describes the movements and summarily outlines the principal characteristics. This is not to say that zoology or botany are superior to chemistry or physics, astronomy or geology, so far as dignity and scientific perfection are concerned. Yet it cannot be denied that, in any order of empirical knowledge whatever, the scientific form can be outlined only after a sufficiently large number of facts have been collected so that some generality and guiding principle may emerge from their comparison. Even the deviations of ordinary rules, when they are compared with one another, manifest a tendency to take place according to certain laws. This is how the appearance of organic monstrosities, which were for so long simply a cause of superstitious terror among the ignorant and then an object of curiosity for the learned, have in the end given rise to a scientific theory which, under the name of teratology, today takes its place within the bounds of the natural sciences.

308. Science is knowledge logically organized. Logical organization or systematization may be summed up under two principal heads: first, the division of the subject matter and the classification of whatever objects scientific knowledge rests upon; second, the logical connection of propositions which reduce as much as possible the number of axioms, fundamental hypotheses, or experiential data, and from which everything has been deduced that may be deduced by reasoning, except to check the reasoning by means of confirming experiments. It follows from this that the scientific form will be more perfect in proportion as it establishes clearer divisions, more decisive classifications, and more clearly marked degrees in the sequence of relations. It also follows from this that to increase our knowledge and to perfect our science are not the same thing. Science is perfected by the conception of an apt idea which introduces a better order in acquired knowledge without increasing the amount of it; while a science which is enriched by the addition of new observations and new facts which are incompatible with previously adopted principles of systematic arrangement and classification will lose with respect to perfection of scientific form. Ordinarily this retrogression is only fleeting; it is the first symptom

of a scientific crisis or revolution; and just as the perfecting of form provokes new researches and an increase of what may be called scientific materials, increasing the materials in this way gives rise to new relationships which suggest other principles of systematic arrangement and classification.   When experience reveals new facts which are in contradiction to one of the fundamental hypotheses, we are led to set up other hypotheses which are in agreement with all the known facts and are sometimes more simple than those which have been given up.   Nevertheless, in order to be justified in affirming that progress in discoveries will ultimately in every case introduce the perfection of scientific form or the perfection of the logical organization of knowledge, it would be necessary to be able to affirm that the conditions of the artificial development of our intelligence are in perfect harmony with those of the ordering of the universe.   This is an assumption which many philosophers will not hesitate to permit themselves, but which will always appear reckless to a circumspect thinker.   Even more temerity would be shown in such an assertion as this: that, when dealing with questions relating to abstract and rational sciences which, like mathematics, human intelligence seems to draw from its own depths, we notice that the arrangement which satisfies best the conditions of logical order, which makes the divisions clearer and more symmetrical, the demonstrations more rapid or more rigorous, is not always the one which furnishes the best reasons for the discovered truths, for their consanguinity and their connections (155).

309. There are sciences, like the abstract sciences, whose subject matter has nothing in common with the chronological order of events, and which, having no historical data to accept, have nothing to borrow from history.   The theorems of geometry, the rules of the syllogism, hold for all times and for all places.   It is clear that no historical element enters into them. Yet, in other respects, these sciences as well as others may have, in so far as they are products of human activity, their own history, which serves to account for their nomenclature and for their external forms. Many of the sciences which deal with natural phenomena have dispensed with the necessary support of a historical basis or idea in their theoretical construction.   Chemistry and physics properly so called, which deal with laws that we regard as immutable and with

properties that we assume have always been inherent in matter, are in this group. As a result, in this respect there is no reason to seek the explanation of the details which the present order of things offers us by taking account of the way phenomena which precede them have succeeded and been connected with one another. Using some examples to explain what we mean, in chemistry the values of the *chemical equivalents* of each of the elementary substances must be obtained by observation. Once we have these values, we can deduce the values of the chemical equivalents of compound bodies from them theoretically. In so far as the theory is unable to give us, in a manner which satisfies everyone, the key to the relations which exist among the chemical equivalents of the various bodies which are regarded as elements, it will be necessary to accept the table of numbers which measure these equivalents as a fact and an empirical datum. But in doing this we do not give up the idea that the relations between these numbers must have a theoretical explanation (which is inseparable from the permanent nature of the bodies), which might have been discovered were this nature better known to us, and for which it would not be necessary to know through what phases the parts of the matter on which we have based our experiments have previously passed; for the same explanation must hold for the other chemically identical although individually distinct bits of matter whose histories are entirely different, or which have passed through entirely different phases. In the same way, each different ray of the solar spectrum has its own *index of refraction* for each refracting material, and, in the present state of theoretical knowledge, experiment alone is able to furnish us with the numerical values of these indices. We nonetheless admit that the causes of unequal refraction rest in the permanent conditions of the constitution of luminous rays. Thus a more penetrating theory would give the reason for these indices in such a way as to make it unnecessary to connect the historical knowledge of the phases through which the world has passed with the theoretical knowledge of the nature of light and of material bodies. The same thing may be said with regard to a great many *constants* or *numerical coefficients* or, even more generally (and without distinguishing between things which can and those which cannot be expressed numerically), in regard to a multitude of facts which are taken as experimental data in the sciences as they are constituted today; and yet no one would think of confusing these facts with historical data, for which only the history of past events, and not the theory of permanent facts, would replace the observation of actual phenomena.

310. On the contrary, in a great many other cases, the data that science accepts and on which it necessarily depends have and can only have a historical proof. For example, celestial mechanics gives us the theory of perturbations in the planetary system and shows us the stability of this system by assigning limits, in one way and another, to the very slow and very small oscillations that the elements of the orbits undergo. But it does not tell us what causes have established certain relations of distance and mass among the members of the system, so that once the order is established it tends to endure. The physical reason and the immediate cause of this very remarkable fact, the most striking of the indications of a directing intelligence (57), is certainly found in the series of phases the world has passed through before reaching that final and stable order whose majestic simplicity we admire.

Theory explains the causes of the tides and of their periodic and regular inequalities to us on the basis of the paths of the moon and the sun. With the help of certain empirical data, it enables us to assign in advance for each point on the coasts the hour and the approximate height of the tides for each day of the year and for many years to come. On the basis of the way in which the water is distributed in the oceans, their average depth and their total volume, theory also teaches us that the permanence of the equilibrium of the seas is guaranteed in the present order of things, so that the oscillations that the attraction of the heavenly bodies produce in them cannot go so far as to bring about the submersion of the continents. But what causes have determined, and have done so in such a way as to satisfy this condition, the average depth and the irregularities of the ocean basins, the volume of water which replenishes them, and that empirical *constant,* characteristic of each locality that is called in the theory of tides the "maximum tide for the port"? In the present state of things, this theory accepts these facts as so many observational data. There are no branches of the natural sciences which do not offer examples of analogous contrasts.

In the study of languages, grammatical structure is the subject matter of a truly scientific theory. With the exception of certain irregularities which must be laid to caprices of the ear or of usage, reasoning and analogy account for syntactical laws and forms. On the other hand, the material composition of words and the lineage of languages can be explained in general only by historical precedents, with regard to some of which we actually possess historical information, while others are lost in the darkness which surrounds the origin of races and peoples.[2] When we lack this historical information, these are for us so many facts, established by observation, that theory accepts and on which grammatical science rests.

The same thing can be said about the system of measurements that peoples and generations hand down sometimes after they have undergone slow and progressive alterations introduced by usage, at other times after sudden reforms due to the intervention of public authority. Among all civilized nations, the different customary systems of measurement have presented some sort of systematic co-ordination which is explained by the suitability of the notation and by other theoretical considerations. But the system of arrangement and of subordination among the parts always leaves the choice of certain fundamental standards an arbitrary matter. The reason for the diversity of these standards among different peoples lies entirely in historical antecedents whose trace is never completely effaced. For example, take the case of the metric system that the French legislators have constructed with the announced intention of offering to all peoples a system the elements of which all have a basis in science, and which would not bear the imprint of any particular nationality. It is not difficult to recognize the influence of a national tradition which, in the series of multiple or submultiple decimals of physical magnitudes, has led to a preference for those which approximate the standards used in an earlier system, in order to arrange, for the people upon whom authority has imposed it, the transition from one system to the other. So, too, the modern map of our departments, which was drawn up with the intention of abolishing the

---

[2] " In this respect *history* is necessary, for by no other means does [the past] come down to us; in the remaining respect, *art*, which needs but a few brief rules." Varro, *De ling. lat.*, VIII, 6.

influence of habits born from historical precedents, presents
a multitude of peculiarities which only the ancient provincial
map can explain. Similar remarks could be made about everything connected with the economic sciences and with the organization of societies (301). The association and the contrast of
historical data with theoretical or scientific data is observed everywhere.

311. In order that the function of the former may be or may
tend to be obliterated before that of the latter, in natural phenomena as much as in matters in which human activity intervenes,
the influence of individual and accidental particulars, for which
history alone accounts, must be of such a nature as to grow
gradually weaker, as in the physical examples mentioned elsewhere (54), and finally to disappear, so that a tangible influence is
left only to the permanent conditions found in the intrinsic nature
of the object to which science or theoretical knowledge is applied.
Otherwise, since the individual and accidental particulars act in
different ways, it is necessary that their effects compensate each
other and cancel each other out and thus tend toward the average
and general results which then become the only objects with which
science is concerned. But it is far from necessary that the conditions be always satisfied. The shock accidentally communicated to
some parts of the mass of a large object does not shake the entire
mass, and all the effects are limited to brief, internal vibrations
from parts adjacent to those which received the shock. Contrariwise, in the case of fermentation, having the process begin at one
point in the mass is sufficient to have it spread throughout the
whole mass, the molecular composition of which will undergo a
complete change. In the case of certain historical causes, time,
far from weakening their influence, extends and strengthens their
effects. The European calendar, which is now in use throughout
the world, is full of irregularities and anomalies that a scientific
editing would have eliminated, and for whose explanation it would
be necessary to go back to the obscurest beginnings of a little town
in Latium. But fate decreed that the calendar of the priests of
this city should become that of the peoples of southern and
central Europe, which later became part of its empire. Fate also
decreed that religious beliefs altogether contrary to those which
directed the original redaction were afterwards adopted by other

European nations that Rome, in the days of its ascendancy, had not been able to vanquish. Finally, the development of European civilization was destined to carry this calendar to the ends of the earth. It is clear that the longer the institution (so arbitrary, weak, and circumscribed when it was born) has lasted and spread, the greater the reason why it should last still longer and spread still more, and even that, all the initial causes having disappeared, the reasons for its existence can now no longer be deduced except from its earlier existence and duration. So it is that, in some way, an individual variety, resulting originally from the fortuitous interaction of external causes, has become stronger in being transmitted from one generation to another, becoming a racial characteristic, or perhaps even acquiring the value of a specific characteristic, and perpetuating itself independently of the influence of external causes, or even because of a force of its own which resists the action of external causes.

312. After all that has been said, we can believe that the distinction between history and science is even more essential than it seemed to Bacon (301), and that it is not exactly due to the presence in the human mind of two faculties, one of which is called memory and the other reason. Men would never have made use of their memory and their reason in order to write history and treatises on the sciences if, in the evolution of phenomena, one part were not composed of permanent and regular laws, susceptible to systematic co-ordination, while another part is left to the influence of anterior facts, products of chance or of accidental combinations among different series of independent causes. The notion of chance, as we have tried to show elsewhere (36), has its basis in nature and is not relative solely to the weakness of the human mind. The same thing must be said of the distinction between historical data and theoretical data. A mind which could penetrate further than ours into the series of phases through which the planetary system has passed would recognize, as we do, facts which are primordial, arbitrary, and contingent (in the sense that theoretical knowledge cannot explain them), and which it would be necessary to accept as historical data, that is to say as the consequence of the accidental concourse of causes which had acted in still more distant times. To presuppose that this distinction is not essential is to admit that time is an illusion, or is to set up an

order of realities in the midst of which time would disappear. But our philosophy undertakes no such bold flight. We are trying to remain within the sphere of ideas that the reason of man can attain while at the same time conserving the capacity to distinguish that which pertains to the particular characteristics of the human mind and that which pertains to the nature of things rather than to the way our faculties are organized.

313. What makes the essential distinction between history and science is that the former includes the succession of events in time, while the latter is concerned with the systematizing of phenomena without reference to the time when they occur. The description of a phenomenon all the phases of which follow one another and are interconnected necessarily according to laws that are made known through reasoning or experience falls within the domain of science and not of history. Science describes the succession of the eclipses, the propagation of a sound wave, the course of a disease which passes through regular phases, and the name of history can only improperly be applied to such descriptions. On the other hand, history necessarily enters into the picture (when there is an inevitable gap in our knowledge resulting from a lack of historical information) where we see not only that, in its present imperfect form, theoretical knowledge is insufficient to explain phenomena, but that even the most perfect theory would still require the collaboration of a historical datum.

If there is, properly speaking, no history where all events necessarily and regularly derive from one another, because of constant laws which govern the system, and without the accidental concurrence of influences which are extraneous to the system encompassed by theoretical knowledge, then a series of events which are not connected in any way has no history in the true sense of the term. Thus the records of a public lottery would show a succession of singular drawings; this may sometimes pique the curiosity, but it would not constitute a history. The drawings would follow one another without being linked together, with the first exercising no influence on those which come later. The situation here is analogous to those annals in which the priests of antiquity sought to record monstrosities and prodigies as these were brought to their attention. All these marvelous events, which had no connection with one another, cannot form a history in the true sense of the term, even though they may follow one another according to a certain chronological order.

On the contrary, the conditions of a historical connection begin to come out in a game such as backgammon. Here each throw of the dice, although brought about by fortuitous circumstances, nevertheless has an influence on the results of the subsequent throws. The requirements of historical connection show themselves still more in the game of chess, in which the reflective determination of the player is substituted for the chance of the dice, yet in such a way that the ideas of the player give rise to a multitude of accidental encounters when crossing those of his opponent. The account of a game of backgammon or of chess, if we should decide to pass the record of it along to posterity, would be a history just like any other, having its crises and its denouements. This is so because the moves not only follow one another, but they are also linked together in the sense that each move has more or less influence on the series of subsequent moves and is influenced by the preceding moves. Should the game become still more complicated, the history of a part of it would become philosophically comparable to that of a battle or campaign, except for the importance of the results. It might even be possible to say without whimsy that there have been many battles and many campaigns whose history no more deserves to be remembered today than does that of a game of chess.

314. Historical connection consists, therefore, in an influence exercised by each event upon other events, an influence which may be more or less extended, but which at least should make itself felt in the neighborhood of the event under consideration, and which, in general, is proportionately greater when it acts more directly on events closer at hand. The characteristic of such a connection is to introduce a certain continuity into the succession of acts comparable to that of which the path of a curve would give us the image in the graphic representation of certain phenomena (46), or, better still, like that which represents for us the outline of the course of a river on a map. In spite of the disorder and confusion of fortuitous and secondary causes with respect to accidents of detail, and without any theory, this sort of connection is all we need to enable us to grasp the general drift of events, to distinguish periods of growth and decline, of progress, of arrest, and of de-

cadence, and periods of formation and dissolution, among nations and social institutions, as for beings to which nature has given a characteristic and individual form of life. The task of the historian who seeks to raise himself above the role of the simple annalist consists in placing in a suitable light, of indicating without indecision as without exaggeration, these dominant and characteristic traits. He must do this without being misled by the role of secondary causes, even when fortuitous circumstances give them an air of grandeur and brilliancy before which the slower and more hidden action of the basic causes seems to fade away. It necessarily follows, and this is quite difficult for other reasons, that the historian takes account of the mutual influence, of the reciprocal penetration of the different series of events each of which has its own principles, ends, laws of development, and, so to speak, its open account in the book of destiny. The historian must unravel, in the very complex woof of historical events, all those threads which are subject to any interlacing and meandering.

315. But does not this historic connection itself indicate how history treated in this manner differs from a scientific theory? Let us assume (in order to follow out the comparison mentioned above) that we are asked to mark by a line on a geographic map the direction of a great watercourse or of a chain of mountains, and that we wish to speak of that general direction which prevails over local and accidental irregularities and meanderings. We contend that there is a great similarity between this problem and those that the philosophical historian proposes to himself. In fact, everyone knows that in order to have a correct idea of the relief of a mountainous country, it must be studied from a point of view which wipes out the innumerable irregularities and the bizarre convolutions that have accumulated as a result of local accidents, in such a way that at first they present to the eyes of the astonished traveler nothing but an inextricable labyrinth. From a higher or more distant point, from which one's view takes in a still greater number of objects at one time, we see outlined those great alignments which are unimpeachable evidence of a dominant principle of regularity and of an order in the disorder. If these lines of inclination, as we now call them, meet one another, it is necessary to look, on the one hand, for an increase of intricacy and of the disorder among the details toward the points where the

junction occurs. On the other hand, and when considering the whole situation, an increased amount of upthrusting is to be expected. The latter is the result of the interaction of the two systems of causes, the average and the general effects of which have had a common tendency in this direction. In order to give a good description of a mountainous country, it will be necessary to indicate as clearly as possible and, more particularly, as accurately as possible the great features to which all irregularities of detail are subordinated. These irregularities need not deceive us. We need not be mistaken as to the exact location of the point at which the greatest upthrust occurred simply because local accidents may have produced in some other part of the chain, where its average height has become appreciably less, a peak which is higher than even the loftiest summits of the section where the upthrust reached its highest point.

It remains to be seen whether or not all these orographic problems are among those which may be mathematically defined and which lead to a technical and exact solution. Now, this is not the case; and all the mathematical formulas which it might be pleasant to devise for this purpose would have the radical defect of being arbitrary. This is so because they would apply equally well to the case in which the fact of a constant and general direction is most striking, and to those cases in which there are many well-marked successive deviations in the general direction, and, finally, to those cases in which the irregularity and amplitude of continual convolutions exclude any idea of a general and dominant direction. Moreover, since the transition from one case to the other must be made through continuous nuances and gradations, logic is clearly unable to distinguish the extreme cases; this distinction can result only from an instinctive evaluation which becomes progressively less clear and less sure in proportion as the extreme terms of the series fall farther apart from each other. Even when the fact of a general direction is beyond question, the ideal line which indicates this direction is not a scientifically defined line; the bearing of the chain is not an angle which can be assigned within any desired limits of approximation, or the determination of which is affected only by errors inherent in every operation of measurement. If we actually want to draw the line on a map or to mark down somewhere the numerical value of the bearing, the choice of the line or of the reading will involve something arbitrary, or something of which we cannot give an exact account.

The same thing must be said of points at which mountain chains and masses, which also are called orographical systems, begin and disappear. In most cases, we cannot find among them such clear-cut breaks in continuity as would rigorously preclude connecting them with one another through some of their ramifications, and to do this in such a way as to abolish ultimately the most natural distinctions. At other times, on the contrary, notable material gaps, such as those which may be the result of the interposition of an arm of the sea, must be rejected as local accidents and should not lead us to fail to recognize the systematic unity of the disconnected parts. In the appraisal of the intrinsic value of all these systematic bonds and in the very conception of systematic unity, therefore, judgments occur throughout the process in which reason does not proceed by way of definition and logical deduction, and the probability of which does not lend itself to an accurate evaluation. Systematizing conceptions of this sort are introduced into the description of natural facts not only for the convenience of the mind, but also to give the key to a sound understanding of the facts in themselves. They must not be confused with truly scientific theories, and still less with the positive aspect of the sciences, which is amenable to the continual control of experiment. On the contrary, they have all the characteristics of philosophical speculation, characteristics on which we have insisted unceasingly throughout this work, and to which we must return in the following chapter, which is devoted to indicating the contrast between philosophy and what is properly called science.

316. Is it not clear by now that all these reflections apply, *mutatis mutandis*, to history when it is handled philosophically, to the picture of historical events when we intend to bring their dominant traits into relief and to forestall the confusion of their details by distinguishing the massive structures and the subdivision of the principal groups? Can this art of penetrating into the deep-seated reason of facts, of discerning their disposition, of laying hold of their guiding threads be reduced to fixed rules? Does it lead to categorical distinctions? Does it throw light equally in all directions? No, there is no doubt that it does not. All the systematic conceptions on which history is based, when dealt with

philosophically, may be more or less disputed, and none lead to what may properly be called demonstration or to positive and experimental verification. Yet no enlightened and impartial mind would hesitate to accept these conceptions as giving to the essential reason of things and to the progressive development of events an expression which is as faithful, as completely free from partiality and arbitrariness and as completely separated from fortuitous accidents as the imperfect means, of which our art makes use in such complex matters, permit. Actually, in portraying his thought, the historian, unlike the geographer, does not have access to graphic and perceptible symbols. He is like the traveler who does not have the advantage of a plan and who must make up for this deficiency by the strength of his memory and imagination and the picturesqueness of his style. Finally, like the philosopher, he is faced continuously with the necessity of using metaphorical language, in spite of the fact that he never ceases to recognize its inadequacy (211).

Thus historical composition is closer to art than to science, even when the historian is clearly less concerned about pleasing and exciting by making his accounts interesting than he is about satisfying the desire of our understanding to know and to comprehend. The historian, like the philosopher, or rather indeed because he is or may be a philosopher, needs those gifts of imagination that we very rightly suspect in the case of purely scientific work, just as does the philosophical painter of nature. According to the trenchant expression of one of the masters of critical literature, " One may say in this sense that he [the historian] needs to be a poet, not simply in order to be eloquent, but in order to be truthful." [3] So it is that history, whose connections with science and philosophy we have been examining, has connections with poetry and art, too, and in this way the three members of Bacon's tripartite division (301) tend to unite, yet without losing their identities.

Moreover, if the historian is an artist, and up to a certain point

---

[3] M. Villemain, *Tableau de la littérature au XVIIIe siècle*, lecture 29. [Abel François Villemain, 1790-1870. An outstanding nineteenth-century French critic, orator, and minister of state. He was the author of numerous books dealing with historical and literary subjects and was active in the political life of France prior to the revolution of 1848.]

a poet, in so far as he has to perceive discerningly a specific character—and the work of carefully perceiving and reproducing a specific character is always the business of art, not of science (193),
it is clear that his composition will have to share the characteristics
of poetic composition to a much higher degree when the dramatic
interest of the account, the importance of the actions, the strong
unity of the subject matter will, in spite of all he can do, place it,
so to speak, on the poet's tripod. Thus Voltaire has said, " An exposition, a plot, and a denouement are as necessary in a history as
in a tragedy ";[4] a maxim that must not be generalized too much
since, among things which have no necessary end, and which like
the sciences and civilization, on the contrary, lead to a process of
continuous perfecting, it is possible to have a strong historical
unity without plot or denouement. But at least it is possible to
say that historical composition, which is susceptible to as many
varieties of kinds as there are different temperaments and proportions among the principal faculties of the human mind, is singularly suited to bring their harmonies and contrasts into relief.

317. On its poetic side, history always has the privilege of exciting interest, however slight may be the importance of the events
related; and what science would neglect as too particular and too
individual is often what is best adapted to art. Moreover, events
which have left behind them no trace may still interest the philosopher, if they are applicable to some general maxim of morals
or politics that cannot be too strongly impressed on the mind and
justified through examples. But in the general system of human
knowledge, and considered as a necessary adjunct to scientific
knowledge, it seems that history should appear only in so far as it
tells of things which are necessary or useful in the explanation of
phenomena and of present facts and those which should happen
later. Everything which has disappeared without leaving any
traces and without having any influence on the order of actually
existing things has no basis, so to speak, on which it can be known
and becomes part of the realm of vague curiosity, that nothing
limits or determines. History will make note of the violent over-
flowing of a river which has filled up its old channels and cut out a
new bed in which it continues to flow, but it will ignore the description of the annual and periodic floods after which the river

4 Letter to President Hénault, January 8, 1752.

returns to its usual course. If the permanent effect of these annual floods is the progressive building up of alluvial soil, history will point out the general result without going into the detailed enumeration of phases which all resemble one another, and whose differences offer no particular feature worthy of interest, since all these differences must compensate for each other in the long run.[5] Thanks to the improvement of which scientific form is capable, the field of the sciences may spread further and further without the human mind's ceasing to include and master it. Conditions of another sort must limit the indefinite accumulation of historical materials; without this all proportion would be lost. It has not seemed fit to define and circumscribe the object of historical researches and traditions more than this, when the force of circumstances makes plain the need of a rule in these matters.

318. It often happens that contemporary historians usurp the name of science for history just as philosophers usurp it for philosophy. This is one of the abuses of modern style and one of the consequences of the illumination the sciences have shed and of the popularity they have acquired. The most serious inconvenience of this confusion is that of suggesting pretended scientific formulas, by means of which the fatalistic historian admirably explains the whole past, but to which he would not care to entrust himself in the prediction of the future. In that respect he is like the authors of those epic stories in which a divine personage reveals in the person of a hero the destiny of his race, provided, of course, that his clairvoyance ceases precisely at the time of which the poet has sung. We have no difficulty in understanding that we can reduce to scientific form certain branches of knowledge which relate to the details of the organization of human societies; for, with observations which have been accumulated by the use of statistics, we succeed in establishing positively permanent laws and relations whose variability itself argues a regular progression of sustained influences. But the same thing may not be said for

---

[5] " The events by which the fate of nations is not materially changed leave a faint impression on the page of history, and the patience of the reader would be exhausted by the repetition of the same hostilities, undertaken without cause, prosecuted without glory, and terminated without effect." Gibbon, *History of the Decline and Fall of the Roman Empire*, Vol. II, Chap. XLVI. [Modern Library Giants edition, p. 413.]

political history. This is so because in the migration of races, in invasions and conquests, in great revolutions of empires, and in changes of mores and beliefs, there are accidental facts and entirely individual forces, which are of such a nature as to exert a perceptible influence over all succeeding ages, or for which periods of time beyond our grasp would be required if their influence were to be blotted out. And, on the other hand, political history is a theatre in which the acts of fortune, however frequent and surprising they may be, are not repeated often enough or are repeated in circumstances too dissimilar for us to be able with certainty, or with a sufficient probability, to disentangle the disturbances of chance from constant and regular laws. Thus, a history may very well have its philosophy, but not its scientific formulation. It may have its philosophy, for philosophical discernment distinguishes causes of different natures, some permanent and others accidental, and it recognizes their tendency to be subordinated to one another, without always being able to give demonstrative evidence for its insights. And while, in regard to things which fall within the field of statistics, it is possible to run back over the series of proofs and thus to confirm the truth of the results already brought to light by earlier observations in such a way as to reach scientific certainty, it would be contrary to the way in which successive historical phases follow one another if they were to lend themselves to this confirmative experience. Consequently, however well the past may be or may appear to be explained, it can never give us more than a singularly indecisive glimmer of the future, not only with respect to accidents of detail but also with respect to general results, that may always be modified and completely changed by unforeseen accidents, like those which have modified the series of antecedent events, taken as a whole.

Yet, if pragmatic history can never become a science, it is perfectly plain that in certain fields of knowledge we find the scientific form connected with a foundation of historical researches. Numismatics, for example, assumes the forms of a science, has its rules, principles, and classifications, although its basis is drawn from history, and although it is of very little use except as it serves to clarify history. Thus one might experience some difficulty in putting it in its proper place in an encyclopedic system, and come to different conclusions, depending upon whether we attach more importance to form or to matter. But that merely proves the insufficiency of our artificial classifications and does not in the least affect the basis of things.

319. We must dwell further on these two expressions, "natural history" and "natural sciences," the simultaneous use of which seems to be traceable to an essential distinction which it is important to clear up. In fact, it is not without reason that up to the present time the name of history has been given to the description, or at least to certain parts of the description of nature. Take physical geography as a term of comparison. When we look at a terrestrial globe on which the contours of the continents and oceans are outlined, as though they had been arrested during the course of the latest revolution of our planet, as a result of the concourse of a multitude of causes having an extremely variable sphere of influence and degree of generality or particularity, we are impressed at first only by the capricious irregularities we see. Then after a more attentive examination certain general traits, certain remarkable conformities or resemblances in the articulations or of the terminations of the continents stand out.[6] To what extent are these resemblances fortuitous; to what extent can they be imputed to the action of a general cause? In its answer to such a question the mind hesitates more or less depending upon whether it is more circumspect or bolder. Facts of the same sort are too few in number to allow the elimination of accidental and disturbing causes to be made with certainty; statistics are inapplicable; truly scientific theory is not possible; but philosophical induction must not be neglected on that account (42). One passes from pure description, which is not a science, which is applicable to a kind of understanding comparable in all respects to historical understanding and to facts which can be explained only through historical precedents; one passes from that, we say, to philosophical speculation, and the force of things leads to it, *omisso medio*, without passing through the intermediary of scientific formulations.

[6] These analogies (*instantiae conformes*), which give rise to what may be called geographic philosophy, were very well pointed out by Bacon (*Novum Organum*, Bk. 2), at the same time as the analogies in the arrangement of the parts of vertebrate animals, which have served as a point of departure in anatomical philosophy.

Descriptive natural science need not be confused with natural history. Descriptive anatomy is a science, for it uses classifications and systematic connections which principally bring into relief the general and constant laws of organic structure, of conditions of organic unity and harmony, and not accidental facts and precedents for which history would furnish the key. If observation of the facts leads to a philosophy of anatomy, it leads to it or should lead to it only mediately, after the data have been subjected to the scientific co-ordination of which they are capable. On the other hand, in matters which relate to the geographic distribution of animals and plants, the grouping of mineral substances in rocks and veins, the distribution of the heavenly bodies and of their orbits in the tremendous expanse of space, and many other subjects that we cannot even mention, there is a multitude of facts the reason for which is purely historical; facts which are connected with one another historically (314) and not scientifically; facts which philosophy arranges in groups, as it does all facts of history properly so called, on the basis of probable inferences, but without being able to submit them to precise laws capable of experimental verification, as is possible in the case of the facts which serve as the basis for the positive sciences. Consequently, a natural history of the heavens and the stars will find its place side by side with the history of universal gravitation; a natural history of the earth, the strata, the rocks, the veins of ore, and the distribution of minerals over the surface of the earth will find its place side by side with physics, chemistry, and crystallography; finally, a natural history of plants and animals will find its place side by side with plant and animal physiology. These types will often be confused in monographs and in didactic writings; but at other times the distinction between them will be well marked. And even though the confusion were inevitable, reason would still have to grasp the principle of generic distinction, if it is to have a philosophical understanding of the whole.

CHAPTER XXI

ON THE CONTRAST OF SCIENCE AND PHILOS-
OPHY, AND ON THE PHILOSOPHY
OF THE SCIENCES

320. Let us put aside for a moment all the theoretical considerations which have been the object of this book up to this
point; let us run over rapidly in our thinking the historical picture
of the development of the human mind. However little attention
we give to the matter, we shall be struck by the contrast that this
picture presents to us between those two orders of speculation that,
in ordinary language, when we are not aiming at positive precision, we agree to designate under the names of " science " and
" philosophy." Science starts from certain elementary notions
common to all men, and combines them to form a body of doctrine
by means of the forces of reason alone, or rather, gathers together
observations and experiments by means of which we are able to
advance to the discovery of the laws to which certain phenomena
are submitted. Philosophy has to do with the origin of our
knowledge, with the principles of certainty, and seeks to make its
way into the reason for the facts upon which the structure of the
positive sciences rests.

The development of the sciences is essentially progressive. New
facts which they establish serve as a point of departure for the
discovery of others. The only things in them that are likely to
perish are their methods and systems; that is to say, the artificial
bonds which are posited to make possible the co-ordination of facts
whose natural connection still escapes us. Except for this nothing
limits the acquisitions they can make; the combinations of abstract notions are without number; the domain of nature will
never be exhausted by man, and the work of observers tends constantly to enlarge our idea of its immensity and infinite variety.
On the contrary, philosophical speculations are hedged in by a
circle of problems which, under diverse forms, always remain
basically the same. Even as they presented themselves vaguely to

thoughtful minds during earlier periods of human history, so they still present themselves, though more clearly expressed, to the enlightened minds of the luminaries of modern science, refined by the cultivation of literature and the arts. Man has it in his blood to pursue incessantly the solution of these mysterious questions, which are all of pressing interest to him; and whether or not he is able to attain his end, there is a secret pleasure connected with the efforts he makes to approach it. His thinking is improved by making a thorough examination of the conditions of an insoluble problem, as it is by actually solving a scientific problem or by discovering a new being or by determining the law of a series of phenomena.

When we speak here of man in general, it is clearly understood that we have in mind only those for whom philosophical meditations constitute an intellectual need. Such men have lived in all periods of the history of culture. No doubt there are many persons for whom this need does not exist, either because of a lack of culture or as a result of natural inclinations. There are some whose philosophy consists in scorning all philosophical speculation, and who repeat, after Montaigne, that it deals with matters which are beyond our reach (that is to say, to speak more precisely, it deals with questions which do not lead to a scientific and positive solution), "ignorance and lack of curiosity are two very soft pillows for a good head." But a person who has spent his life weighing the opinions of the philosophers in the balance of doubt is a long way from furnishing an example of this repose of ignorance and lack of curiosity. As Mme. de Staël put it, " To those who ask you of what use philosophy is, reply confidently: ' Of what use is anything that is not philosophy? ' "

321. Originally the human mind had only an obscure consciousness of its different faculties; it has succeeded only gradually in separating them, classifying them, and at one time recognizing their independence, at another time their subordination. We see that religion, morality, civil legislation, philosophy, and science have been confused during the infancy of all civilizations; and no one would venture to say that the struggle for the emancipation of the domination of one or of the other is definitely ended. Should we then be surprised if certain persons still tend to confuse philosophy with science even in our own day? And, it must be

said, today this tendency should be attributed not to scientists but
to philosophers. The positive sciences have acquired such a great
brilliance, they have rendered such indubitable services to
humanity, that it is natural enough to want to gain public good
will toward the speculations of philosophers by claiming the name
of science for them (318). Some treat philosophy as the science
par excellence; other more modest persons wish at least to think of
philosophy as a science or as a system of independent and autonomous sciences, as certain, as positive, and as progressive in their
nature as others may be, and which will be seen to be such just as
soon as those who cultivate them use the correct method.

These pretensions are not new.[1] It has long been the case that
in their prolegomena, nearly all philosophers have presented
philosophy as a science poorly fashioned prior to their time and
now waiting to be constructed. To them has fallen the task of
pulling down the poor constructions, of clearing the ground, of
setting up the plan, and of laying the foundations; that of their
successors would consist in carrying through the execution of the
plan and of going on with the building. But unfortunately (or, to
speak more accurately, owing to the force of circumstances), those
who follow them have not resigned themselves to this modest role
of continuators. These latter could not resist, any more than their
predecessors could, the temptation to alter the plans and to re-
build the structure from the ground up. So Voltaire was able to
say that *metaphysics* consists of things that everyone knows and of
things that no one will ever know. Yet this did not keep him from
busying himself with metaphysics throughout his whole life. Without being Voltaires, many discerning minds who have not gone
deeply enough into the reason for these continual alterations must
have conceived little esteem for a body of learning which is always
being remade, and the principles of which are perpetually being

---

[1] " And this method of study seems to me to be most appropriate both for
the sagacity of the teacher and for the advantage of the students, in order that
we may not seem to be more eager to destroy than to build and that we may not
be tossed about in constant uncertainty by the snortings of bold geniuses in
continuous changes of doctrine; but that at last, restraining the zeal of sects
(which is stimulated by the empty fame of innovation), mankind, *by establishing fixed doctrines, may proceed further, with unstumbling feet, in philosophy
no less than in mathematics.*" Leibniz, Dutens edition, Vol. III, 316.

brought into question. They have ceased to consider it as a true science, and in this they have been well advised. But they have gone further—they have regarded the field of philosophic speculation with disdain as a sterile field. Seeing clearly that philosophy is not a science, as most philosophers pretend, they have concluded that philosophy is naught. If the errors of man were able to prevail against the laws of his nature, the human mind would tend to be maimed by having one of its nobler faculties condemned to inaction or to impotency.

As a matter of fact, it is by no means necessary to conclude that philosophy is beyond the possibility of general improvement just because it does not have the same sort of progressive development as the sciences. There is no question that the germ of all the most important problems is found in the obscure texts of the Brahmans, beneath the bizarre symbols of the Egyptian priests, in the subtle dialectics of the Greeks, and beneath the dry argumentation of the Scholastics. Nevertheless, philosophy has made some progress at least in this sense, that its problems are more clearly stated, its difficulties better classified, and their subordination better established; that systems are recognized as belonging to generic types, that their mutual connection and affinities can be traced, that their probabilities can be weighed, and that the scope of the conclusions they involve can be estimated exactly. Philosophy still proceeds by the method of exclusion. If it cannot reach the solution of problems directly, it may be able, by an analysis which is often rigorous, to indicate the reason why these problems are insoluble or are amenable to a number of solutions which are either limited or indefinite. It shows the impossibility of certain solutions by establishing their incompatibility either with the data of science, or with the native wisdom and understanding of humanity. In this way it circumscribes the indetermination of a problem that the nature of things has not made amenable to a determinate and truly scientific solution.

322. Another error into which men have fallen by failing to discern the essential characteristics which distinguish philosophical speculation from the positive sciences consists in comparing them from the point of view of the difficulties they present. Thus philosophy, under the name of metaphysics, has been regarded by the unlearned as the realm of those abstractions which are most

wearisome for thought, and many philosophers seem to have taken over the task of vindicating this opinion. On the contrary, others have constantly said that sane philosophy or sane metaphysics is the clearest and simplest thing in the world. Now the positive sciences could not give rise to such contradictions. The method adopted in the exposition of a science, the clarity or the obscurity of the style, the choice of notations or of terminology, may make it more or less difficult to understand. But these things do not completely invert the order of the difficulties, reduce the most important heights attained by the science to the level of elements, or regard as most important those notions which others consider elementary. This clearly shows us that philosophy cannot be compared to science, in the sense that it would form either the first or the last step of the ladder. It is the product of another faculty of the intelligence which, in the sphere of its activity, is exercised and perfected in a way which is peculiar to it. It is also something less impersonal than science. Science is identically transmitted by oral teaching and in books; it becomes the common inheritance of all minds and soon loses the stamp of the exceptional mind which created or enlarged it. In the order of philosophical speculations, developments of thought are awakened only by the thought of others. These developments always preserve a personal character which obliges everyone to set up his own philosophy. Philosophical thought is much less under the influence of the forms of language than is poetic thought, but it still depends upon language, while science is transmitted from one language to another without any modification.

323. When we distinguish religion, morality, poetry, history, philosophy, and science as answering to different faculties in our nature, this does not mean that these faculties can be developed with absolute independence and that there is nothing to mark strong connections or even intimate penetrations among them. The anatomist has reasons for distinguishing the arterial, venous, lymphatic, and nervous systems in the structure of the human body. This distinction is not a pure mental construct, devised to facilitate the description and the study of a complicated whole; it has a natural basis in essential differences of composition, texture, and functions. Nevertheless, these different systems influence each other reciprocally, are interconnected and interpenetrate down to

their ultimate ramifications and can never perform their functions independently of one another. The blood carried by the arteries nourishes and stimulates the nervous tissue, the reciprocal influence of which is indispensable for the formation of the blood and the operation of the arterial system. This analogy shows us how, in the web of intellectual life, it may be possible to have distinctness and dependence among the same faculties simultaneously. Today no one would dream of denying the connections between philosophy and art, or of failing to recognize the well-marked distinction between the genius of the philosopher and that of the artist. In the same way, it must be recognized that the philosophic element and the scientific element, although distinct from one another, are combined or share in the natural and regular development of intellectual activity. Philosophy without science soon loses sight of our real relations with the universe to lose itself in imaginary spaces; science without philosophy would still deserve to be cultivated for the sake of its applications to the needs of life, but beyond that we cannot see that it would offer reason a sustenance worthy of it or that it could be regarded as the ultimate goal of the work of the mind. So it is, to borrow another example, that the philosophic theory of the court of law runs the risk of degenerating into scholastic subtleties if it does not rest on a solid study of the positive laws. The science of the laws, kept apart from any cultivation of philosophy, still finds a useful application in the bosom of the court, but, in the judgment of Cicero, he alone is worthy of the name of legal expert who, while instructing himself in the science of the positive laws, does not stop improving his thought until he has succeeded in reaching the reason of these laws.

324. The writings of philosophers continually make allusion to the findings of the sciences. Philosophers choose examples designed to give a more concrete expression to their ideas and to point out applications and verifications of their theories. Their choices are more or less apt depending upon the adequacy of the philosophers' fund of scientific information. But what is more essential to notice and what explains, in part, the confusion of philosophy with science is that, in the field of speculations which belong naturally within the domain of philosophy, here and there theories are found which are actually reducible to scientific form.

As an example, we have in logic the theory of the syllogism, which may be compared with that of algebraic equations (251). But this accidental interpolation of some scientific matters need not delude us, as it has the philosophers of the Scottish School, to the point of forcing the assimilation of the positive sciences (or, to use the English expression, of " natural philosophy ") with that order of speculations that, in our language and time, is especially designated by the name of philosophy.

The intervention of philosophy in the sciences is even more frequent and essential. The elements of a science cannot be brought to light without grappling with those primary notions by means of which it is connected with the general system of human knowledge, notions whose criticism is the proper field of philosophy. Each author, according to his turn of mind, confines himself more or less to this preliminary and inevitable criticism, although the body of science remains the same, whatever the philosophic system in which the critic has his place. If, so to speak, philosophy lays hold of the sciences at their foundation, it also dominates their highest reaches; and to the extent that the positive sciences have made progress, the mind finds new occasions for returning to the causes, the reason, and the end of things. In this way it is brought back to the position of philosophic speculation.

325. We have been indicating some general observations. Now we must enter into more detailed explanations which, however, will be nothing but the reproduction, from a new point of view, of the principal ideas presented in the preceding chapters. First of all, which of the innumerable meanings that the word " philosophy " has received, in the language of the common man as well as in that of professional authors, shall we accept? Shall we apply it, as was done in the last century, to the secret doctrine of a few scholars, working hand in hand for the subversion of ancient institutions, which they claimed were contrary to the progress of reason and to the perfecting of human nature? Shall we follow the ancients and give the name of philosopher to all those who purport to set themselves apart from the masses by their way of looking at and thinking of things; who retire from the concerns and the encumbrances of practical life, in order to give themselves up to the meditative life? Evidently, the question of contrasting science and philosophy, the philosophic mind and the scientific

mind, does not arise when we give the term philosophy meanings like those just mentioned.

Without stopping these abuses of language, if we separate the fundamental idea from the accessory ideas which have complicated it or sometimes even concealed it, we shall find that we understand essentially by philosophy, on the one hand, the study and investigation of the reason of things, and, on the other hand, the study of the forms of thought and of the general laws and processes of the human mind. Let us recall the names of authors or the titles of books that continue to be highly thought of; let us bear in mind what is now and has always been understood by the philosophy of history, by the philosophy of law, by the philosophy of mathematics, by chemical philosophy, by philosophical anatomy, and so on. By so doing, we shall recognize that there is no branch of human knowledge, whether its object be nature or man, the visible or the invisible, into which the philosophic mind has not claimed to have entered. The thing that always leaves its impress to a greater or lesser extent upon the thought of authors who discuss the unique character of philosophy is this more or less marked tendency to seek out the rational connections among the parts of a single whole, to grasp the profound reason for the phenomena observed or the truths arrived at, which may remain hidden behind immediate causes or logic premises, which are liable to vary, according to the arbitrarily selected system of logical coordination.

But certain things follow from the very fact that the field open to the philosophic activity of the mind is the investigation of the reason of things, since the reason of things is not something which falls under the senses, not something which can be established by means of sensory experience. The judgments that we form in this matter are only judgments which conform to an internal type, to an idea. Therefore, it is perfectly clear that, no matter what their subject matter may be, philosophical investigations lead us back into the world of ideas, and that every philosophic question is intimately connected with the evaluation of certain regulative and fundamental ideas or with the criticism of their representative value.

Reciprocally, certain things also follow when we do not limit ourselves to describing ideas as so many phenomena occurring in the human mind, which is the thing that identifies psychology or a branch of psychology; when, from the presence of these ideas in the human mind, we wish to infer the existence of certain relations among things, certain laws by which attention is called to the phenomena of the external world; in a word, when we wish to take account of the representative value of ideas. We believe that we have more than proved that we can do this only by having recourse to the intervention of the philosophic sense, of that superior sense which grasps the reason of things, and whose judgments, while in no way reducible to the forms of logical demonstration, have a probability which in certain cases entirely excludes doubt, and in other cases is so weakened that the mind is left in a state of complete indecision.

Thus, we see on all sides an intimate connection between investigation into the reason of things, to whatever order of things it is applied, and the criticism of the regulative ideas of the human understanding. It should be possible to present every essentially philosophical question from these two points of view; and, reciprocally, this twofold appearance is what characterizes philosophical questions, to the exclusion of all others.

326. It is wrong, therefore, to depict philosophy as the science or as the aggregate of the sciences which have as their subject matter the human mind and its various faculties, as over against the sciences whose subject matter is the external world and man himself, considered from the point of view of his organic nature and of the functions of his animal life. By the same token, this is also true of the term " metaphysics," the peculiar origin of which we now understand, but which the Scholastics, deceived by an erroneous etymology, always used by opposing metaphysics to physics, as the science of minds to that of bodies. For this reason, we say this term is poorly chosen. Philosophy enters into everything; into physics as into morals, into mathematics as into jurisprudence and into history, into mechanics, which deals with the motions of inert bodies, as into physiology, which deals with the most delicate forces of the organism and of the functions carried out by living beings. Philosophy enters into the sciences formed by arrangements of abstract ideas as much as it does into history which tells of actual events, and as it does into poetry and the arts which constantly use sensory images.

327. Positive history is contrasted with the philosophy of history, when by positive history is understood the totality of well-established historical facts that anyone can verify from the sources, and the certainty of which is independent of opinions that may result from philosophical conceptions designed to bring these facts together. Similarly, we must understand by positive science the collection of facts that anyone can verify, in such a manner as to become certain that they are exact, through one of those procedures which leaves no doubt in the mind, or which at most would give rise to the vain objections of a skepticism against which nature protests, and of which no one seriously takes any stock in the conduct of life. These positive facts are the materials of science, as well as of other knowledge which is non-scientific or not logically co-ordinated (308), but in and of themselves they do not constitute science. It is still necessary that certain ideas intervene in order to bring about their distribution and classification, in order to reduce them to sequential order or methodical arrangements, and especially in order to give us the key to and the reason for them. Therefore, by varying our examples and by considering the different sciences one after another, let us examine for a little while how we can distinguish the positive aspects of the sciences from those which are not, and how the different sorts of elements, whose concurrence is required for the organization of science, are combined.

328. Pure mathematics is found in the highest position. The characteristic which distinguishes it from all the other sciences is that it rests upon exact truths that reason is capable of discovering without the help of experience and which can nevertheless always be confirmed by experience, within the limits of approximation that experience allows (28 and 268). Verification is exact in matters relating to enumeration, computation, and calculation, for which the mind uses the assistance of conventional and discontinuous symbols which have a fixed and determinate value. On the contrary, in matters involving the comparison of continuous and measurable magnitudes, verification will be approached more and more closely in proportion as we work with greater care and with the help of more perfect instruments. For example, given this proposition of geometry—"The number of solid angles in a polyhedron, added to the number of its planes, gives a number

higher by two than the number of the edges "—we can construct as many polyhedrons as we wish, and it will always be found that the theorem is verified exactly because the numbering of the angles, planes, and edges can be carried out with absolute precision, and because whole numbers can pass from one value to another only abruptly. On the contrary, given this other proposition—" The sum of the angles of a rectilinear triangle is equal to two right angles "—it may also be verified on as many triangles as one wishes by carefully measuring the three angles and adding the three measured values. But in this case the verification will never be more than approximate, and the degree of this approximation will depend upon the care taken in the working and in the perfection of the instruments of measurement. In both cases, this possibility of endlessly verifying the conclusions of theory through experiment is what confers on mathematics the character of a positive science. So it is that, resting on both bases of human knowledge, mathematics forces itself irresistibly on the most practical minds as well as on extraordinary minds which are most attracted toward abstract speculation.

On the contrary, this experimental confirmation is not possible for sciences other than pure mathematics, the subject matter of which is ideas and relations that the reason conceives, but that do not fall under the senses (268)—such sciences as jurisprudence, morals, and natural theology. For this reason, unlike mathematics, such sciences cannot be called positive. Thus, set over against natural or rational theology are the positive religions or dogmatic theologies, which draw their authority from a principle other than the rational principle. Similarly, the decisions of the civil legislator are called positive law. These decisions constitute so many facts whose authority subsists independently of all reasonings of legal experts. These cases exist in a manner which is almost analogous to the way historical facts exist, irrespective of the conceptions that reason uses to explain their connection.

Therefore, the oft-repeated statement that the passions and the vices of man may obscure the truths of geometry, as they do those of morality and natural religion, if the same interest is taken in them, is wrong. The most evil passions could not prevent mathematical truths from lending themselves, to the exclusion of those of morality and natural theology, to an experimental confirmation which establishes them as positive facts, and which prevails against all the fallacies of the heart or of the mind.[2] Passions, interests, and prejudices divide mathematicians just as they do other men, even about that which falls within the province of mathematical speculation and the realm of pure reason, as soon as the issue concerns ideas which can no longer be given a sensory form. They differ, too, about theories which are a part of science, but not of that part of science which is positive and compatible with empirical verification.

Moreover, we shall point out later how morality also may be treated as a positive science when regarded from a certain point of view. But this is a question of an empirical morality and not of morality considered as a rational science like pure mathematics.

Certain parts of logic, the theory of the syllogism, for example, can be considered a rational and positive science at one and the same time. In fact, nothing stands in the way of verifying the correctness of the laws of the syllogism by showing, through the use of as many examples as may be desired, that as soon as we violate these rules we are led into arguments which are patently false and absurd. For this reason, no one has ever raised doubts as to the rules of the syllogism, any more than they have about those of arithmetic and geometry.

---

[2] " If you and I were to disagree as to whether one number were more than another, would that provoke us to anger and make us enemies? Should we not settle such a dispute at once by counting? . . . . And if we were to disagree as to the relative size of two things, we should measure them and put an end to the disagreement at once, should we not? . . . . And should we not settle a question about the relative weight of two things by weighing them? . . . . Then what is the question that would provoke us to anger and make us enemies if we disagreed about it, and could not come to a settlement? Perhaps you have not an answer ready; but listen to me. Is it not the question of right and wrong, of the honorable and the dishonorable, of the good and the bad? Is it not questions about these matters which make you and me and everyone else quarrel, when we do quarrel, if we differ about them and can reach no satisfactory agreement? " [Plato, *Euthyphro*. Quoted from the translation by F. J. Church, as revised by R. D. Cumming, " The Library of Liberal Arts " No. 4, p. 7f. (New York: The Liberal Arts Press, 1954.)]

" Nothing can be adequately known except by experience. Argumentation reaches a conclusion but cannot certify the result or remove our doubt, in order that the mind may rest at ease in an intuition of truth, unless the road of experience discovers it." Roger Bacon, *Opus majus*, Part IV, Chap. 1.

329. In the exposition of mathematical doctrines, we come across principles, ideas, and conclusions that cannot be submitted to the test of experiment. We also find in the writings of mathematicians discussions relative to theoretical questions that experience knows no way of settling. Either of these situations is, by itself, sufficient to tell us that these questions do not pertain to positive science; that, properly speaking, they are not mathematical or scientific; that they fall within the field of philosophical speculation from which, however we may try, science cannot be completely isolated, and from which it could become separated, were such a thing possible, only at the expense of its own high position. The intimate union and yet original independence of the philosophical element and of the positive or properly scientific element in the system of human knowledge manifest themselves here through this very remarkable fact, that the mind cannot proceed in an orderly fashion in scientific construction without adopting some philosophical theory, and that nevertheless the progress and certainty of science do not depend upon the solution given to the philosophical question (156 and 215).

Suppose a person wishes to write a treatise on algebra, differential calculus, or mechanics; or suppose he is appointed to give a course of public lectures in these sciences and must clearly set up his method in such a way as to introduce negative quantities, infinitesimals, and the measurement of forces; even then, as far as possible, he should set aside each of these questions by itself, in monographs and in separate studies. A person will impose his method dogmatically, or rather, he will lead his reader or listener to it in a roundabout way, by using a critical discussion, or through the weight of inferences and the authority of examples. But whatever method is used, it will be necessary to come to a decision about the system itself. Yet whatever this indispensable method may be, the same theorems, the same formulas, and the same technical applications will be attained. For example, everyone uses the same rules for finding the negative roots of an algebraic equation, whether he adopts the same way of looking at negative roots as did Carnot, d'Alembert, or someone else. There are no philosophic conceptions which serve as necessary premises in scientific construction, analogous to the foundations of a material structure, which must be made firm if the structure is not to fall to pieces.

This kind of construction stands in spite of the imperfection of its foundations, because at each step it finds so many buttresses and solid supports in verifications both by calculation and by experimentation. So far as satisfying the demands of reason is concerned, this should not hinder us from perfecting without limit the ideas according to which we can take account of the theory philosophically, in so far as these ideas are dependent upon us, and which would be sufficient justification for them in the absence of all verification and all experimental control.

In mathematics, philosophic controversy often rests not on the primary ideas which are the point of departure for the theory, but on the way in which certain results obtained by means of deduction must be taken and on the degree of extension and restriction which they allow. Does the sort of proportion which can be demonstrated (and which can be verified, if need be) for commensurable magnitudes, no matter what the value of the commensurable relation may be, subsist for incommensurable magnitudes, and, if this is the case, in what way must it be defined and understood? This is certainly a question which pertains to mathematical theory, but not to the positive part of mathematics. For, as soon as it is a question of proceeding to actual measurements, we can mean by incommensurable magnitudes only those whose common measure is proportionately smaller when we work with greater care and with more perfect instruments. Therefore, when mathematicians, not content with this simple observation, which would suffice for the practical man and for the empiricist, go to the great trouble of reasoning in order to prove that the proportion established in the case of commensurability still holds when we pass to incommensurables; when they devise different methods for demonstrating this result directly or indirectly; when they admit some and reject others, although experience cannot intervene to end their discussions; they fall back on the analysis and criticism of certain ideas of the understanding which are not susceptible to perceptible manifestations. They place it on the ground of metaphysical speculation. They practice what we are agreed on calling metaphysics, even though men may disparage it, and by no means what may be called positive science.

The philosophy of mathematics still consists essentially in discerning the rational order of dependence of as many abstract truths as the sagacity of inventive minds has successively and laboriously discovered, often by very roundabout means. It also consists in preferring one concatenation of propositions to another (although the latter is just as impeccable from the point of view of logic, or sometimes even more convenient logically), because it satisfies better the condition of exhibiting this order and these connections, just as they follow from the nature of things, independently of the means that we have of demonstrating and knowing the truth. It is evident that this work of the mind must not be confused with that which has as its object the extension of the positive sciences. It is just as evident that the reasons for preferring one order to another are in the category of those which are imposed on us neither by experience nor by the processes of logical demonstration. To deny the philosophy of mathematics because of that would simply be to deny one of the conditions of the rational and regular construction of mathematics; just as to deny philosophy in general would be to deny one of the conditions of the construction of the general system of human knowledge, or one of the essential elements of this system.

330. In physics, the facts and even the laws that anyone can establish by means of concrete experiments, and against which the arguments of skeptics who deny the value of the evidence of the senses would be opposed in vain, make up the positive part of the science, as distinguished not only from premature and dubious hypotheses, but even from conceptions about which all physicists agree, but which are nevertheless incapable of concrete demonstration. For example, when a chemist gives the formula for the composition of a substance, all that experiment can verify is the presence of certain elements associated in certain measurable proportions. This is the positive result of the analysis of the chemist; but this is not sufficient to establish a body of doctrine. But the formula does something more: it also expresses an idea according to which we conceive the elements as united in a certain way to form compound bodies of the first order; the latter as united in such a way as to form compound bodies of the second order, and so forth. When the structure of the compound substances submitted to analysis is conceived in this way, it may happen that all chemists agree, being persuaded by the force of analogies and inferences. But it may also happen, and often does happen, that

their opinions differ. Yet it is not necessary that the controversy arising among them, a controversy which is philosophical rather than scientific in the proper sense of that term (being one which rests upon ideas rather than upon facts), should be capable of being decided by means of experimentation, which gets at only the concrete fact. Furthermore, this difference of opinion need not upset the progress of discoveries in the positive aspects of science or the series of applications made of it daily in meeting the needs of industry and in the practice of the arts.

The substances that today we call "hydrochloric acid" and "chlorine" were still called "muriatic acid" and "oxygenated muriatic acid" in the early years of this century, even after the great discoveries of Lavoisier. Instead of deriving the former from the latter by the addition of hydrogen, the latter was derived from the former by the addition of oxygen. Chemists know that all the sensible results can be explained as logically by the older theory as they can by the newer one, provided that the corresponding quantities of oxygen and hydrogen are present in the same proportion, by volume, which they have when they combine to form water (220). Therefore, when, on the basis of many striking analogies and inferences, it occurred to Gay-Lussac and Thénard to regard Lavoisier's oxygenated muriatic acid as an element, to which they gave the name of chlorine, they changed the philosophy of chemistry completely by this illuminating idea. Yet this did not affect the positive facts of which chemistry was already in possession, and it did not change the nature of the applications that were made up until that time of chlorine and its derivatives in the arts and in industry. On the contrary, when other chemists, whose investigations were stimulated by the researches of these men, later discovered "iodine" and "bromine," substances which have so great a chemical kinship to chlorine, their discoveries not only gave support to the theoretical views which had established chlorine as an elementary substance, but their work, at the same time, also enriched the field of positive chemistry by increasing the number of technical applications, and it was also notable because it opened the way for one of the most unusual inventions of our age, that of photography.

The same chemical analogies which establish the relationship of chlorine, iodine, and bromine, and which constitute the reason

why it is now agreed to regard them as so many elementary substances or chemical radicals belonging to the same group, are likewise the basis for agreeing to admit the existence of another radical
of the same group, to which the name " fluorine " is given, although up to the present time it has not been possible to separate
it from the well-known compounds in which it is thought to be
present.   Everything that has been said about fluorine pertains
therefore to the philosophy of chemistry and not to positive chemistry.   Someone may succeed in making fluorine known [3] in the
same way in which chlorine was made known, but it would still be
possible, strictly speaking and without contradicting any positive
fact, to say of fluorine what was said of chlorine before the theory
of Gay-Lussac and Thénard was adopted.   To do this would be to
carry on an erroneous philosophy, but it would not be contrary to
the positive facts of science.

   If chemists are in agreement with respect to the substances we
have been speaking of, there are others concerning which contrary
theories are still to be found.   Thus the school of Berzelius admits,
under the name " ammonium," a problematic metal which has
never been isolated and which probably is not isolable, but whose
metallic nature becomes evident through the properties of the
products which result from its combination with other substances.
Why, indeed, should there not be some chemical combinations that
purely accidental and accessory circumstances would tend to keep
from occurring except in the presence of reagents which have
brought about their formation?   And why should these accidental
circumstances conceal from the eyes of reason the essential or
fundamental analogies that these combinations would exhibit with
other combinations whose existence in a state of isolation is made
possible by equally accidental circumstances, and which, thus isolated, are able to affect the senses strongly?   Is it not the peculiar
quality and the natural function of reason to arrange the characteristics of things and the things themselves according to their
intrinsic proportion without stopping with appearances and concrete facts?   However, let us say, along with [J. B.] Dumas, that,
" Every time a theory requires the admission of unknown substances, it must be challenged; it is to be assented to only with the

---

[3] [Fluorine was isolated in 1866 by H. Moissan.]

greatest reserve, when it is no longer possible to deny it, or at least when the most forceful sort of analogies are presented by it." [4] But when do these analogies reach this point? This must be decided by the philosophic sense and it can neither be logically determined by means of definitions and precise rules nor established experimentally. This matter falls within the domain of the philosophy of chemistry and in that area in the organization of the theory which remains distinct although not separate from the positive part of our chemical knowledge.

331. Philosophical conceptions play a still greater role in the biological sciences.[5] We find them intervening everywhere: in the classification of species, in the anatomy of organs, in the explanation of regular and of anomalous developments, and in the theory of forces and vital functions. On all sides, when a need appears for facts that are established by observation, and the constantly increasing methodically arranged collection of which is the positive part of the biological sciences, ideas arise designed to express the real order that nature has introduced among these facts so as to link them together rationally, to determine the organs, to indicate relationships and analogies, and to bring out ties of subordination and dependence. The deep questions to which these ideas gave birth, and with which so many illustrious scholars have occupied and still do occupy themselves, are not scientific questions (as is often said with little justification). They do not lead to positive solutions. They are not questions that can be settled by observation and experiment. The way in which they are solved has no direct influence on technical applications, does not tend to extend or to restrict the power of man over the rest of creation, as happens

---

[4] *Leçons de philosophie chimiche*, given at the Collège de France, lecture 9, page 341.

[Jean Baptiste Dumas, 1800-1884, an eminent French chemist and senator. Elected to the Institute in 1832, he became professor of organic chemistry in the School of Medicine. In addition to holding important public offices, he rendered France significant public service. His writings include a *Treatise on Chemistry Applied to the Arts*, published in eight volumes, 1828-1845, and the work referred to above, published in 1837.]

[5] [Since we commonly use the term " natural science " in a manner which includes both the physical and the biological sciences, the term *les sciences naturelles* is translated by either " the biological sciences " or " the life sciences." See note 7, page 506.]

in the case of questions of fact, that is, with respect to those which belong to the positive part of the sciences. But, on the contrary, these are the questions which most enhance thought and toward which the mind feels invincibly drawn by the need of exercising the highest faculties nature has placed in it.

A celebrated anatomist of our times [6] has sought to set up a distinction and a contrast between what he calls "descriptive" sciences and those to which he gives the name "general" sciences. It is our opinion that this is nothing but the distinction and the contrast between the positive part of the biological sciences and the philosophy of these same sciences. This will be judged by the passage that we quote below and by the comments we shall make about it, in order to make our own ideas clearer.

The unity, harmony, and purpose of man and nature would be poorly known, if our knowledge were limited to what comprises the descriptive sciences. However indispensable it is to know the truth about the details which make up these latter sciences, we are aware that these truths are not detached from one another; we know they impinge upon and are connected with each other by diverse and numerous relations, by some common element which serves them as it were as a source and point of departure. Therefore, the study of these relations, the examination of these principles forms a new series of general facts which are discovered to be the key to all these particular traits and to bring them together as a body of doctrine. This is the purpose of the general sciences. . . . In the descriptive sciences, we are always in search of the characteristics which differentiate facts from one another; in the general sciences we search for the characteristics which relate them to one another. In the former, nature is dissected and facts are isolated; in the latter, they are connected and joined by the force of analogies. The study of the analogy between organic beings, therefore, forms the essence of the general sciences, as that of their differentiating characteristics forms the essence of the descriptive sciences. That is the source of their differences, and of their subordination, of the simplicity of the descriptive sciences, and of the complexity and the extension of the general sciences. . . .

Since the sole purpose of the descriptive sciences is to make known a

[6] Serres, *Principes d'organogénie*, Part 1.
[Etienne Renaud Augustin Serres, 1786-1868. A French physiologist. He was appointed professor of comparative anatomy at the Jardin des Plantes in 1839. In addition to the work referred to above, he published *The Laws of Osteogeny* in 1815, and *The Comparative Anatomy of the Brain in the Four Classes of Vertebrate Animals,* in two volumes, 1824-1826.]

given object, a series of organs or bodies, their work is in some ways entirely mechanical, entirely material. . . . On the other hand, because the general sciences propose to establish the conditions necessary for the existence of the organs of living bodies, and because they propose to make known how they came to be what they are, either taken by themselves, or in relation to one another, their work is necessarily superior, more intellectual; it is concerned entirely with reflection and comparison. . . . If it is permissible to put it in this way, the descriptive sciences are the body of nature, while the general sciences are its spirit.

Therefore, there can be no doubt that facts may be the basis for the general sciences, as they are of the descriptive sciences, but they differ in their quality. The facts which make up the descriptive sciences are elementary; those in the general sciences are raised to the second, third, and fourth power. As far as the facts are concerned, that is the whole difference, and the certainty will be the same for the one as for the other.

But if the certainty is the same in these two types of sciences, it must be recognized that the causes of error are both much more numerous and much stronger in the general sciences than in the descriptive sciences. The latter have only one danger to avoid, that of saying too much. Because of the desire to go into details, we become prolix; we smother the salient characteristics under a mass of insignificant ones; we describe without making known. Descriptive human anatomy has often shown this failing. This is the source of its dryness, but it is also the source of its invariable certainty. The situation is just the opposite in the general sciences. Just as soon as the mind has grasped a relation, a characteristic which is common to many facts, it aspires to understand everything; it invents instead of explaining; it wanders instead of directing. If generalizations are to be useful in the sciences, it is necessary to know how to limit them. So we see here, in the abuse of details and in the abuse of generalities, the two sources of dangers in the biological sciences. . . .

Now, with all due respect for an eminent naturalist, let us ask what can be meant by "facts raised to the second, third, and fourth powers"? Very general facts admit of as much certainty as individual facts and may, for that very reason, furnish the materials for a positive science. Thus it would be as certain that all mammals have a spinal column, although we have not dissected all the species of mammals, as it is that all horses have one, although not all the individual members of the species have been dissected. What makes a fact positive is the possibility of indefinite verification in all individual cases, although the number of individual cases is never exhausted. Furthermore, general facts are decribed in the same way as special or particular facts. For

example, along with Bichat we may describe the general properties
of different tissues which enter into the structure of organs, exactly
as we describe the structure of each organ in the particular anatomy. In this sense, the sciences which the author we have quoted
calls " general " are descriptive like the others and also as positive
as the others. Perhaps they are more difficult, but they are amenable in the same way to uniform and continuous progress in the
hands of gifted observers and faithful historians. On the other
hand, a purely descriptive science, in which facts lack any natural
or artificial connection, in which they cannot be submitted to any
classification, because no mention has been made of the relations
which unite them and of the generalities which summarize them,
such a science, we say, would not deserve the name of science
(308), and nothing of this sort is to be found within the limits of
the biological sciences. Finally, it must be recognized that simplicity is no more likely to be found in sciences which deal with
details and particular facts. On the contrary, they should be note-
worthy for their complexity, whereas the sciences which aim at .
ascending to generalities and principles tend by that very fact
toward simplicity and unity.

But it is incontestable that in this work of progressive abstraction and generalization, philosophical thought, which discerns
analogies and goes back to the reason of phenomena, moves further and further away from facts which are susceptible of rigorous
verification and concrete demonstration. As a result, when experimental control is lacking, the individual reason more and more
runs the risk of being left to its own fancy. In other words, the
philosophical element predominates in the sciences that the author
calls general, in those which, even because of this predominance
of the philosophical element, must, according to him, be considered
as the *spirit* of the biological sciences. Inversely, the positive or
strictly scientific element for the most part enters into what he
calls the descriptive sciences, which he considers as the material
part or as the *body* of the biological sciences. The other remark,
that the characteristics which differentiate them may be determined with an altogether different precision than the resemblances,
is perfectly correct and comes back to our theory of continuous
and discontinuous transitions. When we face the problem of
differentiating beings, and for that purpose take them in a fully

developed condition in which they are most unlike one another (131), the work of the mind rests on perfectly distinct ideas, to which our artificial nomenclatures and logical methods are very well adapted. On the contrary, in order to perceive resemblances, it is necessary to fill in gaps, to come back to continuous transitions which by their nature rebel against logical distinctions and classifications and present many obstacles to the perfecting of the scientific form.

332. Nevertheless, there is one sense in which this expression (which seems so strange when one first runs across it), "facts raised to the second, third, and fourth powers," can be clearly understood. In all branches of statistics we collect great numbers of particular facts for the purpose of establishing, through the calculation of the mean values, not exactly facts but tendencies or aptitudes which are due to the action of constant causes, whose nature we usually ignore, but whose existence can always be revealed when a sufficient number of particular observations have been accumulated to compensate, almost exactly, for effects due to accidental anomalous causes (35). Such tendencies and such influences, which do not fall directly under the senses and which the numbers alone imply, are facts of another sort than those which are particular and concrete. Nevertheless, they too are positive facts in this, that it is possible to repeat the experiment and to confirm by means of a new experiment what had been concluded from the earlier one. Since these facts are derived from the observation of a great number of particular and concrete facts, through the elimination of what is sensible and particular in them, a person might say, if he so desired, that they are positive facts of a different order or of a different degree than concrete facts. Furthermore, since we can compare the averages of a great number of partial series and take the average of these averages, in order to draw from them what is common to all these series or what is independent of the conditions peculiar to each series in particular, we see that in this sense it is permissible to admit positive facts of different orders or degrees. Thus individual varieties are with reason regarded as being of another order than racial varieties, the latter of a different order than specific varieties, and so forth.

But as phenomena the nature of whose laws we are studying become more complex, the more cases there will be in which we can make our experiments only under a form other than that to which statistical investigation is suited. Even now the physical sciences offer us many examples of this in some of their branches, such as meteorology. The situation is something altogether different in the biological sciences in which, as is so often said (not without some exaggeration), there are no absolute rules. We take as a rule in these sciences only what happens most commonly or what has a very great tendency to happen, always leaving room for anomalies and exceptions. Whatever is to be thought of this maxim, which is itself not to be taken in an absolute sense, it is certain that the number of cases in which experimental and positive confirmation is practicable only by means of statistical procedure should be singularly increased as we go from the physical to the biological sciences, simply because of the complication of causes. Thus the positive part of the medical sciences can be little more than that part which rests upon experiments instituted on a large scale, after the manner of statistical experiments. Nevertheless, this is not the sort of knowledge the sick person asks of his doctor, and it makes little difference to him that the doctor knows whether, with a certain type of treatment, the proportion of cures is sixty or sixty-five out of every hundred. The patient is concerned with the particular and concrete fact; and the accidental complication may have much more importance for him than the constant tendency. Thus there has always been for the doctors not only an art and a bringing together of bits of knowledge, but also a methodical science that was taught and learned before anyone dreamed of collecting and publishing any such statistical numbers. Only this science is not positive, if by positive science must be understood (as we believe we have established) that which rests upon facts capable of being established through experience. Indeed, wherever medicine is not in possession of absolute rules, the experience of an individual fact proves nothing; and what is called " long run experience " is simply statistical experience divested of the numbers which give it a precise meaning and conclusive value.

333. When we enter into the sciences which deal with the moral nature of man and with the organization of human society, the growing complexity of the objects which they include makes it much more necessary to appeal to observations whose great number compensates for the anomalies of chance, if we wish to give

experiential confirmation to these sciences and to support them on the basis of positive data. Hence the very name "statistics," which, in the correct and restricted sense, is synonymous with "political arithmetic," but which is often extended, as we ourselves have extended it in this work, by analogy to all investigations whose subject matter is the accumulation of great numbers of particular facts, for the purpose of discovering constant influences and tendencies which run through the effects of accidental and variable causes. Furthermore, we should be mistaken if we believed that the growing complexity of causes makes it necessary to pile up a greater number of particular facts in order to keep the average results perceptibly uniform. It is rather the opposite which happens. For example, in our country there is clearly greater lack of uniformity from one year to another in the mean temperature, in the mean direction of the winds, and in the amounts of rainfall, than in the relation between the number of persons accused of some crime to the total population, or in the proportion of the number of those accused who are convicted or acquitted. It is primarily in regard to things which depend more upon man's free will that, since each individual case is increasingly independent of those which are contiguous, there are fewer irregular influences of the sort which (as in the meteorological phenomena cited as an example) could affect a whole series of particular observations, only to make way subsequently for contrary tendencies. Consequently, it is here that compensation for all transitory and anomalous influences should take place most quickly.

334. But, in recognizing the utility and even the necessity of statistical experimentation if certain parts of moral, political, and economic doctrines are to be given the character of positive sciences, we must be on our guard against reducing these doctrines to the positive features which they may thereby acquire. We must even be on guard against considering the part that has already been made positive, or which is capable of becoming positive, as that which contains the more important and essential portions of these doctrines. Let us imagine questions like these: should capital punishment be continued or abolished? should the institution of slavery be continued or abolished in certain countries? I admit that we may be in possession of statistical documents which prove quite clearly that following the abolishment of the death

penalty, there has been a certain increase in the number of persons accused of those crimes which formerly carried the death penalty; that following the suppression of slavery, there has been a certain decrease in the population, in the production and consumption of goods of various sorts. But shall we regard these questions as settled for these reasons? Will the very slight difference be sufficient to make us regard them as settled; and, in the contrary case, how great a difference will be required to entail the solution? If the difference is nonexistent or insignificant, must the questions be regarded as insignificant, and their solutions thought to be immaterial or arbitrary? Obviously nothing of that sort can be admitted. And even if we are fortunate enough to find, through the use of statistics, positive facts which give support to theories in morals and politics which are satisfying to our reason; even if, in certain cases, these positive results may silence dangerous agitators, nevertheless, in a great many cases, reason feels that above all it should take account of other principles, other rules and other ideas, which alone are capable of bringing order and light into theories, of obtaining conscientious acquiescence, and even of directing statistical investigation in such a way as to make it fruitful and conclusive. Just imagine the morality of conscience, civil jurisprudence, and the law of nations being obliged at each step to cite mathematical or statistical proof that a certain solution provokes more misdemeanors and frauds, gives rise to more lawsuits, brings discord into more households, divides more families, ruins more people, enriches more rogues, is injurious to population, agriculture, industry, and commerce, results in more wars, makes them more bloody, and so on! It would not only be necessary to renounce morals, jurisprudence, and the law of nations for want of ability to produce such positive proofs; but also the positive facts, like most of the facts of history, would most often give only confused and contradictory results, in so far as reason had not interpreted them and arranged them according to certain rules which are within its power, by respecting all the principles of human nature, those which do not call for sensible and direct manifestation as much as those which reveal themselves in concrete facts.

Even the economic sciences, which have to do more especially with things from the point of view of social utility and of the well-being of individuals, find it necessary for reason to fix, independently of experience, the elements and the conditions of the problems which they discuss. Statistics will clearly teach us that the population has increased, that the price of merchandise has been raised or lowered, that the proceeds from taxes has been increased, that we have harvested more wheat, drunk more beer, or spun more cotton; but, with all that, have the people become more robust, wiser, and more healthy? Is society better off, the State more peaceful within and more respected without? These are far more important questions, and still deserving of more attention. It is to be hoped that the progress of civilization will not put an end to these questions although they cannot be settled peremptorily, scientifically, and positively, that is to say, by a process of exact calculation or by a specific experiment which could have here no other form than the statistical. The happiness of individual persons, the well-being of the State, are not defined by Aristotle's rules of logic, are not written in numbers and algebraic formulas, are not established by statistical numbers, although some of their indirect effects may be established in this way. It is, however, always necessary that the philosophical sense intervene in order to compare the fate of individuals and the order of the State to an ideal type, and to evaluate the relations of concrete and measurable effects with the intelligible principle of order or of disorder which has produced them.

335. So it is, therefore, that throughout the sciences we find philosophical speculation intimately connected with the part which is positive or scientific in the proper sense of the term, that is, with the part which permits of indefinite progress, technical applications and the control of concrete experiments. We shall have to establish everywhere this twofold fact: that the intervention of the philosophical idea is necessary as a guiding thread, and to give science its authoritative and regular form; and that, nevertheless, the progress of positive science is not suspended by the unsettled condition of philosophical questions. Reciprocally, it would imply that we might be able to expect the positive and experimental solution of a philosophical question from the progress of scientific knowledge. If we were unable to discern the philosophical character of a question *a priori*, we should recognize it *a posteriori* and through inference, by seeing that the progress of positive knowledge keeps the question in its state of scientific indefiniteness.

Thus, as it is manifest that the immense progress made since the time of Newton and Leibniz in one of the higher branches of mathematical analysis known as infinitesimal calculus does not hinder us from discussing, as we have been doing for almost two centuries, the principles of this calculus themselves, every discerning mind is sufficiently aware that such discussions do not rest on a point of doctrine that may be resolved scientifically, but on a philosophical question necessarily connected with the exposition of the doctrine.

336. We must not confuse with truly philosophical questions hypotheses which have to do with facts which are inaccessible to observation, whether in the provisional state of our knowledge, or because of the limits that circumstances place on the extension of means of observation and experimentation that we have at our disposal. It is more than probable that observation will never settle what we ought to think about the ingenious hypothesis of the plurality of worlds, and that we shall not be able to carry through the work that would enable us to know on empirical grounds the composition of the interior and deep strata of our earth. The obstacles which make such observations impracticable always come from accidental and accessory circumstances rather than from essential reasons. They result from the limits of our physical powers and from the imperfection of the material instruments which we are able to set up, rather than from limits imposed fundamentally on all knowledge based on sensory perception. On the contrary, it would be repugnant to reason that, by sufficiently increasing the power of our telescopes, we should be able to solve experimentally the problem of knowing whether the universe is or is not limited in space; that, by sufficiently increasing the magnifying power of our microscopic instruments, we should be able to lay hold of the primary elements of matter, to settle by experimental means the question of life, of atoms, and of action at a distance. Positive science is unconcerned with what the nature of the obstacles may be which stand in the way of the extension of our knowledge, as soon as they are recognized to be humanly insurmountable. But in philosophy they are distinguished because, on the one hand, in philosophy one is clearly less preoccupied with facts than with the reason for facts and with their subordination, and because, on the other hand, one's purpose

is to disentangle the hierarchy of the intellectual faculties of man, a hierarchy which shows itself in the explanation of the causes of our ignorance, so well as in that of the causes of our knowledge.

337. If we go to the trouble of bringing together all the scattered observations in this chapter and the preceding one, I think we shall no longer be led to distinguish clearly in the intellectual and moral nature of man *three* principal faculties (logically and, so to speak, anatomically distinct), as Bacon thought (301), but rather *five* principal forms of development, appropriate to as many *idiosyncrasies* or different *temperaments,* and corresponding to as many general ideas, rubrics, or categories, that may be designated thus—Religion, Art, History, Philosophy, Science—setting them forth in the order which indicates their union well enough and which conforms with what we know of the general progress of civilization. Indeed, every civilization has begun with religion and is at first entirely bound up in it; art and poetry are born in the shadow and under the influence of religion; the history of nature and of man is separated later from the mythological and poetic wrappings; and everywhere philosophy, by being connected at first with the symbols of religion and art, has preceded science, which seems to be the last conquest of man's mind and the product of an advanced civilization in all its maturity. History calls in art and philosophy; science can rarely isolate itself from philosophy and history; but the mutual action and combination of different principles is no reason for confusing them. All the efforts that have been made to bring them into opposition to one another have never succeeded in eradicating them from the human mind, because they result essentially from its nature and from the nature of its relations with external objects. This has been said many times of religion and philosophy, of poetry and science; it must similarly be said of science and philosophy. Therefore, we insist on this important point that we have kept before us throughout: namely, that philosophy is not a science, as is so often said, but that it is nevertheless something which human nature, if it is to be complete, can no more ignore than it can ignore science and art. If we should have succeeded in putting this truth in a new light, we shall believe that, for our part, we have contributed some small bit to the correcting of certain prejudices and to the general progress of reason.

CHAPTER XXII

ON THE CO-ORDINATION OF HUMAN
KNOWLEDGE

338. We are now in a position to examine how Bacon applied the principle of his tripartite division to the encyclopedic classification of human knowledge, to evaluate the criticisms and the changes to which his system has been subjected, and to propose for ourselves an attempt at synoptic co-ordination, however imperfect a co-ordination of this sort must necessarily be for reasons that we have stated (243). After having laid down his three main rubrics: HISTORY, POETRY, and SCIENCE—which correspond to three distinct faculties: MEMORY, IMAGINATION, and REASON—Bacon divided HISTORY into *natural* and *civil*; POETRY into *narrative, dramatic,* and *allegorical*; SCIENCE into *philosophy* and *revealed theology.* Philosophy deals with *God, nature,* and *man*; this led Bacon to reintroduce (through a duplication of entries) *revealed theology* in the first subdivision of *philosophy*, whereas at first he seemed inclined to exclude revealed theology from this rubric, the only one for which the need of a synoptic table had developed. At this time we shall go into details only in so far as this is necessary to give a summary of the views of the great English philosopher.

1. DOCTRINE OF GOD

Theology $\begin{cases} \text{Natural} \\ \text{Revealed} \end{cases}$

(Appendix) Doctrine of Angels and Spirits

2. DOCTRINE OF NATURE

Speculative $\begin{cases} \text{First philosophy} \\ \text{Special physics} \\ \text{Metaphysics} \end{cases}$

Operative $\begin{cases} \text{Mechanics} \\ \text{Magic} \end{cases}$

| (Main Appendix) *Mathematics* | Pure | Geometry
Arithmetic, algebra |
| | Applied | Perspective
Music
Astronomy
Cosmography
Architecture
Engines (arts of the
    engineer) |

3. DOCTRINE OF MAN

Generalities on the nature of man.

| *Philosophy of humanity* | Sciences relative to the human body | Hygiene
Medicine
Cosmetics
Athletics
Painting
Music |
| | Sciences relative to the human mind | Logic, grammar
Rhetoric, etc.
Ethics |

| *Civil or political doctrine* | Principles of conservation
Principles of affairs
Principles of government or
    of the State |

The incoherencies of the details and the peculiarity of certain relations are too apparent for us to insist upon them. To tell the truth, the only part of Bacon's work which deserves attention is the fundamental idea of the tripartite division. Let us see to what extent d'Alembert has accepted and modified it.[1]

339. To begin with he changes the order of the principal faculties, by systematically doing violence to all psychological and historical inductions, and arranges them thus:

MEMORY, REASON, IMAGINATION;

[1] The tables developed by Bacon and d'Alembert are found, with the *Discours* by d'Alembert and the explanations by Diderot, at the beginning of their *Encylopédie*. M. Bouillet has reproduced the whole thing in his edition of Bacon's *Philosophical Works*, Vol. 1, 489ff., and in his *Introduction*, he has added some meticulous historical details.

the corresponding rubrics are:

HISTORY, PHILOSOPHY, POETRY;

but the substitution of the word PHILOSOPHY for the word SCIENCE is only a matter of style, and at bottom, for d'Alembert as for Bacon, these two terms have the same value. HISTORY and POETRY are subdivided almost as in the Baconian table, but with considerable additions: for *technology* (or, as we should say, the *arts, crafts,* and *manufactures*) is made a part of *natural history*; while the *fine arts* (*music, painting, architecture,* and so on) are brought together in the same rubric, IMAGINATION, as *poetry* properly so called. Here is the abridged table of the rubric PHILOSOPHY:

1. GENERAL METAPHYSICS OR ONTOLOGY

2. SCIENCE OF GOD

Natural and revealed theology.   Divination.   Magic.

3. SCIENCE OF MAN

Universal pneumatology   Science of the mind   { reasonable / sensitive

Logic .... { Art of thought   (Ideology) / Art of recording   (Writing, hieroglyphics, heraldry) / Art of communication   (Grammar, pedagogy, philology, rhetoric, etc.)

Morals .... { general / particular   (Natural jurisprudence, economics, politics)

4. SCIENCE OF NATURE

Metaphysics of bodies or general physics   { Extension, movement, velocity, etc.

Mathematics   { Pure   { Arithmetic / Geometry

Mixed   { Mechanics / Geometrical astronomy / Optics / Acoustics / Pneumatics / Art of conjecture

Physico-mathematics

		Anatomy
		Physiology
	Zoology	Medicine
		Veterinary medicine
Particular		Hunting and fishing
Physics	Physical astronomy—Astrology
	Meteorology
	Cosmology
	Botany, agriculture
	Chemistry

Natural affinities are by no means less violated in this table than they are in Bacon's. It is particularly surprising to find botany between cosmology and chemistry, and so far from zoology; to see the art of conjecture, or the mathematical theory of chances, put in following acoustics and pneumatics. Moreover, in placing *natural history* with civil history under another heading than that which contains the *natural sciences,* such as zoology, cosmology, and botany, d'Alembert does not sufficiently explain what it is that distinguishes astronomy from the history of the heavens, zoology from the history of animals, botany from the history of plants. To put the matter in a word, he gives no valid reason for this dislocation which was not involved in the original scheme of Bacon, which was worked out in a period during which the knowledge of nature had not progressed so far toward scientific co-ordination as in the century during which d'Alembert wrote.

340. Following the attempts of Bacon and d'Alembert, we shall mention only two more of them, because of the fame of their authors, Bentham and Ampère, who, moreover, have treated this subject not incidentally, but *ex professo,* and who have made it the subject-matter of special treatises.[2] In their encyclopedic tables, both Bentham and Ampère abandon Bacon's principle of tripartite division, and both propose to apply rigorously the principle of dichotomous classification, which Ampère in particular thinks is based strictly on the nature of things, as if the very form of the rule

[2] a. *Essay on the Nomenclature and the Classification of the Principal Branches of Art and Science,* taken from *The Chrestomathia of Jeremy Bentham,* by G. Bentham. Paris, 1823.

b. *Essay on the Philosophy of the Sciences, or an Analytical Exposition of a Natural Classification of All Human Knowledge,* by A. M. Ampère, Paris, 1834.

and the applications that have already been made of it in the different branches of the natural sciences had not indicated the extent to which it should be considered artificial. We shall not insist on this point further at the present time since no one has contested, or will contest it. Bentham particularly elaborates the abuse of dichotomous ramifications with fatiguing excess, and the fabrication of strange words, designed to express the series of bifurcations. A first bifurcation gives him on the one hand *metaphysics* (*cœnontology*), on the other hand the science of particular beings (*idiontology*), which is then divided into the science of bodies (*somatology*) and the science of minds (*pneumatology*). This is nothing but the application of ideas commonly held by medieval scholastics. *Somatology* is broken up into the science of quantities (*posology,* mathematics) and the science of qualities (*poiosomatology*); while *pneumatology* is subdivided into *noology* (logic and ideology) and *anoopneumatology,* which includes *pathoscopy* and *ethics.* But it would be tedious to follow Bentham through the subsequent ramifications of his encyclopedic table. A great number of forced and artificial subdivisions would be found in it, and it does not even have the value of an artificial classification, namely, that of gaining a clearer view of the ensemble of the classified objects.

341. This is not true of Ampère's classification because his sagacity and his vast knowledge did not allow him to fail to recognize the true affinities of the sciences on this point. Furthermore, he had to violate his artificial rules of bifurcation in order finally to reach a series which, as a matter of fact, represented the natural relations of the different bodies of scientific doctrine better than those which we have previously examined. This can be judged from the following extract, in which the first column designates what Ampère called "branches," the second column, "subbranches," and the third, *sciences of the first order* which the author then bifurcates into *sciences of the second* and *of the third order.* But without following through these last ramifications, in which the contrivance of bifurcation becomes more and more apparent, we shall limit ourselves to joining by brackets the names of some sciences of the third order with the names of the sciences of the first order, whose position in the general order it is also advisable to indicate.

First realm—COSMOLOGICAL SCIENCES

Mathematical Sciences	pure	Arithmology / Geometry
	physico-mathematics	Mechanics / Uranology
Physical Sciences	physical properly called	General physics (chemistry) / Technology
	geological	Geology (mineralogy) / Oryctotechny
Natural Sciences	phytological	Botany / Agriculture
	zoological	Zoology / Zootechny
Medical Sciences	physico-medical	Physical Medical (pharmaceutics) / Hygiene
	medical properly called	Nosology (therapeutics) / Practical medicine (diagnostics)

Second realm—NOOLOGICAL SCIENCES

Philosophical Sciences	philosophical properly called	Psychology (logic) / Metaphysics (ontology, natural theology)
	moral	Ethics / Thelesiology
Dialectical Sciences	dialectics properly called	Glossology / Literature (bibliography, literary criticism)
	eleuthero-technics	Technesthetics / Pedagogy
Ethnological Sciences	ethnological properly called	Ethnology / Archeology
	historical	History (chronology, philosophy of history) / Hierology (symbolics, disputation)

Political Sciences	ethnocritical	Nomology (legislation, jurisprudence) Military art
	ethnogenetic	Social economy (statistics, theory of wealth) Politics (international law, diplomacy)

In going over this table, it must seem strange that chemistry and logic are no more than sciences of the third order, while orycto-technics and military art are of the first order. Moreover, we are struck throughout by the arbitrariness and even falsity of the principles of classification. Surely, the relation of the zoological sciences to the botanical sciences and the physical sciences is not the same as that found between the geological sciences and the physical sciences or between the historical and ethnological sciences. The same thing must be said of the relation of geometry to arithmology, as compared with those of hygiene to medical physics, or of literature to glossology. In order to show, by a single example, what can be done to these discordances in regard to sciences of the second and third order, we shall select the branch with which Ampère was most familiar because of the nature of his work, that of the *mathematical sciences,* which he breaks up as follows:

Arithmology	Elementary arithmology	Arithmography Mathematical Analysis
	Megethology (theory of magnitudes)	Theory of functions Theory of probability
Geometry	Elementary geometry	Synthetic geometry Analytic geometry
	Theory of forms	Theory of lines and surfaces Molecular geometry
Mechanics	Elementary mechanics	Cinematics Statics
	Transcendent mechanics	Dynamics Molecular mechanics
Uranology	Elementary uranology	Uranography Heliostatics
	Uranognosis	Astronomy Celestial Mechanics

It is very well to distinguish pure arithmetic or the theory of numbers from the theory of magnitudes which have become quantities and are expressed by the series of numerical values, both whole and fractional. But the theory of numbers par excellence, which deals with the properties of numbers in themselves, abstracted from every mathematical notation and of every method of calculation, and which is by no means part of *elementary* arithmology, should be placed neither in arithmography nor in the theory of equations, which is what Ampère means by mathematical analysis. Furthermore, algebra, of which the theory of equations constitutes the most essential part, belongs, together with the theory of functions, in megethology and not in pure arithmetic. Finally, the theory of probability is primarily a part of the theory of combinations and numbers. The classification of geometry is unsound at every point. The theory of lines and surfaces cannot be isolated from geometry, whether synthetic or analytic; and there is no molecular geometry, since crystallography, which Ampère calls by that name, is only a branch of three-dimensional geometry when it is considered from a mathematical point of view. On the contrary, Ampère had a happy idea when he invented a new word, "cinematics," in order to distinguish that theory which marks the natural passage from geometry to mechanics properly so called, and in which the properties of motion are considered apart from the forces which produce it and the time during which it takes place. We shall not dwell on the subdivisions of uranology, which are all very arbitrary. We shall now proceed at last to the statement of the ideas which have served to guide us in attempting a new classification.

342. At first, common sense wishes to distinguish knowledge which has been brought together into bodies of principles for a technical or practical purpose only, from those that are of special interest to speculation, that we cultivate for their own sake, for the satisfaction of our reason and of the instinct of curiosity which is an integral part of our nature. No doubt we can deduce from all the sciences some consequences having practical utility and find a law or a theorem in regard to a question of application. Reciprocally, our technical learning may be used profitably in the advancement of the speculative sciences. Facts observed in metallurgical operations, in agriculture, or in the practice of medicine

may contribute to the clarification of certain theoretical points in chemistry, botany, or animal physiology, just as, in general, the theories with which the chemist, botanist, and physiologist are concerned will be applied by the engineer who works on the development of mines, by the scientific agriculturalist, and by the physician. But the profound causes for distinction, which gives a separate existence and a kind of autonomy to agriculture and medicine, do not allow them to be considered as pure applications or as simple appendages of botany and physiology. The philosophical value of their characteristics makes it still less permissible to regard chemistry and physiology as appendages or as accessory sciences in relation to agriculture or medicine. The same distinction is more or less perceptible everywhere; and in other respects the importance and the development of the technical sciences depend upon different characteristics of the state of civilized nations and are by no means proportional to the importance and the philosophic rank of the sciences to which they must be attached. Therefore, we are naturally led to arrange our technical knowledge in a particular serial order, parallel to the sequence or the sequences in which the speculative sciences are drawn up. These interest us especially because of the notion they give us of the laws of nature, as well as of the facts which have determined the ordering of the world and the destinies of humanity.

343. This is the point at which the opportunity for a more subtle distinction makes itself felt, a distinction founded upon a more abstract principle and on more profound reasons. There are sciences whose particular characteristic is to bring together into a system the eternal truths or the permanent laws of nature, which pertain to the essence of things or to the qualities with which the creative power has been pleased to endow the objects of creation. On the other hand, there are some sciences which rest upon a series of facts which have successively produced one another and which are explained by one another by going back as far as possible towards primordial facts, which must be admitted without explanation, since we lack knowledge of the preceding facts which would explain them. As a result of this distinction, "nature" and the "world" are two terms which do not express the same idea.[3] One

---

[3] See the book by Mr. Alexander von Humboldt, entitled *Cosmos*.

is the point of view of the physical scientist who systematizes the laws of chemical combinations, and for whom iodine and bromine are just as important elements as chlorine because they play perfectly analogous roles in chemistry; the other is the point of view of the geologist, who seeks to discover how the different chemical substances are distributed over the surface of the earth and enter into its composition. In the next to the last chapter we developed the truly distinctive characteristics of history and science properly so called, or natural history and the natural sciences, of physics and cosmology. Here we return to them only in so far as it is necessary to justify an attempt at classification and to furnish through the classification itself, if it appears natural and clear, a sort of counterproof of the theory.

344. Finally, this analysis leads us to arrange the table of human learning into three parallel series (see the table on pages 500 and 501): the THEORETICAL series, the COSMOLOGICAL and HISTORICAL series, and the TECHNICAL or PRACTICAL series. And at the same time the manner in which the facts, laws, and phenomena are subordinated and connected (as it has been set forth throughout this work), by going from the more fundamental to the more special, and from the more simple to the more complex, gives us the basis for establishing a series of *levels* or of groups. The combination of the divisions by levels and by series constitutes a *table of double entry* (237), that is to say, the most convenient and the least defective arrangement for representing clearly and as faithfully as possible a system of complicated relations.

The distribution of levels enables us to distinguish five groups or families, namely:

The first group, composed of the MATHEMATICAL SCIENCES;
The second group, composed of the PHYSICAL and COSMOLOGICAL SCIENCES;
The third group, composed of the BIOLOGICAL SCIENCES and NATURAL HISTORY properly so called;
The fourth group, composed of the NOOLOGICAL SCIENCES and all the branches of SYMBOLICS;
The fifth group, composed of the POLITICAL SCIENCES and HISTORY properly so called.

As far as the order of the levels and the principal divisions are concerned, our classification squares with that of Ampère and it

SKETCH OF SYNOPTIC CLASSIFICATION OF HUMAN KNOWLEDGE

	THEORETICAL SERIES	COSMOLOGICAL & HISTORICAL SERIES	PRACTICAL & TECHNICAL SERIES
I. MATHEMATICAL SCIENCES	Elementary Arithmetic Theory of numbers, Theory of combinations { Logistics, Algebra, Theory of functions Geometry { Elementary Geometry, Trigonometry, Transcendental Geometry Theory of chances and of probabilities / Rational Mechanics { Cinematics, Statics, Dynamics	. . . .	Calculus, Metrology Surveying — Geodesy, Descriptive geometry, Stereotomy, Perspective, etc. Industrial mechanics { Machines, Motors, Industrial use of machines & motors Hydraulics — Navigation Nautical astronomy, Gnomonics, Measurement of time, Calendar
II. PHYSICAL AND COSMOLOGICAL SCIENCES	Physics properly called { General properties of bodies, Universal gravitation, Physical mechanics, Acoustics, Imponderable agents, Light, Heat, Electricity and magnetism Physico-chemical sciences { Molecular actions, Internal structure of bodies, Crystallography Chemistry { Mineral, Organic	Astronomy Terrestrial physics, Meteorology Geology, Physical geography Geognosy, Oryctognosy, Mineralogy	Application of physics to industry Architectonics (Arts of the engineer) Industrial chemistry, Docimasy, Metallurgy, etc.
III. BIOLOGICAL SCIENCES AND NATURAL HISTORY PROPERLY —	Vegetable { Anatomy, Embryology, Teratology, Physiology Animal { Anatomy, Embryology, Teratology	Botanical, Classification & distribution of vegetables, Botanical paleontology Zoological, Classification & distribution of animals	Phytotechnics, Agricultural sciences Zootechnics, Breeding of animals — veterinary art, etc.

III *Continued*	Anatomy, Embryology, Teratology, Physiology } Human — Phrenology, Physiognomy — Empirical psychology	— Anthropology, Classification & distribution of human races — Ethnology, Linguistics	Hygiene, Gymnastics, Physical education — Pedagogy — Medical Sciences — Pathology, Clinical, Surgery, Pharmaceutics
IV **NOOLOGICAL AND SYMBOLIC SCIENCES**	Ideology — Logic — Esthetics — Natural Theology — Ethics	Hieroglyphics — Paleography, Philology — Mythology and religious symbol, Sacred traditions, Dogmatic theology — Ethnography	Mnemonics — Grammar & Prosody, Rhetoric & Poetry — Music, Plastic Arts, Religious monuments, Religious rites, Casuistry — Natural law, or conscience
V **POLITICAL AND HISTORICAL SCIENCES PROPERLY CALLED**	Theory of Institutions { Religious, Political, Administrative, Juridical, Military } — Social economy — Statistics — Chrematology or Theory of Wealth	Archeology, Iconography — Numismatics — Historical chronology — Political geography — History { Ecclesiastical, Political, Civil, Military, etc. } — History of civilization, commerce of the arts and sciences — Biography — Bibliography	Legislation & Jurisprudence, International, Ecclesiastical, Political, Administrative, Civil, Commercial } Law — Military arts — Commerce — Finances, Manufactures — Arts & Crafts

could not have been otherwise, whether we had set out from theoretical considerations or had been guided by the relations that the conformity or the analogy of studies have established between the different categories of scientists and scholars.[4] For, in the light of theoretical considerations, it is impossible to confuse mathematics with the physical and cosmological sciences and to ignore their immediate dependence upon mathematics. The sciences whose subject matter is the world of living things presuppose knowledge of the general properties of bodies and of the general arrangement of things in the world, while they carry us, through the natural history of man and through empirical psychology, which is closely connected with physiology, at least up to the edge of the domain of speculation and of all the sciences which deal with different aspects of the laws of the human understanding and of the moral nature of man. Finally, the sciences whose subject matter is the organization of societies or of political bodies can spring only from those which deal as much with the physical nature of man as they do with his intellectual and moral nature.

In saying that this order is imposed upon us, we are by no means saying that it is perfect. For example, between mathe-

---

[4] See, at the end of the *Nouveaux essais sur l'entendement humain*, Leibniz's reflections on the division of the sciences, and notably on what he calls the " civil division " of the sciences according to the *faculties* and the professions. Notice his views on the *economic faculty* " which would contain the arts of mathematics and mechanics, and everything relating to the detail of the subsistence of men and the commodities of life, in which agriculture and architecture would be included." On this point, as on many others, Leibniz was ahead of his times, and he broached questions which are now being discussed. Over and above the division according to professions and *faculties* which, according to Leibniz's own statement, principally concerns the practical or technical sciences (since institutions of public instruction, like the professions, should have adapted themselves to the needs and usages of society), there is a sort of official division for the theoretical sciences, which had not been born, so to speak, at the time of Leibniz and of which he does not speak. This is the result of the official establishment of the academies and of the relation of scientists according to convictions that they themselves have as to the affinities between the sciences they cultivate. It is clear that we must mistrust any systematic classification which would too overtly disturb an arrangement of which a philosophical account has not always been taken, and which may contain some defective parts, but which, on the whole, is established by the assent of learned bodies and of the enlightened public.

matics and logic, and between physics properly so called and certain branches of the economy of societies, to which we have proposed to give the name "social physics," there are affinities which the arrangement of the table does not indicate, although they are expressed faithfully enough by adopting the *schematism* or the arrangement which follows, and which conforms still more to Ampère's principle of classification:

	THEORETICAL SERIES	COSMOLOGICAL SERIES	TECHNICAL SERIES		THEOLOGICAL SERIES	HISTORICAL SERIES	PRACTICAL SERIES
Mathematical sciences	"	"	"	Noological and symbolic sciences	"	"	"
Physical and cosmological sciences	"	"	"	Political and historical sciences	"	"	"
Biological sciences and natural history properly so called	"	"	"				

But then the study of man would be divided into two parts, and we should no longer take into account the continuous transitions by means of which we proceed from the study of the functions of the nervous system and of animal sensibility to the study of the superior faculties of the intelligence (as we shall explain in the following chapter). If this chain were broken, it would disturb relations still more intimate and more essential than those that the adoption of this new arrangement or another analogous one would seek to express. It must be concluded from this that while one system is preferable to another, it is impossible to express exactly and completely by any sensible scheme the relations of which we catch glimpses, and which form so many natural affinities among the different parts of the system.

345. We may ask ourselves why it is that, when we use the distinction between history and science properly so called as a principle of classification, we do not apply the same distinction that we have so eagerly sought to establish between science and philosophy. The reason for this is that the philosophical element, which comes to

be allied with all the branches of our positive knowledge, in history as in science, cannot be separated or anatomically distinguished with as much clarity as science can be from history, although in other respects the separation of the two historical and theoretical elements is far from being absolute. But the schematism works on the whole, and that is all that can be asked in making classifications of this kind. As regards anatomy properly so called, the skeletal, arterial, lymphatic, and nervous systems can easily enough be isolated and dealt with separately. After being prepared in this way these systems can be perfectly described in terms of regions and organs. A different anatomy, based upon the distinction between the tissues which function as the elementary materials for all the parts of the organism, does not lend itself in the same way either to the isolation of materially separated systems or to the topographical description of the organs. The same thing is true of the philosophical element of our knowledge. Furthermore, we have not proposed distinct divisions in our table for philosophy properly so called, but only for branches of our knowledge which can be more or less reduced to scientific form, and in which the influence of philosophical speculation predominates more than in the others.

Likewise, although religion, art, and poetry may not be sciences, and although this may have strangely curtailed or weakened their role as compared to such sciences as botany or chemistry, all these things have their places in an encyclopedic table of human learning when dealt with from the point of view from which they are considered by the scientist. That is to say, they must be included in such tables in so far as they pertain to the history of the human race and are the manifestations of certain faculties of the human mind, or of instincts and needs the study of which forms a part of the study of our nature. It does not follow that because of this it is necessary to compare the Bible, the *Iliad,* or the *Laocoön* to an algebraic table or medical treatise (which would be to fall into profanation or barbarism). But a table of human knowledge is neither a method of bibliographical classification, nor a catalogue of the productions of man's genius or of the inspirations of a supernatural wisdom. Bibliographical divisions must be adjusted to the relative abundance of the productions in each branch of literature, philosophy, history, the sciences, and the arts. This is a relative abundance which changes with the state of civilization, mores, institutions, and beliefs. The rational subordination of the diverse parts of our knowledge cannot be regulated by such conditions.

346. In order to complete the commentary on our table, we have to add to the preceding generalizations some explanations concerning details. In the family of the *mathematical* sciences, the column under the second series is necessarily empty (309). In the first column we distinguish two parallel series (241), because in fact, beyond the elements of arithmetic, the science divides into two branches. One branch deals with the properties which numbers possess in so far as they are numbers, and independently of any application to the measurement of magnitudes, by means of a unit which is arbitrary and capable of being fractionated. This application is precisely the object of the other branch, and constitutes what has been called "logistics" [5] from the time of Viète [6] and Descartes, and before modern algebra underwent the developments which made it a science that can be considered by itself, independently of every application to other parts of mathematical theory. Logistics, algebra, and the theory of functions are so many sections in Ampère's theory of abstract magnitudes or megethology. From that we generally go on to geometry and rational mechanics, in which is found the basis for the applications of the calculus of magnitudes to the explanation of the phenomena of nature; while, by means of the theory of combinations, which has the clearest sort of relationship with that of pure numbers, we go on to the calculus of chances and of mathematical probability which is the other source from which the applications of numbers to all the natural phenomena are derived, phenomena introduced by means of a complexity of causes, some dependent upon and some independent of one another (36).

[5] "The ancients distinguished arithmetic and logic, assigning the consideration of whole numbers to the former, and to the latter the consideration of fractions and every sort of calculation and *logos*." Leibniz, Dutens edition, Vol. III, 133.

[6] [François Viète (or Vieta), 1540-1603. A celebrated French mathematician and public servant. He published several works on mathematics and related subjects. He contributed greatly to the development of algebra by making it a purely symbolic science, and also to trigonometry. It is said he was the first person to represent the known quantities in algebra by symbols.]

In the systematic arrangement of theoretical sciences, rational mechanics forms the natural transition from mathematics to general physics, just as, in the technical arts, industrial mechanics forms the transition from the mathematical arts to industrial physics. But yet, in spite of the fact that they borrow some principles or some data from experiments, the mathematical form is so strongly predominant in them, we need have no hesitation about including them in the family of the mathematical sciences.

347. We have no particular remark to make about the division which includes the family of the *physical* and *cosmological* sciences. The distinction of the two series, physical and cosmological, is sufficiently self-evident. A similar distinction in the group which includes the *biological* sciences [7] and *natural history* properly so called, requires more attention. It involves the question of distinguishing in the structure of living beings, plants or animals, which are known to us, that which follows from the general laws of organic beings which are independent of time and space, from that which follows from the succession of facts and accidental causes which have differentiated the races and the species, determined their geographical distribution, created or maintained some and annihilated others, and given to the world, *parte in qua,* the appearance under which we know it. How obscure these recondite questions as to origins are! And how, in the discussion of particular cases, can the departure from the general law and the special facts, the accidental and the essential, be made exactly? However, from the point which the sciences have now reached, scholars no longer confuse the work of the naturalist who describes, compares, and classifies the species, with the experimental researches of the physiologist or with the laws that the anatomist discovers and formulates. In the opinion of all competent judges, zoological characteristics are one thing, anatomical and physiological characteristics something else. On the basis of zoological characteristics, the fauna of Aus-

---

[7] We have adopted here this expression that others have proposed, for the purpose of ending the confusion that, according to etymology, the expressions " physical sciences " and " natural sciences " present. The ambiguity of these expressions goes back to the time when the domain of the naturalist and that of the physicists were not distinguished. Among the English, who have preserved the traditions of the Middle Ages in all things better than we have, a doctor (*un médecin*) is still called a " physician." [See also the note on p. 479.]

tralia are strongly contrasted with the fauna of other continents. Yet, the laws of anatomy and of physiology apply to this fauna just as they do to all others. And should another continent and another fauna be discovered, we shall be sure from now on that the same theoretical laws would apply to new zoological forms found there. The progress of each science in a manner appropriate to it will certainly make the separation still more marked. But from now on it will be enough to show that in this respect the family of biological sciences does not constitute an exception to the general system of the sciences, whatever difficulty we may find in carrying out in detail a rigorous separation. For, once again, nature does not bind herself to the absolute precision of our logical rules; and from the fact that the distinction of the two systems has a natural basis, it does not follow that the two systems may not be combined and be connected through some of their ramifications.

The level of the biological sciences separates clearly into four members or subordinate groups, although crossovers and minglings are found in the passage from one group to another. The vegetable and animal kingdoms will not be confused by anyone even though certain creatures exist which cannot be unequivocally identified as either plants or animals. Still less should empirical psychology be confused with human physiology, although there is an intimate connection between them, both in relation to the abstract theory of ideas, and as a result of the fact that psychology is found to be connected with the whole group of noological sciences. The reasons (already given in Chapter IX, and to which we shall return in the next chapter) which lead us not to separate empirical psychology from the family of natural sciences, militates equally against placing ethnology and linguistics in the second series, following anthropology or the natural history of the human species, or in the third series, pedagogy, the relations of which to empirical psychology are of the same nature as those of hygiene and physical education to human physiology. Moreover, it is clear that these are questions of analogies and affinities that can be evaluated differently, and concerning which there would be little reason for making any dogmatic pronouncement.

348. Since man is destined by nature to social life, in the study of his intellectual and moral nature we have to take account continuously of the facts and ideas which arise out of the relations of

man to his fellow beings, in the midst of civil and political society. Therefore, it is not necessary to try to establish a rigorous separation between the group of *noological* sciences and the group of *social* or *political* sciences. Ethics or morality has relations with the theory of legislation and with all the branches of positive law. Grammar, literature, and the fine arts presume the existence of cultured societies and the dealings of the individual with society. To tell the truth, the division of the two groups is purely artificial, but it is convenient. This is especially so because it corresponds well enough to the distinction that it is convenient to make between the sciences which hardly permit of continuous and indefinite progress, because the kind of observations on which they rest has long furnished nearly everything it can furnish, and the sciences which, in most of their parts, at least of those which deserve to be called positive (334), should be strengthened and extended, in proportion as the progress of observation and experimentation brings additional facts to light and gives greater certainty or precision to the facts already known or foreseen.

CHAPTER XXIII

ON THE SCIENTIFIC CHARACTER OF PSYCHOL-
OGY AND ON ITS PLACE AMONG
THE SCIENCES

349. For half a century, philosophers have had much to say about psychology and psychological observation. The relations between the psychological study of man and philosophical speculations are so close that we feel obliged to follow up the rough draft of the general table of human knowledge by going further into the development of matters which concern psychology—its principles, its methods, and its connections with the other branches of human knowledge. With this done, we shall come to the end of the task we proposed for ourselves when we undertook this *Essay*.

350. What strikes us first as we move from physiology to psychology and from the phenomena of animal life to those of the intellectual life (127) is the impossibility of fixing precisely the point at which the one is superimposed upon the other or the exact point at which the series of psychological phenomena originates. Those psychologists who have pretended to stick closest to nature, to describe the gradual development of the functions of intelligence most cautiously and clearly, have all taken the phenomenon of "sensation" as the point of departure in their descriptions and as the first stage in their theoretical construction. But how many steps, how many modifications in the sensibility, and how many varieties in these affections there are which we lump together under the generic and abstract term "sensation," since we lack the ability to distinguish them clearly! At the lowest level we divine rather than establish, in elementary tissues, an obscure sensibility which is often indicated by the special name of "irritability," in order to draw attention to the great gap that lies between it and that perfected sensibility adapted to the system of sensory organs, as a result of which sensory perceptions occur. But even those who think it necessary to use different terms to designate things so remote from each other regard irritability, which is emi-

nently a vital force and one very distinct from the physical properties of tissues, as simply the rudimentary manifestation of the capacity for sensing which becomes perfected in proportion as the organism becomes perfected and complicated, always tends toward centralization and systematic unity, and which, on the contrary, is gradually weakened in proportion as the organism wastes away and breaks up into its primordial elements.

Sensibility of the sort found in the lower forms of animal life which have no nerve centers, or in which a centralized pattern of nerve structure is relatively imperfect, is certainly to be put above the purely organic sensibility of the elementary tissues or of the organs which are not connected with the general pattern of animal life, although it is still far removed from the general sensibility found among the higher animals. This intermediate group may be represented by insects which lack a brain, but which nevertheless perform so many marvelous acts that indicate a sort of instinctive sensibility and perception of which our customary ways of getting knowledge can give us no idea. It would surely be as contrary to reason to resort to the hypothesis of Cartesian mechanism in order to explain the acts of the ant and the bee as it would be to explain those of the dog and the elephant in this way. From another point of view, it is also impossible that anyone who has delved a little into the laws of organic structure should admit that the way in which the most industrious insect perceives resembles that of the animal whose structure predisposes it to functions of a higher order, whose natural aptitudes have been perfected, since becoming domesticated, through its contact with man.

351. Thus sensation, the phenomenon taken as first term in the psychological series, is not really a primitive fact, or one from which it is possible to set out as from a well-established point of reference with which all subsequent facts could theoretically be connected. On the contrary, it is taken as a point of departure arbitrarily, and no one would know how to fix it with precision in the midst of a continuous series of phenomena whose true origin will always escape observation and knowledge. It is impossible to say how many nuances this so-called "obscure" sensibility passes through in going down the series from higher animals to the ultimate animalcules; among the higher forms, and in man himself, it seems at one time to be localized in certain organs, and at another

time to belong to the system of phenomena that consciousness brings together and centralizes, according to the progressive evolution of organic structures and functions.

352. Nor is the transition from sensation to judgment any more abrupt. In order to convince ourselves of this, we need only to glance over what naturalists and philosophers have written concerning illusions of the senses. Education and habit cannot explain the more or less obscure or distinct judgment which is implied in every sensory perception; and psychologists extricate themselves only by arbitrary hypotheses, often surcharged to such an extent with details of pure invention that they deserve to be called philosophical fictions. The imagination has been stimulated with respect to these matters by the fact of the long period of infancy in human beings, by the long time it takes the infant to acquire the faculties which give him knowledge of things outside himself. It has always been noticed that the human species is a unique exception in this respect. It seems that nature was able to satisfy the conditions of the birth of the child only by curtailing the period of pregnancy at the expense of foetal development. It also seems that our species reproduces, to a certain extent, the anomaly observed in the order of marsupials, in which, since parturition is always immature, special means of protection preserve the life of the young until they have attained the degree of organic perfection which is found at birth among the young of other species as a result of their prenatal development. Now, if we consider how contact is established between the newly-born and the external world among these species (which in this regard are more favored than we) we shall find nothing resembling the laborious apprenticeship, the slow education of the sense organs, that the systematic explanations of the psychologists should lead us to expect to find, in order to establish the transition from sensations to judgments concerning distances, forms, and other properties of bodies. The movements of young birds which run about and find their food after they break out of their shells must not be compared to the movements of the newborn infant who seeks the breast for his nourishment; the latter are instinctive movements and are not accompanied by perception, or are accompanied by only an obscure perception of external objects; the former presuppose a clear and distinct perception of distances and forms of a type which must

remain almost unchanged from the beginning throughout the whole life of the animal, and which the child acquires only a long time after its birth.

353. If there were not, even for us, some judgments that are originally spontaneous and that nature had intimately connected with sensation, although this connection cannot be logically explained; if the apparent spontaneity of these judgments were the result of nothing but training and habit; then another sort of training and other habits would have to have the power of depriving us of these, after the development of our understanding had made us aware that such judgments are erroneous. Now, we observe just the opposite of this, and the illusions that judgment destroys must not be confused with those that judgment or reason corrects but does not destroy. It may be, as some authors have said, that a person born blind, but who has been given his sight as the result of an operation, believes, at first, that he is able to touch the walls of his rooms; that all raised surfaces would seem flat to him; and that later on, and when illusions have been overcome, the trees bordering a roadway, which, because we have become familiar with the laws of perspective, appear to us to be of equal height and arranged in two parallel lines, seem to him to be smaller and smaller and to be arranged in two converging lines. But over and above these illusions that are cleared up through habit, there are others which persist. The professional astronomer, like anyone else, sees the sky in the form of an inverted bowl, the moon as a plane disk, larger when at the horizon than when at the zenith. When we go to see a cyclorama a certain amount of time is needed for the illusion to be produced, and it has not been noticed that the habit of frequenting a cyclorama lengthens or shortens the time needed for the occurrence of this sensory illusion, an illusion in which reason surely has no part. The apparent diameters of the fixed stars seem to contract in a high-powered telescope. This is an illusion that is explained on the basis of the fact that the apparent movement of the stars becomes so much more rapid when they fall within the field of the telescope. And this explanation squares with the phenomenon everyone has noticed of the apparent foreshortening of houses, trees, and other objects which border a railroad over which one is traveling at high speed. Now, no one can establish that the habit of making astronomical

observations or of traveling on a train has any effect upon the persistence of these illusions. In the experiment in which two fingers are crossed over a moving ball, the frequent repetition of the experience, plus the intervention of the sense of sight, do not destroy the illusion of the sense of touch although it may be corrected by them. If there are persistent illusions in sensory perception, as over against other illusions that are corrected by the concurrence of the other senses and by habit, it is clearly necessary that the sensations of men, like those of animals, imply a different sort of judgment than the superior judgment of which the animal is certainly incapable, and by means of which our reason corrects these persistent mistakes.

354. But these are principally voluntary movements and, in general, rapid decisions of the will. They imply in the most marvelous manner, if not a series of judgments and of arguments, in the logical sense of these terms, at least a continual activity of the intelligence which makes it perceive with rapidity relations of convenience or inconvenience. These are all at least as complicated as, and often even more complicated than, those that we grasp only slowly in the process of logical deduction. All games of dexterity and reckoning, all physical and mental skills offer wonderful examples of this. In order to formulate a scientific theory of perceptions and judgments of this nature, we like to imagine that the act has been broken up originally into distinct moments, and that afterwards habit has progressively diminished the intervals of time which separate them, until every trace of them in consciousness is effaced. This is a device of the same sort as that by means of which, in the calculus, we first choose intervals which later on we submit to a process of indefinite subdivision in order to come back to the continuity that we can reach only in this roundabout way (201). But this is simply a device adapted to our manner of conceiving things and not to the nature of things themselves. No doubt experience confirms that where things taught and learned methodologically are concerned, it is necessary to proceed pretty much in this way, that is to say, to break up the continuous action so that it may lend itself to methodical transmission. But since nothing of this sort is observed in any actions of animals and in a great many of our own actions, it must be concluded that habit does not only reduce the intervals to the point of making consecu-

tive actions indiscernible, but rather that the effect of habit is to endow us with special aptitudes, by stimulating and directing our capacity to grasp relations toward a particular end, without being required to work our way through a series of logical deductions. We become skilled at the game of billiards not by studying carefully the mechanical problems, the theory of which the player should know thoroughly in order to give a logical account of his game, but rather through practice. That is to say, we become skilled by cultivating through habit and by directing toward this recreational purpose the ability, which we have in varying degrees, of sizing up at a glance the relations between movements of parts of our body, the intensity and the direction of the movements impressed on the balls, and the modifications that these movements should undergo as a result of friction and of striking other objects. The skillful player is aware of all these things in his own way, not in that of the mathematician; he lets himself be guided by the lessons of experience, without being able to abstract, as does the physicist, the fundamental data of experience or to give a scientific account of these data. No one will dispute the fact that the player is continually using the faculty of judgment, although in a manner closer to the spontaneity of sensory perception than to the methodical and reflective procedure upon which the attention of logicians and psychologists is exclusively focused, when they wish to present a scientific theory about the operations of thought.

355. When these operations are judged on the basis of the knowledge they have produced, an unmistakable gap exists between judgments based upon the relation between the senses and those that reason conceives as absolute and necessary. Kant has worked out more rigorously than anyone else the impossibility of deducing anything but a relative and conditional judgment from a sensory experience, unless a higher faculty intervenes. But, on the other hand, let us remember that the general ideas of reason acquire their maximum clarity only in certain select intelligences that are either predisposed toward them or are placed in circumstances favorable to the stimulation and development of the forces of the mind. Let us also remember that among other persons we find these general ideas confused and in all degrees of ambiguity, to such an extent that it would be risky to assert that they do not exist germinally in the most primitive sensory perception and in

what occurs in the infant or the idiot. With these things in mind we should recognize that this line of demarcation (which is very useful in logic, where we are preoccupied with the intrinsic value and with the form of ideas rather than with the forces and the sources that nature brings into play to produce them) loses some of its fixity when carried over into the field of psychology. In the latter, on the contrary, we have in view the natural development of the forces of the mind rather more than the intrinsic character-istics of the products of thought. So, coming back again to a com-parison which is no doubt remote, a certain kind of classification of rocks, based upon the chemical nature of their principal con-stituents, may be perfectly obvious in mineralogy where rocks are studied according to their composition and structure, rather than on the basis of the circumstances under which they are formed and their position in the earth. But this classification is of no use to the geologist who, on the contrary, observing rocks where they are found and having the history of their formation as the principal object of his study, establishes transitions, mixtures, and substitu-tions of elements, which confuse or unite (from his point of view) that which the classification of the mineralogist has distinctly separated.

356. The development of activity, will, and freedom is parallel to that of sensibility, intelligence, and reason; and the continuity of transition which is observed in one series will also be observed in the parallel series. These two orders of phenomena correspond to and suppose one another mutually and are connected with one another after the fashion of mechanical action and reaction (168). Corresponding to the obscure sensibility of organic tissues and of lower animals is a similarly obscure activity, a mysterious force of which not only the cause or the essence, but the mode of action is incomprehensible to us. In proportion as the affections of the sensibility are co-ordinated and centralized so that they tend to-ward more distinct knowledge, voluntary direction also comes to be determined with greater clarity. Finally, it can no longer be denied that the phenomenon of desire and will exists in the higher animals, in very young children, idiots, insane persons, and in men who are lost in their dreams or are intoxicated. However, in these cases, that higher degree of the active power which constitutes free will either cannot exist, or no longer exists, or is momentarily

suspended. And then, what of the degrees of this self-possession, which is the condition of eminently free and fully responsible action! When does the voluntary act begin to entail the responsibility of the agent? St. Augustine wonders if a child in a cradle, in a fit of temper, has not sinned; and the question he presents to the theologians we present to the philosophers. We recognize in the action and in the remarks of a demented person a malicious will that we repress by means of threats and corporal punishment and that nevertheless does not entail responsibility in the eyes of the expositors of civil and religious laws. We attribute our passions to the animals; we call them cruel, obstinate, timid, and cowardly; and if it is not permissible to take these expressions literally, as attributing to the appetites and inclinations of animals a character of morality which belongs only to man, nevertheless one cannot refuse to see in these affections of the animal nature the basis of the appetites and the passions of human nature. Not only are the organic conditions analogous, but the analogy persists in the psychological affections; and the powers of the human mind, which sometimes master these in the name of a higher principle, are subject to lapses, to gradual weakenings, the consequence of which is that no one whose eye does not penetrate into the depths of man's being has the right to set himself up as the judge of the absolute values of man's actions.

If nature had introduced clear-cut distinctions in the series of psychological phenomena, an order of succession that the mind might have been able to grasp clearly, the language of psychology would not have been so slow in becoming established, and the precision of the ideas would have introduced precision into the language. Had this happened, the impossibility of defining things for want of the knowledge of their essence or of the power to break up the idea into simpler ideas, would not have hindered the reaching of an agreement on the terms by which to designate them, from the moment they had been distinguished by thought. On the contrary, when we see that after so many attempts the language of psychology is continuously being reconstructed and is always in its infancy, that the meaning of terms varies from one author to another, or rather that each author tries vainly to convey the same idea by using exactly the same word, and so gives rise to endless distinctions and contradictions on the part of critics, we are bound to infer that the indecisiveness of language is the counterpart and the mark of the indecisiveness of ideas. Consequently, we should no longer be surprised that the psychologists, starting from obscure and indecisive beginnings, have been unable to give their language and their systems a truly scientific precision, absoluteness, and logical sequence.

357. These considerations naturally lead to the discussion of the meaning of the famous proposition which, as has been said, sums up the whole psychology of Condillac and his school: "Attention, judgment, reasoning, memory, imagination, desire, will, the passions, in a word, all the faculties of the human mind *are only transformed sensation.*" Since the philosophy of Condillac, which was so disdainful of the past, has ceased to prevail in France, it has been repaid for that contempt by contempt; it has been given a sophistical form under which it would be unworthy of the attention of serious minds. Nevertheless, the influence it has had on eminent men forces us to believe that this doctrine does not run counter to the laws of reason; that it could have been better interpreted and defended for a longer time, when, by one of those whims of which we find so many examples in the history of opinions, after having enjoyed exclusive domination it passed completely from sight. But our problem here is not to examine all the aspects of Condillac's philosophical system. We wish to call attention incidentally to the meaning of the formula which has been understood to sum up his psychology, and, to begin with, to bring together some examples whose comparison seems to us to be well-suited to do away with its ambiguity.

358. When an engineer proposes to make use of the *kinetic* energy developed by some natural agent, such as a stream of water, in order to produce some mechanical effect, he conceives of a device which first collects the energy that is produced and then stores it up so that it may be meted out uniformly later on as it is needed, even though the motive force may act intermittently. Other parts of the installation are designed to distribute and transmit the kinetic energy in various directions until it reaches the machines whose construction and whose operating parts are specifically adapted to the production of the different mechanical effects to perform which the kinetic energy is received and utilized. Now the study of the laws of mechanics teaches us that in this case the

engineer manipulates the power put at his disposal by nature in the same way that the sculptor works the mass of clay that he molds to his liking, and with which he may successively make, destroy, and remake such forms as he pleases. After having been momentarily present in the main shaft of the drive wheel, the kinetic energy passes, by means of intermediate gears, to levers, flywheels, and pistons; here it produces rotary motion, there rectilinear motion; here the motion is intermittent, there it is continuous. It makes little difference how the kinetic energy is obtained in the first place, for, at least theoretically and except for the fact that there is a loss of energy in the operation of the machine, a loss which the skill of the mechanic constantly tends to reduce, the engineer is always able to introduce the kinetic energy in the form best suited for the purpose he has in mind. This can be conveniently expressed by saying that the machine can only transform a given quantity of energy, or that the kinetic energy transmitted from one part of the machine to another is simply the same energy transformed.

Similarly, when a banker exchanges certain kinds of money for others, silver for gold or gold for paper, demand notes for long term notes, sight drafts for notes that are payable on maturity, checks which are payable in a given place for bank drafts payable at some far distant point, he simply transforms, in terms of the needs of his business, a value which is fundamentally always the same. The form in which he receives this value makes little difference to him, for he will always be able, through banking operations, to obtain this value in the form best suited to his actual needs, except for the diminution of value or the loss which results from the cost of exchange and discount, a cost that commercial competition and the perfecting of the machinery of commerce constantly tend to reduce to their *minimum*. Not only the banker but also the merchant and the manufacturer who work with actual goods see in the goods only an absolute and homogeneous quantity, namely, commercial value, brought into being under different forms. A commodity may have been materially used up, but its value survives and has gone into another commodity which has been produced. From the point of view of the economist and the sort of ideas and facts he analyzes, it is accurate to say that the value or part of the value of the goods produced is simply a transformation of the value of the commodity that has been consumed.

359. To go back to the terms of our first example, let us suppose that the factory for whose operation the kinetic energy of a stream of water is used is a powder works. In this case, after it has gone through all the parts of the mechanical system in a variety of forms, the kinetic energy is finally spent and used up in the shock of the pestles which grind and mix the materials out of which the powder is made. This powder is itself potentially a very powerful substance. It will be used for shattering objects, for clearing out pieces of rock, for producing powerful explosions. In a word, it contains a motive force which requires only a spark to set it off. But can it be said that the motive power used up by the pestles has passed into the powder; that the potential energy of the powder is only the transformation of the energy originally furnished by the motive force and distributed through the different parts of the machine used in manufacturing the powder? To say this would be a mistake; for there is no proportionality, no relation between the force dispensed in the mill and the mechanical power of the powder produced. One is not the result of the other; the original energy has been used up in grinding and mixing absolutely inert materials. In order to explain the mechanical properties of the powder we must have recourse to certain laws of chemistry and physics, which are not related to the use of power in the manufacturing of the powder. The using up of the original energy has been simply one of the *conditions* for bringing together and into intimate contact the materials of the explosive mixture, in order to make possible the ultimate use of mechanical force that nature has given the chemical reaction of the elements of the mixture when it is activated by some device for exploding it.

Similarly, when powder is used in mining operations, it is necessary to use a certain amount of energy to drill holes in the rock, to put the powder into them and pack it down, and to fire it. But except for this there is no relation whatever between the amount of energy used up in this work and the mechanical effects produced by the detonation. The preparatory work will not determine the ultimate production of power which is the result of the expansion of the gases created by the burning of the powder; it will simply be one of the preliminary conditions which must be fulfilled before the potential energy of the powder, the explanation of which is found in physical laws of a particular sort, can be utilized in the production of a useful effect. One sort of energy will have been used up, another will have been created, but the latter will not be the first transformed. The transition of the former to the latter cannot be explained simply by the laws of mechanics, at least not by those which are sufficient to explain the operation of a machine itself. A natural action must intervene which has its own laws and its special reason for being.

A steam engine which is fired by coal is used to extract coal from the bowels of the earth and to raise it to the surface of the earth. In one sense this machine reproduces the commodity it consumes. It absorbs force and it creates force since, by virtue of the properties of gaseous substances, every source of heat is equivalent to a source of mechanical force. But in such a case it cannot be said that there has been a transformation of the force used up. The relation which exists between the consumption of the machine and its useful effect can be traced to accidental or fortuitous circumstances. The same expenditure of combustible material and of power would have had to be applied in raising to the same height the same weight of materials that could in no way have the property of reproducing heat or power.

360. When, instead of applying his muscular force directly to the production of a mechanical effect, man acts to overcome resistances by using a machine, the entirely passive function of the machine consists in transforming the kinetic energy that man possesses through his nature as a mechanical agent, by building it up or by dissipating it at certain points in space and at certain moments in time, but without fundamentally altering it, and above all without adding anything to it. On the contrary, when a sailor uses his muscular force to spread and to give the right direction to the sails of his ship, to direct the rudder, or to work the ropes, instead of acting himself upon the obstacles to be overcome, he uses forces of nature, which are incomparably more powerful than his own, to act for him. It would be vain to attempt to explain the effects produced, if we did not take account of this intervention of outside forces which the work of the sailor directs and puts into play, but of which he is not the productive source.

Man himself and the animals, considered from the point of view of their bodily structure, may be likened to mechanical structures in which lever arms, fulcrums, and all the elements of a machine are found. In structures of this sort, a contraction of groups of muscles is the source of the kinetic energy which then, by being transformed and distributed according to the laws of mechanics, comes to the organs by means of which the animal acts directly upon external bodies. But whatever obscurity may exist concerning the causes and the particular form of muscular contraction, we fail to see how it is possible to escape the conclusion, that here, as in the burning of the explosive mixture, or as in the combustion of coal, a special action of nature intervenes by means of which it not only transforms but also creates mechanical force out of nothing. And if we consider the series of still more subtle phenomena which intervene between the action of external stimulants and the nervous reaction of the muscular fiber, we shall judge that there exists among the various terms of the series a heterogeneity which does not allow us to conceive the passage from one to the other by means of a simple transformation, but which, on the contrary, obliges us to admit the intervention of natural forces *sui generis*, of which the above examples, however, give the idea by their simplicity and to some extent by their relative crudeness.

361. Now, in what sense can we say with Condillac that sensation is transformed to become attention, judgment, reason, memory, desire, will, and so on? Does this mean that the phenomena of sensation alone make understandable to him everything that goes on in the understanding and in the will after sensation has occurred? Does it mean that, once this first fact is given, all the others are found to be virtually included in it, and that the highest insights of reason, the most reflective determinations of the will, contain nothing more than what was in the phenomenon of sensation without either additions or subtractions, without profound modifications due to the intervention of forces whose cause and controlling principle would exist elsewhere than in sensation itself? If such were Condillac's notion (and it is true that his language lends itself only too readily to this interpretation), it would be difficult to understand how a doctrine which runs so counter to common sense could have found its place among those extreme aberrations to

which philosophers have been led by yielding to the desire to twist the facts of nature to fit the artificial unity of their systems.

But if Condillac simply wanted to describe the sequential order among psychological phenomena, which makes it possible to consider them as proximate causes of one another, he would only have done what is done in all branches of the study of nature, what all physiologists do without the import of their language being misunderstood. For example, when they explain the general unity of the functions of nutrition, by taking edible material from the time it is put in the mouth, and following its transformation in the tissues where assimilation takes place, they do not pretend that the act of grasping and of masticating foods is sufficient to explain digestion, nor that assimilation is nothing but a prolonged and modified digestion. They admit the intervention of a variety of forces and principles whose concurrence is the necessary condition of the accomplishment of the total function. But they lack the ability to trace the phenomena back to their true causes, by means of scientific observation and analysis, in such a way as to isolate them and to assign each of them its exact part in the process. Therefore, for purposes of explaining the phenomenon, and with the limits of possible observation, they consider the description of the circumstances in which the phenomenon is produced and they regard as the cause of a given phenomenon other observable phenomena in whose absence the former would not have occurred.

Condillac's formulation would no longer seem to be anything more than the correct expression of the continuity that prevails in the series of psychological phenomena which develop from one another and proceed from one another through an endless operation of vital and creative energy.[1] Thus the pure teachings of Condillac have been succeeded by modified doctrines out of a desire to give a larger place to the activities of the mind. After this, however, the difficulty, or, better, the impossibility of determining precisely the extent of the division has introduced an indecision into the exposition of these doctrines which have sought to explain and not merely to describe phenomena. The artifice of style may sometimes conceal this indecision, but criticism cannot fail to bring it to light again as soon as it examines the matter thoroughly.

362. By making sensation the foundation, so to speak, of the

---

[1] " Nature brings forth nothing piecemeal." Pliny, *Natural History*, XVII, 22.

whole series of affections of which the human mind is capable, Condillac continued the profound separation, laid down by the Cartesians, between material phenomena, which, according to them, are all equally reducible to pure mechanism, and spiritual phenomena, which are all equally incompatible with the essential properties of matter, from the vaguest sensation to the highest acts of intelligence and freedom. He opposed Buffon when that great writer granted to animals the affections of sensibility, since he felt that these are compatible with corporeal nature, even though Buffon was too genuine in his portrayal of nature to fall into the absurd hypothesis of Cartesian automatism by completely denying to animals a soul, that divine principle of freedom and reason. Buffon's opinion was a protestation by his common sense, supported by science and genius, against the paradox into which the predilection for systematic constructions and abstract speculations had led the whole Cartesian school. His opinion was a return to the idea admitted in antiquity, of a sensitive and animal soul and of a rational and free soul (127ff.), an idea which Maine de Biran, one of the most original thinkers of this century, has reproduced under forms which are very much more settled but which for that reason give evidence of being an arbitrary and systematic contrivance.[2] According to him, the essence of the human *self* or of the human mind consists in the ability to make free choices. The affections of the sensibility, the images of the imagination, the violence of the passions belong to animal nature and fall within the province of the organism; the human person, the self, a *hyperorganic* force (*compos sui*), the essence of which is self-possession and free self-determination, is united in man, and in man alone, with animal life. The mystery of this union replaces the mystery of the union of thinking substance and extended substance in the Cartesian system. The hyperorganic force of the self is suspended in dreams, in drunkenness, and in insanity, while the animal life continues to function, almost as, according to Bichat's theory, organic life goes on during anomalous or periodic suspensions of the animal life. When the person recovers himself, finds himself once more in the presence of the phenomena of sensibility, imagination, and passion, which are realized in the sphere of functions of animal life, it is always by a free and voluntary

[2] *Philosophic Works*, Vol. III, *passim*.

determination that his strength is revealed and his authority manifests itself. In the absence of the self, there are sensations, but not knowledge; passion, but not will: for knowledge and will presuppose the consciousness of freedom and personality, of self-possession. Consequently, neither will nor knowledge is present in any degree whatever, either in animals or in men when they are dreaming or delirious.

363. But these consequences themselves show wherein the ingenious system whose principal characteristics we are outlining is excessive and contrary to fact. Our own experience proves to us that in dreams, although we may or may not experience a vague feeling of our inability to act or to co-ordinate our action, we know the persons and things of which the imagination reproduces images; that the faculties of comparing, judging, remembering, and anticipating are brought into play as a result of these deceptive perceptions. It is plain to see that when we pass from sleeping to waking, these functions are carried out with greater regularity or co-ordination, but that their nature does not suddenly change. We have every reason to believe that the same thing is true of them when the infant, by a sort of awakening or, rather, progressive attentiveness, comes little by little into possession of the faculties of the intellectual life. It is as contrary to a legitimate inference to refuse to concede that animals have any knowledge and even any feeling of their individuality as it is to accord to them a knowledge and a personality similar to our own. The human individual is the highest form of self-consciousness, or, at least, we have no idea of any higher form of it; but this consciousness has its degrees and its progressive development as do the other phenomena of life.

364. If a sharp line of demarcation may no more be drawn where Buffon and Maine de Biran have placed it than where it was placed by Descartes, it must be recognized that the thought of the former tends toward a more faithful expression of the natural hierarchy of phenomena and functions, and the real distinction that the creator has placed between humanity and animality; a distinction of an altogether different sort than that which separates one species of animals from another, and corresponding to destinies that have nothing in common. In fact, reason tells us that there are two levels in the series of phenomena which can all be called psychological in the sense that they all affect our internal senses:

a lower level, including the affections which are also found in animals, or, if this way of putting the matter is preferred, those to which man is subject in so far as he shares in the nature of animals; and a higher level, including all parts of the series in which it is certain animality does not share. This is a fact that no theory can upset, and a theoretical conception will be preferable for the very simple reason that it sets off so important a fact to better advantage. This is true even though elsewhere the separation of the levels, which is clearly manifest when we take in at a single glance the unity of the strata of which they are composed, loses some of its clarity as we approach the median region in which they shade off into one another and precludes neither the interdependence of the parts of the system, nor the reaction of the parts on one another. It is pretty much after this fashion that the distinction of geological strata does not hinder us from observing the confusion and mixture of rocks as we pass from one stratum to another on the one hand, and, on the other hand, that the lower strata often carry marks of profound alterations, resulting from an influence exerted by the strata which have been deposited on top of them.

365. To the degree that we move toward the higher level of the phenomena of the psychological series, or toward exclusively human psychology, the assistance we can draw from anatomical and physiological observations becomes increasingly inadequate. There are many obvious reasons for this. In the first place, gross anatomy teaches us nothing or nearly nothing about the organization of the brain which tends to explain or make conceivable the functions of that organ of thought. Another more subtle or finer anatomy, the first attempts toward which did not appear before the beginning of this century, has still advanced too little to throw any light on this very difficult and complicated subject. Moreover, although psychological aptitudes do not cease to be connected with organic characteristics, it is an important consideration that the connection rests upon organic characteristics whose importance decreases in proportion as psychological capacities of a higher order are concerned. These are no longer characteristics of genera and species, but of varieties of races, or more often still of individual varieties, characteristics in whose transmission there is no fixity and which are linked to differences of aptitude of the greatest im-

portance, intellectually and morally.   Consequently there is a well-marked contrast between the study of facts from the point of view of the naturalist and the physician, and the study of the same facts or of related facts from the point of view of the psychologist and the moralist.   Without exceeding the limits of a legitimate inference, we can predict that this contrast will be maintained, whatever progress may be made in the future in the study of organic characteristics in their relation to intellectual and moral aptitudes.

It is likewise incontestable that the progressive development of human thought, although never independent of the organism, tends to be directed more and more in terms of laws which are appropriate to it, and which have only more and more indirect relations to organic dispositions.   Everything in the sensibility of animals is adapted to the perception of space and relations of position in space (139).   There is homogeneity and direct connection between the organic arrangements of the apparatus of sensation and the relations for the perception of which it must serve as an intermediary between the animal and exterior objects. But man finds the idea of duration and of the co-ordination of things in time in the intimate consciousness of his personal existence.   This is an idea that no organic structure in itself has the capacity to set up, because this is a matter of relations of which no organic structure can present the imprint and immediate representation.   Therefore, if the consecutive production of the idea of time and of duration must be attributed to the energy characteristics of the force which produces the consciousness of personal existence, then there is no reason for requiring that the organism preserve the imprint of all the affections of the mind which are the result of previous impressions and which should still exert such a fundamental influence upon later resolutions and actions.

366. A man is present at a lecture that captures his attention and involves him in serious reflection.   At this moment his senses are stimulated, his brain at work; and if a sufficiently penetrating eye were able to read into the most intimate details of the organism, it would discern a thousand traces of this work of the organs. In addition, the age and the temperament of the subject, the state of his health, and his dietetic habits, all being things which certainly reflect themselves in the organic dispositions, exercise, concurrently with the memory of the past and the concern for the

future, a less contestable influence on the impression that the lecture produces at the moment. Meanwhile other circumstances supervene; the senses again become quiet or experience excitations of another sort; all that remains of a very lively impression is a memory which is lost in a mass of other memories. But, behold, twenty years later, as a result of one of those associations of ideas which reason discerns and to which the senses contribute nothing, as a result of one of those moral analogies which relate facts physically most disparate, an unforeseen event revives this memory, recalls the forgotten reflections and obliterated emotions. Henceforth these reflections and emotions will come to exercise a decisive influence over all the conduct of the person. They will lead him to give up his habits, his most cherished affections; they will lead him into solitude; they will persuade him to impose privations and penances on himself. Here is an interesting subject for the study of the moralist and an occasion for applying the knowledge he has acquired of the strength of the human heart. But to imagine that all this can be explained by the convolutions of fibers or by vibrations of molecules, to require that organic structures retain indefinitely the imprint of all the affections which later have an influence on the decisions of the mind through the power of memory, is to fall into one of those systematic exaggerations that reason resists, even when it does not have the means at its disposal for logically demonstrating its absurdity or of experimentally proving its falsity.

What we are saying of the ideas connected with the nature of time or of duration may, of course, be said of all the conceptions of reason which are not of the physical and sensory type. It is true that we are unable to deal with abstract ideas without the aid of sensible symbols, and for that reason animal sensibility and the organism always retain a place in the work of thought. Therefore, let us consider two men who are working on the same abstract speculations but who think in two different languages. If the microscopic anatomy of their brains could be carried out far enough, it is very probable that differences would be found in them corresponding to the diversity of vocal sounds which are made by both during this imaginative work. Other organic modifications would correspond to the degree of attention, to intellectual effort, and would vary from one individual to the other, according to innate or acquired aptitudes; but they might be identical, either in the same individual or in different individuals, although their thoughts bear on very diverse subjects. Now let us admit that these organic modifications, imperceptible to actual observation, and, so to speak, infinitesimals, which should differ principally with respect to the nature of the symbols serving to support thought and with respect to the intensity of the intellectual effort, nevertheless have I know not what in common. And does this " I know not what," this infinitely small thing of another order, represent the abstract truth, the identical theorem for the two intellects, in whatever language they think them, and whatever difficulty or facility they may have had in grasping them? Common sense forbids so complicated and so gratuitously constructed a scaffolding, a hypothesis which, far from explaining anything, does not even give an inkling of a possible explanation.

367. All that we have been able to do up to the present time has been to attempt to establish the connections of certain organic characteristics with certain intellectual and moral aptitudes, without at all going into the why of these connections; whereas, in the case of animal sensations, we catch a glimpse of the relations between the construction of the apparatus of the sense [3] and, if not the nature, at least the order and the intensity of the sensations produced. Gall's theory on the correlation of aptitudes with the development of the various regions of the cerebral hemispheres, developments which he believes can be interpreted by means of the bumps on their bony covering, is even now a remarkable example of the results which the empirical study of the connections in question may lead to. This is true even though the celebrated author of this doctrine allowed himself to be led prematurely, as always happens, to the construction of a system whose consequences the best minds have not been able to accept, to the extent that they go beyond the observed facts. We clearly understand that experimentation properly so called, that process which artificially sets up conditions in which phenomena are produced in order to establish their independence or to show their connection, becomes impossible in the higher phases of psychology. We must depend on the observation of the facts as they present themselves to us in all their

---

[3] [*la construction de l'appareil sensible*]

complexity. We can imagine how prodigious this complexity is in comparison to the simplicity of astronomical phenomena, which also lie beyond experiment and which we must be content to observe as they unroll themselves before us. It will not be difficult to understand how very arduous it must be to discover the true order, the actual subordination of psychological phenomena through the complication of the appearances. However, we may move toward the solution of this great problem, and no doubt it will be approached through a minute, patient, and intelligent study. Organic lesions and alterations resulting from morbid or unnatural anomalies will be compared with the nature of the general or partial disorders they have introduced into the functions of the intellectual life, such as the destruction of memory for certain kinds of facts or words, the perversion of certain judgments or tastes, and hallucinations of various kinds. And, at the same time that the relationship between the organism and the functions is established, we shall have a less imperfect knowledge of the subordination of the functions, which is a strictly psychological question, more important in itself than the relations of the functions with the organic structures.

In this comparative study we are concerned, on the one hand, with organic varieties of characteristically decreasing importance, which for that very reason are less clearly indicated and somewhat transient, and, on the other hand, with very complex phenomena in which many causes concur (in such a way that each cause in particular *tends* toward the production of a certain effect, rather than producing it efficaciously and constantly). Therefore, this study is a case in which the procedure has to be statistical, that is to say, it is one in which a great number of observations are collected in an attempt to unravel the constant influences by counterbalancing the effects which are due to fortuitous and variable causes. This is how we shall succeed in giving precision to the vague evaluation that each person makes for himself because of the influence that age, temperament, dietetic habits, and racial characteristics exert on moral and intellectual capacities. These are influences which would often not be manifest in a comparison of one individual with another, although the comparison strikes everyone's attention, and although it has given rise to proverbial statements, when, even without having recourse to statistical tables, one race is compared with another, one sex to the other, young persons to adults, and so forth, in an over-all view.

368. The same method is applicable when we wish to study not the immediate influence of the organism on the production of psychological phenomena, but the influence of these phenomena on one another, in such a way that mediately or immediately they are dependent upon the organism. For example, if statistical documents can establish a connection between age and criminality or the tendency to commit a certain kind of crime, they will also clearly be able to establish a connection between criminality and the level of education. But in neither case will the establishment of the relation carry with it the explanation of the causes or of the kind of influence, although a series of statistical observations, suitably directed, may in the long run throw some light on the nature of the influence itself. In certain cases it is no longer impossible to set up a system of experiments which reveal the influences we wish to study and to isolate them sufficiently from accidental and disturbing influences, without for that reason failing completely to take account of numerous series of observations like those which statistics records. For example, pedagogy is a science or an art the relations of which to general psychology are almost the same as those of medicine to psychology. In pedagogy, as in medicine, we are not strictly limited to the observation of phenomena as they occur of themselves. Direct experimentation is not impossible, although the respect due to human nature and the very purpose of the pedagogical art add to the intrinsic difficulties of experimentation and restrict its field. Well-conducted pedagogical experimentation is very appropriate for clarifying the operations of the mind and the inclinations of the heart and the connection of aptitudes and dispositions. Take a single example, that of relaxation through a variety of tasks; is it not clear that we can attain a sufficient precision in the conditions of the experiments which might be devised to study this matter to enable us to determine indirectly what the faculties are among which nature has given the greatest amount of independence by the very constitution of the organs of thought, and to do this in such a way as to confirm or to overthrow a theory established on considerations of another kind, such as the phrenological system of Gall?

369. But much time and effort will be necessary before psychology, like medicine, can be brought to a truly scientific form.   Even though the human mind may never be capable of carrying out this conquest, it would not follow that the precept of the oracle, " Know thyself " (Γνῶθι σαυτόν), had been a vain one, for science is only one of the forms of " gnosis " or " knowledge," in the broadest meaning of that term (305).   Therefore, to suppose that psychology has been destined never to assume the form of a scientific system by no means implies that it is necessary to conclude that nothing can be gotten out of common proverbs and the meditations of philosophers on the subject of psychological facts.   It would simply be found that since psychology eludes the indefinite progress which pertains only to knowledge which is linked together in scientific systems, the wisdom of the ancients would long ago have brought together everything true and useful that may be deduced from the reflective observation of psychological phenomena, everything that this observation yields when it is not twisted to meet the needs of arbitrary hypotheses and premature systems.

When psychology is thought of in this way, do we not find in it the prototype of that Socratic philosophy which was so highly regarded by the Greeks and yet which was so rapidly replaced by the profound or subtle systems of the celebrated men who came immediately after the Socratic school?   Judging the matter from the tradition of antiquity, would it not be possible to compare the intellectual and moral study of man, as Socrates appears to have conceived it, to the medicine of Hippocrates, devoid of all system, created well before the period in which it became possible to believe in the scientific co-ordination of pathological facts which, even in our day, give rise to so many ephemeral theories, and nevertheless are already so rich in profound aphorism, in judicious diagnostic skill, and directions for treatment that their very wisdom has made them survive all the revolutions of science.

370. If the nature of psychological data is, in fact, better suited to be translated into aphorisms than into theorems; if, at least up to the present time, the superiority of the ideas of the enlightened man over those of the ignorant, where they touch upon phenomena of this order, has been a product of wisdom rather than an acquisition of science; then this practical conclusion may be inferred, that the literary studies of the young child should prepare

him for the observation of the moral and intellectual nature of man, but that psychology cannot really be taken as the subject of an elementary and authoritative course. Experience establishes that oral teaching is useful for young minds only on the condition that it rests on exact ideas which are rigorously interconnected. Where the nature of things or the imperfection of our knowledge have presented an obstacle to the exact definition of ideas and to their systematic concatenation, teachers and pupils must necessarily be content with words and with empty and arbitrary formulas.

On the contrary, logic is a true science which fits easily into the program of instruction. The same thing may be said of those philosophical problems which, although not susceptible to a scientifically demonstrated solution, nonetheless admit of a rigorous definition and an exact analysis (321).

Those who propose to introduce psychology into the curriculum by placing it at the beginning of the course of philosophical studies naturally give the reason that it is in order to study the faculties of the human mind before proceeding to an analysis of the ideas that these faculties furnish. The ancient scholastics cited an equally plausible reason for attributing priority to logic when they said that logic is the instrument that serves us in the search for truth, and that the instrument should first be known before it is used. But happily nature has accorded us more liberty in the use of our faculties and in the exercises which should perfect them. It is possible to study logic and to make a thorough study of a great mass of questions which belong to general philosophy without giving our attention to psychological functions, just as it is possible to understand gymnastics without a previous study of the anatomy of the human body, and just as a music teacher may give lessons which are profitable to his pupil without first having to teach him the theories of acoustics or the anatomy of the ear, which he himself almost always ignores.[4]

371. However, let us now leave this question, which is after all of interest in the practice rather than in the theory of teaching. Let us move on to the discussion of the very principle from which philosophers have set out who pretend to attain scientific demon-

----

[4] [This reference to the content and method of instruction reflects Cournot's persistent interest in pedagogical matters. See his *Des Institutions d'instruction publique en France* (Paris: Librairie de L. Hachette et Cie, 1864).]

stration as fully as physicists and chemists, and perhaps even better than they; to be like the latter the continuators of Bacon's reform, by basing their philosophical systems upon psychology, and psychology upon observation. The fundamental principle of these philosophers is that there are two kinds of observation, corresponding to two distinct or rather contrasting kinds of facts. There is sensory observation which applies to phenomena of the external world and to the study of man himself, considered from the point of view of his bodily nature. Then there is an introspective observation which is simply the attentive contemplation of the *facts of consciousness*, that is to say, the psychological phenomena which take place within consciousness and which are immediately known to us through our consciousness of them.[5] This distinction is undoubtedly justified, in that some of the materials of our study and of our ideas are furnished us by the relations of the senses properly so called, others by the internal affections of our sensibility, and, finally, still others by a superior faculty to which we give the name "reason." But all our faculties depend on each other, and all our knowledge is interconnected. If the physicist and the naturalist observe with their senses, they observe still more with reason; and no one can make use of reason and of the senses without a sort of inward observation of the evidence that consciousness gives us

[5] " The term ' consciousness ' is ever respectable when it signifies the lively and profound feeling of our duty, in whatever way these duties are made known to us, whether a moral instinct reveals them to us immediately, or whether experience and reflection have indicated them to us; but to extend this name consciousness and that of observed facts to metaphysical abstractions, to mental intuitions, to secret inspirations, is to substitute illusions for realities, enthusiasm for study, and belief for science." Daunou, *Cours d'histoire* [a series of lectures], given at the Collège de France in 1828. We cite this severe judgment without approving it at every point by a good deal and without trying to explain what in it is true.

[Pierre Claude François Daunou, 1761-1840. Politically active in the late eighteenth century in France, he was one of the framers of the constitution of the year III (1795), the first president of the Council of Five Hundred, and a member of the committee appointed to draw up the constitution of the year VIII (1800). He was editor of the *Journal des Savants* from 1816 to 1838. Early in this period he was appointed professor of history in the Collège de France. Among his numerous writings the *Cours d'histoire*, cited here by Cournot, was published in twenty volumes, the first appearing in 1842, two years after Daunou's death.]

of the impressions of the senses and the conceptions of reason. When the term "observation" is used in the language of the sciences, when the sciences of observation are set over against those which are based upon calculation and reasoning, it is always understood that we speak of observation which is regularly organized and systematically conducted, which leads to the discovery of hidden phenomena by means of connections that reason conceives between sensory phenomena, and sometimes with the artifice of methods, as in statistical research, sometimes with the artifice of instruments, as in astronomy and physics. No one would think of saying that mathematics and geometry are observational sciences, because, if we become introspective, and if we observe in this way what are called the facts of consciousness, we shall find that we possess the idea of number, the idea of space, the idea of a straight line, and the idea of mathematical axioms; for example, that the whole is greater than the part, that a straight line is the shortest distance between two points. Psychology would come no closer to being an observational science, because introspective observation, that is to say, attention given to the evidences of consciousness, may have taught philosophers, like everyone else, that man experiences sensations and desires, that he has ideas and a will, that he makes judgments and deliberates, and that, depending upon the circumstances, he feels himself constrained or free in his decisions to act. If, when they reason about these elementary data which are common to all men, philosophers reach conclusions that laymen ignore and conclusions on which philosophers are often far from being in agreement, it would be an abuse of words to rank their theories among the observational sciences, whatever their scientific value may be. Therefore, the whole matter boils down to this question: Can the introspective observations philosophers tell us about be carried beyond these elementary notions and even extended indefinitely in such a way that a group of diligent and patient observers might be gotten together, each of whom would methodically avail himself of the work of his predecessors, not just in the formal solution of a multitude of problems raised at the present time, but even in the discovery of a mass of facts of which we have no idea at present? Or, on the contrary, in anticipation of the possibility that it may some day be brought to such a stage of development that it can take its place among the observational sciences, should not psychology be introduced by means of a system of empirical investigations, if not exactly like that found in such or such a branch of the physical or natural sciences (since in this respect all observational sciences offer something particular), at least analogous to it and differing from it only with respect to specific varieties, instead of forming a distinct and contrasting genus?

372. Now, when the problem is put this way, it is almost resolved. Apart from the fact that experience informs us that this introspective observation, this solitary contemplation of phenomena which occur in the privacy of consciousness, has never produced anything which resembles a body of scientific doctrine, the reason for this result or the absence of a result is easy to understand. It is not without reason that for so long a time the consciousness of the psychologists has been compared to the eye which sees objects outside of itself and which is not able to see itself.[6] Were it not that the device of the mirror, which allows the eye to see its own image, has no analogue in questions dealing with the examination of consciousness, then, because of the very intervention of reflection on the facts of consciousness, the phenomena which we wish to observe are found to be necessarily complicated by a new phenomenon and often modified or weakened. Astronomers and naturalists skilled in micrography, who observe with eyes aided by powerful instruments, know how often one is likely to have an illusion in delicate observations, by believing that one sees what one expects and wants to see as a result of preconceived opinions. If thought can react to this degree on sensation, the organic and physiological conditions of which are much better established, how much more reason is there for intellectual phenomena of a higher order, which have repercussions in consciousness, to be disturbed by the attention that is given to them. They may be disturbed to the point that it becomes difficult or even impossible to grasp them by means of introspective observation, as they are or as they would be without the inevitable interference of this disturbing factor. Here (to use figuratively the

---

[6] " The soul cannot see itself, but, like the eye, the soul discerns other things." Cicero, *Tusculan Disputations*, Bk. I, Chap. 28.

language of astronomers) the effects of the perturbations are of the same order of magnitude as the effects of the principal causes which we wished to separate.

Not only does the attention given to the facts of consciousness modify and alter them, but it also often makes them pass from non-being into being; or, to speak more accurately, it brings to the state of " facts of consciousness " psychological phenomena which would not have been echoed in consciousness unless attention had been given to them, and which may pass through a multitude of phases before reaching that of which we have a clear consciousness, which is the only one capable of becoming the object of introspective observation. What idea would we have of the way we appeared on the earth and of the manner in which we existed during the early years of infancy, if we had been abandoned in solitude from an early age, if the sight of what happens to our fellows and the accounts of our parents did not inform us about what happened to us in that early period of infancy, no traces of which are retained in our memory? Now, each psychological phenomenon goes through a period of early infancy, so to speak, in its progressive evolution, a phase that consciousness is unable to lay hold of, or memory to retain, and of which we form an opinion only through induction, analogy, and the indirect observation of external manifestations that we have good reason to believe are connected with inner phenomena which are hidden from direct observation.

373. Some other comments still remain to be made. In order that an observation may be called scientific, it must be capable of being carried out and repeated in circumstances which permit an exact definition. This must be done in such a way that in each repetition of the same circumstances it is always possible to establish the identity of the results, at least within the limits of error which inevitably affect our empirical determinations. Besides this, it is necessary that, in defined circumstances and within the limits of error which we have just indicated, the results be independent of the temperament of the observer; or that, if there are exceptions, they be traceable to a peculiarity of temperament which manifestly renders such a person unsuited for that kind of observation, without shaking our confidence in the constancy and in the intrinsic truth of the observed fact. But nothing of this sort seems

to be encountered in the conditions of introspective observation on which we might wish to establish a scientific psychology. On the one hand, we have the problem of fleeting phenomena that cannot be grasped because of their perpetual metamorphoses and of their continuous modifications; on the other hand, these phenomena are essentially variable with the individual in whom the role of observer and that of the subject under observation are united. The phenomena change, often completely, as a result of varieties of temperament which have the greatest degree of mobility and instability and have the least characteristic value or importance in the general plan of the works of nature. Of what importance to me are the discoveries that a philosopher has made or that he believes he has made in the depth of his consciousness if I do not read the same thing in mine or if I read something else entirely different in it? Can this be compared to the discoveries of an astronomer, a physicist, or a naturalist who urges me to see what he has seen, to feel what he has felt, and who, if I do not have sufficiently good eyes or a sufficiently delicate touch, addresses himself to other persons better endowed than I am, who see or feel the same thing so exactly that it will be clearly necessary for me to yield to the truth of an observation to which all those agree in whom the qualities of the witness are found?

Thus we see that the most useful observations concerning the intellectual and moral nature of man are gathered not by philosophers favorably disposed toward theories and systems, but by men who are truly endowed with the spirit of observation and capable of grasping the practical side of things, by moralists, historians, statesmen, legislators, and teachers of young persons. Generally speaking, useful observations have not been the fruit of solitary contemplation and introspective study of the facts of consciousness. They are rather the result of an attentive study of the conduct of men placed in diverse circumstances and subjected to passions and influences of all sorts, from which the observer takes great care to free himself as much as possible. So it is that here, as elsewhere, direct observation is principally of sensible facts, which, it is true, the testimony of our consciousness teaches us to relate to internal affections which escape the senses. In that the observations of which we are speaking resemble those of the physiologist, who makes judgments about the sensitivity of certain tissues in an animal on the basis of the convulsive movements the animal makes when these tissues are injured. They have a resemblance, although a more remote one, to the observations of the physicist, who judges the relative velocity of the vibratory motions of two sounding bodies by means of the musical interval of the two sounds produced.

374. Nevertheless, if we have not succeeded in giving to facts gathered in this way concerning the intellectual and moral nature of man a scientific co-ordination comparable to that which connects facts revealed through a methodical observation of inanimate nature and of the living organism, the reason for this is evident and pertains to the nature of the facts observed, rather than to that of the instruments of observation. We have been endowed with certain senses of exquisite perfection and admirably adapted to our natural relations with external objects in everything that concerns the preservation of animal life in the individual and in the species. Nevertheless, these senses would soon be found deficient for the scientific investigation of sensible phenomena as the evidence of consciousness is deficient in the scientific investigation of internal phenomena, unless reason perceived such relations of dependence among the perceptible facts that some can be regarded as the manifestation and measure of others. Thus, through indirect observation, we reach facts which we should be unable to reach through direct observation. In this way we measure magnitudes out of all proportion to the organs of sense, for example, the very great velocity of light and the minuteness of light waves. All this is a consequence of the inexhaustible variety of the combinations to which the co-ordination of phenomena in space gives rise and of the fruitfulness characteristic of mathematical conceptions (141). But when we enter directly upon the study of intellectual phenomena of the highest order, so that we lose all trace of the relations with the organism, the phenomena do not lend themselves to any co-ordination in space. They are not for that reason exempt from co-ordination in time; but, as far as empirical investigation is concerned, this condition is the same as nonexistence. It seems that we have no means of estimating the time required for the occurrence of a phenomenon of this nature, the interval of time which necessarily separates two determined phenomena or two determined phases of the same phenomenon.

Phenomena may be superposed or entangled without our being able to distinguish them, either through their co-ordination in space or through their co-ordination in time.

We can hope to surmount these difficulties only by fastening the broken chain together again, if this is possible. If this is to be done we must proceed gradually in the regular and methodical study of intellectual phenomena, beginning with those whose connections with the conditions of organic structure are most evident, and proceeding thus step by step, in such a way as to profit from the arrangement already introduced into phenomena of a lower order, for the purpose of attempting the scientific analysis and arrangement of phenomena of the next higher order.

Finally, where all representation in space becomes impossible, where every measure fails us, the means of precision and of scientific control, drawn from the use of numbers, still find their place. The conception of order, of combinations, and of chance is even more abstract and general than the ideas of space and time. Consequently, numbers govern the intellectual and moral world just as they do the physical world (36). And statistical numbers, skillfully employed, may again bring to light combinations and a regular arrangement that the complexity of causes and effects does not allow us to recognize in the observation of individual cases. Often statistical procedures can only give greater exactness and clarity to notions that are gained through the experience of life, and that occur in the form of general maxims or aphorisms in the writings of philosophers and moralists, but that become elements of scientific research and comparisons only when they have been fixed by numbers. Thus, however empirical psychology is envisioned, we fail to see how it can be based upon processes of observation essentially different from those used in other scientific researches. Empirical psychology is in the same position as the other sciences—all of which have their peculiar difficulties—to which methods of observation must be accommodated by subjecting them to different modifications. Only in this case the difficulties are greater, so that progress is slower and the approximation to truly scientific forms is much more belated.

375. Beyond this empirical psychology, which is a branch of anthropology and which may be regarded in some ways as the consummation of the biological sciences, there is doubtless another

psychology which does not require this apparatus of observation, this slow accumulation of detailed facts. This psychology should no more be put among what are called the observational sciences than arithmetic or geometry should be, however applicable they may be to some observable and observed facts, a condition without which any science would be chimerical. It is clear that it is possible to study the conditions of a conclusive argument, to classify our ideas into different categories, to set forth the rules of good method, to discuss the value of different kinds of proof and inferences, and to invoke principles of morality, by continuing their application to given species, without examining how, under what conditions, by virtue of what forces, because of what natural springs the notions, ideas, rules, and principles in question have made their appearance in the mind. When thus conceived psychology is nothing but logic and theoretical ethics. Far from having its basis in the study of the organism, of the functions of life, and of the aptitudes and the natural needs of man, as does empirical psychology, it continually tends to exclude all these things. It seeks to bring to light relations and general truths that men lay hold of because of their nature as intelligent beings and reasonable creatures, as would be the case with any other creature on whom God had bestowed intelligence and reason in the same degree, but by different means, and in a physically different order of things.

If it has been necessary for man's senses and his brain to be organized in one certain way rather than another in order to bring his intelligence to the point of being able to conceive mathematical ideas and arguments; if, too, a very complicated concatenation of historical events has been necessary to bring certain human societies to a state of intellectual culture which might permit the development of the sciences and of mathematics in particular, we see very clearly that the essence of mathematical truths does not depend upon the progress of events which have led to the clearing of forests, the construction of cities, the invention of writing and printing, and, finally, the establishment of professions and academies. Neither does the essence of mathematical truths, as we see with equal clarity, depend upon the arrangement of ganglia and nervous plexuses, on the composition of the blood and the body fluids, and on the properties of heat and electricity. If it has pleased the Creator, in the harmonious arrangement of the world which is the object of our observations, to use so many extraordinary means for revealing to a Newton, and through him to other men, the fundamental truths which had been hidden from them, the disparity between the result and the means, the simplicity and the great generality of the one, the complexity and striking singularity of the others, force us to believe that the same result might have been obtained in other ways. Or at least we are forced to believe that the legitimacy of our judgments concerning the results is not subordinated to the state of our knowledge about the nature of the means. What we have been saying apropos of mathematical conceptions may be said in regard to the still more abstract and general ideas which are the subject matter of logic. Likewise, the notion of justice and injustice and that of the obligation to respect the rights and the persons of our fellow men present themselves to our minds as not dependent on the nature of the causes which have perfected our species to the point of making the individual living being a person, any more than on the instincts and needs which have created in the midst of human societies different sorts of property. So the same regulative principles ought to govern beings that, although having equally attained the dignity of moral persons, yet had done so in circumstances physically and historically unlike those in the midst of which the destinies of humanity are worked out (169ff.).

376. Thus, we must be on our guard against confusing psychology, which is the empirical knowledge of intellectual facts in their natural relations with the organization and the constitution of the thinking subject, and logic, which deals with the relations between ideas as they follow (as reason has made us see) from the nature of ideas themselves, independently of the way they are elaborated and appear in the human mind.[7] The same dis-

---

[7] " That logic has already, from the earliest times, proceeded upon this sure path is evidenced by the fact that since Aristotle it has not [been] required to retrace a single step, unless, indeed, we care to count as improvements the removal of certain needless subtleties or the clearer exposition of its recognized teaching, features which concern the elegance rather than the certainty of the science. It is remarkable also that to the present day this logic has not been able to advance a single step, and is thus to all appearance a closed and completed body of doctrine. If some of the moderns have thought to enlarge it by introducing *psychological* chapters on the different faculties of knowledge

tinction always recurs between the knowing subject and the object of knowledge, between the things which depend upon the nature of the thinking subject and those which depend, on the contrary, only on the characteristic qualities of the object of thought. Moreover, this profound distinction would not be impeded except that, in dealing with abstract and rational logic, we are often led to touch on questions whose empirical solution belongs to what is properly called psychology, considered as a branch of anthropology or of the natural history of man. It would be still less possible to enter into the application of the sovereign principles of morality without entering into a circle of abstractions in order to consider man as nature has made him, with his appetites, his instincts, and his needs, which belong to the physical constitution of individuals and of the species.

We have tried to indicate clearly the contrast between psychology and logic, to point out the true character of empirical psychology, and to show what psychological observation consists in, because it has seemed to us that modern doctrines have introduced confusion on all these points. But we do not lose sight of the fact that all our classifications are somewhat artificial, and that, finally, (as we have so often reminded ourselves) all our knowledge is interconnected because all our faculties aid one another.

(imagination, wit, etc.), *metaphysical* chapters on the origin of knowledge or on the different ideas of certainty according to difference in the objects (idealism, scepticism, etc.), or *anthropological* chapters on prejudices, their causes and remedies, this could only arise from their ignorance of the peculiar nature of logical science. *We do not enlarge, but disfigure sciences, if we allow them to trespass upon one another's territory."* Kant, *Critique of Pure Reason*, Preface to the second edition [translated by Norman Kemp Smith (London: Macmillan and Co., Ltd., 1929), pp. 17, 18].

CHAPTER XXIV

AN EXAMINATION OF SOME PHILOSOPHICAL SYSTEMS, AS THEY ARE RELATED TO THE PRINCIPLES PRESENTED IN THIS WORK. PLATO, ARISTOTLE, BACON, DESCARTES, LEIBNIZ, AND KANT.

377. We certainly do not propose to undertake a summary of all the innumerable systems of philosophy, nor even to sketch a few of them without omitting any of their essential characteristics. So our purpose in bringing this work to a close with some historical considerations is and can be only to point out very quickly, for the purpose of clarifying our own principles, the comparisons which it seems natural to make between them and some of the celebrated doctrines toward which, as toward so many prototypes, all cultivated minds readily turn.

Ab Jove incipiendum.[1] It is very clear why we give our attention to the oracle of ancient wisdom, the sublime Plato, whose writings have had an influence on Greek philosophy, and, consequently, on all philosophy down to modern times, comparable to that which the Homeric poems have exercised on Greek letters, and, consequently, on all the literatures of western Europe—an influence which persists in spite of all the destruction wrought by the barbarians, and of all the revolutions which religions, languages, and customs have undergone. Now, if we try to draw from Plato's vast system of dialectical subtleties and poetic conceptions that which bears most directly upon our problem, we shall find, first of all, that his great genius has enabled him to express by means of the most striking image—the famous myth or symbol of the "Cave"—the abstract conditions of the fundamental problem of critical philosophy. This problem consists in moving from the idea to the object, from appearances to things, from phenomena to relative or absolute reality. Some prisoners are shut in a cave[2] in such a way that they

---

[1] [" Let us begin with Jupiter."] Quintilian, X, 1.
[2] Plato, *The Republic*, Vol. II, Bk. 7 [" The Loeb Classical Library," (Cambridge, Mass.: Harvard University Press, 1946), pp. 119ff.].

are forced to turn their backs to the pale light of a fire which illuminates them from afar. Puppets pass and repass in front of the mouth of the cave, their shadows being thrown on the back of it. These shadows are the only things the prisoners are able to see and are the sole source of their ideas of the nature of things. What a difference there is between these phantoms and the wooden or clay figures which are their source, and between those figures and living animate nature! And what precautions must be taken when the prisoners are freed and their fetters loosened little by little, in order to accustom their weakened eyes first to the appearance of the fire, to which the cave owes the little light that it enjoys, then to the reflected light of the sun, and, finally, to the sun itself! From beginning to end, the image is strikingly appropriate (83). But Plato, who so clearly conceives the relation of appearance or phenomenon to reality, does not think to ask himself how, in spite of their chains, the prisoners of the cave would be able to distinguish the shadows which are thrown before their healthy and normal, although weakened eyes, from the spots and the empty phantasms which, without an external cause, would confuse their eyes and their disordered imaginations. He does not consider that criticism which, proceeding by way of probable inference and not by positive demonstration, distinguishes in a complex impression between what is affective and what is representative, between what is traceable to the sentient subject and what is traceable to the nature of the object of knowledge, between what deceives us and teaches us nothing and what gives us true knowledge, and does so even though the known truth is still not or may never be the absolute truth, but a relative truth which is probably the only sort to which created beings can attain.

Or, more likely, this criticism did not escape such a mind as Plato's, but he disregarded it. He himself is careful to teach us this in the *Theaetetus,* in which Socrates, after having raised the eternal objection drawn from dreams and madness, and after having made sport of Protagoras' doctrine which makes sensation the criterion of knowledge, and man *the measure of all things,* by asking why the pig and the baboon might not also legitimately make the same pretense, puts into the mouth of Protagoras or of his partisans this remarkable reply:

You say the sort of thing that the crowd would readily accept—that it is a terrible thing if every man is to be no better than any beast in point of wisdom; but you do not advance any cogent proof whatsoever; you base your statements on probability. If Theodorus, or any other geometrician, should base his geometry on probability, *he would be of no account at all.*[3]

And Socrates, or rather Plato, who shares strongly Protagoras' feeling in this matter, sets about in search of arguments that may be fairly presented. But since the nature of things does not admit of categorical demonstration, he has no alternative but to resort to dialectical subtleties to which the genius of the Greeks was so inclined, and whose sterility they can be allowed to have failed to recognize more than we. Socrates is in the right when a little further on he says, " Then knowledge is not in the sensations, but in the process of reasoning about them; for it is possible, apparently, to apprehend being and truth by reasoning, but not by sensation." [4] He is wrong when he rejects the only kind of arguments and proofs, or rather of inductive inferences, that reason may use in criticizing our sensation, and in showing the weakness of the judgments they suggest to us.

378. Moreover, how could Plato have been satisfied, as being knowledge of the highest sort, with inferences and probabilities which do not even satisfy the needs of geometry—a science which in his eyes is very inferior to that whose object is the search for primary truths and the essence of things? In fact, according to his theory, there are four ways of knowing, of which the first and the most perfect, which goes back to principles and to the reason of things, is the only one which properly deserves to be called *knowledge.* The instrument of this knowledge *par excellence* is dialectic [5] and its subject matter,

---

[3] [Plato, *Theaetetus,* " The Loeb Classical Library," (Cambridge, Mass.: Harvard University Press, 1928), p. 81.—Cournot's italics.]

[4] [*Ibid.,* p. 165.]

[5] " The dialectic of Plato is the search for the general in the particular, the absolute in the relative; the search for the scientific ideal. It is a method which tends upwards, which, separating the complex, changing, and individual from our various perceptions, carries up to the essence, the permanent, the one. It is an analysis in the sense that it breaks things up for the purpose of excising the accessory and attaining the essential, or that which in each thing subsists in

546 ON THE FOUNDATIONS OF OUR KNOWLEDGE

the other section of the intelligible [by which] I mean that which the reason itself lays hold of by the power of dialectics, treating its assumptions not as absolute beginnings but literally as hypotheses, underpinnings, footings, and springboards so to speak, to enable it to rise to that which requires no assumption and is the starting point of all, and after attaining to that again taking hold of the first dependencies from it, so to proceed downward to the conclusion, making no use whatever of any object of sense but only of pure ideas moving on through ideas to ideas and ending with ideas.[6]

The second way of knowing, which belongs to the mid-ground between opinion (δόξα) and true knowledge (ἐπιστήμη), and of which, according to Plato himself, geometry or arithmetic offer the clearest prototypes, is what he calls διάνοια, a word very difficult to translate, and that we shall translate as " theoretical knowledge." The object of this theoretical knowledge is still intelligible things or ideas, but of a different class than those mentioned above.

This then is the class that I described as intelligible, it is true, but with the reservation first that the soul is compelled to employ assumptions in the investigation of it, not proceeding to a first principle because of its inability to extricate itself from and rise above its assumptions, and second, that it uses as images or likenesses the very objects that are themselves copied and adumbrated by the class below them, and that in comparison with these latter are esteemed as clear and held in honour . . . . Students of geometry, reckoning and such subjects first postulate the odd and the even and the various figures and three kinds of angles and other things akin to these in each branch of science, regard them as known, and, treating them as absolute assumptions, do not deign to render any further account of them to themselves or to others, taking it for granted that they are obvious to everybody. They take their start from these, and pursuing the inquiry from this point on consistently, conclude with that for the investigation of which they set out . . . . And do you not also know that they further make use of the visible forms and talk about them, though they are not thinking of them but of those things of which they are the likeness, pursuing their inquiry for the sake of the square as such and the diagonal as such, and not for the sake of the image of it which they draw? And so in all cases. The very things which they mould and draw, which have shadows and images of themselves in water,

---

eternal reason; it is a synthesis in the sense that, from complex and variable phenomena, it seems to form, by virtue of intelligence, something which is not a phenomenon." *Abélard*, by M. de Rémusat, Vol. I, 300.

[6] Plato, *The Republic* [*op. cit.*, pp. 113, 115].

these things they treat in their turn as only images, but what they really seek is to get sight of those realities which can be seen only by the mind.[7]

After " true knowledge " (ἐπιστήμη) and " theoretical knowledge " (διάνοια), included within the common rubric of "intelligence," or rather of " intellection " (νόησις), comes " belief " (πίστις) and " probability " (εἰκασία), which are so many degrees of opinion (δόξα). The world of nature particularly belongs to the world of opinion, and Plato does not hesitate to hazard conjectures concerning such material where he is satisfied

[to adhere strictly] to what [he] previously affirmed, the import of the " likely " account, [and] will essay (as [he] did before) to give as " likely " an exposition as any other (nay, more so), regarding both particular things and the totality of things from the very beginning.[8]

Here, certainly, is a well-established hierarchy, a clear and well-connected system, and one very naturally suggested by the state of knowledge in Plato's time. What he calls knowledge or " science " is what we call " philosophy," while διάνοια is what is universally referred to as "science " today. He is on solid ground when he subordinates geometry to philosophy, as physics is subordinated to mathematics. But he is wrong in not recognizing the value of confirmation by a possible experiment, which gives to geometry the character of a positive science, a character denied to philosophic speculation. He falls into an even more important error (although one fully excusable in his time), when he concludes on the basis of the place of philosophy in the hierarchical series that probable inferences are inadmissible in philosophy, although the nature of things makes it necessary for us to be content with inferences of this sort in physics but not in mathematics. And for that reason he condemns his philosophy to remain empty and sterile except as he becomes, through a happy infidelity to his principles, as so often happens in his case, the ingenious and eloquent expositor of those opinions or philosophical beliefs which are based, not upon rigorous demonstrations, *more geometrico*, but upon probabilities

[7] [*Ibid.*, pp. 111, 113.]

[8] [Plato, *Timaeus*, " The Loeb Classical Library," (Cambridge, Mass.: Harvard University Press, 1929), p. 111.]

and likelihoods, upon inductive inferences and analogies of the very same sort as those which he theoretically disdains.[9]

And notice to what extent a preconceived abstraction may influence the mind. Plato, this great artist, the best prepared of all philosophers for sensing the resources furnished to the expression of philosophical thought by art and poetry, where it bears upon matters which are incompatible with the precision and the dryness of logical forms; Plato sets about to underrate art and poetry and to set them almost at naught. For the artist and the poet are only imitators of sensible objects, which in their turn are nothing but images of intelligible things. Consequently the artist and the poet are still further from pure truth, or from that which is the object of true knowledge, than are those who, after the fashion of the physicists, deal directly with perceptible objects.[10]

379. Instead of providing what it promised and what it was impossible for it to provide, the philosophy of Plato was promptly modified. It was better organized than any other to lend itself to two contrary tendencies: to incline toward mysticism and toward skepticism. The mystical tendency predominated when Hellenic civilization, returning in its old age to the fables that had charmed its youth, and subject to all sorts of foreign influences, probed deeply the symbols with which Plato's poetic imagination so often succeeded in clothing his thought. But this degeneration of Platonism, whatever curiosity it may have from the point of view of the history of the human mind, is of no interest for our purposes. On the contrary, the tendency toward skepticism showed itself early even in the gardens of the Academy, even in the halcyon days of Greek philosophy. On the basis of the division so clearly established by Plato, what was more natural than to reject in opinion

---

[9] " However various were the significations in which the ancients used ' dialectic ' as the title for a science or art, we can safely conclude from their actual employment of it that with them it was never anything else than the *logic of illusion* (*die Logik des Scheins*)." Kant, *Critique of Pure Reason* [translated by Norman Kemp Smith (London: Macmillan and Co., 1929), p. 99]. The *Schein* of Kant bears a great resemblance to the Δόξα of Plato, and the German philosopher is correct in the sense that the Platonic dialectic made use of probable inferences rather than of rigorously concluded demonstrations. Compare this judgment of Plato's dialectic by Kant with that of M. de Rémusat, mentioned above.

[10] *The Republic*, Bk. 10 [*op. cit.*, pp. 419ff.].

(δόξα) everything incapable of being brought within the limits of science (ἐπιστήμη) and that which really falls within the realm of opinion? This does not mean that it is impossible to support the affirmative and negative sides of an argument indifferently (which skepticism exaggeratedly maintains); it means rather that it is necessary to be content with probable inferences and arguments which convince the reason, although they are not secure from sophistical objection, and although they do not have the kind of certainty which is characteristic of mathematical demonstrations (διάνοια). We are not sufficiently well acquainted with those schools which the ancients call the " Second " and " Third Academy " to be able to evaluate clearly their notion of philosophical probability and of the nature of probable knowledge. The writings of the philosophers of these schools have not come down to us as have those of Plato and Aristotle, or those of the Neo-Platonists of Alexandria; we know them only secondhand through references and inadequate extracts. Cicero himself is only an imitator; and although the admirable lucidity of his mind and his Roman common sense led him in general to take from the Greeks what was more substantial and to cut down the scholastic subtleties, it is not completely certain that he always grasped those parts of the teachings of his masters which would most interest us. It seems that a Greek named Arcesilaus or Carneades, whose discernment had been sharpened by the study of geometry, which was flourishing at that time, as had been the case with Plato, was in a position to give to the theory concerning opinion and philosophical probability a more definitive form than we find in the writings of Cicero and other ancients.[11] However, when we consider that the doctrine of mathe-

---

[11] " As a result of the dual relation of the representation (φαντασία) to the object (τό φανταστόν) and to the subject (ὁ φαντασιούμενος) Carneades, concluding that objective knowledge is impossible, thought that neither the senses nor the intellect offer a sure witness (κριτήριον) to objective truth, and he permitted only probability to subsist (τὸ πιθανόν, probabilitas) in three different degrees (ἐμφασίς or πιθανὴ φαντασία, ἀπερίσπαστος, διεξωδενμένη ἤ περιωδενμένη φαντασία). This is what is called the probabilism (εὐλογιστία) of Carneades." * Tennemann,† Manuel de l'histoire de la philosophie, § 168 in the translation by Cousin.

* [Arcesilaus (or Arcesilas), who lived from 316 to 241 B.C., was the founder of the " Middle " or " Second Academy; " Carneades, circa 214 to 129 B.C., was the founder of the " Third Academy."]

matical probability first appeared only in very modern times, and that so many mistakes had been made by the most skillful philosophers and mathematicians as to how it was to be understood and applied, we are strongly tempted to believe that the Greeks in their heyday did not have truly definitive ideas on this subject and were unable to express precisely that of which they could not give an exact account to themselves.

380. Plato fused together in his own way Pythagorean doctrines and the teachings of Socrates, and his system bears the mark of his taste for mathematical abstraction. Aristotle, wishing to do something different and better than Plato, constructed his on the basis of his genius as a naturalist, observer, and classifier. As an observer, he attached the greatest importance to sensory experience; he even wished that it might be used if necessary in all abstract and theoretical reasoning, and he expressed his position on this matter in terms that Bacon would not have disclaimed.[12] He distinguishes with perfect clarity deductive or syllogistic reasoning and reasoning by inductive inference (ἐπαγωγή): the first of which moves from the general to the particular, and is better adapted to the nature of things; the second of which moves from the particular to the general, and which, proceeding immediately from sensation, is better adapted to our nature.[13] We might think we were listening to Locke or Condillac, when he describes how sensation gives rise to memory, memory to experience, and the comparison of experience to common or general ideas, which are the principle of art or knowledge.[14] He even goes much too far in this direction, when he re-

---

† [Wilhelm Gottlieb Tennemann (1761-1819). A German philosopher who was an exponent of Kant's doctrine. He was professor of philosophy at Jena and Marburg. His writings include a translation of Locke's *Essay on the Human Understanding*, a *History of Philosophy* in eleven volumes, and an abridged version of this work. It is the latter to which reference is made here.]

[12] " The facts, however, have not yet been sufficiently grasped; if they are, then credit must be given rather to observation than to theories, and to theories only if what they affirm agrees with the observed facts." *De generatione animalium*, Book II, 10 [760b 30. *The Works of Aristotle* (Oxford: The Clarendon Press, 1910), Vol. V].

[13] " In the order of nature, syllogism through the middle term is prior and better known, but syllogism through induction is clearer to *us*." *Analytica Priora*, Book II, 23 [68b 35. *Op. cit.*, Vol. I].

[14] " So out of sense-perception comes to be what we call memory, and out of frequently repeated memories of the same thing develops experience. From ex-

gards it as obvious that, if we had one less sense, we should necessarily have one less science; [15] for it is possible to prove (108) that doing away with several of our senses would not necessarily involve doing away with any body of knowledge, because the sensations which are suppressed, having no true representative value, do not contribute, so far as they exist, to the formation of science or knowledge properly so called.

But in spite of such well-marked tendencies toward what we should call empiricism and sensualism, tendencies which he owes no doubt to the positive direction of his works in the natural and descriptive sciences, Aristotle has the same ideas as Plato concerning the hierarchy of knowledge. For him as for Plato the highest sort of knowledge is that of first principles and of the reason of things,[16] and the syllogism is its proper form. Physics would have the same relation to mathematics as mathematics has to first philosophy. In physics we must seek only probability; and therefore it should be considered to fall within the realm of opinion ($\delta\acute{o}\xi a$) and not of science ($\dot{\epsilon}\pi\iota\sigma\tau\acute{\eta}\mu\eta$); for the subject matter of science [17] is exclusively the things which are necessarily of such and such a nature, and which cannot be otherwise; while the field of opinion includes all the things which sometimes appear in one way, sometimes in another, so that we can reason in regard to them only through probability and conjecture. It is true that elsewhere [18]

perience again—i.e. from the universal now stabilized in its entirety within the soul, the one beside the many which is a single identity within them all—originate the skill of the craftsman and the knowledge of the man of science." *Analytica Posteriora*, Book II, 18 [100a 4. *Op. cit.*, Vol. I].

[15] " It is also clear that the loss of any one of the senses entails the loss of a corresponding portion of knowledge, and that, since we learn either by induction or by demonstration, this knowledge cannot be acquired." *Analytica Posteriora*, Book I, 15 [81a 38. *Op. cit.*, Vol. I].

[16] " The first principles and the causes are the most knowable; for by reason of these, and from these, all other things are known, but these are not known by means of the things subordinate to them." *Metaphysica*, Book I, 2 [982b 2. *Op. cit.*, Vol. VIII].

[17] " So though there are things which are true and real and yet can be otherwise, *scientific knowledge* clearly does not concern them: if it did, things which can be otherwise would be incapable of being otherwise." *Analytica Posteriora*, Book I, 27 [88b 33. *Op. cit.*, Vol. I].

[18] " For it is when a man believes in a certain way and the starting-points are known to him that he has scientific knowledge, since if they are not better known

Aristotle consents to regard as being included within the field of scientific knowledge not only what always and necessarily happens, but also what happens ordinarily and most frequently (ἐπὶ το πολὺ), by excluding only what happens accidentally (τὸ συμβεβηκός), or by anomaly. It follows from all this that Aristotle foresees, but in the most confused sort of way, the applications of the doctrine of chances and probabilities and the future science of statistics, although he is not sure whether he should put it in ἐπιστήμη or in δόξα. The philosopher of Stagira obviously had no notion of philosophical probability and of the principles of analogy and induction. In his eyes, if induction (ἐπαγωγή) is valid and if it is possible to draw conclusions by proceeding from the particular to the general, it is only because all the particular cases have been exhausted. In fact, this alone can give *positive* certainty (327) of the generality of the principle.

381. This is the place to insist on the distinctive trait of Aristotle's genius and system, namely, the tendency to classification. His comprehension of natural history, although as varied and as profound as was possible in his time, could not result in classifications of the sort modern naturalists have established. But it was perfectly plain that his learning fixed his attention by preference on the relations of the individual to the species and of the species to the genera, and that it suggested to him the idea of extending further and further, by abstraction, this hierarchical progression of beings, until the supreme genus, abstract Being as such (τὸ ὄν), is reached. From this source come two corresponding inventions: that of the theory of the syllogism, which would be sufficient to immortalize the name of Aristotle, and that of ontology, which contributed so much to the discrediting of peripateticism. Everything hangs together in the doctrine of peripateticism; the rules of definition are derived from the principle of classification, and the theory of the syllogism is connected with the rules of definition. Although all our knowledge comes to us through individual beings, which alone make impressions on our senses, the supreme category of being or substance is the source of every axiom, of every syllogistic argument, and, consequently, of all scientific or rational

---

to him than the conclusion, he will have his knowledge only incidentally." *Ethica Nicomachea*, Book VI, 4 [1139b 33. *Op. cit.*, Vol. IX].

knowledge.[19]   Every logical proof rests finally upon the principle of contradiction, that is, the principle that a thing cannot both be and not be at the same time.   This amounts to the same thing as Condillac's principle that all our judgments are nothing but a series of identities.   This principle is the *quid inconcussum* which suffices for Aristotle against the skeptics; for since we have sensations, it must be that they come from something and that that something exists; [20] as though that were sufficient to permit us to affirm that they represent something and to enable us to recognize what they represent!

Thus Aristotle, by exaggerating the resources of the principle of identity and of syllogistic deduction, by wishing to make all principles or axioms proceed from a single principle or axiom, failed to recognize the active and continual intervention of the forces of the intellect in those special judgments, based upon ideal constructions, that Kant has so well characterized and indicated by the name of synthetic judgments *a priori* (262).   From these judgments, much more than from syllogistic deduction, we obtain mathematical truths, and their role and importance in geometry had been very clearly perceived by Plato, as can be inferred even from the passages that we have cited above.   This is so because Plato was a geometrician and, from all appearances, Aristotle was not, any more than Condillac was, when he confused the science of numbers with the "language of calculation" (*la langue du calcul*), and the mathematical spirit with the spirit of analysis.

But if Aristotle misconstrues or neglects in his system that which makes mathematical speculation fruitful, he falls into a still more serious omission, so far as its harm to philosophical speculation is concerned, by not taking account of the genus of probability

---

[19] " So, too, certain properties are peculiar to being as such, and it is about these that the philosopher has to investigate the truth. . . . So, too, there are many senses in which a thing is said to be, but all refer to one starting-point; some things are said to be because they are substances, others because they are affections of substance." *Metaphysica*, Book IV, 2 [1004b 15 and 1003b 5. *Op. cit.*, Vol. VIII].

[20] " For sensation is surely not the sensation of itself, but there is something beyond the sensation, which must be prior to the sensation; for that which moves is prior in nature to that which is moved." *Metaphysica*, Book IV, 5 [1010b 35. *Op. cit.*, Vol. VIII].

which is its unique support. In doing so, he weakens at one and the same time induction (by reducing inductive inference to the status of being nothing but a summary of particular observations) and deduction, which can never be anything more than a means of classifying and not of extending our knowledge, if the general idea is nothing more than an abstract epitome of particular ideas, that is to say, nothing but something that inductive probability alone can give (47). Every distinction between artificial and natural abstraction, based upon the reason of things, is thus found to be lost from sight. By wishing to give greater exactness and scientific precision to Plato's teachings, Aristotle aggravates in two ways the mistake that his master had made. This mistake was the result of an entirely excusable illusion when we bear in mind that Leibniz later shared it, and that, even in our own day, many distinguished minds fall into it. The mistake consists in not recognizing that philosophy is something other than science, and that it lies within the realm of δόξα and not of ἐπιστήμη. For, on the one hand, Plato, under cover of the flexibility of the forms of his dialectic, was able to belie his theory in carrying it out and to make continual use of probable inferences, which no longer lent themselves to the exactness, or, rather, to the rigidity of the forms of the Aristotelian syllogism. He was able to use the poetic forms of language (that we know to be useful or necessary), even though they are incompatible with the dryness of syllogistic argumentation. On the other hand, Aristotle substitutes an ontological subordination, a hierarchy of categories, genera, and species, for a rational subordination among truths and facts, according to which they are the basis or reason for one another. This is the very illuminating rational hierarchy that Plato had always had in view, although not always clearly enough. Aristotle's subordination hinges on the idea of being or substance. This idea may be regarded as fatal for the human mind, in the sense that it is always precipitated into inscrutable mysteries without settling anything with respect to which it wished to make a thorough examination.

To the degree that the domination of peripateticism extended and consolidated itself, these causes of error were bound to be more pronounced. It is especially during the Middle Ages that the syllogism was put into practice in the scholastic tournaments of the bachelors and doctors. It is also during the Middle Ages, at the time of the dispute between realism and nominalism, that we see the partisans of Aristotle finally approach the fundamental problem of philosophy—the problem which has to do with the representative value of our ideas—but without being able to succeed in freeing it of the obscurities inherent in the ontological hierarchy (167). Thus, their long quarrels were bound to result only in abstruse subtleties of a sort repugnant to good minds and to cause the true meaning of the question to be increasingly lost sight of.

382. Two great reformers, Bacon and Descartes, made their appearance a few years apart; and their philosophical daring, seconded by the admirable discoveries of Galileo and Kepler and by the advance of civilization, gave the human mind the impulse which initiated a movement that still endures. Bacon follows Plato in that he, like Plato, has the inspiration of the poet and the apostle. Furthermore, his philosophy is aphoristic rather than dialectical. Above all, he has that feeling for the grandeur and the majesty of nature which was missing in the genius of the Greeks, who were inclined to take man as the measure of all things, in philosophy as in religion, art, and poetry. *Homo, naturae minister et interpres* [21]. . . . These few words are sufficient to characterize

[21] " Man, the servant and interpreter of nature, understands and controls so much of the regular successions of nature as he has worked with or will mentally observe; he neither has accomplished, nor can accomplish more." *Temporis partus masculus.* And, in another variant of the same aphorism, " Man, the servant and interpreter of nature, understands and controls so much of the regular succession of nature as he will mentally observe while guided by the laws of nature." *Novum Organum*, I, Aphorism 1.

[Throughout this work it has been impossible to locate a number of Cournot's references, at least in their exact expression. In many cases, as a glance at the various French editions of the *Essai* will show, Cournot either gave no explicit reference for materials he cited or his source was an edition of the works of various authors to which we have been unable to gain access. The above variants of one of Bacon's aphorisms are a case in point. The Latin text of neither variant, as it appears on page 567 of the third edition of the *Essai*, agrees with the text of the aphorism as found in Volume I, 241 of *The Works of Francis Bacon*, collected and edited by Spedding and Ellis; the text of the second variant is to be found in *De Interpretatione Naturae, Sententiae XII*, 1, Volume VII, 365, of this collection; the text of *Temporis partus masculus* may be found in Volume VII, 7 through 32 of the same collection.

Where it has not been possible to find texts used by Cournot, because we have been unable either to identify or to gain access to an edition used by him, new translations have been made of materials he quoted.]

the whole of Bacon's philosophy and to fix the point of view from which he looks at things. He is the priest of the new cult of nature, which spurns the excessive and mystical phantoms of the oriental imagination as well as the extravagance of Greek subtlety and of scholastic controversy. Far from taking the human understanding as the measure of all things, he affirms that from the very first he finds in its constitution the causes of error and illusion.

The Idols of the Tribe have their foundation in human nature itself, and in the tribe or race of men. For it is a false assertion that the sense of man is the measure of things. On the contrary, all perceptions as well of the sense as of the mind are according to the measure of the individual and not according to the measure of the universe. And the human understanding is like a false mirror, which, receiving rays irregularly, distorts and discolours the nature of things by mingling its own nature with it.[22]

Elsewhere he says,

The great deceptiveness of the senses traces the pattern of external nature [from analogy] to man (*ex analogia hominis*) and not [from analogy] to the universe (*non ex analogia universi*); and *this is not to be corrected except by reason and universal philosophy* (*per rationem et philosophiam universalem*).[23]

Then come the other types of *idols* or illusions, and in the second group those which he calls, in an allusion to the metaphor of Plato, the "idols of the cave," which affect, not the species, but the individual varieties, as though Plato had not had in mind humanity as a whole rather than only a few men placed in special and exceptional conditions in his simile of the prisoners of the cave. However, it is of little importance that Bacon's "idols of the tribe" are really Plato's "idols of the cave;" what would be important for us would be to know how Bacon thinks it possible for the human mind to correct these illusions which are common to all men and which are an integral part of their nature. He certainly indicates it in the short and emphatic lines just quoted, *ex analogia universi . . . . per rationem et philosophiam universalem*, but indicate it is all that he did. Elsewhere (82), by returning to the simile of an interposed mirror which deflects luminous rays and

[22] *Novum Organum*, I, Aphorism 41. [*Bacon Selections*, edited by M. T. McClure, "The Modern Student's Library" (New York: Charles Scribner's Sons, 1928), pp. 288, 289.]

[23] *Novum Organum*, II, Aphorism 40. [*Op. cit.*, p. 467. (Cournot's italics).]

deforms images, he seems to admit the impossibility of clearing away illusions due to such a cause. Bacon ceaselessly appeals to observation, to experience, to analogy, and to induction; but he does not give the philosophical theory of induction and analogy; he does not grasp the rational principle on which philosophical probability is based and which authorizes us to draw from experience more, even infinitely more, than there is in experience itself. After having thrown down peripateticism, he unwittingly falls again into peripatetic formalism. By presenting the inductive procedure, at least in its execution, as approximately the same as Aristotle's idea of ἐπαγωγή, he seems to see in it only a means for carrying out the separation or sorting of generalities and particularities.[24] Immediately all his attention is given to the invention of a sort of sieve, capable of performing this separation by mechanical means, so to speak. This results in prolix enumeration of *instances* or *forms* of induction, to which he attaches as much or more importance as the Scholastics had attached to the forms of the syllogism, and of which no one who has come after him has ever made the slightest use. The rapid progress of physics has precluded our being led into error in this connection by the example of a great man; and the high truths whose eloquent interpreter he was, being unmixed with error, contributed to the education of his century and to the lasting conquests of the human mind.

383. It must not be thought that Bacon, who was singularly

[24] In order to give an idea of Bacon's procedure in carrying out this defective part of his work, we quote the following passage: " We must make therefore a complete solution and separation of nature, not indeed by fire, but by the mind, which is a kind of divine fire. The first work therefore of true induction (as far as regards the discovery of Forms) is the rejection or exclusion of the several natures which are not found in some instance where the given nature is present, or are found in some instance where the given nature is absent, or are found to increase in some instances when the given nature decreases, or to decrease when the given nature increases. Then indeed after the rejection and exclusion has been duly made, there will remain at the bottom, all light opinions vanishing into smoke, a Form affirmative, solid and true and well defined. This is quickly said; but the way to come at it is winding and intricate. I will endeavor however not to overlook any of the points which help us towards it." Bacon, *Novum Organum*, II, Aphorism 16. [*Op. cit.*, pp. 410, 411.]

It is especially necessary to recognize that there is some possibility of making these considerations a good basis for and the means of interpreting statistical tables.

preoccupied with discoveries made in the realm of nature, regarded the study of the human mind and its faculties as vain. Far from it; in his eyes this study is the most important of all, for without it the pageant of nature would be nothing but a descriptive phantasmagoria.[25] On the other hand, we have seen that he said, and with reason, that the illusions of the human mind can be corrected only through the attentive study of nature, through the conception of the general order of the world, *ex analogia universi*. Apparently, this is a vicious circle from which we must escape, and Bacon has not been careful to indicate clearly the way out. The fact is that, if not in theory, at least in practice and in the works which are his principal claims to glory, Bacon is particularly concerned with the explanation of the physical world and with the extension of the positive sciences, which are destined to increase man's power over nature. Now we do not have to gain a precise idea of the rational principles of analogy and induction to be assured that the continual control of sensible experience, in the physical and the biological sciences, must sooner or later result in our rejecting the prejudices with which we are imbued and the false ideas which would follow from innate tendencies or acquired habits, as incompatible with the orderly explanation of observed facts. But there is more to it than this. The absolute incomprehensibility of certain natural facts or the absolute irreducibility of one order of phenomena to another should attest to the falsity or the inadequacy or the incoherence of certain fundamental data of our understanding, in so far as it applies to the comprehension of the external world. But Bacon did not undertake the description and the criticism of the human understanding from this point of view. Still less does he seem to have proposed to himself the philosophical study of the human mind, considered in itself, and not in so far as it is applicable to the knowledge of external things. In this inner world, there is without doubt a general order to be discovered, a hierarchy to be established among faculties and functions, which makes it possible for some to control others, and for the function of each to be judged, *ex analogia universi* . . . . *per rationem et*

[25] " He who first and before anything else has not explored the working of the human mind and has not accurately charted the channels of knowledge and the hazards of error will find everything masked and, as it were, spell-bound; if he will not use the proper tools, he will not understand." *De interpretatione naturae* [ed. by Spedding and Ellis, Vol. VII, 366].

philosophiam universalem. Bacon's great principle, which he is content to indicate in such a summary manner, would therefore still be applicable here. But Bacon neglected to carry it out, or, rather, we see that he was unable to carry it out without being in possession of a more complete theory of the principles and of the nature of philosophical induction. For here the criterion of sensory experience is lacking, at least if this name is reserved for experience which leads to a precise determination and to exact experiments (372). It is no longer possible continually to confront theoretical conceptions and positive facts with one another; the intelligence which knows and the intelligible relations which are the objects of knowledge tend to be constantly confused. In place of those irresistible probabilities, which have the legitimacy of the proof called " common sense," we are most often forced to be content with possibilities that different minds may evaluate differently. At this point, therefore, the Baconian aphorisms are no longer sufficient, or at least it is necessary to give a development and interpretation of his principles that Bacon himself has not given. Nor was it given by the Scottish school when it proposed to carry out, in the philosophy of the human mind, the reform attributed to Bacon in what the Greeks called " physics," and which the English call " natural philosophy."

384. What, as a matter of fact, is the fundamental dogma of the Scottish school? Let us listen to what the leader of this school has to say:

All reasoning must be from first principles; and for first principles no other reason can be given but this, that, by the constitution of our nature, we are under a necessity of assenting to them. Such principles are a part of our constitution, no less than the power of thinking: reason can neither make nor destroy them . . . . A mathematician cannot prove the truth of his axioms . . . . A historian, or a witness, can prove nothing, unless it is taken for granted that the memory and senses may be trusted. A natural philosopher can prove nothing, unless it is taken for granted that the course of nature is steady and uniform. How or when I got such first principles, upon which I build all my reasoning, I know not; for I had them before I can remember: but I am sure they are parts of my constitution, and that I cannot throw them off . . . . If we are deceived in it, we are deceived by Him that made us, and there is no remedy.[26]

---

[26] " Inquiry into the Human Mind," Chap. V, Sect. 7 [*The Works of Thomas Reid,* collected by Sir William Hamilton (Edinburgh: MacLachlan, Stewart, and Co., 1846), p. 130].

. . . [The several original principles of belief] make up what is called "the common sense of mankind " . . . . A clear explication and enumeration of the principles of common sense, is one of the chief *desiderata* in logic [27] . . . . There seems to be no remedy for this . . . unless the decisions of common sense can be brought into a code in which all reasonable men shall acquiesce,[28] and so forth.

Thus Bacon had been wrong in supposing that there were inherent illusions in the make up of the human mind ("idols of the tribe") and in thinking that they could be corrected *ex analogia universi*. For, if such illusions occur, they are without remedy. Even those illusions which would affect only individuals ("idols of the cave") are for these individuals as irremediable as the others; for the principles which they would destroy are so many articles in the general code that would not be admitted by individuals among whom such illusions are found and that no one would have the right to impose upon them. They would be shut out from communion with the great body of men, and that is the end of the matter. All criticism of our intellectual powers (*facultés*) by one another is impossible; all have the same right to infallibility; all the articles of the code enjoy an equal authority, and antinomies, if by chance they are found (as we know some are found) are absolutely insoluble. It is necessary to agree with the ignorant that the earth is immovable; for nothing is more contrary to common sense than for us to believe that we are actually being carried through space with a velocity of about eighteen and three-quarters miles a second. What one person regards as a first principle or as a principle of common sense will by no means have this character in the eyes of another person. For example, we shall not agree entirely with Reid in regard to the axiom that "the course of nature is uniform and invariable." This is not true in certain respects (48) and could not be true in any sense without making physical research sterile for that reason. In a word, the Scottish school, in pretending to extend and to complete the work of Bacon, took for its fundamental maxim one directly opposed to the aphorism of the master. Thus, with all the talent that it has been able to bestow upon details, and in spite of the wise moderation which preserved

[27] [*Op. cit.*, Chapter VII, p. 209.]

[28] " Essays on the Intellectual Powers of Man," Essay VI, Chap. 2 [*op. cit.*, p. 422].

it from excesses into which other schools have fallen, it has not led to any hierarchical subordination, to any rational classification of the faculties of the intellect, or to any evaluation of the representative value of the facts which it describes.[29]

385. During the past thirty years, so much has been said in France about Descartes, his influence, and his method, that we inevitably feel apprehensive about dwelling on a shopworn topic. He was careful not to fall into the error into which Reid later fell, and he did not place on the same level the different faculties whose organization and operation constitute the human understanding. We should always allow ourselves to be persuaded by the evidence of our reason, but we are not bound to believe in that of our imagination and our senses. And why must we allow ourselves to be persuaded by the evidence of our reason? It is because we know that God exists, that he is perfect, and because " our ideas or notions, being real things which come from God insofar as they are clear and distinct, cannot to that extent fail to be true." [30]   But, on all these points, the importance of the question and the prestige of Descartes' name require that he be allowed to speak for himself, yet without needing to reproduce and discuss here the well-known arguments the concatenation of which comprises the Cartesian proof of the existence of our soul, of God, and of the divine perfections:

Finally, if there are still some men who are not sufficiently persuaded of the existence of God and of their soul by the reasons which I have given, I want them to understand that all the other things of which they might think themselves more certain, such as their having a body, or the exist-

---

[29] " If the Scottish school has rendered a service and an important one to philosophy, it is surely to have established in human minds once and for all, and in such a way that it cannot be forgotten, the idea that there is a science of observation, a science of facts, *in the way in which this is understood by the physicists,* whose object is the human mind, and whose instrument is the inmost sense, and whose result should be the determination of the laws of the mind, as that of the physical sciences should be the determination of the laws of matter." Jouffroy's preface to his translation of *The Works of Thomas Reid,* p. cc. The fundamental error of the Scottish doctrine consists in this very comparison, as we believe we have proved (371ff.).

[30] Descartes, *Discourse on Method,* Part IV [translated by Laurence J. Lafleur, " The Library of Liberal Arts " (New York: The Liberal Arts Press, Inc., 1950), p. 25].

ence of stars and of an earth, and other such things, are less certain. Even though we have a *moral assurance* of these things, such that it seems we cannot doubt them without extravagance, yet without being unreasonable we cannot deny that, as far as metaphysical certainty goes, there is sufficient room for doubt. For we can imagine, when asleep, that we have another body and see other stars and another earth without there being any such. How could one know that the thoughts which come to us in dreams are false rather than the others, since they are often no less vivid and detailed? Let the best minds study this question as long as they wish, *I do not believe they can find any reason good enough to remove this doubt unless they presuppose the existence of God.* The very principle which I took as a rule to start with, namely, *that those things which we conceive very clearly and very distinctly are all true,* is known to be true only because God exists, and because he is a perfect Being, and because everything in us comes from him. *From this it follows that our ideas or notions, being real things which come from God insofar as they are clear and distinct, cannot to that extent fail to be true.* Consequently, though we often have ideas which contain falsity, *they can only be those ideas which contain some confusion and obscurity, in which they participate in nothingness.* That is to say, they are confused in us only because we are not wholly perfect. It is evident that it is no less repugnant to good sense to assume that falsity or imperfection as such is derived from God, as that truth or perfection is derived from nothingness. But if we did not know that all reality and truth within us came from a perfect and infinite Being, *however clear and distinct our ideas might be,* we would have no reason to be certain that they had the perfection of being true.

After the knowledge of God and the soul has thus made us certain of our rule, it is easy to see that the dreams which we have when asleep do not in any way cast doubt upon the truth of our waking thoughts . . . . For in truth, whether we are asleep or awake, we should never allow ourselves to be convinced except on the evidence of our reason. *Note that I say of our reason, and not of our imagination or of our senses* . . . for reason does not insist that all we see or visualize in this way is true, but it does insist that all our ideas or notions *must have some foundation in truth,* for it would not be possible that God, who is all perfect and wholly truthful, would otherwise have given them to us. Since our reasonings are never as evident or as complete in sleep as in waking life, although sometimes our imaginations are then as lively and detailed as when awake or even more so, and since reason tells us also that all our thoughts cannot be true, as we are not wholly perfect; *whatever of truth is to be found in our ideas will inevitably occur* in those which we have when awake rather than in our dreams.[31]

---

[31] [*Op. cit.,* pp. 24-26. Cournot's italics, except for the word " reason " italicized in the last paragraph.]

We must guard ourselves against reproaching Descartes for not having gone farther into the nature of *moral certainty*; for a great man who is in advance of his time cannot go entirely beyond it. He should, however, be reproached for being content with a reason of the most obscure sort, when he rejects as false that which is confused and obscure in our ideas, *because to that extent they partake of negation.* What's that! we place between our eyes and visual objects lenses which deform the images, distort the lines we see, so that what was clear, regular, and well ordered, becomes involved and confused; how can the interposition of the lenses be made to resemble a participation in nothingness? This is simply a disturbing cause, as real as the others, producing effects also as real; a disturbing cause whose existence reason can discover, through its feeling for the order of things, and the effects of which it can discern in some cases, just as, in other cases, reason is also able to give quasi-certain assurance that our perceptions and our ideas are not affected by disturbing causes of this sort. If the perfection of God guarantee us nothing more than that our ideas or notions must have *some foundation in truth,* are we not still left to discover what this foundation is? And in order to separate what is clear and distinct, and consequently true, from what is confused and obscure, and consequently false, must not reason allow itself to be guided by its innate feeling of order and disorder, harmony and confusion? If such is the case, the criterion used by Descartes will differ in no way from ours. And if the distinction in question can be made only by means of probable inferences and not by means of *infallible* arguments, it will make little enough difference, as far as the value of the conclusions drawn by means of critical examination is concerned, that the idea of order may have been connected with the notion of the existence and the perfection of God by demonstrative reasons. On the contrary, these supposedly demonstrative reasons do not by any means have the virtue of convincing all minds. Furthermore, they imply and will always imply transcendent problems which do not admit of a positive solution. Therefore, their problematic and transcendent character is improperly reflected with respect to the solution of purely logical questions which we have sought to connect with them, and the independence of which must, on the contrary, be clearly demonstrated.

386. Observe now that if Descartes seems to subordinate the

certainty of his axiom—that all our ideas are true in so far as they are clear and distinct—to the notion of Divine veracity, on the other hand, by his stratagem with respect to the existence and the attributes of God, he infers the existence of a perfect being from the idea of a perfect being. From this it follows that it is fundamentally an axiom for Descartes that all things must be such as our understanding clearly conceives them to be. That one of his disciples who has followed most rigorously all the consequences of the Cartesian teaching, Spinoza, is not deceived on this point, since he tells us, "The *properties of the understanding* which I have principally noted and which I clearly understand are these:—I. *That it involves certainty,* that is, that it knows things to exist formally just as they are contained in it objectively." [32]

Thus Spinoza admits from the outset that the conceptions of our intelligence are the infallible criteria of the truth of things; that in this mirror of the human mind there can be only images exactly like the objects which produce them and in no way altered by the character of the reflecting mirror itself or by the medium which surrounds it. That is dogmatism in its extreme form, so extreme, indeed, that there is no longer room for the criticism of the human understanding.

For the rest, as Descartes (without falling into the excesses of Spinozism, for which it would be unjust to make him responsible) finds no notion more indelibly printed in the human mind than that of substance, no fact more incontestable than that of thought, no notion more clear than that of extension, he traces the distinc-

---

[32] *On the Correction of the Understanding,* § 108 [*Spinoza's Ethics and De intellectus emendatione,* "Everyman's Library" (New York: E. P. Dutton and Co., Inc.), p. 262. The italicizing of the second phrase is by Cournot].

See also Malebranche, *Recherche de la Vérité,* Book IV, Chap. 11; and Bossuet, whom no one will accuse either of Spinozism or of letting himself be carried away by his imagination, like the celebrated priest of the Oratory [Malebranche], gives this precept in his *Logique* (Book I, Chap. 64): "To know the differences of things by means of ideas, that is to say *not to doubt* that there is a difference among things when we have different ideas." Locke says the same thing: "Our simple ideas *are all real* in the sense that they always agree with the nature of things." This is the reason for contending that Locke and Condillac, as well as Spinoza, Malebranche and Bossuet, follow Descartes in that, while adopting systems wholly opposed in regard to the origin of ideas, they agree on the principle that every clear and distinct idea is necessarily true or conforms to objective **reality.**

tion between thinking and extended, spiritual and corporeal substances with a steadfastness unknown before him, and follows intrepidly all the extreme consequences of his premises and of the uncompromising classification to which they led him.   In the physical world there are only extension and movement; there are only impulses received and transmitted, and no actions at a distance and no forces properly so called; animals are pure machines and are or should be explained, so far as their bodily nature is concerned, by the most passive and the crudest sort of mechanism, just as everything in the spiritual realm should be explained by means of the purest and most exalted principles.   Middle terms are proscribed as obscure.   The metaphysics of Aristotle gives way to a new metaphysics, its premises so well adapted to the constitution of our intelligence, that from its inception it captivated all the serious minds of the seventeenth century because of its clarity.   But soon, because of the consequences that result from it and which contradict the suggestions of common sense, no less than the discoveries of the sciences, this metaphysic lost its influence little by little, although during the period when it was most influential it could not be compared in any way with the powerful system whose place it had wished it might take.

387. As he himself tells us,[33] it is in such circumstances that

---

[33] " Although I am one of those who have worked very hard at mathematics I have not since my youth ceased to meditate on philosophy, for it always seemed to me that there was a way to establish in it, by clear demonstrations, something stable.   I had penetrated well into the territory of the Scholastics when mathematics and modern authors induced me while yet young to withdraw from it. Their fine ways of explaining nature mechanically charmed me; and, with reason, I scorned the method of those who employ only forms or faculties, by which nothing is learned.   But afterwards, when I tried to search into the principles of mechanics to find proof of the laws of nature which experience made known, I perceived that the mere consideration of an ' extended mass ' did not suffice and that it was necessary to employ in addition the notion of ' force,' which is very easily understood although it belongs to the province of metaphysics.   It seemed to me also that the opinion of those who transform or degrade animals into simple machines, notwithstanding its seeming possibility, is contrary to appearances and even opposed to the order of things.

" In the beginning, when I had freed myself from the yoke of Aristotle, I occupied myself with the consideration of the void and atoms, for *this is what best fills the imagination;* but after many meditations I perceived that it is impossible to find principles of true unity in mere matter, or in that which is only

Leibniz undertook the reform of the idea of substance, by positing in principle that there is no substance which is not endowed with action or force, and even which does not actually tend to exercise this action or this force (*conatum involvens*). That is to say, while conceding completely to the ancient schools that there is a basis for the notion of substance, he declares it sterile if the idea of force is not incorporated in it and does not give it life. Actually he undertook to deduce *a priori*, from the idea of force, everything the peripatetic and the Cartesian schools hoped to deduce from the notion of substance. Now the term " force," used with the degree of generality that Leibniz attaches to it, can be applied to mechanical force, to vital and organic force, to the free determination of the self, and even lends itself (vaguely, to be sure) to the conception of an infinite number of intermediate modalities. It should follow from this that the system of Leibniz is better adapted than any other, if not to a precise and scientific explanation, at least to a philosophical and general conception of the ensemble of the phenomena of nature in their inexhaustible variety. Thus, as Maine de Biran has very well said: [34]

This reformed metaphysics no longer makes room for only two great classes of beings, entirely separate from one another and excluding every intermediary; but one and the same series includes and joins all the beings of creation. Force, life, perception are everywhere distributed among all the degrees of the series. The law of continuity suffers neither interruption nor break in the passage from one degree to another and fills without gaps, without any possibility of empty space, the immense interval separating the ultimate monad from the supreme intelligent force from which everything emanates.

388. However, we must not concern ourselves here with this *philosophy of nature*, however curious it may be in itself, and however superior it may appear to all the efforts of this sort that have been attempted before and after Leibniz. Another part of his doc-

---

passive. . . . It became necessary, therefore, to recall and, as it were, reinstate the *substantial forms*, so decried now-a-days, but in a way to render them intelligible, and distinguish the use which ought to be made of them from the abuse which had befallen them." *A New System of the Nature and of the Interaction of Substances* [from *The Philosophical Works of Leibniz* translated by G. M. Duncan (2nd ed.; New Haven: The Tuttle, Morehouse and Taylor Company, 1908), pp. 77, 78].

[34] Maine de Biran, *Oeuvres philosophiques*, Vol. IV, " Exposition de la doctrine de Leibniz," published by Cousin.

trine is of much more interest to us. In promulgating his principle of sufficient reason and opposing it to the principle of contradiction, which Aristotle had set up as the fundamental axiom or the major support of every scientific truth (381), Leibniz is, among all philosophers, the first to indicate clearly the essential purpose of all philosophical study, namely, to conceive things in the order according to which they constitute reasons for one another,[35] an order which must not be confused either with the chain of causes and effects, or with that of logical premises and consequences (18 ff.).

The great foundation of *mathematics* is *the principle of contradiction or identity,* that is, that a proposition cannot be *true and false* at the same time; and that therefore *A* is *A,* and cannot be *not A.* This single principle is sufficient to demonstrate every part of arithmetic and geometry, that is, all *mathematical principles.* But in order to proceed from *mathematics* to *natural philosophy,* another principle is requisite, as I have observed in my *Theodicaea:* I mean, *the principle of sufficient reason, viz.,* that nothing happens without a reason why it should be *so,* rather than *otherwise.* And therefore Archimedes being desirous to proceed from *mathematics* to *natural philosophy,* in his book, *De aequilibrio,* was obliged to make use of a particular case of the great principle of *a sufficient reason.* He takes it for granted, that if there be a *balance,* in which every thing is alike on both sides, and if equal weights are hung on the two ends of that balance, the whole will be at rest. 'Tis because no *reason* can be given, why one side should weigh down, rather than the other. Now, by that single principle, viz., that *there ought to be a sufficient reason why things should be so, and not otherwise,* one may demonstrate the being of a *God,* and all the other parts of *metaphysics* or *natural theology;* and even, in some measure, those principles of *natural philosophy,* that are independent of *mathematics:* I mean the *dynamic* principles, or the *principles of force.*[36]

We think we have shown in many passages in this book (28, 265) that the expression of Leibniz would be incorrectly understood if the use of the principle of contradiction were made the essential

---

[35] " Our reasonings are founded on *two great principles, that of contradiction,* in virtue of which we judge that to be *false* which involves contradiction, and that *true,* which is opposed or contradictory to the false.

" And *that of sufficient reason,* in virtue of which we hold that no fact can be real or existent, no statement true, unless there be a sufficient reason why it is so and not otherwise, although most often these reasons cannot be known to us." [Leibniz, *The Monadology,* §§ 31, 32, *op. cit.,* p. 313.]

[36] " Leibniz's Second Paper: Being an Answer to Dr. Clarke's First Reply." [See Leibniz, Letters to Samuel Clarke, § 1, *op. cit.,* pp. 330, 331.]

characteristic of mathematical speculation, and the use of the principle of sufficient reason the essential characteristic of speculation in physics and metaphysics; while it must be said that the continual application of the idea we have of the reason of things is what essentially characterizes philosophical speculation, whether it rests upon abstract relations, such as those which are the subject matter of mathematics, or on the interpretation of natural facts, or on the laws of our understanding. What ought particularly to invite our attention here, and what we have already had occasion to indicate, is the negative form under which Leibniz puts to work the principle or the idea that we have of the reason of things in such a way as to make it a means of rigorous demonstration, *more geometrico,* through the device of reduction to an absurdity; but in such a way as to restrain its application singularly and even to leave aside the most important applications of the principle, those to which it owes its prerogative as a regulative and dominating principle. We have the idea that everything must have its reason, apparently in order to find out positively the reason of things that we know and why the knowledge of some leads us to knowledge of others. We do not have the idea only in order to yield this negative judgment, that things cannot exist, in this case, in such a necessarily very greatly limited manner, in which, because of certain peculiar conditions of symmetry, we can affirm that there is no reason why they might receive a certain determination rather than some other contrary or symmetrical determination.

389. Finally, in order to make clearer the sense in which we use these expressions of "positive" and "negative" judgment, let us take one of the examples Leibniz has given us in his theory of *indiscernibles.* Many people have said that there are not two perfectly similar objects in nature, for example, two leaves exactly like one another. And they have had reason for saying this, because, since the arrangements which occur in this case and produce the varieties of individuals and the particularizations of the specific type are innumerable, it is, if not utterly impossible, at least infinitely improbable that we should happen upon two absolutely identical arrangements. This is a positive application of the notion we have of the reason of things. We find in the simplicity of the arrangement which would lead to exact resemblance, among an infinite number of arrangements, that the play of independent

causes may introduce the reason why this arrangement is not and may not be realized physically, although it implies no contradiction and is not in this sense mathematically or metaphysically impossible (33). But this is not the way Leibniz understands it. According to him, if the two objects A and B were rigorously identical, they would be indiscernibles, and there would be no sufficient reason why A should not be in place of B and B in place of A. God himself could not resolve upon a choice which would not have a sufficient reason. Therefore, any hypothesis which would imply the existence of indiscernible things is inadmissible. This is an instance of what we understand by the application of the principle of the reason of things under its negative form, and by way of the negation or exclusion of a hypothesis; for evidently the mind does not work in the same way when it judges that a certain thing should receive the determination A, because it sees the reason for it, or when it judges that a thing cannot receive a determination other than A, for example the determination $+A'$, because it sees no reason why it should receive the latter rather than the contrary determination $-A'$. This is the same distinction that mathematicians establish between direct demonstrations, which instruct the mind, that is to say, which show it the reason for the demonstrated truth, and demonstrations which are indirect or by absurdity, which *compel the mind* (often better than the others, because of a certain logical prerogative attached to negative forms) but do not instruct it. Thus in the example we are considering, our affirmation neither has nor can have the kind of certainty which characterizes mathematical demonstrations. The fact affirmed is certain only in the sense that the contrary fact is infinitely probable, the adverb " infinitely " here being taken in its exact meaning and not improperly, as it sometimes is, to take the place of some other adverbial superlative. The mistake that Leibniz makes is in not being satisfied with this kind of certainty, which should be enough for us, since it is due to the nature of the matter in question. He seeks another certainty; and he attains it or thinks he attains it, but at the cost of what daring! And who can persuade himself to follow him in this daring enterprise of assigning to the Divine power the limits beyond which it cannot go?

However, let us continue the Leibnizian deductions. If the co-existence of indiscernible things is inadmissible, so is a succession of indiscernible degrees, and an indiscernible motion is also an impossible motion. Thus it would be impossible to suppose two concentric spheres of equal density and perfect homogeneity, of which one (the inner sphere) would be impelled by a rotational motion; for neither man nor angel (nor God himself, if one dares to say it) would be able to distinguish the system in which such a movement is taking place, from a system which is identical in all respects except that this motion is lacking.[37] Now, in order to refute this consequence, it is sufficient to imagine that the inner sphere has a density at first less and then greater than that of the sphere which encloses it. No matter how small this difference may be, in one direction or the other, the movement, according to Leibniz, will be discernible and consequently possible. Therefore, at the limit, and in the continuous passage from the least to the greatest density it is clearly necessary that the movement remains possible, although momentarily indiscernible. This must be possible because of the principle of continuity that Leibniz invokes everywhere and upon which he himself has based his great discovery of the differential calculus. Furthermore, let us conceive of a series of such systems, in which the density of the internal sphere tends to be increasing, by being first smaller and then larger than that of the sphere which surrounds it, and that in the series a sphere is found in which the difference in density is nil. In that case the intelligence of an angel would not be necessary, but that of a man would be sufficient to judge, *by inductive inference,* that the movement discerned in all the systems which are part of the series, with the single exception of this one, belong also to this intermediary system, and ceases to be distinguishable only because of a particular and accidental relation. Thus it is that after having observed a rotary motion in all

[37] " If we contrive two perfectly concentric spheres, perfectly similar each to each as well as similar in all their parts, one to be included within the other so that there is not the slightest gap, then, whether we assume the inner sphere to revolve or to remain at rest, neither an angel, *nor, I may say, any greater being,* could ever make a distinction between its states at different times nor have the judgment to discern whether the inner sphere rests or revolves, and by what law of motion." Leibniz in the *Acta Eruditorum* for 1698; also in the collection of Desmaizeaux, Vol. I, p. 212; [and Erdmann edition, p. 159, col. 1].

But even in Leibniz's hypothesis, there would be mechanically discernible effects. However, for our purpose we have no need to enter into this discussion. It is simpler to reason from his hypothesis and grant him his premises for it.

stars which have marks that make this motion discernible, we shall not hesitate to infer that a star whose surface presents no mark that permits its rotation to be discerned also has a similar motion.

390. Leibniz, with the idealistic genius of Plato, but with a more solid logic and with an incomparably vaster knowledge, has therefore made the same mistake as Plato; he has sought to deduce everything, *more geometrico*, from an *a priori* principle, and he was not able to bring himself to admit a kind of inductive reasoning in philosophy, based upon probability or δόξα. Consequently, it was necessary for him to use the device of *reductio ad absurdum* to give a negative form to the statement of the principle of the order and reason of things (whose role and importance in philosophy he appreciated better than anyone before him), and, as a result, virtually to restrict the application of the principle, and always, through the desire to attain the end, to yield to the temptation of going beyond its legitimate applications.

Not that Leibniz did not often appeal to the notion of philosophical probability and to inductive judgment based upon the order and reason of things. For example, we find among his writings these passages:

Nevertheless it is reasonable to attribute to bodies real movements, according to the supposition which explains phenomena in the most intelligible manner [38] . . . . The reality of phenomena is marked by their connection, which distinguishes them from dreams [39] . . . . The truth of sensible things [consists] only in the *connection* of phenomena, which must have its reason, and that it is this which distinguishes them from dreams; but that the truth of our existence and of the cause of phenomena is of another kind, because it establishes substances.[40]

These passages are explicit, and the last especially is notable in that it clearly indicates the distinction between the idea of reason and the idea of causality and because it indicates the origin of the contrast between the clarity of philosophy which contemplates the rational order of things and the obscurities of ontology which glosses the notion of substance. But if the passages cited contain

[38] Leibniz, *A New System of the Nature and of the Interaction of Substances*, § 18 [*op. cit.*, p. 86].

[39] Leibniz, *Examen des principes du R. P. Malebranche*, Dutens edition, Vol. II, 210; [also Erdmann edition, p. 695, col. 2].

[40] Leibniz, *The New Essays on the Human Understanding*, § 14 [*op. cit.*, p. 237].

in germ the whole doctrine that we have undertaken to develop, it must be recognized that at no point does Leibniz explain the nature of this judgment based upon the order and the connection of phenomena. Neither does he tell us why it is reasonable to propose one hypothesis rather than another about true, that is to say, about real motions, nor whether this reasonable judgment is probable or certain, and of what kind the probability or certainty is. In this respect Leibniz adheres to the primary inspirations of common sense; this is not the problem that he loves to probe deeply and on which he gives himself free rein. He reserves for other uses the power of his constructive genius.

391. The story of the scientific and philosophic works of Leibniz presents two special features well worth attention. His first work as a young man, his inaugural dissertation,[41] is devoted to the theory of combinations. He treats in his own way and from a much more general point of view this problem with which Pascal and Newton were concerned in their turn, but in some ways incidentally, in connection with the solution of certain problems in arithmetic and algebra, while Leibniz, with reason, saw in algebra only a particular application of the theory of combinations, and a branch of his *universal language* or *calculus of combinations*—or *combinative symbolism*—(*caractéristique universelle ou combinatoire*).[42] In the immense variety of his works, he never loses sight

[41] *Disputatio arithmetica de complexionibus*, inaugural thesis for the celebrated Leipzig Academy, March 6, 1666.

[42] "Only one part of this algebra to which we give so much attention is of general usefulness. . . . Indeed, I do not know to what extent the whole system of algebra can be commended unless it benefits that higher science which I am wont to call the *calculus of combinations* (*combinatoriam characteristicam*)." Leibniz's *Letter to Oldenburg*, Paris, December 28, 1675. [See *Die philosophischen Schriften von G. W. Leibniz*, 7 vols., ed. by C. T. Gerhardt (Berlin, 1875-90), VII, 5.]

[Two entries in André Lalande's *Vocabulaire technique et critique de la philosophie* are helpful in fixing the meaning of the phrases *caractéristique universelle* ou *combinatoire* and *combinatoriam characteristicam*. The first entry is on page 113 of the first volume of the 1926 edition of the work, under the heading *Combinatoire:* " A. The mathematical science whose object is to arrange by order all the possible combinations of a given number of objects, to enumerate them, and to study the properties and relations of them; B. For Leibniz this science applies to concepts of all sorts, and thus constitutes the synthetic part of logic, so that it merges into the art of invention." The second entry is found on

of this fundamental idea. He gives evidence in many passages in
his correspondence of his high regard for the mathematical study
of games of chance. He knew and appreciated the discoveries of
Pascal, Fermat, and Huygens in regard to the matter of chances
and mathematical probabilities and even the applications that had
already been made of them to statistics by J. de Wytt and Hudde.[43]
But he did not cultivate this branch of science on his own account.
And so the mathematician-philosopher who first conceived the
generality and the importance of the theory of combinations seems
to neglect the most philosophical use that can be made of it in the
application of the notion of chance to the interpretation of natural

---

page 96 of the same volume, under the heading *Caractéristique:* " A. The art of
representing ideas and their relations by symbols of ' characters '; B. A system of
such symbols: the *Caractéristique universelle* of Leibniz (also called *Spécieuse—*
that is to say, Algebra—*générale*) is intended to be at one and the same time a
universal philosophic language and an algorithmic logic."]

[43] " Being a great gambler, he (the chevalier de Méré) devoted his first works
to the estimation of averages, which gave birth to the excellent ideas *On Chance,*
by Fermat, Pascal, and Huygens, of which P. Roberval was unable or did not
wish to understand anything. M. le Pensionnaire de Wytt pushed the matter
still further and applied it to other more important uses in connection with life
annuities, and M. Huygens has told me that M. Hudde too had had some excellent meditations thereon which he, like many others, had suppressed. Thus
games of chance themselves deserve to be examined; and if any mathematician
were to carry his meditations over into that field, he would find many important
considerations, for men have never shown more spirit than when they play."
Réplique aux réflexions de M. Bayle, at the end of the collection by Desmaizeaux.
    On the basis of the date of the covering letter sent to Desmaizeaux, this
passage appears to have been written in 1711, and, considering the relation of
Leibniz to the whole Bernoulli family, it is surprising that he gave no evidence
of having any knowledge of the important works by Jacques Bernoulli, who
died in 1705, and whose *Ars conjectandi* had appeared in 1713, through the
assistance of his nephew, three years before the death of Leibniz in 1716.
    [John de Witt or de Wytt, 1625-1672: one of the most eminent statesmen in
the history of the Dutch nation. The title " M. de Pensionnaire," by which he is
referred to above, is that of First Magistrate of the Republic of Holland. At the
age of twenty-three he gave evidence of his profound knowledge of mathematics
in writing a *Treatise on Curved Lines.* Jan Hudde, circa 1636-1704: filled public
offices in his native city of Amsterdam, including that of Burgomaster. He did
some writing in mathematics, a field in which he was well versed. Personne or
Personier de Roberval, 1602-?: a French mathematician. He made several significant discoveries in mathematics and in 1632 was appointed to a chair of
mathematics at the Collège de France.]

phenomena and to the criticism of our ideas. This is certainly a curious circumstance that it is not necessary to impute to a systematic bias (as the very terms Leibniz uses show), but it is to be greatly regretted, since he has deprived us of the light that his able mind would certainly have shed on a subject so deserving of attention, and which fell so completely within the scope of his work.

The other special feature that we wish to point out is the following. As we have already remarked, Leibniz ceaselessly invokes the law of continuity.

There is always a presentiment in animals of what is to happen because the body is in a continual stream of change, and what we call birth or death is only a greater or more sudden change than usual, like the drop of a river or of a waterfall. But these leaps are not absolute, and are not the sort I disapprove of; e.g., I do not admit of a body going from one place to another without passing through a medium. Such leaps are not only precluded in motions, *but also in the whole order of things or of truths.*[44]

His system of monads and forces admits of innumerable gradations in perceptions, some being more obscure, some more distinct; each monad representing, from its point of view, the entire universe, and the representation varying without discontinuity with each change of aspect of the monad. Moreover, since

the perceptions that are found together in the same soul at the same time include a truly infinite number of small *indistinguishable* feelings, that the future might unfold, we must not be astonished at the infinite variety that should result from them in time. All that is only one consequence of the representative nature of the mind, which should express what is taking place, and even what will take place in its body, and to some extent in all bodies, as a result of the connection or communication between all parts of the world.[45]

Well and good; but we cannot avoid being astonished that Leibniz thought it possible to distinguish and to define these feel-

[44] Leibniz, Letter to Rémond de Montmort, Feb. 11, 1715 [*Leibniz Selections,* " The Modern Student's Library " (New York: Charles Scribner's Sons, 1951), p. 188. Cournot's italics].

[45] Dutens edition, Vol. II, 78. [From Leibniz's *Lettre à Basnage,* i.e., *Lettre à l'auteur de l'histoire des ouvrages des savants,* 1698, published in *Ouvrages des savants,* July, 1698, p. 329. Also in the Erdmann edition, p. 153, col. 2. Cournot's italics.]

ings, these perceptions, and these *indistinguishable* ideas by means of one characteristic. Nor can we avoid being surprised that, when he advanced the view of the continuity *of every order of things or of truths,* he had believed this continuity to be compatible with the construction of language which would represent by combinations of elementary symbols all the perceptions of the intelligence, in such a way that, in going from the simple to the complex and by returning from the complex to the simple, it would be as easy and possible to find as it is to demonstrate all sorts of truths [46] and to do this with the exactness which is peculiar to the method of geometricians and to algebraic calculation.[47] Elsewhere, by a sensible contradiction, he recognizes " a taste, distinguished from the understanding, which consists in confused perceptions, of which not enough is known to make them reasonable, and which is something approaching instinct." [48] It must be fully recognized that Leibniz wished to preserve his two favorite ideas in his system without seeking to reconcile them and without realizing that they are irreconcilable; and the mathematician to whom we are most indebted for introducing the art of expressing continuity in the variation of measurable quantities was not disposed to notice the consequences which resulted from the impossibility of expressing or of characterizing continuity in non-measurable things and qualities.

392. Leibniz had been the Plato of Germany; Kant became her Aristotle, and a veritable new era was opened by him. His will always be the glory of having indicated in the description of the phenomena of knowledge, with a rigor unknown before him, the distinction between form and substantial basis, between mold and matter, between what is adventitious and can be traced to the way things are influenced from without and what can be traced to the very nature of the mind endowed with the capacity of knowing.

---

[46] *Historia et Commendatio Linguae Characteristicae Universalis.* Raspe's collection, pp. 535ff. See also the article cited above from Maine de Biran. [Erdmann edition, pp. 162ff.]

[47] " Therefore some special mode of statement is necessary, and moreover one like a thread in a labyrinth, by the force of which, no less than by the Euclidean method, questions may be resolved after the fashion of calculus." *De primae philosophiae emendatione.* Dutens edition, Vol. II, 19; [also in Erdmann's edition, p. 122, col. 1].

[48] Dutens edition, Vol. I, 46. See again the passage cited above in § 283.

There can be no doubt that all our knowledge begins with experience. . . . But though our knowledge begins with experience, it does not follow that it all arises out of experience. For it may well be that even our empirical knowledge is made up of what we receive through impressions and of what our own faculty of knowledge (sensible impressions serving merely as the occasion) supplies from itself . . . Such knowledge is entitled *a priori*, and distinguished from the " empirical," which has its sources *a posteriori*, that is, in experience . . . . Experience teaches us that a thing is so and so, but not that it cannot be otherwise. First, then, if we have a proposition which in being thought is thought as *necessary*, it is an *a priori* judgment . . . . Secondly, experience never confers on its judgments true or strict, but only assumed and comparative *universality* (*angenommene und comparative*) . . . . Empirical universality is only an arbitrary extension of a validity (*willkürliche Steigerung der Gültigkeit*) holding in most cases to one which holds in all . . . . When, on the other hand, strict universality is essential to a judgment, this indicates a special source of knowledge, namely, a faculty of *a priori* knowledge. . . . Now it is easy to show that there actually are in human knowledge judgments which are necessary and in the strictest sense universal, and which are therefore pure *a priori* judgments. If an example from the sciences be desired, we have only to look at any of the propositions of mathematics; if we seek an example from the understanding in its quite ordinary employment, the proposition, " every alteration must have a cause," will serve our purpose . . . . Even without appealing to such examples, it is possible to show that pure *a priori* principles are indispensable for the possibility of experience, and so to prove their existence *a priori*. For whence could experience derive its certainty, if all the rules according to which it proceeds, were always themselves empirical, and therefore contingent? . . . Such *a priori* origin is manifest in certain concepts (*Begriffen*), no less than in judgments. If we remove from our empirical concept of a body, one by one, every feature in it which is [merely] empirical, the colour, the hardness or softness, the weight, even the impenetrability, there still remains the space which the body (now entirely vanished) occupied, and this cannot be removed. Again, if we remove from our empirical concept of any object, corporeal or incorporeal, all properties which experience has taught us, we yet cannot take away that property through which the object is thought of as substance or as inhering in a substance. . . . Owing, therefore, to the necessity with which this concept of substance forces itself upon us, we have no option save to admit that it has its seat in our faculty of *a priori* knowledge.[49]

Certainly this is an exposition of remarkable lucidity and precision. But, in what Kant says of induction, we already see ap-

[49] Kant, *Critique of Pure Reason*, Introduction to the second edition, §§ I and II [translated by Norman Kemp Smith (London: Macmillan and Co., Ltd., 1929), pp. 41-45. German insertions are Cournot's].

pearing the germ of an error or the first traces of a gap that is to constitute the most important deficiency in his system. It is clear from the passage cited, and even more in the brief and terse paragraph in his *Logic* [50] which he has given over to the consideration of judgment by analogy and induction, that the philosopher of Königsberg, as well as the Stagirite, sees in induction only a logical recapitulation of particular experiences. Everything that follows from the above, being based neither on experience nor upon reason (as he conceives it), is for Kant only a *presumption* or a probability without scientific value and the origin of which he has no special interest in scrutinizing. This diffidence about probability or δόξα, from which Kant was no more free than were Plato, Aristotle, and Leibniz, kept him from grasping its eminently rational meaning and value. For evidently the rational principle of induction and analogy is among those data of reason of which Kant so ably tells us, which make experience possible, and which above all make the interpretation and discussion of experience possible. If Kant had undertaken to submit this datum of reason to analysis, as he submitted others to it, he would have been in possession of a criterion by means of which we should be able to judge the representative value not only of the empirical elements of knowledge, but also of the other elements *a priori* which make it up.

393. Let us follow the course of his deductions further. Our knowledge is extended by experience; that is incontestable; and if

mathematics gives us a shining example of how far, independently of experience, we can progress in " a priori," [it is because] it does, indeed, occupy itself with objects and with knowledge solely insofar as they allow being exhibited in intuition (*als sich solche in der Anschauung darstellen lassen*) ; [51]

that is to say, only in so far as we are able to imagine or to represent them by means of sensible symbols or forms, whether natural or artificial, but susceptible in the last analysis of a precise value and

---

[50] " Every rational argument must yield necessity: therefore, induction and analogy are not rational arguments but only logical *presuppositions* or empirical arguments. One clearly obtains general propositions by induction, but not universal propositions." *Logic,* Chap. III, § 84, in the translation by M. Tissot. [Cournot's italics.]

[51] Kant, *Critique of Pure Reason.* Introduction [to the second edition, § III. *Op. cit.,* pp. 46, 47. German insertion Cournot's].

a rigorous definition. This is the point at which he makes his illuminating distinction between synthetic and analytic judgments (262). Logic is indebted wholly to Kant for this. By reducing the principle of identity or contradiction, which the ancient logicians praised so highly, to its proper role, he gives the true explanation of the fruitfulness of mathematical speculation with respect to the extension of knowledge and of our inability, properly speaking, to extend our knowledge or to acquire truly new knowledge in the field of metaphysics. The purpose of intellectual activity in the realm of metaphysics should be limited to defining and putting in order the " a priori " elements of our knowledge, invariably fixed by the constitution of the human mind, by assigning to each its role and its range. Such is the definition of the " critique of pure reason." Its role, among the other sciences, is comparable to that of a magistrate in charge of the police of a city; [52] the work of this magistrate is not, as the economists would say, a productive one; but by preventing all kinds of disorder and usurpation of rights, he favors the productive work of the other citizens and turns it in the direction most advantageous for themselves and for the city.

Thus criticism has a purely negative role according to Kant. And it is necessary that its role remain negative, since Kant intends to proceed in all matters with the rigor of logical determinations, and since nothing is better suited to the precision to which logical rigor gives rise (as we have had numerous occasions to recognize) than to proceed by the method of limitation, exclusion, and negation. But, just as induction and analogy make it possible for us legitimately to go beyond the logical and rigorous consequences of experience and to extend our affirmations and our knowledge, which are not absolutely certain but extremely probable, far beyond the limits of actual experience; just as, and as a result of the same principle, it is permissible to believe that we can, in the field of pure reason, effectively extend our affirmations and our knowledge; so for that sort of synthetic judgment " a priori," philosophical probability has a quality analogous to, if not identical with, that which ideal construction has for the extension of our knowledge in pure mathematics. This being the case, criticism in philosophy is no

---

[52] Kant, *Critique of Pure Reason*, Preface [to the second edition. *Op. cit.*, pp. 26, 27].

longer necessarily reduced to the negative role assigned to it by Kant. It shares the fate of historical criticism, which no doubt frequently ends in nothing but negative conclusions, but which, as a result of the order it introduces among scattered fragments, is also often able to reconstruct, if not with an absolute certainty at least with a high probability, that which time has destroyed. Its affirmations are finally accepted by all reasonable persons, although it can refer neither to a possible experience nor to a geometrical demonstration in order to convince them.

394. The most important question that we can put to ourselves, concerning the subject of the conceptions or judgments *a priori* that we find in the human mind, is without question that of knowing whether these judgments or conceptions correspond to the reality of external objects and teach us something of this reality, or whether we can validly conclude nothing from them as to the nature of the existence of things, and whether our study of them has no other result than causing us to know how our intelligence is organized. The response that Kant made to this question is dictated by the premises that we have set forth:

Hitherto it has been assumed that all our knowledge must conform to objects. But all attempts to extend our knowledge of objects by establishing something in regard to them *a priori*, by means of concepts, have, on this assumption, ended in failure. We must therefore make trial whether we may not have more success in the tasks of metaphysics, if we suppose that objects must conform to our knowledge. This would agree better with what is desired, namely, that it should be possible to have knowledge of objects *a priori*. . . . We should then be proceeding precisely on the lines of Copernicus' primary hypothesis. Failing of satisfactory progress (*nicht gut fort wollte*) in explaining the movements of the heavenly bodies on the supposition that they all revolved round the spectator, he tried whether he might not have better success if he made the spectator to revolve and the stars to remain at rest.[53]

But this is still simply a hypothesis, only a presumption, and Kant wishes an *apodictic* demonstration. He finds it, or thinks he finds it, in what he calls the *antinomies of reason* (145), that is to say in the contradictions into which we fall when we attribute an objective value to ideas *a priori*, as reason conceives them.

[53] Kant, *Critique of Pure Reason*, Preface to the second edition. [*Op. cit.*, p. 22. German insertion Cournot's.]

If, then, on the supposition that our empirical knowledge conforms to objects as things in themselves (*als Dingen an sich selbst*), we find that the unconditioned (*Unbedingte*), *cannot be thought without contradiction,* and that when, on the other hand, we suppose that our representation of things, as they are given to us, does not conform to these things as they are in themselves, but that these objects, as appearances (*Erscheinungen*), conform to our mode of representation (*nach unserer Vorstellungsart*), the contradiction vanishes; and . . . we thus find that the unconditioned is not to be met with in things, so far as we know them, that is so far as they are given to us, but only so far as we do not know them, that is, so far as they are things in themselves.[54]

The French reader may well find something obscure and barbaric in this technical language; but we have explained it (Chap. I), and it is perfectly accurate when correctly understood. Yes, we agree with Kant that, if the human reason is surprised in glaring contradictions, as a result of the tendency which leads it to attribute an absolute truth to the manner in which it conceives things, this is the demonstrative proof and not simply the probable indication that this inclination deceives it and that things are, at bottom and speaking absolutely, different than it conceives them to be. We are already able to regard it as very probable that the human mind is not so constituted as to attain absolute truth in all things, nor perhaps even in anything, and the contradictions of philosophers sufficiently attest to the weaknesses and to the gaps in the human mind. But no one has, like Kant, with a dialectic as concise as it is profound, given demonstrative evidence of these contradictions and of the necessity for them.[55]   However, because

---

[54] Kant, *Critique of Pure Reason,* Preface to the second edition. [*Op. cit.,* p. 24. German insertions Cournot's.]

[55] However, Pascal has said, " All principles are true, those of the sceptics, of the Stoics, of the atheists, and so on; but their conclusions are false, *because the opposed principles are also true."* This remarkable fragment, printed for the first time by Prosper Faugère,* is to be found on page 92 of the second volume of his edition. The first editors, being Jansenists and Cartesians, apparently suppressed it because they did not wish to admit that opposed principles (the thesis and antithesis, in Kant's terms) might have the virtue of being equally true and of imposing themselves on reason with the same evidence. The whole basis of

* [Armand Prosper Faugère : 1810-1887. A French *littérateur* who is said to have published the first correct and complete edition of Pascal's *Pensées* in his two volume edition of the *Pensées, fragments et lettres de Blaise Pascal* (1844). He three times won prizes for eloquence offered by the French Academy, one of these being for his " Eulogy on Pascal " in 1842.]

Copernicus would have been mistaken had he wished to maintain against Kant that his ideas about space and about the movements of bodies were truths with an absolute verity, does it follow that he was mistaken in maintaining against Ptolemy that the sun is at rest and that the earth is in motion? We may very well believe, contrary to the assertion of Kant, that our representations are patterned after phenomena and not phenomena after our representations; that is to say, that the order found in our representations can be traced back to the order which exists in phenomena and not inversely, without for that reason believing that the faculties which bring us into touch with nature had been set up in such a way as to grasp the first principles and the fundamental reason for the order of phenomena. Induction must intervene in order to make allowance for what pertains to the nature of things perceived and what pertains to the organization of our faculties. It is necessary to judge, as Bacon said, *ex analogia universi*; it is necessary to give up apodictic demonstrations, in the use of the principle of contradiction; it is necessary to admit a kind of proof that Kant excluded, even the idea of which was not suggested to him by his inflexible and formalistic logic, and the omission of which, in spite of all his efforts, involved him in an absolute skepticism.

395. We must bring to a close this very rapid, and nevertheless already very long, excursion into the immense history of the efforts and the systems of the human mind concerning a question from which all others arise. We are not undertaking to review the more modern systems that have produced the philosophic movement in Germany, following Kant's reform, and the adventurous boldness of which is in such marked contrast with those promises of severe regulation and careful repression of any rash undertaking that the *Critique of Pure Reason* contains. Kant had proposed to show the impossibility of passing legitimately from the description of the

---

Pascal's thoughts revolves around the contradictions of reason; but it does not necessarily follow that he used as much care and precision in defining and classifying them as Kant used a century later. There are many analogies between the philosophy of Pascal and that of Kant. The latter *saved himself*, that is to say prepared himself against *annoying doctrines*, by opposing the *practical* reason to the speculative; while Pascal opposed religious faith to reason and hoped that faith might be reached through experience. In that (it must be recognized) he shows a truer and more profound feeling for the conditions and the needs of human nature.

laws and the forms of the human understanding to affirmations concerning the way in which things-in-themselves exist; he set out especially to prove categorically that the absolute escapes us. Yet after him, all the efforts of the metaphysicians have had as their purpose what they call the passage from the subjective to the objective, and the comprehension of the absolute. Some have exhausted themselves with analyses which are always subtle, often obscure, sometimes profound, in order to deduce the *not-self* from the *self,* in order to identify intelligence and nature, in order to create the world through the force of logic and the efficacy of ideas. Others have thought it necessary to return to psychological observation, to describe the phenomena of "consciousness" more completely than did Kant, and in such a way as to find in it what he declared it was impossible to find in it, namely, that greatly desired passage and that means of concluding validly from an order of internal phenomena to an order of external truths and realities. But, once more, it is not our intention to enter into the discussion of these recent theories.[56]   Every discussion must come to a close, under penalty of trying the patience of the most favorably disposed reader.   Moreover, an unknown author always displays poor taste when he pits his personal opinions directly against those that his contemporaries have sustained with all the vigor of their talent and the authority of their name.   He has only the right to propose them with modesty and to present in the best way he can the reasons which have persuaded him.   We have already developed them in this work, not without interjecting some digressions along with them.   There remains nothing more for us to do now to dispose the judgment of the reader toward them favorably except to draw them together again in a quick résumé.

[56] We refer the reader to the judgment Jouffroy passes on them.   (Preface to the French translation of *the Works of Reid,* p. cxcii.)

CHAPTER XXV

SUMMARY

396. If, following all the developments into which we have entered and all the applications that we have tried to make, we wish to summarize in a few pages the body of principles which make up the substance of this book, we must first of all recall the brief quotation that we have already made from Bossuet (17): "Order and reason are very closely related. Only reason can introduce order into things, and order can be understood only by reason: order is the friend of reason and its subject matter." In fact, one has been able to see in all that precedes that relations of the most intimate sort exist between the idea of order and the idea of the reason of things, or rather that it is the same idea under two different aspects. To understand that one fact is the reason of another, that one truth proceeds from another truth, is nothing more or less than to grasp the relations of dependence and subordination, that is to say, to lay hold of an order among the diverse objects. This dependence impresses us and is perceived by us only because we have the faculty of comparing and preferring one arrangement to another as being more simple, more regular, and consequently more perfect; in other words, because we have the idea of what constitutes the perfection of order, and because it is the essence of our reasonable nature to believe that nature has introduced order into things and to believe ourselves proportionately closer to the true explanation of things when the order in which we come to arrange them seems to us to satisfy better the conditions of simplicity, of unity, and of harmony, which, according to our reason, constitute the perfection of order.

This idea of the order and reason of things must not be confused with the idea of the chain of causes and effects. The former finds its application in things and for truths which do not depend upon one another in the same way that an effect depends upon its active or efficient cause. Therefore, we are not able to account for

the idea we have of the order and reason of things by means of the
kind of observation and by the evidences of consciousness which
the notions of cause and effect suggest to us. This idea is the very
source of all philosophy, the final and supreme end of all philosophical speculation, that which pre-eminently characterizes the
spirit of philosophic curiosity and which gives, in different degrees,
a philosophical imprint to all the work of thought, in matters of
taste and imagination, as in those which arise out of erudition and
knowledge.

397. The idea of the order and the reason of things is above all
the basis of philosophical probability, of induction, and of analogy.
To assign a law to phenomena is to draw from a single principle
the reason for the variegated and multiple appearances which first
strike our attention; it is to introduce order into the confusion of
appearances. So the idea of law, in its most general aspect, as it
was grasped by Montesquieu's genius in the beginning of his im-
mortal work, is still, from another point of view, nothing but the
idea of order or of the reason of things. But what ground have we
for believing in the existence of such a determined law and for
going beyond the immediate consequences of observation and experience by affirming in a general and absolute way what experience has been able to establish only in particular cases and approximately? This ground follows from the character of simplicity
inherent in the presumed law and from the improbability that, in
the innumerable mass of fortuitous combinations, chance had
caused us to fall upon some observations which are capable of being
connected by means of so simple a law, if this law did not have an
intrinsic existence independent of the chance observation of it, and
if it did not also connect facts of the same nature that we have not
observed. The character of simplicity may be so striking, the number of observations may be such, the approximation may fall between such narrow limits, that not the least doubt remains in the
mind, in spite of sophistical objections that can always be made
against any proof which does not have the character of mathematical demonstration. In other circumstances, probability is weakened, because it shades off through insensible degrees, in conformity
with the law of continuity, the most striking image of which is
furnished always by the gradual diminution of light and the transi-

tion of colors.[1] Different minds may be affected differently by it
and yet it may not be possible either to assign a precise point at
which conviction ceases, or indecision begins, or the point at which
indecision gives place to a contrary conviction, namely, the conviction that we are ignorant of the law of the phenomenon.

398. The probability which comes from the feeling for the order
and the reason of things and which is the true basis of most of the
judgments we hold in speculations of the highest sort, as well as in
the most ordinary affairs of life, the probability that we call philosophical probability, resembles in many ways the mathematical
probability which results from the evaluation of the chances favorable or contrary to the occurrence of an event. Both are connected, although in different ways, with the notion of chance, which
is at bottom (as we believe we have established) nothing other
than the notion of the independence and of the interdependence of
causes. Both are capable of being increased and decreased imper-
ceptibly, without sudden modifications which give rise to sharp
demarcations. But their dissimilarities are no less notable. It is
especially important to understand clearly that philosophical probability is always incompatible with the possibility of numerical
evaluation, for the important reason that we are able neither to
enumerate all the possible laws or all the forms of order, nor to
classify them, nor to arrange them in such a way as to determine
the characteristics of the simplicity of the laws and the perfection
of the forms and the relative importance of these characteristics by
means of a determination entirely free from arbitrary elements and
numerically expressible. This is the case even though we have a
faculty for laying hold of the contrasts between simplicity and complexity, between harmony and discords, between regularity and
confusion, and between order and disorder.

399. The criticism of our knowledge or the discussion of the
representative value of our ideas is an immediate application of the
principles of philosophical probability. From the beginning we
check our own judgments against one another, in such a way as to

---

[1] " And though a thousand different gradations gleam in it (the sky) the
change from one to another itself eludes the eyes as they watch. So much alike
are the adjacent steps in the process of the change, though the extremes are
clearly to be distinguished." Ovid, *Metamorphoses*, Book VI, § 6, lines 65ff.

distinguish what can be traced to accidental and abnormal circumstances from what pertains to the very basis and to the habitual and permanent conditions of our individual make-up. Then we check our personal impressions, ideas, and judgments against those of other men in such a way as to take into account the individual idiosyncrasies, the particularities which are associated with race, nationality, and the prejudices of an education that varies according to the time and the place, disentangling all this from that which has its roots in the very foundation of human nature itself and which should be regarded as belonging to the normal form of the species. But philosophical criticism goes further and asks itself whether this common basis of impressions and ideas, inseparable from the constitution of the species, does not depend upon it in such a way that we have no means of judging, in accordance with the general order of the world and the analogies that it suggests, at what point our ideas are in conformity with external reality. It asks itself if there may not be, among the diverse functions or faculties which are in operation in the economy of our intellectual organization, a hierarchy such that one of them may serve to control or to regulate the others; so that, for example, reason may criticize the worth of the testimony of the senses or of the instincts of consciousness, in the same way that it criticizes the value of judicial or historical evidence. Now, if, contrary to the evidence of the senses and in spite of natural instincts, reason clearly succeeds in convincing itself that it is necessary to explain the apparent movements of heavenly bodies by our own movement, what is to prevent it from utilizing inductive inferences of the same sort, with the result that it can distinguish, among the impressions that affect our sensibility, that which goes back to the constitution and the order of external things from that which can be traced to the organization and the economy of the sensible being? Actually we succeed very well in distinguishing what follows from the circumstances under which we observe things and from the particularities of our nature as observers from what pertains to the very constitution of the things observed not only in the example which we have selected as the most striking but also in a number of others; and we always notice, as we should, that the effect of this distinction is to introduce into things an order and a harmony which our first perception of them does not yield and to reduce the laws of phenomena to a more simple expression by separating from them everything which complicates them or masks them in immediate observation. When, because of clear and distinct conceptions, we succeed in connecting phenomena by means of simple laws, by submitting them to a precise co-ordination, it is repugnant to reason to admit that such laws are imaginary; that the conceptions which they express are not bound up with the nature of our ideas and have no basis in the nature of external things; that in reality the laws of phenomena are more complicated. Finally, it is repugnant to admit that in being combined with the laws peculiar to our intelligence, such simple laws introduce, as a result of a prodigious chance, an impression of order and simplicity, instead of an increase of complexity and confusion. Without doubt, this is only a probable induction and the contrary hypothesis is not rigorously demonstrated to be impossible. But the probability may be of the order of those with which no right mind ever refuses to agree. And in other cases in which the probabilities would no longer appear irresistible, it would still be worth while for reason to consider them carefully and to compare them, not only in order to make up our minds with respect to practical matters, but still more to bring to bear speculatively on things the best judgment possible with the rudiments which are given us.

Inversely, what if, in the explanation we attempt to give of phenomena, the system of our ideas presents incoherencies and disagreements; what if there are gaps which we are unable to close, certain connections that we are unable to establish, contradictions that we are unable to resolve; and what if the progress of discoveries, by perfecting certain branches of our knowledge in isolation, by developing them and simplifying them, always leaves in existence the same fundamental gaps, the same irreducible contradictions? Should this be the case, on the basis of it there would be reason to infer with a high degree of probability, not that these apparent incoherencies belong to the nature of things, but, on the contrary, that they come essentially from the nature of our intelligence, which is not accommodated, *parte in qua,* to the exact perception of the order of the world and the harmony of nature.

400. In fact, if we set aside purely abstract consideration for an instant in order to glance at the order of the world and the economy of nature, we are immediately struck by the admirable

harmony which reigns there in general, and by the abnormal facts which detract from the general harmony. Our reason distinguishes the different principles to which it can have recourse and between which it is possible to choose, with more or less probability according to the case, in order to explain the harmonies which are so striking to us. But even without being obliged to make a choice, we understand very well that the causes, whatever they may be, which have introduced harmony into nature should have introduced it into the disposition of external things, the impressions they make on us, and the notions that these impressions suggest to us. There are opposed systems acting and reacting on one another which should tend to be reconciled and not to be ceaselessly opposed. After all, man is himself part of the world, and the veracity of his faculties is, in certain respects, only a consequence of that same necessity which produces the harmony of the world, and which forces nature to be put in accord with itself.[2] Yet, the essential harmony in the general plan of the world admits of certain deviations, allows certain exceptions, and sometimes even leads to or presupposes certain partial discordances. Where harmony is not necessary to maintain the existing order, where the influence of mutual reactions ceases to extend, the discordances may be perpetuated. Consequently, however needful it may be that man's faculties square with the nature of external things to the extent that this is required for the accomplishment of his role in the world and for the conservation of individuals and of species, the need for an accord of this sort is no longer felt in matters relating to ideas that give birth, in the womb of a civilized society, to an increase of culture which appears to be an accidental and abnormal fact in the general order of the world in the first place. Here is rooted a distinction of which one must not lose sight: the distinction between beliefs, on the one hand, that nature has charged itself to inculcate in us, and to accomplish which it takes all possible care to resist every sophistical inclination; and ideas, on the other hand, the critical examination of which nature has left to the philosophers for the reason that these ideas do not bear upon that which is fundamental and essential in the functions which nature has assigned to individuals and to species.

401. To judge that in certain respects our ideas conform to the

---

[2] "The principle of uniformity which constrains the nature of things to be self-consistent." Pliny, *Natural History*, II, 13.

reality of things is to affirm that the true relations of things are not falsified or complicated by the nature of our perceptions. But this is not to pretend that it is within our power, be that as it may, to attain absolute truth. The astronomer makes known the laws of the movements of the planetary system, and he is very sure that he has separated them from everything which results from the position of the observer. Nevertheless, these movements, the laws and theory of which he formulates, are still only relative to the system of which the sun, the earth, and the other planets make up the parts, just as the movements that we observe from a ship are relative to the system composed of the ship and the bodies on it. Our notion of these relative movements, of their causes and their laws, is not tainted with illusion; they are not simply apparent but fully real; nevertheless this reality is only relative. For, in so far as these systems are parts of a more general system, they actually describe in space more complicated movements which result from a movement common to the whole subordinated system and from the internal movements of the system. Now, it is not possible for us to reach the final term of this series, nor to have in space points of reference which are absolutely fixed or of whose fixity we may be absolutely certain. Consequently the notion we have of a phenomenon which consists of movements or combinations of movements is never true (that is, it never conforms to external reality) except in a relative sense, although it may tend more and more toward absolute reality the higher we go in the hierarchy of systems. A clear and decisive example of this is a sort of *schema* well suited to making understood how we are able to have notions freed from every internal cause of error or illusion, which conform perfectly in this sense to external reality and which, nevertheless, do not reach absolute reality, to which we can only gradually approach.

402. Thus the different faculties through whose functioning the knowledge of things is possible call attention to a superior faculty that directs and controls them, that strengthens or weakens our prejudices and our natural beliefs, which are fruits of acquired or transmitted habit and of the prolonged action of external causes.[3]

---

[3] " It is the prerogative of reason to be the highest source of all certainty and to contain a system of principles and consequences which may be true on the basis of reason alone and through the harmony which is proper to it." Tennemann, *Manuel de l'histoire de la philosophie,* Introduction, § 45 of the translation by Cousin.

This superior faculty is what apprehends things or seeks their reason, order, law, unity, and harmony. Its means of criticism or control are not categorical and peremptory demonstration but inductive judgment or philosophic probability, the force of which in certain cases is not less irresistible. It is possible to say that this faculty which controls the others controls itself and that in this sense it, to the exclusion of all others, is truly autonomous. For, if nothing outside of it corresponds to the idea of order (as it is found in us) how could it happen that by penetrating further and further into the knowledge of the external world we might increasingly find that everything in it occurs in conformity with this regulative idea? On the basis of this we should necessarily be led to fall into all the excesses of the skeptical schools and to suppose that all the notions we believe we have of an external world are only fantastic creations of our own mind, which in reality contemplates no existence other than its own. Without doubt this constitutes an irrefutable but only slightly contagious skepticism, and one which does not need to be taken account of by the most responsible minds.

403. The principles of philosophic criticism being thus presented (and these are none other than those which guide us in every type of criticism), the problem is to apply them to the principal ideas which may be regarded as the support of the entire system of our knowledge; to the ideas which come to us most immediately from the senses; to those which the mind elaborates as a result of its power of comparing, combining, generalizing, and abstracting; and, finally, to the ideas of morality and esthetics. By reconsidering the analysis of our sensations from this point of view, it is not difficult to distinguish those which have a representative value from those which are unable to represent anything, at least directly, although they may have the virtue of informing us, through the persistence of certain impressions, of the presence of external objects that other sensations depict to us and make it possible for us to know. Consequently our sensations offer this characteristic difference: some cannot be suppressed without destroying or upsetting the system of our knowledge, while others may be successively abolished without giving rise to any necessary alteration or mutilation of our knowledge, whether common or scientific. If from sensations we pass on to abstract and general ideas, we similarly distinguish those which have a basis only in the constitution of our

mind, in the nature of its instruments, and in the needs of our methods, from those which represent the arrangement, the relations of subordination, the principles of unity, and the harmony that nature has put into its works. These are the principles and relations which the human mind could not invent, but which it must grasp and express in the best way it can, although most often still with imperfections of which it is aware and which it strives ceaselessly to remedy. Finally, in the domain of morality and esthetics, in which it should be expected that a more influential role would be left to the needs of nature, to the customs and conveniences of social life, to the influence of education, of imitation, and of prejudices which have their roots in historic precedents, reason does not relinquish the function of discovering laws of a higher order and more general rules to which man is subject in so far as he is an intelligent and reasonable being, and not for the satisfaction of any appetite or instinct of his sensible nature or because of any particularity of his specific constitution.

404. Far from having to argue from the illusions and errors to which our natural dispositions or acquired habits sometimes incline us in order to sanction an absolute skepticism, it is precisely because of our experience that it is within the power of reason to discover such illusions, notwithstanding the force of natural dispositions, that we have more reasons for adhering to our natural beliefs when reason confirms them. A similar experience is, so to speak, the touchstone of theory; and we shall be more sure of the impartiality of a judge who does not always decide in favor of the same side. There are, therefore, good reasons for insisting especially upon the applications of philosophical criticism, in so far as their concern is to point out imperfections inherent in our intellectual make-up or in the artificial instruments that the intellect uses, as well as the insurmountable obstacles that they produce in the exact perception of the order and relations of things. The results of such a criticism, which seems to be purely negative, come marvelously to the help of reason in its positive and theoretical assertions. Now, it should be recognized that the human understanding is so constituted, because of its union with animal nature, that it has been able to have direct representation or intuition only of extension and of the forms of space. For everything else, it is obliged to rely on the artifice of symbols and language or on the symbolic use of the forms of space,

as an auxiliary means of indirect representation and expression. That is the reason for the very important disparity between the things to be represented and the means of representation. This results from the fact that, while nature generally follows the law of continuity in all things and follows it in such a way that discontinuity is seen only as a peculiar or accidental case, conventional symbols, like those of language, on the contrary, lend themselves naturally only to the expression of clearly distinct ideas and relations, among which there are no continuous transitions or indiscernible nuances. On the other hand, spatial figures lend themselves fully to the representation of continuous things only in the very special case in which the attribute of continuity is associated with that of measurable magnitude. Another important disparity comes from the linear form of discourse, which does not allow us to make perceptible or to fix by means other than that of imperfect images, borrowed from the figures of geometry, the infinite variety of forms the idea of order may assume, and which are actually offered to us in the study of the relations of beings. From this it follows that logic, which draws its name and its form from the name and form of language, is often a rebellious and natively defective instrument, as much for the perception as for the explanation of the true relations of arrangement and subordination among things. The insufficiency of logic has been felt and proclaimed many times, by catching it in contradiction with the indications of good judgment, that is to say, with the judgment of that superior and regulatory faculty that we call reason. Yet, even after the researches on the philosophy of language by the metaphysicians of the last century, there still remained the problem of showing the fundamental reasons for these discordances, of setting forth the influence they exercise, not only with respect to the speculative works of philosophers and scientists, but also with respect to the procedures of the practical arts and even the mechanisms of social institutions.

405. What do we look for, what should we look for in speculation as in application? The truth; that is to say, apparently, the conformity of the notion that we form of things with the things themselves, the resemblance of an image to its prototype. But if there are cases in which the truth consists in grasping a precise point, an exact number, from which it is impossible to deviate with-

out committing a demonstrable error, how much more frequently does it happen that calculation gives place to an *estimate* which could be exactly known only as a result of a chance whose probability is infinitely small and for which we do not even possess means of regular approximation? Are there not insensible transitions and gradations from the image whose representation is the most striking to the image which offers the least conformity with its prototype? And why should we suppose that the analogue of the transitions which are rendered perceptible to us in the forms of the extended is not found in the sphere of purely intelligible ideas and relations? To grasp the intelligible relations of things in all their truth, to the extent that lies within man's power, to choose the sensible images which are least imperfectly fitted to the expression of such relations, will, therefore, most often no longer be the work of the calculator who advances by sure and reckoned steps, applying methods, combining or developing formulas, connecting propositions. It will be the work of an artist, in whom a particular sense, given by nature, perfected by use and study, guides and sustains the hand in making the rough draft of the plan as well as in sharpening the details. In his own way, the philosopher will be a poet or a painter. This is why a mark of personal individuality is stamped on the productions of his mind; this is also the source of the occasions of inferiority which do not permit philosophical speculation to have a development parallel to that of the sciences. There is no continuous progress except when the condition of an identical transmission from one intelligence to another is fulfilled, nor is there identical transmission except under the conditions of a rigorous definition of ideas and a logical connection of propositions. Now, in most cases, the objects of philosophical speculation are of such a nature that these essential conditions cannot be satisfied or can be satisfied only very imperfectly, with too rough an approximation for the mind to be content.

406. If it is true that in the absence of natural divisions, of sharp distinctions, and of invariably fixed points of reference, our knowledge is unable to take on the logical organization which gives it the character of science and which is the principle of indefinite progress, we have the very simple explanation of the imperfect state of our knowledge concerning the subject which most interests us, that is to say, the operation of our intellectual faculties. Be-

cause nature has not furnished us with lines of demarcation and points of reference concerning which everyone is forced to fall into agreement, so that each person supplements it according to his fancy, language can never be fixed nor can the ground of discussion be affirmed. On the other hand, we are powerless to determine with precision the circumstances of observation and the conditions of experience, simply by invoking experience and observation ceaselessly. This is the principal reason why psychology, considered as a branch of anthropology, which in turn falls within the vast system of the sciences whose subject matter is the functions and forms of life, has still not attained the truly scientific state. And this is why we are justified in censuring as chimerical the pretension of raising, upon the basis of a pretended introspective observation of psychological phenomena, a body of scientific teachings which may be like the counterpart of the physical and natural sciences, based upon the observation of external phenomena and of facts which fall under our senses.

But, it must be added, this empirical psychology, whether or not it is able to get out of the rudimentary state in which it finds itself and so deserve some day to be counted among the experimental sciences, is neither the proper study of philosophy nor the necessary introduction to philosophical studies. The understanding has its laws, the knowledge of which does not necessarily presuppose knowledge of the ingenious processes to which nature has had recourse in order to make us intelligent beings. It is no more necessary to understand these laws than it is to know the theory of the vibratory motions in sounding bodies and to know the anatomy of the ear and the physiology of the auditory nerve in order to grasp and to make use of the laws of music. The principles of philosophic criticism have the special virtue of putting us in a position to make judgments about the representative value of our ideas that in no way require that we know how our ideas come to be formed, nor what phases our impressions and perceptions pass through before taking the definite forms that the laws of our nature assign to them.

407. Philosophy has as its object the order and reason of things. Consequently the philosophic spirit permeates the sciences which deal with abstract truths or with the arrangement of the material world, as well as those which refer to man considered as an intelli-

gent and moral being. We may be able to understand a little of the
nature of man and his role in the world only by observing the connection of all the phenomena of nature and their hierarchical progression; from those which are most simple, constant, and universal
and which, because of all these characteristics, function virtually as
a foundation or framework for all the others, up to those which
exhibit more complexity and organic perfection, and which, because
of this, must admit of more particular and less stable combinations.
On the other hand, there is no order of phenomena in the physical
world that we may not explain by our ideas and that consequently
may not lead to a critical examination of the value of some of the
fundamental ideas with which all our theories are connected. Thus,
by the very act of examining the order and reason of things, we
discuss the laws and the forms of our understanding, by discerning
harmonies or contrasts in this relationship. And while it is permissible for what is properly called science, in so far as it seeks only
to arrange facts methodically, to study man and nature, the subject
and the object of knowledge, independently of one another, philosophical speculation, which rests upon the relation of these correlative terms, is never at liberty to deal with them separately.

408. Positive facts, that is to say, those of which we are able
to acquire certain proof by calculation or measurement, by observation, by experiment, or, finally, by a concurrence of evidence of
various sorts which leaves no room for reasonable doubt, serve as
the materials for the sciences. But a collection of facts of this sort,
even in large number, is suitable for the establishment of a science
only when they can be arranged according to a certain logical order,
appropriate to the nature of the instruments of thought, which
makes up the essence of the scientific form. As a result of the logical organization and the systematic classifications of our knowledge,
when these are possible, we deduce conclusions from premises, we
compare and combine well-defined ideas, and we discover new
truths by the force of reasoning alone. And if the truths or the
facts thus presented or discovered are confirmed by observation or
experiment, we obtain at one and the same time both the highest
certainty that it is possible for us to attain, and the most brilliant
evidence of the power of our intellectual faculties. If the nature
of the facts is not suited to this logical order of which we speak, but
if it may be supplied by artificial definitions and classifications, the

scientific form will still be possible.   However in this case it will have the value only of an artificial scaffolding for sustaining and for conveniently transmitting our knowledge.   In general, when this condition holds, instead of being able to serve in the discovery of new truths without the aid of new observations, we shall have to expect that the series of observations, by revealing new facts, will upset the old scaffolding.   Finally, if it is not possible to submit known facts to a logical arrangement, even to an artificial one, scientific form becomes impossible.   This neither hinders these facts from being perfectly certain, nor reason from distinguishing in them an order and a group of relations worthy of its full attention.   Thus it happens that there is a positive history and a philosophy of history, even though the connection of historical facts, which results from the influence of prior facts on succeeding ones, cannot in any way resemble logical connection or the methodical arrangement of facts which serve as material for science in the proper sense of the term.   Philosophy, which more or less permeates the warp and woof of all the sciences, which intermingles there (in different proportions according to the subject matter) with the positive part of our knowledge, should never be confused with science, since it still manifests itself where the conditions of the organization and of the scientific schematism fail.

409.   It is of the very nature of things that philosophical truth can neither be put beyond the possibility of dispute, as is the case with positive facts, nor be categorically demonstrated by means of reasoning, calculation, or reduction to absurdity, as may be done with abstract truths which are the object of what we call the exact sciences.   After the sciences have been enriched by a sufficiently large number of positive facts, the assent of good minds is able to make an idea, a philosophic conception, prevail which places these facts in a more illuminating order, which better takes account of their connections and their dependence.   But the idea itself is not a fact which falls within the domain of sensible experience, a result that calculation is able to make manifest, or a theorem capable of categorical proof.   We propose it and sometimes welcome it, but we do not impose it.   Philosophical probability, in whatever degree it may be found, never excludes paradoxical or sophistical contradiction.   We are no more able to measure this probability than we are to express numerically the degree of resemblance be-

tween the intelligible relations of things and the idea we have of
these relations, between this internal idea and the expression we
attempt to give of it with the help of symbols of language and other
sensible forms to which we try to reduce it. The feeling for the
true in philosophy is not more capable of being rigorously broken
up and analyzed than is the feeling for the beautiful in the arts.
And the upsetting of common sense, like the perversion of taste,
does not, properly speaking, constitute a refutable error.

410. Therefore, it is clearly essential not to confuse the sciences
and philosophy, and, in the ultimate alliance which often operates
between scientific work and philosophical speculation, to distinguish clearly what pertains to the one and what to the other. Any
confusion in this regard would be prejudicial to the progress or
to the dignity of the human mind. Philosophy is especially interested that this confusion should not occur; for, as it will always
be easy to prove that philosophy is not a science, that it neither
develops nor advances in the way the true sciences do, the inanity
of philosophy would be concluded from this, despite constant
protests of the human mind, if it were not possible to succeed
in the other direction and fully establish the fact that philosophy
has its proper domain and that it issues from a special faculty; and
if, by knowing fully the true character of it, we have at one and the
same time the explanation of the superiority of its role and of the
inferiority of its resources.

If the human mind needs to make this distinction in order to
make its penchant to philosophize legitimate, it must with still
more reason be invoked as an excuse by an author who has risked
writing a book on philosophy, that is to say, of writing about
questions which have been debated over and over again since man
first took it into his head to write. Great as would be the extravagance of pretending to reveal primary principles which God has
intended to hide from the sight of men; [4] equally as great would
be the presumptuous folly of attempting to solve within the sciences one of those questions for which a host of minds, many of the
great minds, have sought a solution without finding it, and of
attempting to settle in a doctrinaire fashion an issue which the
centuries have left open; nevertheless we may be allowed, without

---

[4] " To whom has the root of wisdom been revealed? " Ecclesiasticus 1: 6.

offending the rules of wisdom and modesty, to propose some new
clarifications, which tend on the contrary merely to prevent doctrinaire decisions and absolute dogmatism in which—to use the words
employed twenty-three centuries ago by the father of history—
" It seems unlikely, however man may desire it, that he will ever
reach knowledge that is absolutely certain." [5]

[5] Εἰδέναι δὲ ἄνθρωπον ἐόντα, ὅπως χρὴ, τὸ βέβαιον, δοκέω μὲν οὐ δαμῶς. Polymn.,
50. [It has been impossible to identify this author definitely. He may be Polym-
nestor or Polymnastus of Colophon who lived about 660 B.C.]

INDEX

Abelard, P., 330
Absolute, the idea of the, 22, 581, 582; certainty, 106, 107; existence, 212; knowledge, 598; laws, 39; reality, *11–15*, 179, 221-223, 543; scepticism in Kant, 129, 218; truths, *21, 22, 104*, 105, *133*, 544, 580, 589
Abstract concepts and ideas, 22, 23, 33, 217, 226–233, 237, 241, 247, 270, 314, 327, *343, 347*, 348, 371, *383*, 393, 527, 590ff.; logic, 542; notions, 301; relations, 154; sciences, 227, 396, 445; truths, 73, 234, 235, 475, 476
Abstraction, *28*, 50, 56, 188, *226–251*, 264, *291*, 301, 343, 364, 394, 413, 482, 550, 554
Accessory causes, 259, 269
Accidents, 57, 103, 329, 420, 459
Accidental anomalies, 116; causes, 29, 30, 34, 35, 45, 244, 267, 268, 426, 485, 506; characteristics, 172, 173, 242; circumstances, 144, 260, 478; colors, 12, 14, 16; concurrence, 451; deviations, 308; disturbances, 440; events, 41; facts, 260, 450, 461; forces, 459; influences, 261; occurrences, 265; phenomena, 30; relations, 98, 173
Action at a distance, 186, 223
Actions, instinctive, 203; moral, 275, 276; reciprocal, 32, 33; vital, 199
Active causes, 26; faculties, 253; forces, 26, 27; principle, 203; touch, 159, 163, 164
Adaptation, harmonious, 102, 270; in animals and man, 83, 85, 95, 102
Affections of organic life, 283, 284; sensory, 134
Affective sensations, 155, 284
Alembert, J. d', 3, 221, 237, 435, 474, 491ff.

Algebra, 233, 236, *298, 299*, 309, 310, 316, 317, 320, 337, 339, *366–368*, 381, 505, 572
Ambiguity, *19–21*, 101, 102, 109
Ampère, A., 493–499, 505
Analogy, *68–72, 106, 172*, 217, 231, 261, 281, 384, *393–396*, 482, 536, *552, 557*, 577, *578, 584*
Analysis, 135, 163, 167, 179, 180, 185, 218, 301, 302, 359, *371–389*, 395, 396, 539, 577, 578
Analytic judgments, 371-389, 577, 578; methods, 374–383
Animal functions, 156, 167; life, 86, 88, 135, 142, 163, *170, 192–195*, 203, *224*, 283, 284, 319, 320, 374, 509, 510, 538; nature, 86, 134, *135*, 164, *167*, 209, *214*, 252, 524, 525; sensibility, 284, 503; species, 127, 214
Animality and humanity, 524, 525; characteristics of, 164
Animals, 18, 19, *83–85, 95, 102*, 169, 192, 193, 213, 214, 374, 516, 524
Anomalous causes, 78, 79, 115
Anthropology and psychology, 539, 542
Antinomies of pure reason, 222, 223, 579
Apparent motions, *8–10, 17*, 117, 442, 443
Appearance, *10, 11, 15, 115*, 178, *543, 544*
Approximation, *65, 66*, 267, 280, 301, 302, 313, 317, 472
A priori belief, 176; conceptions, 579, 580; constructions, 211; elements of knowledge, 577; ideas, 579, 580; judgment, 215; synthesis, 382–387; synthetic judgments, 553
Arbitrary definitions, 20, 21, 25
Arcesilaus, 549
Archimedes' principle, 386
Aristotle, 30, *33, 34*, 160, 188, 192, 193,

ganization, 402, 410–434; powers, 400, 410; principles, 396; proofs, 402; questions, 395

Jurisprudence, 388, 390–410, 412, 428, 429, 472

Jury, the institution of the, 253, 254, 257, 302, 352, 399, 411, 412, 416–419, 421, 422, 425, 433

Justice, 253, 302, 411, 413, 419, 421

Kant, I., 6, 10, 11, 20, 30, 68, 69, 104, *129–133*, 199, 212, 217, *218, 222, 223*, 375, 381–385, 387, 514, 541, 542, 548, 553, *575–581;* criticism of the philosophy of, 576ff.

Kepler, J., 56–61, 232, 297, 555

Kléber, J., 42

Knowledge, *5–17*, 18, 34, *48*, 50, 68, *71, 73*, 77, 84, *98–104, 106–133, 134–170*, 171, 180, 182, 196, 203, *206, 207*, 210, 218, 219, 223, *224*, 226, 231, *235, 236*, 239, 242, 247, *251–254, 262*, 269, 296, 297, *315*, 330, 331, 334, 374, 376, *377*, 382, *384, 434, 437–441*, 444, 445, 449, 457, 458, *462, 469*, 471, *472, 474, 476*, 488, *490–508*, 531, 541, 542, *544–553*, 575, *577–579, 585–598*

La Bruyère, J., 265, 270

Lalande, A., 572

Lambert, J., 45

Language, *19, 20*, 32, *48, 146*, 154, 179–181, 188, 207, 209, 213, 227, 235, 236, *251*, 261, 281, *306–323*, 324, 326, 328, 338, 342, *343, 345, 346, 356, 358*, 371, *372*, 381, 384, 427, 442, 448, 456, 466, 516, 517, 572–575, *591, 592*

Laplace, P., 221

Lavoisier, A., 297, 319, 477

Law, 10, 15, 16, *28–30*, 32, 39, *49, 54–70, 73, 74, 78*, 79, 81, *84–87*, 90, *92, 93*, 95, *100, 101*, 103, 110, 112, *113–118*, 120, *121*, 131, 135, 137, 149, 150, 152, 170, 178, *180–183, 186–192*, 196, 199, 207, 209, 213, *217–221*, 227, *231–235*, 254–258, *260*, 261, 264, *265*, 271, *283, 284*, 287, *288, 291*, 292, *293*,

299, *301*, 308, 310, *314, 317*, 365, *384*, 385, *390–410, 410–434*, 435, 436, *438–440*, 448, *450, 459, 461*, 462, *467, 469, 472, 473, 497–499, 502–507, 510, 517*, 519, 520, *526*, 574ff., 584ff., 591, 592

Lawyers, 378, 387, 388, 396, 397

Legal experts, 393, 398–400, 426, 467; fictions, 293, 394–396; philosopher, 28, 29; principles, 396, 397; reasoning, 390; texts, 400

Legendre, A., 345, 362

Legislative forms, law and, 397; power, 402–404, 408, 410

Legislators, 110, 293, 297, 397, 398, 403, 408, 414, 417–419

Leibniz, G., 3, 20, 25, 35–38, 77, 111, 179, *195*, 208, *212, 219–221*, 223, 234, 274, 284, *298–301*, 316, 317, 350, *385*, 388, 397, 406, 464, 488, 502, 505, 554, *565–575;* criticism of his philosophy, 566ff.

Life, *86–89*, 123, 124, 163, *191–209*, 224, 283, 284, 304, *319, 320, 466, 467*, 509, 510, *524, 529*, 538

Light, 137, 149, 182

Limits, method of, 299, 300; of experience, 208; of observation, 54, 63–67, 70, 71, 142; of possible error, 46; of precision, 303, 304

Linear order of discourse, *347–370*, 592; series, 336, 344, 348–359

Linguistic roots, 324

Linguistics, 247

Linnaeus, C., 192, 328, 352–354

Literature, 436, 509

Living things, *50*, 82, *83, 87, 88, 91–96, 183, 191, 195–200, 204–206*, 214, 287, 288, 336, 374

Locke, J., 550, 564

Logic, 6, *23, 68, 71*, 227, 233, 241, 242, 253, 257, 293, 294, 329, 337, 338, 346, *361*, 363, *368*, 371, *378, 379, 383*, 384, 396, *426, 473*, 483, 487, 532, *540–542, 592*

Logical abstraction, 207, 229, 231, 232,